RECENT ADVANCES IN DYNAMICAL ASTRONOMY

ASTROPHYSICS AND SPACE SCIENCE LIBRARY

A SERIES OF BOOKS ON THE RECENT DEVELOPMENTS
OF SPACE SCIENCE AND OF GENERAL GEOPHYSICS AND ASTROPHYSICS
PUBLISHED IN CONNECTION WITH THE JOURNAL
SPACE SCIENCE REVIEWS

VOLUME 39

RECENT ADVANCES IN DYNAMICAL ASTRONOMY

PROCEEDINGS OF THE NATO ADVANCED STUDY INSTITUTE IN DYNAMICAL ASTRONOMY HELD IN CORTINA D'AMPEZZO, ITALY, AUGUST 9–21, 1972

Edited by

B. D. TAPLEY AND V. SZEBEHELY

Department of Aerospace Engineering and Engineering Mechanics,
University of Texas at Austin, Austin, Tex. 78712, U.S.A.

D. REIDEL PUBLISHING COMPANY

DORDRECHT-HOLLAND / BOSTON-U.S.A.

Library of Congress Catalog Card Number 73–83571

ISBN-13: 978-94-010-2613-0 e-ISBN-13: 978-94-010-2611-6
DOI: 10.1007/978-94-010-2611-6

Published by D. Reidel Publishing Company,
P.O. Box 17, Dordrecht, Holland

Sold and distributed in the U.S.A., Canada and Mexico
by D. Reidel Publishing Company, Inc.
306 Dartmouth Street, Boston,
Mass. 02116, U.S.A.

Softcover reprint of the hardcover 1st edition 1973

TABLE OF CONTENTS

PREFACE VII

LIST OF PRINCIPAL SPEAKERS IX

LIST OF PARTICIPANTS XI

I. REGULARIZATION

E. STIEFEL / A Linear Theory of the Perturbed Two-Body Problem (Regularization) 3
J. WALDVOGEL / Collision Singularities in Gravitational Problems 21
D. C. HEGGIE / Regularization Using a Time-Transformation Only 34
J. BAUMGARTE / Stabilization of the Differential Equations of Keplerian Motion 38
F. NAHON / The Particular Solutions of Levi-Civita 45
O. GODART / Example of Integration of Strongly Oscillating Systems 53
W. BLACK / The Application of Recurrence Relations to Special Perturbation Methods 61
D. G. BETTIS / Numerical Solution of Ordinary Differential Equations (*Abstract*) 71

II. THE THREE-BODY PROBLEM

V. SZEBEHELY / Recent Advances in the Problem of Three Bodies 75
R. F. ARENSTORF / Periodic Elliptic Motion in the Problem of Three Bodies (*Abstract*) 107
G. KATSIARIS and C. L. GOUDAS / On a Conjecture by Poincaré 109
G. KATSIARIS / The Three-Dimensional Elliptic Problem 118
P. G. KAZANTZIS / Second and Third Order Variations of the Three Dimensional Restricted Problem 135
C. G. ZAGOURAS / Planar Periodic Orbits Using Second and Third Variations 146
E. RABE / Elliptic Restricted Problem: Fourth-Order Stability Analysis of the Triangular Points 156
P. GUILLAUME / A Linear Description of the Second Species Solutions 161

III. THE *N*-BODY PROBLEM AND STELLAR DYNAMICS

G. CONTOPOULOS / Problems of Stellar Dynamics 177
W. T. KYNER / Invariant Manifolds in Celestial Mechanics 192
S. J. AARSETH / Theoretical Aspects of *N*-Body Computer Simulations 197
S. J. AARSETH / Integration Methods of the *N*-Body Problem (*Abstract*) 208

R. WIELEN / Numerical Studies of the Gravitational Problem of *N*-Bodies (*Abstract*) 209
J. R. DORMAND / Star Encounter Simulations 211
O. GODART / Yukawa Potential and Stellar Encounter 216
C. AUST / An Investigation of the Rôle of Turbulence in the Formation of Star Clusters by a Point-Mass Simulation 222

IV. THE THEORY OF GENERAL PERTURBATIONS

G. HORI / Theory of General Perturbations 231
J. HENRARD / The Algorithm of the Inverse for Lie Transform 250
U. KIRCHGRABER / The Property of Covariance in Hori's Non-Canonical Perturbation Theory (*Abstract*) 260
J. RAE / Subdynamics in Classical Mechanics 262

V. THE SOLAR SYSTEM AND ORBITAL RESONANCES

G. A. WILKINS / Astronomical Systems of Units, Co-ordinates and Time-Scales 273
R. L. DUNCOMBE / Masses of the Planets and Satellites (*Abstract*) 287
R. L. DUNCOMBE / Planetary Ephemerides 289
P. J. MESSAGE / Orbits of the Natural Satellites. Lectures on Some Types of Orbital Motion Amongst Them 309
G. COLOMBO and F. A. FRANKLIN / On the Origin of the Titan-Hyperion System 312
M. W. OVENDEN / Planetary Distances and the Missing Planet 319
R. R. ALLAN / Change of Inclination in Passing Through Resonance 333

VI. TRAJECTORY DETERMINATION AND THE MOTION OF RIGID BODIES

M. S. DAVIS / Review of Non-Numerical Uses of Computers 351
P. HERGET / Elementary Dynamical Astronomy 392
B. D. TAPLEY / Statistical Orbit Determination Theory 396
M. SCHNEIDER / Orbit Determination by Solving a Boundary Value Problem 426
J. E. COCHRAN / Long-Term Motion in a Restricted Problem of Rotational Motion 429
M. VITINS / Uniform Theory of a Rotating Rigid Body with Dynamical Symmetry 454
W. J. ROBINSON / Attitude Stability of a Satellite in the Restricted Three-Body Problem 457
B. L. STANEK / Analytic Treatment of Straight-Line Flights into the Sun after Planetary Fly-By 461

PREFACE

This volume contains the text of the lectures and papers presented at the NATO Advanced Study Institute in Dynamical Astronomy, held in Cortina d'Ampezzo in August 1972. The main sponsorship of the Scientific Affairs Division of NATO is gratefully acknowledged. Additional support received from many other sources, especially those from the University of Texas and from the United States National Science Foundation, made wider participation at the Institute possible.

The Director of the Institute enjoys this opportunity of expressing his appreciation to his co-workers and especially to Dr E. Stiefel, Co-Director of the Institute. The speakers and the audience formed an indivisible group of students at this Institute. The audience delivered short research papers during the afternoon sessions and the principal lecturers of the morning became the audience. We were all students of celestial mechanics and what we learned is summarized in this volume.

The two weeks Institute concentrated on a few subjects of special interest to present-day celestial mechanics and dynamical astronomy. Emphasis was placed on the fundamental mathematical and astronomical aspects rather than on the immediate applications. Teaching and the presentation of new research results were combined during the Institute.

The major areas of interest were: (i) *regularization*, (ii) *resonance* and (iii) *the many-body poblems.*

(i) Since one of the modern trends in celestial mechanics is the use of regularization, in a number of areas, several lecture series and papers were dedicated to this subject. Indeed, the interest of the participants was such that no day went by without questions raised or references made to regularization – independently of the subject at hand at that particular time. The concept of regularization is fundamental in the analytical and numerical aspects of the solution of those differential equations which dominate the science of celestial mechanics. Consequently, it is not surprising that during the two weeks of the Institute regularization was an ever-recurring subject. The elimination of singularities and consequent existence proofs are the original raison d'être of regularization. In modern numerical analysis, the stability of numerical integration techniques depends fundamentally on the introduction of the proper regularizing transformation. Consequently, the subject of regularization covers existence problems basic in mathematics as well as essential numerical methods without which certain solutions could not be obtained at all. The principal lecturers on this subject were Drs Baumgarte, Bettis, Nahon, Scheifele, Stiefel, and Waldvogel.

(ii) The second subject which met with considerable interest during the Institute was *resonance* in the solar system. The non-linear differential equations governing dynamical astronomy show non-linear resonance phenomena not unlike those occurring in the other non-linear physical sciences. The near resonance orbital conditions occurring in the solar system, the couplings between the orbital and the rigid-body motion of satellites and planets were the fundamental examples for application of non-linear resonance theories during the Institute. Cosmological theories, concerning the origin of the solar system and its past and future development, based essentially on close resonances, were treated in detail. The principal contributors to this subject were Drs Allan, Colombo, Duncombe, Kyner, Message, Ovenden and Wilkins.

(iii) The great, celebrated unsolved problem of celestial mechanics is the gravitational *many-body problem.* The introductory chapter to this field was the problem of two bodies which was immediately followed by the presentation of general and special perturbation methods and classical and statistical orbit determination techniques. The modern non-numerical uses of computers offering an exciting impetus to this field were also discussed. Those problems which cannot be handled by perturbation techniques, such as the stellar three-body problem and the gravitational n-body problem received considerable emphasis during the Institute. Modern numerical and analytical techniques were discussed and the presentation of several new results made this subject one of central interest. The principal speakers were Drs Aarseth, Arenstorf, Contopoulos, Davis, Goudas, Guillaume, Henrard, Herget, Hori, Kirchgraber, Rae, Szebehely, Tapley and Wielen.

The material is organized into six basic areas: (I) Regularization, (II) The Three-Body Problem, (III) The N-Body Problem and Stellar Dynamics, (IV) The Theory of General Perturbations, (V) The Solar System and Orbital Resonances and (VI) Trajectory Determination and the Motion of Rigid Bodies.

Some of the didactic aspects of the lectures are only summarized in this volume, especially when the material is easily accessible elsewhere in the open literature. The unpublished didactic contributions, the main research lectures and the short original communications of the participants delivered at the Institute appear in this Volume and are offered to our colleagues.

LIST OF PRINCIPAL SPEAKERS

Aarseth, S. J.	(Norwegian)	Institute of Astronomy, Madingley Road, Cambridge, England
Allan, R. R.	(British)	Space Department, Royal Aircraft Est., Farnborough Hants, England
Arenstorf, R. F.	(U.S.A.)	Dept. of Mathematics, Vanderbilt Univ., Nashville, Tenn. 37235, U.S.A.
Baumgarte, J.	(German)	Universität Braunschweig, Braunschweig, West Germany
Bettis, D. G.	(U.S.A.)	Dept. of Aerospace Engineering & Engineering Mechanics, Univ. of Texas at Austin, Austin, Tex. 78712, U.S.A.
Colombo, G.	(Italian)	Istituto di Meccanica, Università di Padova, Padova, Italy
Contopoulos, G.	(Greek)	Univ. of Thessaloniki, Thessaloniki, Greece
Davis, M. S.	(U.S.A.)	Dept. of Physics, Univ. of North Carolina, Chapel Hill, N. C. 27514, U.S.A.
Duncombe, R. L.	(U.S.A.)	Nautical Almanac Office, U.S. Naval Observatory, Washington, D.C. 20390, U.S.A.
Feagin, T.	(U.S.A.)	NASA Goddard Space Flight Center, Code 552, Greenbelt, M., U.S.A.
Garfinkel, B.	(U.S.A.)	Yale Univ. Observatory, New Haven, Conn. 06520, U.S.A.
Goudas, C.	(Greek)	Dept. of Mechanics, University of Patras, Patras, Greece
Henrard, J.	(Belgian)	Facultés Universitaires de Namur, Namur, Belgium
Herget, P.	(U.S.A.)	Cincinnati, Observatory, Observatory Place, Cincinnati, Ohio 45208, U.S.A.
Hori, G.	(Japanese)	Dept. of Astronomy, Univ. of Tokyo, Bunkyo-Ku, Tokyo, Japan
Kirchgraber, U.	(Swiss)	Seminar für Angewandte Mathematik, Eidgenössische Technische Hochschule, Clausiusstrasse 55, Zurich, Switzerland
Kyner, W. T.	(U.S.A.)	Dept. of Mathematics, Univ. of New Mexico, Albuquerque, N. M. 87106, U.S.A.
Message, P. J.	(British)	Dept. of Applied Mathematics, Univ. of Liverpool, Liverpool L69 3BX, England

Nahon, F.	(French)	Université de Paris VI, 11 Rue M&P Curie, Paris 5e, France
Ovenden, M. W.	(British)	Dept. of Geophysics and Astronomy, Univ. of British Columbia, Vancouver 8, British Columbia, Canada
Rae, J.	(Belgian)	Chimie Physique II, Faculté des Sciences, Univ. Libre de Bruxelles, Ave Roosevelt, 1050 Bruxelles, Belgium
Scheifele, G.	(Swiss)	Dept. of Aerospace Engineering & Engineering Mechanics, Univ. of Texas at Austin, Austin, Tex. 78712, U.S.A.
Stiefel, E.	(Swiss)	Seminar für Angewandte Mathematik, Eidgenossische Technische Hochschule, Clausiusstrasse 55, Zürich, Switzerland
Szebehely, V.	(U.S.A.)	Dept. of Aerospace Engineering & Engineering Mechanics, Univ. of Texas at Austin, Austin, Tex. 78712, U.S.A.
Tapley, B. D.	(U.S.A.)	Dept. of Aerospace Engineering & Engineering Mechanics, Univ. of Texas at Austin, Austin, Tex. 78712, U.S.A.
Waldvogel, J.	(Swiss)	Seminar für Angewandte Mathematik, Eidgenossische Technische Hochschule, Clausiusstrasse 55, Zürich, Switzerland
Wielen, R.	(German)	Astronomisches Rechen-Institut, Moenchhofstr. 12–14, D-69 Heidelberg, West Germany
Wilkins, G. A.	(British)	H. M. Nautical Almanac Office, Royal Greenwich Observatory, Herstmonceux Castle, Hailsham, Sussex, England
Wissler, E.	(U.S.A.)	College of Engineering, Univ. of Texas at Austin, Austin, Tex. 78712, U.S.A.

LIST OF PARTICIPANTS

Abu El Ata, N.	(Egypt)	Faculté des Sciences, Dept. de Mécanique Tours 66 65, II Quai St. Bernard, Paris 5°, France
Aust, C.	(British)	Dept. of Physics, Univ. of York, Heslington, York YO I 5DD, England
Benest, D.	(French)	Observatoire de Nice, Nice 06, France
Bennet, A.	(U.S.A.)	543 Auburn Dr., Auburn, Ala. 36830, U.S.A.
Bernetti, L.	(Italian)	Istituto de Geodesia e Geofisica, Univ. di Trieste, Trieste, Italy
Bianchini, G.	(Italian)	Istituto di Meccanica, Università di Padova, Padova, Italy
Black, W.	(Scottish)	Dept. of Astronomy, Univ. of Glasgow, Glasgow, Scotland
Calvo, M.	(Spanish)	Dept. of Astronomy, Univ. of Zaragoza, Zaragoza, Spain
Cochran, J. E.	(U.S.A.)	Aerospace Engineering Dept., Auburn Univ., Auburn, Ala. 36830, U.S.A.
Dormand, J. R.	(British)	Dept. of Mathematics, Teesside, Polytechnic, Middlesbrough, Teesside TSI 3BA, England
Eisner, S.	(Swiss)	Witikonerstr. 248, CH 8053, Zurich, Switzerland
Fitzpatrick, P. M.	(U.S.A.)	Dept. of Mathematics, Auburn Univ., Auburn, Ala., U.S.A.
Forti, G.	(Italian)	Astrof. di Arcetri, Largo E. Fermi 5, Firenze, Italy
Godart, O.	(Belgian)	Univ. de Louvain, 3030 Heverle-Louvain, 27 de Croylaan, Belgium
Guillame, P.	(Belgian)	Institut d'Astrophysique, Cointe, 4200, Ougree, Belgium
Hagedorn, P.	(German)	Institut für Mechanik, Univ. Karlsruhe, 75 Karlsruhe, Kaiserstr. 12, West Germany
Heggie, D.	(British)	Institute of Astronomy, Madingley Road, Cambridge CB3 OEZ, England
Hiller, J.	(German)	5307 Wachtberg-Pech, auf dem Reeg 8, West Germany
Jaeger, K.	(German)	53 Bonn – Bad Godesberg 1, Postfach 301, West Germany

Kazantzis, A.	(Greek)	Dept. of Mechanics, Univ. of Patras, Patras, Greece
Kunz, R.	(Swiss)	Bruggaecher 2, CH-8617 Mönchaltorf, Switzerland
Losco, L.	(French)	4, rue Querret, 25 Besancon, France
Manar, A.	(Italian)	Osservatorio Astronomico de Brera, Milano, Italy
Mavraganis, T.	(Greek)	Dept. of Mechanics, Univ. of Patras, Patras, Greece
Cid Palacios, R.	(Spanish)	Univ. de Zaragoza, Dept. Astronomia, Zaragoza, Spain
Poma, A.	(Italian)	Osservatorio Astronomico, Via Ospedale 72, 09100 Cagliari, Italy
Rabe, E.	(U.S.A.)	Cincinnati Observatory, Observatory Place, Cincinnati, Ohio 45208, U.S.A.
Robinson, W. J.	(British)	Dept. of Mathematics, Univ. of Bradford, Bradford, Yorkshire, England
Schmidt, D.	(German)	1827 Avenel Rd, Adelphi, Md. 20738, U.S.A.
Schneider, M.	(German)	Technische Univ., Muenchen, West Germany
Sperling, H. J.	(U.S.A.)	Marshall Space Flight Center, S & A, Aero. Huntsville, Ala. 35812, U.S.A.
Vitins, M.	(Canadian)	Schaffhauserstr. 89, CH 8057, Zurich, Switzerland
Stanek, B. L.	(Swiss)	Sagenbrugg, CH 6318 Walchwil, Switzerland
Zagar, F.	(Italian)	Milano, Via Brera 28, Italy
Zagouras, C.	(Greek)	Dept. of Mechanics, Univ. of Patras, Patras, Greece

Participants of the 1972 NATO Advanced Study Institute in Dynamical Astronomy.

PART I

REGULARIZATION

A LINEAR THEORY OF THE PERTURBED TWO-BODY PROBLEM (REGULARIZATION)

E. STIEFEL
Seminar für Angewandte Mathematik, Zürich, Switzerland

Abstract. Discussion of the numerical solution of Keplerian motion. Singularities and instability of the classical differential equation. Regularization and stabilization by the KS-transformation. Abbreviated description of that transformation and application to the perturbed two-body problem. Orbital elements.

1. Preliminaries

Our main problem will be to describe pure two-body motion by *harmonic oscillations* governed by linear differential equations with constant coefficients. These linear differential equations are everywhere *regular* in contrast to the classical Newtonian equations, which are singular at the collision of the two bodies. Furthermore there is a possibility of mutual stimulation between the methods of celestial and the resulting oscillatory mechanics. Perturbing forces will be included.

Consider a particle of the mass m attracted by a central body of the Mass M and perturbed by a perturbing force P and a perturbing potential V (P contains the forces which cannot be derived from a potential). Denoting the position of the particle by the vector x with respect to a coordinate system centered at the mass M, the Newtonian equations of motion are given by

$$\ddot{x} + \frac{K^2}{r^3}\, x = P - \frac{\partial V}{\partial x}, \qquad K^2 = k^2 (M + m), \tag{1}$$

where dots represent differentiation with respect to the time t, r is the distance between the masses, k^2 is the universal gravitational constant, and

$$\frac{\partial V}{\partial x}$$

is the gradient of the scalar function $V(x, t)$.

The kinetic energy of the particle taken per unit mass is given by

$$T = \tfrac{1}{2}v^2 = \tfrac{1}{2}|\dot{x}|^2 = \tfrac{1}{2}(\dot{x}, \dot{x}),$$

where v is the velocity of the particle and (,) denotes the scalar product of the two inserted vectors.

The total negative energy h, defined by the *energy relation*

$$h = -\tfrac{1}{2}|\dot{x}|^2 + \frac{K^2}{r} - V, \tag{2}$$

B. D. Tapley and V. Szebehely (eds.), Recent Advances in Dynamical Astronomy, 3–20.

leads to the following *law of energy*

$$\dot{h} = -\frac{\partial V}{\partial t} - (P, \dot{x}), \tag{3}$$

where the first term on the right-hand side denotes the partial derivative of the function $V(x, t)$ with respect to t.

In the special case where the potential V does not depend on time explicitly, i.e. $V(x, t) = V(x)$, and no other perturbing forces act on the particle, h is evidently a constant during motion. Such a potential is said to be *conservative*.

2. Regularization and Stabilization

Let us study the above equations of motion from the point of view of numerical integration. At vanishing distance $r = 0$ there is a singularity in the equation of motion (1). If there is no singularity in the perturbing terms, the value of the potential V and the negative total energy h will be finite at this vanishing radius. Consequently, we derive from the energy relation (2), that the velocity $|\dot{x}|$ must tend to infinity as r approaches zero. We are thus confronted with the fact that the solution of the Newtonian set of equations is singular at $r = 0$. These equations are therefore not suitable for numerical integration for small values of the distance r.

A. FICTITIOUS TIME

In order to obtain *regular solutions*, we must multiply the velocity vector $\dot{x}$ by an appropriate factor which compensates the growth of the velocity near collision. Choosing the distance r as the scaling factor we have

$$r \frac{\mathrm{d}x}{\mathrm{d}t} = \frac{\mathrm{d}x}{\mathrm{d}s},$$

where s is a fictitious time, which we will regard as a new independent variable. The physical time t may be recaptured by integrating the expression

$$\frac{\mathrm{d}t}{\mathrm{d}s} = r, \tag{4}$$

which follows immediately from the above equation.

The basic set of Equations (1), (2), (3) may be rewritten in terms of s as the independent variable

$$x'' - \frac{r'}{r} x' + \frac{K^2}{r} x = r^2 \left(P - \frac{\partial V}{\partial x}\right), \qquad t' = r, \tag{5}$$

$$h = -\frac{1}{2r^2} |x'|^2 + \frac{K^2}{r} - V, \tag{6}$$

$$h' = -r \frac{\partial V}{\partial t} - (P, x'), \tag{7}$$

where the prime denotes differentiation with respect to s. Since V is to be regarded as a function of the four independent arguments x_1, x_2, x_3, t, its partial derivative with respect to t in Equation (7) cannot be replaced by differentiation with respect to s.

Equations (5) and (4) determine the unknowns x_1, x_2, x_3 and t as functions of the independent variable s. The physical time t is to be considered as a fourth coordinate of the particle.

The Equations (5) are still singular since the denominator r appears. Nevertheless, the first step of our regularization procedure was successful since the unperturbed *solutions* are regular functions of the independent variable.

We will make this clear by first discussing one-dimensional motion. Putting $r=x$, $V=0$, $P=0$, we have

$$x'' - \frac{x'^2}{x} + K^2 = 0$$

$$xh = -\frac{x'^2}{2x} + K^2, \qquad h' = 0.$$

Solving the energy relation for x'^2/x, we obtain

$$-\frac{x'^2}{x} = 2hx - 2K^2,$$

which is a well-determined quantity at collision. Substituting this energy relation into the first equation yields

$$x'' + 2hx = K^2.$$

This inhomogeneous linear differential equation with constant coefficients obviously has solutions which are perfectly regular functions of s.

In two-dimensional motion the above argument cannot be reproduced because of the undetermined quotients. However, it is very important to realize that singularities in a differential equation need not necessarily imply singularities of its solutions.

In order to make this point clear, let us consider the following linear differential equation of second order for the scalar function $x(s)$

$$(1 - \cos s)\, x'' - \sin s\, x' + x = 0.$$

Dividing this equation by $(1-\cos s)$ in order to obtain the standard form, we find that $s=0$ is a singularity of the equation. On the other hand, we can readily verify that the regular functions $(1-\cos s)$ and $\sin s$ are particular solutions of the differential equation. Since the general solution is a linear combination of these solutions, it is regular for every s.

It must be stressed that singular differential equations are numerically badly behaved near the singularity, even if the solution is perfectly regular. Therefore the next step of a regularization procedure must be to eliminate all singularities in the differential equation. For the differential equations at hand this will be performed in

Sections C and D. Proceeding to the discussion of two-dimensional motion, it is readily verified that a solution of Equations (5) in the unperturbed case, i.e. $V=0$, $P=0$, is given by

$$\begin{aligned} E &= \frac{K}{\sqrt{a}}\, s\,, \\ x_1 &= a\,(\cos E - e)\,, \qquad x_2 = a\sqrt{1-e^2}\,\sin E\,, \qquad x_3 = 0\,, \\ r &= a\,(1 - e\cos E)\,, \end{aligned} \tag{8}$$

where a and e are constants of integration. Since the solution is given by perfectly regular sine and cosine functions, it is not advisable to compute the solution by means of the singular differential Equations (5).

The Solution (8) is the general solution up to a rotation in the x_1, x_2-plane and an additive constant in E.

B. STABILIZATION

From the point of view of numerical integration the stability of a differential equation is of fundamental importance.

Stability in the sense of Ljapunow is defined as follows: Let a reference solution, whose stability is to be determined, be given by specifying its initial values at some instant. By variation of the initial values by a small amount we obtain a second solution. The reference solution is called stable, if the distance between the two solutions for any time onwards can be made to be smaller than any prescribed positive quantity ε by an appropriate choice of the variation of the initial conditions.

It is well known that every elliptic solution of the classical Newtonian Equation (1) is unstable. The reason for this instability can be understood by considering the circular motion of a particle in the x_1, x_2-plane. Furthermore we will assume that the neighbouring solution, obtained by modifying the initial conditions, is also circular. The equations of motion are given by

$$\ddot{x} + \omega^2 x = 0\,, \qquad \omega = \frac{K}{r^{3/2}}\,,$$

where r is the radius of the circle. The period of revolution is $2\pi/\omega = (2\pi/K)r^{3/2}$. If the radii of the two circles differ, then the periods of revolution on the reference orbit and the modified orbit will also differ. Thus the two particles rotate about the origin like the two hands of a clock. No matter how small the difference of the two radii is chosen, sooner or later t reaches an instant, where the two particles are opposite to each other with respect to the origin, which is incompatible with Ljapunow stability.

Recapitulating our criticism of the Newtonian equations of motion from a numerical standpoint, we recall that their solutions are singular and unstable and that the differential equations are singular. The change of the independent variable from t to s produced regular solutions, however the resulting equations of motion are still singular and the solutions are still unstable. The instability of the solutions is due to the fact that the frequency of revolution in circular motion depends on the distance.

Our final goal will be to describe pure Keplerian motion by harmonic oscillator equations with fixed frequency, thus obtaining regular differential equations and solutions as well as stability.

C. MOTION IN A PLANE (Levi-Cività-transformation)

The equations of plane motion are given in Equations (8). As the eccentricity e approaches 1, the elliptic shape of the orbit degenerates into a straight line segment on which the particle oscillates. The position vector makes a sharp bend of the angle 2π at the origin (see Figure 1).

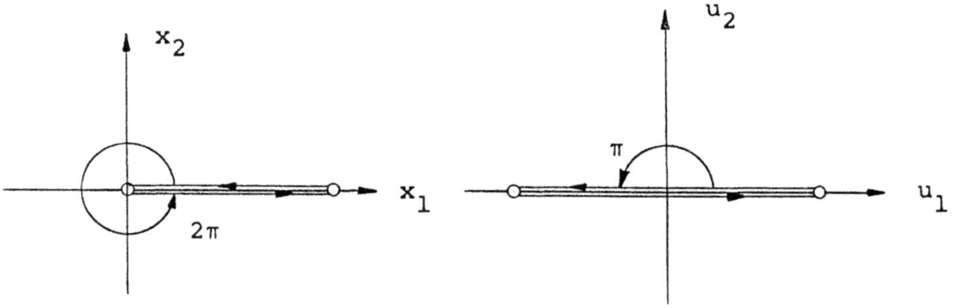

Fig. 1

In order to remove this singular behaviour we will map the physical x-plane onto a parametric u-plane in such a way that the image of the particle passes through the origin after collision. Hence the angle 2π in the physical plane must be transformed to the angle π in the parametric plane. This goal can be achieved by the conformal mapping

$$u_1 + iu_2 = \sqrt{x_1 + ix_2}$$

which, indeed, does halve angles at the origin. Squaring this transformation and separating into real and imaginary parts leads to

$$x_1 = u_1^2 - u_2^2, \quad x_2 = 2u_1u_2, \quad r = u_1^2 + u_2^2, \tag{9}$$

and by differentiation the matrix relation

$$\begin{pmatrix} \dot{x}_1 \\ \dot{x}_2 \end{pmatrix} = 2 \begin{pmatrix} u_1 & -u_2 \\ u_2 & u_1 \end{pmatrix} \begin{pmatrix} \dot{u}_1 \\ \dot{u}_2 \end{pmatrix}$$

is obtained. The 2×2-matrix appearing on the right-hand side is known as the Levi-Cività matrix. It also appears in the coordinate transformation (9) if the transformation is written in the matrix notation

$$\begin{pmatrix} x_1 \\ x_2 \end{pmatrix} = \begin{pmatrix} u_1 & -u_2 \\ u_2 & u_1 \end{pmatrix} \begin{pmatrix} u_1 \\ u_2 \end{pmatrix}. \tag{10}$$

The Levi-Cività matrix has the following three properties:

(1) It is orthogonal.

(2) Its elements are linear and homogeneous functions of the parameters u_1 and u_2.

(3) The first column is the u-vector.

It can be shown that pure elliptic Kepler motion is represented in terms of the parameters u_1 and u_2 by a harmonic oscillation, provided the fictitious time s is used. Thus regularization and stabilization is achieved. (Stiefel and Scheifele, 1971)

D. MOTION IN SPACE

Proceeding to the regularization of the laws of motion in space it is natural to construct a square matrix satisfying the three properties of the Levi-Cività matrix listed above. The construction of such a matrix is intimately connected with the theory of hypercomplex numbers (algebra).

For 2×2 matrices the algebra of complex numbers is very helpful, as we saw in the derivation of the Levi-Cività matrix. Similarily, 4×4 matrices are related to the algebra of quaternions and 8×8 matrices have been studied by Cayley. It turns out that no 3×3 matrices exist which satisfy the three required properties. Thus we are forced to use a 4×4 matrix. Denoting the four parameters by u_1, u_2, u_3 and u_4 our choice is

$$L = \begin{pmatrix} u_1 & -u_2 & -u_3 & u_4 \\ u_2 & u_1 & -u_4 & -u_3 \\ u_3 & u_4 & u_1 & u_2 \\ u_4 & -u_3 & u_2 & -u_1 \end{pmatrix}. \tag{11}$$

This matrix is known as the *KS-matrix* because it was proposed by Kustaanheimo and Stiefel (1971). Note that the Levi-Cività matrix appears in the upper left-hand corner.

Bearing the Levi-Cività transformation (10) in mind, we shall define the transformation of a point in the four-dimensional u-space onto the corresponding point in the three-dimensional physical space by

$$x = Lu\,, \tag{12}$$

where x is the usual position vector in the physical space (considered as a matrix consisting of one single column) augmented by a fourth component which should vanish. In explicit form this transformation is given by

$$\begin{aligned} x_1 &= u_1^2 - u_2^2 - u_3^2 + u_4^2, & x_2 &= 2(u_1u_2 - u_3u_4), \\ x_3 &= 2(u_1u_3 + u_2u_4), & x_4 &= 0, \end{aligned} \tag{13}$$

and it is seen that the fourth component of x does indeed automatically vanish. For the sake of convenient notation we will adopt the convention that a vector in the physical space is supplemented to a four-dimensional vector by adding a fourth component of zero value.

The transformation (12) is called the *KS-transformation*. The KS-matrix is orthogonal,

$$L^T L = L L^T = (u, u)\, E\,, \tag{14}$$

where E is a unity matrix and L^T is the transposed matrix of L, hence the first basic property is satisfied. Obviously the required properties two and three also hold true.

The derivation of the differential equations for the parameters u_i is most conveniently carried out using matrix notation. Using Equation (14), we have

$$\begin{aligned} r^2 &= (x, x) = x^T x = u^T L^T L u = u^T (u, u)\, u = (u, u)^2 = |u|^4, \\ r &= |u|^2\,, \end{aligned} \tag{15}$$

thus

$$L^{-1} = \frac{1}{r} L^T\,. \tag{16}$$

Differentiating the transformation (13), we obtain

$$\begin{pmatrix} \dot{x}_1 \\ \dot{x}_2 \\ \dot{x}_3 \end{pmatrix} = 2 \begin{pmatrix} u_1 & -u_2 & -u_3 & u_4 \\ u_2 & u_1 & -u_4 & -u_3 \\ u_3 & u_4 & u_1 & u_2 \end{pmatrix} \begin{pmatrix} \dot{u}_1 \\ \dot{u}_2 \\ \dot{u}_3 \\ \dot{u}_4 \end{pmatrix}.$$

The matrix appearing on the right-hand side of this equation can be identified as the KS-matrix without its last row. Adding the fourth row of the KS-matrix to the above relation, we obtain as the fourth component of the right-hand side vector

$$u_4 \dot{u}_1 - u_3 \dot{u}_2 + u_2 \dot{u}_3 - u_1 \dot{u}_4\,,$$

but this component should vanish since $\dot{x}$ is a vector in the physical space. Thus we have to *impose* the condition

$$u_4 \dot{u}_1 - u_3 \dot{u}_2 + u_2 \dot{u}_3 - u_1 \dot{u}_4 = 0\,, \tag{17}$$

which is called the *bilinear relation.* This leads to

$$\dot{x} = 2L\dot{u} \quad \text{or} \quad x' = 2Lu'\,, \tag{18}$$

if the fictitious time s is used.

It must be stressed that the use of Equation (18) can only be justified if the bilinear relation is satisfied, that is, if motion in four-dimensional space really does satisfy this relation, which is a non-holonomic constraint.

The *Lagrangian formalism* is a suitable method for deriving equations of motion for generalized coordinates q_i. Let $T(q, \dot{q})$ be the kinetic energy of a mechanical system and let $U(q, t)$ be its potential energy. If the solution is restricted by a non-holonomic constraint of the form $l = \sum_{(k)} a_k(q) \dot{q}_k = 0$, the equations of motion are given by

$$\frac{\mathrm{d}}{\mathrm{d}t}\left(\frac{\partial T}{\partial \dot{q}_i}\right) - \frac{\partial T}{\partial q_i} = -\frac{\partial U}{\partial q_i} + \lambda a_i\,,$$

where λ is a Lagrangian multiplier, which still has to be determined.

In our case the generalized coordinates are u_1, u_2, u_3, and u_4. Using Equation (18) we obtain

$$T = \tfrac{1}{2}|\dot{x}|^2 = \tfrac{1}{2}\dot{x}^T\dot{x} = 2\dot{u}^T L^T L\dot{u} = 2|u|^2|\dot{u}|^2 = 2r|\dot{u}|^2,$$

and taking the derivatives with respect to $\dot{u}_i$ and u_i we obtain

$$\frac{\partial T}{\partial \dot{u}_i} = 4r\dot{u}_i, \qquad \frac{\partial T}{\partial u_i} = 4u_i|\dot{u}|^2,$$

where Equation (15) was used.

With a potential of the form*

$$U = -\frac{K^2}{r} + V$$

and the constraint

$$l = u_4\dot{u}_1 - u_3\dot{u}_2 + u_2\dot{u}_3 - u_1\dot{u}_4 = 0$$

we finally find

$$4(r\dot{u}_i)^{\cdot} - 4|\dot{u}|^2 u_i + 2\frac{K^2}{r^2}u_i = -\frac{\partial V}{\partial u_i} + \lambda a_i,$$

where

$$a_1 = u_4, \qquad a_2 = -u_3, \qquad a_3 = u_2, \qquad a_4 = -u_1.$$

Introducing the fictitious time s as the independent variable by means of

$$(\ )^{\cdot} = \frac{1}{r}(\ )',$$

we obtain

$$u_i'' + \frac{\tfrac{1}{2}K^2 - |u'|^2}{(u, u)}u_i = -\frac{r}{4}\frac{\partial V}{\partial u_i} + \lambda\frac{r}{4}a_i. \tag{19}$$

Since

$$|x'|^2 = x'^T x' = 4u'L^T Lu' = 4|u|^2|u'|^2,$$

the energy relation (6) transforms to

$$h = \frac{K^2 - 2|u'|^2}{(u, u)} - V. \tag{20}$$

Substituting this expression into the left-hand side of Equation (19), we obtain

$$u_i'' + \tfrac{1}{2}hu_i = -\frac{r}{4}\frac{\partial V}{\partial u_i} - \frac{V}{2}u_i + \lambda\frac{r}{4}a_i. \tag{21}$$

The Lagrangian factor λ can be determined in the usual manner by differentiating the constraint, i.e. bilinear relation, and inserting the equations of motion (21). Using

$$l = u_4u_1' - u_3u_2' + u_2u_3' - u_1u_4' = 0$$

* The additional force P will be taken care of later on.

we obtain

$$l' = u_4 u_1'' - u_3 u_2'' + u_2 u_3'' - u_1 u_4'' = 0,$$

hence

$$\begin{aligned} l' = & - \tfrac{1}{2}(h + V)\left[u_4 u_1 - u_3 u_2 + u_2 u_3 - u_1 u_4\right] - \\ & - \frac{r}{4}\left(u_4 \frac{\partial V}{\partial u_1} - u_3 \frac{\partial V}{\partial u_2} + u_2 \frac{\partial V}{\partial u_3} - u_1 \frac{\partial V}{\partial u_4}\right) + \\ & + \lambda \frac{r}{4}\left[u_4 u_4 + u_3 u_3 + u_2 u_2 + u_1 u_1\right] = 0 . \end{aligned}$$

The first square bracket vanishes identically. The value of the second round bracket is the fourth component of $L(\partial V/\partial u)$. Since

$$\frac{\partial V}{\partial u_i} = \sum_{j=1}^{3} \frac{\partial V}{\partial x_j} \frac{\partial x_j}{\partial u_i},$$

we have

$$\frac{\partial V}{\partial u} = 2L^T \frac{\partial V}{\partial x},$$

where the fourth component of the gradient $\partial V/\partial x$ is zero. Thus the second round bracket is equal to the fourth component of the vector $L(2L^T(\partial V/\partial x))$, which in turn is equal to the fourth component of $2r(\partial V/\partial x)$, which vanishes by definition. We are left with the expression

$$l' = \lambda \frac{r}{4}(u, u) = \lambda \frac{r^2}{4} = 0$$

from which we conclude that λ must vanish, since in general $r \neq 0$. Hence the equations of motion do not need to be modified by terms stemming from the restrictive bilinear relation. The bilinear relation is *no essential constraint.*

The resulting equations of motion

$$u'' + \tfrac{1}{2}hu = -\frac{r}{4}\frac{\partial V}{\partial u} - \frac{V}{2}u \tag{22}$$

are of the type of a *perturbed harmonic oscillation* with slowly varying frequency, if the perturbing potential is small.

It is of importance to note that not only the gradient of the potential appears but also the potential itself. Of course the arbitrariness of the additive constant in the potential does not influence the motion of the particle, since the total negative energy h contains the same additive constant as V, thus compensating this arbitrariness. However, the choice of the additive constant does influence the frequency appearing on the left-hand side of the oscillator equation.

Before collecting the final set of the equations of motion, we proceed to generalize our equations to include non-conservative forces. This can be done in a heuristic way by considering the work done per unit of time by the perturbing force

$$\begin{aligned}(P, \dot{x}) &= 2\,(L\dot{u}, P) = 2\,(L\dot{u})^T P = 2\dot{u}^T L^T P = 2\,(\dot{u}, L^T P)\,,\\ (P, \dot{x}) &= (2L^T P, \dot{u})\,.\end{aligned} \tag{23}$$

On the other hand, the work done per unit of time by a four-dimensional force Q in the u-space is given by $(Q, \dot{u})$. Comparing this expression with (23), we see that the force Q is given by

$$Q = 2L^T P\,.$$

Thus the gradient $\partial V/\partial u$ must be augmented by the force Q, hence

$$u'' + \frac{h}{2}\,u = -\,\frac{V}{2}\,u + \frac{(u, u)}{2}\left[-\,\frac{1}{2}\frac{\partial V}{\partial u} + L^T P\right], \tag{24}$$

or written in a slightly different way

$$u'' + \frac{h}{2}\,u = -\,\frac{1}{4}\frac{\partial}{\partial u}\,[(u, u)\,V] + \frac{(u, u)}{2}\,L^T P\,.$$

These are the general equations of motion for the coordinates u_1, u_2, u_3 and u_4 as a function of the fictitious time s. The law of energy (7), which is straightforwardly transformed by use of Equation (23), states that

$$h' = -\,(u, u)\,\frac{\partial V}{\partial t} - 2\,(L^T P, u')\,. \tag{25}$$

Finally the physical time t is evaluated from the equation

$$t' = r = (u, u)\,. \tag{26}$$

The total order of this system (24), (25), (26) is 10, in contrast to the order of the original Keplerian equations which is 6. This increase is no disadvantage from the numerical point of view since it will be seen that these equations have the desired properties of regularization and can be stabilized.

COMMENTS

(1) Assume that a conservative perturbing potential $V(x)$ is given and that no additional force P occurs. Since $V = V\big(x(u)\big)$ does not depend on time t explicitly, h is constant and thus the frequency of the oscillator is a constant. This is one of the most important properties of the above set of equations of motion in practical applications. An example is the motion of a satellite under the influence of the J_2-oblateness potential of the earth.

(2) It is necessary to make some remarks concerning the initial conditions. As previously mentioned the matrix transformation of velocities of Equation (18) is only valid if the bilinear relation holds true and vice versa. Thus by inverting Equation (18) we obtain u',

$$u' = \frac{1}{2r}\,L^T x' = \tfrac{1}{2} L^T \dot{x}$$

as a function of u and x' and enforce the bilinear relation to be satisfied.

The inversion of the KS-transformation (12) is, however, not unique; it is found that all points lying on a given circle about the origin in the u-space lead to the same point in the physical space (Stiefel and Scheifele, 1971). See also the following collection of formulae. Provided the initial conditions are chosen as just described, it can be shown that all solutions of the general equations of motion (24) satisfy the bilinear relation during motion (Stiefel and Scheifele, 1971).*

(3) In case of unperturbed elliptic motion the Equations (24) are reduced to the equations of a harmonic oscillator

$$u'' + \frac{h}{2} u = 0 .$$

The constant frequency is $\omega = \sqrt{h/2}$. These equations are regular and it is easy to verify that these differential equations are stable in the sense of Ljapunow, provided h is held fixed.

In contrast, Equation (26), $t = \int r \, ds$, is unstable. A small change in the radius by a constant amount ε gives rise to an erroneous time which can differ by an arbitrarily large amount from the reference solution, since

$$t = \int (r + \varepsilon) \, ds = \int r \, ds + \varepsilon \cdot s .$$

This instability will be removed in the next section by introducing a time element.

COLLECTION OF FORMULAE

The following arrangement of equations should facilitate the construction of a computational program. Matrix relations are written in explicit notation.

	Symbol
Time	t
Independent variable	s
Differentiation with respect to t	dot
Differentiation with respect to s	prime
Gravitational constant	k^2
Central mass	M
Mass of moving particle	m
$k^2(M+m)$	K^2
Position vector	x
Its three components	x_i
Distance from the origin	r
Perturbing potential	$V(t, x_i)$
Additional perturbing force	P
Its three components	P_i
Total energy of the particle	$-h$
KS-position vector	u
Its four components	u_j
KS-additional force	$L^T P$
Its four components	$(L^T P)_i$
Scalar product	parenthesis, comma

* Furthermore the KS-mapping of such a solution satisfies the three-dimensional Equation (5).

(1) KS-transformation.
(a) Of the position.

$$x_1 = u_1^2 - u_2^2 - u_3^2 + u_4^2, \qquad x_2 = 2(u_1 u_2 - u_3 u_4),$$
$$x_3 = 2(u_1 u_3 + u_2 u_4). \tag{I}$$

(b) Of the distance.

$$r = (u, u) = |u|^2 = u_1^2 + u_2^2 + u_3^2 + u_4^2.$$

(c) Of the velocity.

$$\dot{x}_1 = \frac{2}{r}(u_1 u_1' - u_2 u_2' - u_3 u_3' + u_4 u_4'),$$
$$\dot{x}_2 = \frac{2}{r}(u_2 u_1' + u_1 u_2' - u_4 u_3' - u_3 u_4'), \tag{II}$$
$$\dot{x}_3 = \frac{2}{r}(u_3 u_1' + u_4 u_2' + u_1 u_3' + u_2 u_4').$$

(d) Of the potential. Insert (I) into $V(t, x_i)$.
(e) Of the additional force. Insert (I) and (II) into P_i, then

$$(L^T P)_1 = u_1 P_1 + u_2 P_2 + u_3 P_3,$$
$$(L^T P)_2 = -u_2 P_1 + u_1 P_2 + u_4 P_3,$$
$$(L^T P)_3 = -u_3 P_1 - u_4 P_2 + u_1 P_3,$$
$$(L^T P)_4 = u_4 P_1 - u_3 P_2 + u_2 P_3.$$

(2) Differential equations for u, t, h as functions of s.

$$u'' + \frac{h}{2} u = -\frac{1}{4} \frac{\partial}{\partial u} (|u|^2 V) + \frac{|u|^2}{2} (L^T P)$$

or written in terms of the components

$$u_j'' + \frac{h}{2} u_j = -\frac{1}{4} \frac{\partial}{\partial u_j} (|u|^2 V) + \frac{|u|^2}{2} (L^T P)_j, \qquad j = 1, 2, 3, 4,$$
$$h' = -|u|^2 \frac{\partial V}{\partial t} - 2(u', L^T P), \qquad t' = |u|^2.$$

(3) Initial conditions. Assume $x, \dot{x}$ given at instant $t=0$. Compute u, u', t, h at instant $s=0$ from

$$u_1^2 + u_4^2 = \tfrac{1}{2}(r + x_1), \; u_2 = \frac{x_2 u_1 + x_3 u_4}{r + x_1}, \; u_3 = \frac{x_3 u_1 - x_2 u_4}{r + x_1} \; \text{if } x_1 \geqq 0,$$

else

$$u_2^2 + u_3^2 = \tfrac{1}{2}(r - x_1), \; u_1 = \frac{x_2 u_2 + x_3 u_3}{r - x_1}, \; u_4 = \frac{x_3 u_2 - x_2 u_3}{r - x_1},$$

and

$$u_1' = \tfrac{1}{2}(u_1\dot{x}_1 + u_2\dot{x}_2 + u_3\dot{x}_3),$$
$$u_2' = \tfrac{1}{2}(-u_2\dot{x}_1 + u_1\dot{x}_2 + u_4\dot{x}_3),$$
$$u_3' = \tfrac{1}{2}(-u_3\dot{x}_1 - u_4\dot{x}_2 + u_1\dot{x}_3),$$
$$u_4' = \tfrac{1}{2}(u_4\dot{x}_1 - u_3\dot{x}_2 + u_2\dot{x}_3),$$

finally

$$h = \frac{K^2}{r} - \tfrac{1}{2}|\dot{x}|^2 - V, \qquad t = 0.$$

(4) Checks during integration.
Energy equation

$$h = \frac{K^2 - 2|u'|^2}{|u|^2} - V.$$

Bilinear relation

$$u_4u_1' - u_3u_2' + u_2u_3' - u_1u_4' = 0.$$

3. Regular Elements

Elements are quantities which, during unperturbed motion, vary linearly with respect to the independent variable. The total negative energy h is an example of an element, which remains constant in unperturbed motion. In the classical treatment of the two-body problem 6 elements are found using the physical time t as the independent variable. Elements have the advantage that they vary almost linearly if the motion is influenced by weak perturbations.

We will begin by introducing a time element τ which is defined by

$$\tau = t + \frac{1}{h}(u, u'), \qquad (h > 0).$$

Differentiating this equation we obtain

$$\tau' = r + \frac{1}{h}(u, u'') + \frac{1}{h}(u', u') - \frac{h'}{h^2}(u, u').$$

Inserting Equation (24) and replacing (u', u') by

$$(u', u') = \tfrac{1}{2}(K^2 - r(h + V)),$$

obtained by solving Equation (20), the above equation becomes

$$\tau' = \frac{1}{2h}(K^2 - 2rV) - \frac{r}{4h}\left[\left(u, \frac{\partial V}{\partial u} - 2L^TP\right)\right] - \frac{h'}{h^2}(u, u').$$

In the unperturbed case we have

$$\tau' = \frac{K^2}{2h},$$

hence

$$\tau = \frac{K^2}{2h} s.$$

Therefore τ is indeed an element. Let us note furthermore that, if the energy h is held fixed, the solution τ is stable. Thus we have stabilized the time equation without increasing the order of the system. However the general principle is the same as in the stabilization of the u-equations: the introduction of elements into differential equations can produce stabilization. The element h was introduced into the u-equations.

COLLECTION OF FORMULAE

The differential equations for the vector u, the total negative energy h and the time element τ are given by

$$u'' + \frac{h}{2} u = -\frac{1}{4}\frac{\partial}{\partial u}[(u, u) V] + \frac{|u|^2}{2} L^T P \tag{27}$$

$$h' = -|u|^2 \frac{\partial V}{\partial t} - 2(u', L^T P) \tag{28}$$

$$\tau' = \frac{1}{2h}(K^2 - 2rV) - \frac{r}{4h}\left(u, \frac{\partial V}{\partial u} - 2L^T P\right) - \frac{h'}{h^2}(u, u'). \tag{29}$$

COMMENT. In the canonical theory (Stiefel and Scheifele, 1971), t is canonically conjugated to the the negative energy h. This is the reason why we started by introducing a time element rather than elements for the coordinates u_i.

Our final step is to introduce elements attached to the coordinates u_i. Note that the frequency of oscillation $\omega = \sqrt{h/2}$ is no longer constant in perturbed motion. In order to overcome this difficulty in an elementary way, we will define a new independent variable E by the differential relation

$$E' = 2\omega.$$

In pure elliptic motion we have $E = 2\omega s$, which is the definition of the *eccentric anomaly*. The equations of motion may be rewritten in terms of E, the generalized eccentric anomaly,

$$\frac{d^2 u}{dE^2} + \tfrac{1}{4} u = PT,$$

where

$$PT = -\frac{1}{4\omega^2}\left[\frac{V}{2}u + \frac{r}{4}\left(\frac{\partial V}{\partial u} - 2L^T P\right)\right] - \frac{1}{\omega}\frac{\mathrm{d}\omega}{\mathrm{d}E}\frac{\mathrm{d}u}{\mathrm{d}E}.$$

The frequency term appearing on the left-hand side now has a fixed constant value of $\frac{1}{2}$. Assuming a solution of the form

$$u = \alpha \cos\frac{E}{2} + \beta \sin\frac{E}{2},$$

where α and β are 4-vectors depending on the dependent variable E, and requiring that

$$\frac{\mathrm{d}u}{\mathrm{d}E} = -\tfrac{1}{2}\alpha \sin\frac{E}{2} + \tfrac{1}{2}\beta\cos\frac{E}{2},$$

we obtain the differential equations for α and β in the usual manner

$$\frac{\mathrm{d}\alpha}{\mathrm{d}E} = -2PT\sin\frac{E}{2}, \quad \frac{\mathrm{d}\beta}{\mathrm{d}E} = 2PT\cos\frac{E}{2}.$$

The quantities α and β are indeed elements because, in the unperturbed case where $PT=0$, they are constant.

The two further element Equations (28) and (29) still have to be rewritten in terms of the independent variable E, however this will not be carried out here. The reader is referred to the book from Stiefel and Scheifele (1971). The order of the resulting element equations is ten.

COMMENTS

(1) Note that the derivation of the element equations is much easier than in the classical case where the elements a, e, I, Ω, ω, M, are used. This is a consequence of the linearity of the differential equations. Furthermore, we were able to introduce the elements step by step; beginning with the elements h and τ and ending with the elements α and β.

(2) The elements α, β, h and τ are always well defined regardless which type of pure elliptic Kepler motion is considered. Provided no singularities appear in the perturbing terms, the resulting element equations are absolutely free of singularities even if collision occurs. The elements are indeed *regular*. It should be recalled that the classical Keplerian elements are undetermined if the orbit is circular, if it lies in the x_1, x_2-plane or if collision or ejection occurs. The corresponding classical element equations consequently have singularities if any of these cases occur.

(3) In contrast to the classical element equations which are of the order six, the regular element equations are a system of tenth order. Further on, we will show that it is not possible to find any set of 6 regular elements and that it is necessary to use redundant elements.

4. Final Remarks

In this section we will discuss and clarify problems which inevitably arose during the foregoing description of the regularizing procedure.

A. CONTINUOUS FIELDS

We recall that the essential feature of the KS-transformation was that we could find an orthogonal matrix whose elements are linear homogeneous functions of the parameters u_i and in which the first column is the state vector u. These matrices have been investigated by Hurwitz and are shown to exist for two, four and eight dimensional matrices. The reason for the non-existence of a 3×3 matrix of this type is easily seen as follows. Consider the following 3×3 matrix

$$\begin{pmatrix} u_1 & f_1(u) & g_1(u) \\ u_2 & f_2(u) & g_2(u) \\ u_3 & f_3(u) & g_3(u) \end{pmatrix},$$

where f_i and g_i are linear homogeneous functions of u_i. Putting

$$u_1^2 + u_2^2 + u_3^2 = 1,$$

we note that the state vector u lies on a 2-dimensional sphere S^2. The vectors f and g are perpendicular to the state vector u and to themselves. Hence a local cartesian frame of two tangential vectors is attached to any point of S^2. This field of frames is continuous over the whole field. As follows from a theorem of Poincaré and L. E. J. Brouwer, however, no such field can exist; thus proving the non-existence of a Levi-Cività matrix in three dimensions. In passing we note that only the spheres S^1, S^3 and S^7 allow continuous fields of frames. The sphere S^1 is used in the Levi-Cività matrix whereas the sphere S^3 is attached to the KS-matrix.

B. REGULARIZING TRANSFORMATIONS

It is natural to ask, whether a different more general approach than a generalization of the Levi-Cività matrix can lead to regular differential equations without increasing dimensions. B. Gagliardi has recently shown that it is impossible to regularize without increasing the number of dependent coordinates. He proved that it is not possible to regularize the unperturbed Kepler motion by transformations satisfying the following general assumptions:

(1) The transformation of the physical coordinates x_i to the new coordinates u_j has the form

$$x_i = x_i(u_j), \qquad i, j = 1, 2, 3$$

and its inverse is assumed to exist.

(2) A time-transformation is admitted

$$\frac{dt}{ds} = \mu(u_i, \dot{u}_i, s),$$

where μ is an arbitrary positive density function. The transformation of velocities thus has the form

$$\frac{\mathrm{d}x_i}{\mathrm{d}t} = \frac{1}{\mu} \sum \frac{\partial x_i}{\partial u_\varrho} u'_\varrho .$$

From Equation (15) we see that the KS-transformation is a mapping of the 3-dimensional sphere $S^3 : u_1^2 + u_2^2 + u_3^2 + u_4^2 = 1$ onto the 2-dimensional sphere $S^2 : x_1^2 + x_2^2 + x_3^2 = 1$.

Another question of importance is whether some other mapping of S^3 into S^2 which is simpler than the KS-transformation can also achieve regularization.

About thirty years ago H. Hopf classified mappings, and it can be shown that the KS-mapping is contained in Hopf's simplest class. Hence the KS-transformation is indeed a good choice from the geometrical point of view.

C. THE MANIFOLD OF KEPLER ORBITS

We now turn our attention to the question whether it is possible to find 6 elements which are regular.

Five elements are needed for characterizing the shape and position of the orbit and the sixth is attached to the position of the particle in its orbit. Without loss of generality we may restrict ourselves to the discussion of the five 'geometrical' elements. In classical celestial dynamics they are a, e, I, Ω and ω.

Let us consider the family of elliptic orbits which have a fixed value of the semi-major axis a. To each orbit we will attach a pair of 3-vectors p and q. The vector p is the position vector of the empty focus and its length is $|p| = 2ae$, where e is the eccentricity of the orbit. For circular orbits $p = 0$, and for collision orbits we have $|p| = 2a$. The vector p is uniquely and unambiguously defined for all elliptic orbits. It is called the Laplace vector. The vector q is normal to the orbital plane, has the magnitude $2a\sqrt{1-e^2}$ and its sense of direction builds a right-handed orientation together with the sense of revolution of the orbit. Although there is no sense of revolution in the collision orbit, q nevertheless remains well determined since it vanishes altogether. Hence the q vector is also uniquely and unambiguously defined.

The two vectors p and q are orthogonal. Let us define two other vectors y and z by

$$y = p + q, \qquad z = p - q .$$

Due to the orthogonality of p and q we have

$$|y| = \sqrt{4a^2 (e^2 + 1 - e^2)} = 2a ,$$
$$|z| = 2a ,$$

thus y and z may be considered to be points on two-dimensional spheres of radius $2a$. Any independent choice of any two vectors y and z on the spheres define two orthogonal vectors p and q which in turn determine a Kepler orbit. In topological language this fact is expressed by saying that the structure of the manifold of Kepler orbits with fixed semi-major axis is the product of two-dimensional spheres.

Defining elements is thus equivalent to determining coordinates of two vectors on two spheres. Clearly this cannot be done by adopting on each sphere, two coordinates without introducing singularities (polar coordinates, for example, have singularities). Hence a total set of 6 regular elements does not exist. We have thus shown that regular element require the use of a redundant set of elements.

Acknowledgements

The author is indebted to M. Vitins for establishing the manuscript according to the author's lectures delivered in Cortina d'Ampezzo and for his suggestions and comments.

Reference

Stiefel, E. L. and Scheifele, G.: 1971, *Linear and Regular Celestial Mechanics*, Springer-Verlag, Berlin, Heidelberg, New York.

COLLISION SINGULARITIES IN GRAVITATIONAL PROBLEMS

J. WALDVOGEL

Seminar für Angewandte Mathematik,
Swiss Federal Institute of Technology, Zürich, Switzerland

Abstract. The rotating-pulsating coordinate system commonly used in the elliptic restricted problem of three bodies is modified to be useful also in the limit of rectilinear motion of the primaries. In these coordinates the homographic solutions of the problem of n bodies are equilibrium solutions, and so are, in particular, the homothetic solutions leading to a collision of n bodies. General collision solutions are thus obtained as solutions which asymptotically approach these equilibriums. In an example of a continuable triple collision the expansion valid at the singularity is given explicitely.

1. Introduction

In the motion of n point masses under their mutual Newtonian attraction close encounters and collisions between two or more bodies are very important events. From a practical point of view the handling of close encounters is more important than the investigation of actual collisions. For obtaining a deeper understanding of the problem of n bodies, however, a good knowledge of its singularities is indispensable.

A singularity occurring at a finite time t_0 is called a collision singularity if the position vectors of all n mass points have finite limits as $t \to t_0$. Only for $n \leqslant 3$ it is known that the only possible singularities are collision singularities. For $n > 3$ it cannot a priori be excluded that non-collision singularities occur at a finite time.

In the problem of two bodies ($n=2$) the situation is comparatively simple: the only possible singularities are binary collisions. It is well known that these (even in the presence of other bodies) correspond to algebraic branch points in the solutions. The singularities due to binary collisions may be removed by various regularization techniques, and the motion can always be continued beyond the singularity.

For $n=3$ there was shown by Painlevé and Sundman that the only possible singularities at finite times are binary and triple collisions. Furthermore, a triple collision is excluded when the total angular momentum $\mathbf{C} \neq 0$, but it needs not occur when $\mathbf{C}=0$. The triple collision was first investigated by Sundman (1907) and later in an exhaustive article by Siegel (1941). The result is that a triple collision corresponds to a transcendental singularity of the solutions (branch point of infinite order), and the motion cannot be continued beyond the collision, except in a few special cases.

Collisions of n bodies in general were first considered by Chazy (1918). He found that the n bodies approach a central configuration as they collide. The nature of the singularity is similar for all $n \geqslant 3$, but for $n > 3$ it is not known whether other types of collisions may occur, too. The best results in this respect have recently been established by Sperling (1970). He proved that in a general collision singularity the n bodies separate into clusters of n_i members ($1 \leqslant n_i \leqslant n$, $\sum n_i = n$), and all clusters collapse

B. D. Tapley and V. Szebehely (eds.), Recent Advances in Dynamical Astronomy, 21–33.

simultaneously. The individual cluster shows the behaviour of a collision of all points in a problem of n_i bodies.

In the present paper collisions of n bodies will be described in slightly modified rotating-pulsating coordinates commonly used in the elliptic restricted problem of three bodies. No completely new results are derived, but considerable insight into the theory of homographic solutions and into the behaviour at a collision is obtained, and also expansions valid in a singularity will be given explicitly.

2. The Problem of *n* Bodies

The motion of n point masses m_i subject to their mutual Newtonian attraction is described by the differential equations

$$m_i \frac{\mathrm{d}^2 \mathbf{x}_i}{\mathrm{d}t^2} = U_{\mathbf{x}_i}, \qquad i = 1, \ldots, n, \tag{1}$$

where $\mathbf{x}_i$ is the position vector of m_i in a three-dimensional inertial coordinate system Γ,

$$U = \sum_{i<j} \frac{m_i m_j}{|\mathbf{x}_i - \mathbf{x}_j|} \tag{2}$$

is the force function, and $U_{\mathbf{x}_i}$ is the gradient of U with respect to $\mathbf{x}_i$.

The coordinate system Γ is chosen such that the center of mass of the system $m_i (i=1, \ldots, n)$ is at rest at the origin $\mathcal{O}$ of Γ:

$$\sum_{i=1}^{n} m_i \mathbf{x}_i = 0. \tag{3}$$

3. The True Anomaly in Kepler Motion

Consider the Kepler motion governed by the differential equation

$$\frac{\mathrm{d}^2 \mathbf{x}}{\mathrm{d}t^2} + \mu \frac{\mathbf{x}}{r^3} = 0, \qquad r = |\mathbf{x}|, \tag{4}$$

where $\mathbf{x}$ is a 2-vector and μ is the gravitational parameter of the central body. The angular momentum integral of (4) is

$$r^2 \frac{\mathrm{d}\vartheta}{\mathrm{d}t} = \sqrt{p\mu}, \tag{5}$$

where ϑ and p are the true anomaly and the semi-latus rectum, respectively. Here ϑ will be measured counterclockwise from the *apocenter* of the Keplerian orbit (in contradiction to the customary definition of the true anomaly), in order to exclude that the point $\vartheta=0$ falls into a collision. Now the variable σ is introduced by the differential relation

$$\mathrm{d}\sigma = \frac{\mathrm{d}t}{r^2}\sqrt{\mu}\,; \tag{6}$$

it can be made proportional to ϑ:

$$\sigma = \vartheta/\sqrt{p}\,. \tag{7}$$

σ will be called the *scaled true anomaly*. During half a revolution σ varies in the interval $0 \leqslant \sigma \leqslant \pi/\sqrt{p}$, the upper boundary increasing unlimited as $p \to 0$. In terms of σ the well-known polar equation of the Keplerian orbit is

$$r = \frac{p}{1 - e\cos(\sigma\sqrt{p})}, \tag{8}$$

where e is the eccentricity of the orbit. Denoting the semi-major axis of the Keplerian orbit by a, we now shall examine the Kepler motion in the limit

$$p \to 0\,, \qquad e \to 1 \qquad (a \text{ fixed}) \tag{9}$$

of rectilinear motion. Obviously, the true anomaly ϑ can no longer be used as a parameter in this limit.

However, it is possible to take the limit (9) of Equation (8):

$$r^*(\sigma) = \lim r(\sigma) = \lim \frac{1}{\dfrac{1-e}{p} + \dfrac{e}{2}\sigma^2 + \mathcal{O}(p)}\,. \tag{10}$$

From geometry one easily obtains

$$\lim \frac{1-e}{p} = \lim_{e\to 1} \frac{1-e}{a(1-e^2)} = \frac{1}{2a} = \frac{c}{2},$$

where $c = a^{-1}$ is introduced as an abbreviation of the reciprocal semi-major axis. Hence, Equation (10) yields the relation

$$r^*(\sigma) = \frac{2}{\sigma^2 + c} \tag{11}$$

which proves that σ is a good parameter also for rectilinear Kepler motion. Equation (11) can be shown to be valid for all 3 types of rectilinear motion. The collision takes place at $\sigma=\infty$; $\sigma=0$ corresponds to the apocenter in the elliptic case ($c>0$), and $\sigma=\sqrt{-c}$ represents the particle being at infinity in the hyperbolic or parabolic case ($c\leqslant 0$). For each revolution between two collisions in the elliptic case σ runs from $-\infty$ to $+\infty$.

The timing of the rectilinear motion is obtained from Equation (6) which may be written in the following two forms:

$$t = \frac{4}{\sqrt{\mu}} \int \frac{\mathrm{d}\sigma}{(\sigma^2 + c)^2} \tag{12}$$

$$t_0 - t = \frac{4}{\sqrt{\mu}} \left(\frac{1}{3} \sigma^{-3} - \frac{2c}{5} \sigma^{-5} + \frac{3c^2}{7} \sigma^{-7} - \cdots \right).$$

Carrying out the integration in closed form yields

$$t_0 - t = \frac{2}{c\sqrt{\mu}} \left[\frac{1}{\sqrt{c}} \operatorname{arc\,cot} \left(\frac{\sigma}{\sqrt{c}} \right) - \frac{\sigma}{\sigma^2 + c} \right]$$

for the elliptic case ($c > 0$).

4. Scheibner's Coordinates

Scheibner (1866) observed that the elliptic restricted problem of three bodies is best described in a rotating and 'pulsating' coordinate system (where the primaries remain at fixed positions) with the *true* anomaly of the primaries as independent variable. This idea – in a generalized form – will be used in order to transform the equations of motion of the problem of n bodies.

Consider a fictitious massless particle moving on a reference Kepler orbit under the influence of an equally fictitious central body (with gravitational parameter μ) located at the origin $\mathcal{O}$ of the inertial coordinate system Γ introduced in Section 2. The motion of the particle is governed by Equation (4). We assume without loss of generality the 3-axis of this coordinate system to be perpendicular to the orbital plane of the Kepler orbit.

Consider now a Cartesian frame Γ' rotating about the 3-axis and pulsating such that the fictitious particle occupies a fixed position with respect to Γ'. The transformation of the position vectors $\mathbf{x}_i$ is given by

$$\mathbf{x}_i = rD\boldsymbol{\xi}_i, \qquad i = 1, \ldots, n, \tag{13}$$

where $\boldsymbol{\xi}_i$ is the position vector of m_i with respect to Γ', and D is the rotation matrix

$$D = \begin{pmatrix} \cos\vartheta & -\sin\vartheta & 0 \\ \sin\vartheta & \cos\vartheta & 0 \\ 0 & 0 & 1 \end{pmatrix}. \tag{14}$$

As a new independent variable the *scaled true anomaly* σ, defined in Equation (6), will be used contrary to Scheibner in order to allow taking the limit (9). Such coordinates will be referred to as *Scheibnerian coordinates* (associated with a certain reference Kepler motion), which therefore may be characterized by a rectangular coordinate system Γ' 'pulsating' and rotating about its 3-axis such that a fictitious masseless particle, whose inertial position vector $\mathbf{x}$ describes the reference Kepler motion governed by

$$\frac{\mathrm{d}^2\mathbf{x}}{\mathrm{d}t^2} + \mu \frac{\mathbf{x}}{r^3} = 0, \qquad r = |\mathbf{x}|$$

remains at a fixed position with respect to Γ'. The *scaled* true anomaly $\vartheta/\sqrt{p}$ of the reference Kepler motion is used as independent variable.

Denoting derivatives with respect to σ by primes, differentiation of Equation (13) yields

$$\begin{aligned}\frac{d\mathbf{x}_i}{dt} &= \frac{\sqrt{\mu}}{r^2}(r'D\xi_i + rD'\xi_i + rD\xi_i') = \\ &= \sqrt{\mu}(r'r^{-2}D\xi_i + r^{-1}D'\xi_i + r^{-1}D\xi_i')\\ \frac{d^2\mathbf{x}_i}{dt^2} &= \mu r^{-2}(r''r^{-2}D\xi_i - 2r'^2r^{-3}D\xi_i + \\ &\quad + r^{-1}D''\xi_i + 2r^{-1}D'\xi_i' + r^{-1}D\xi_i'')\\ D^T\frac{d^2\mathbf{x}_i}{dt^2} &= \mu r^{-3}[\xi_i'' + 2D^TD'\xi_i' + [(r^{-1}r'' - 2r^{-2}r'^2)I + D^TD'']\,\xi_i], \end{aligned} \tag{15}$$

where D^T is the transpose of D and I is the unit matrix. By using (14) and (7) the matrix products in (15) may be written as

$$\begin{aligned} D^TD' &= \sqrt{p}C \quad \text{where} \quad C = \begin{pmatrix} 0 & -1 & 0\\ 1 & 0 & 0\\ 0 & 0 & 0\end{pmatrix},\\ D^TD'' &= p(F - I) \quad \text{where} \quad F = \begin{pmatrix} 0 & 0 & 0\\ 0 & 0 & 0\\ 0 & 0 & 1\end{pmatrix}. \end{aligned} \tag{16}$$

Hence, the first two diagonal elements of the factor of ξ_i in Equation (15) are

$$r''r^{-1} - 2r'^2r^{-2} - p = -r\left[\left(\frac{1}{r}\right)'' + \frac{p}{r}\right].$$

Equation (8) implies

$$\left(\frac{1}{r}\right)'' + p\cdot\frac{1}{r} = 1\,. \tag{17}$$

Thus, Equation (15) becomes

$$D^T\frac{d^2\mathbf{x}_i}{dt^2} = \mu r^{-3}[\xi_i'' + 2\sqrt{p}C\xi_i' + (pF - rI)\,\xi_i]\,. \tag{18}$$

In order to transform Equation (1) to Scheibner's coordinates notice that

$$U_{\mathbf{x}_i} = r^{-2}DV_{\xi_i}, \qquad i = 1, \ldots, n, \tag{19}$$

where

$$V(\xi_1, \ldots, \xi_n) = \sum_{i<j}\frac{m_im_j}{|\xi_i - \xi_j|}. \tag{20}$$

The equations of motion for the problem of n bodies in Scheibner's coordinates are therefore

$$m_i(\xi_i'' + 2\sqrt{p}C\xi_i' - r(\sigma)\xi_i + pF\xi_i) = \frac{r(\sigma)}{\mu} V_{\xi_i}, \quad i = 1, \ldots, n, \tag{21}$$

where the matrices C and F are defined in (16), and $r(\sigma)$ is given by

$$r(\sigma) = \frac{p}{1 - e\cos(\sigma\sqrt{p})}.$$

In the limit $p \to 0$, $e \to 1$ we immediately obtain

$$\frac{\sigma^2 + c}{2}\xi_i'' = \xi_i + \frac{1}{\mu m_i} V_{\xi_i}, \quad i = 1, \ldots, n. \tag{22}$$

5. Homographic Solutions

A homographic solution $\mathbf{x}_i(t)$ of the problem of n bodies is a solution such that the configuration of the n point masses remains similar to itself. It is immediate that every equilibrium solution $\xi_i(\sigma) = \text{const.}$ of Equation (21) corresponds to a homographic solution of Equation (1). In these homographic solutions the point masses describe similar Keplerian orbits with one focus at the origin $\mathcal{O}$.

Pizzetti (1904) showed that there are no other homographic solutions to the gravitational problem of n bodies. The proof is rather lengthy and will not be reproduced here; one would have to consider a coordinate system with a variable axis of rotation.

Consequently the homographic solutions may be discussed by determining the equilibrium solutions $\xi_i(\sigma) = \text{const.}$ of Equation (21). Putting $\xi_i'' = \xi_i' = 0$ yields

$$\begin{aligned} r(\sigma)\frac{\partial V}{\partial \xi_{i1}} + r(\sigma)\cdot\mu m_i \xi_{i1} &= 0 \\ r(\sigma)\frac{\partial V}{\partial \xi_{i2}} + r(\sigma)\cdot\mu m_i \xi_{i2} &= 0 \qquad i = 1, \ldots, n \\ r(\sigma)\frac{\partial V}{\partial \xi_{i3}} + (r(\sigma) - p)\cdot\mu m_i \xi_{i3} &= 0 \end{aligned} \tag{23}$$

where ξ_{ij} is the jth component of the vector ξ_i. Necessary for the third of these equations to be satisfied is $\xi_{i3} = 0$ $(i = 1, \ldots, n)$ (which implies $\partial V/\partial \xi_{i3} = 0$) or $p = 0$.

In the first case all points ξ_i lie in the plane of rotation, and the necessary and sufficient conditions for the ξ_i are

$$\begin{aligned} &p \text{ arbitrary}; \quad \xi_{i3} = 0 \\ &V_{\xi_i} + \mu m_i \xi_i = 0, \quad i = 1, \ldots, n. \end{aligned}$$

In the second case we have

$$\begin{aligned} &p = 0 \\ &V_{\xi_i} + \mu m_i \xi_i = 0, \quad i = 1, \ldots, n. \end{aligned} \tag{24}$$

Note that Equation (13) implies the condition

$$\sum_{i=1}^{n} m_i \xi_i = 0 . \tag{25}$$

Hence, in homographic solutions the mass points describe similar Keplerian orbits of arbitrary excentricity lying all in a fixed plane, or they describe rectilinear Keplerian orbits. These latter homographic solutions are called *homothetic solutions*.

A set of points ξ_i satisfying Equations (24) and (25) is said to form a *central configuration* associated with the set of masses m_i. The theory of central configurations was developed by Dziobek (1900); a good account of it may be found in Wintner (1941).

Only a few facts needed later will be mentioned here. Condition (24) requires the resultant force V_{ξ_i} onto m_i to be proportional to m_i and to its position vector ξ_i. The factor of proportionality μ depends on the individual central configuration.

(1) Any two points, 3 points at the corners of an equilateral triangle or 4 points at the vertices of a regular tetrahedron form a central configuration for any set of masses.

(2) To every numeration of n masses on a straight line there exists exactly one central configuration (Moulton, 1910).

(3) Many planar central configurations of 4 masses have been calculated, but a complete survey has not yet been done (MacMillan and Carty, 1932).

6. Collisions

The homothetic solutions ($p=0$) of the problem of n bodies represent examples of motions involving collisions of n bodies. It was shown by Chazy (1918) that just prior to a collision of all bodies in the n-body problem the mass points approach a central configuration. Therefore, it is natural to seek other collision solutions among those asymptotically approaching homothetic solutions. This will be done by determining the solutions asymptotic to equilibriums in a Scheibnerian coordinate system with $p=0$.

This method yields all collision solutions only under the hypothesis that the central configurations associated with a particular set of masses do *not* form a continuum. The hypothesis, however, is still unproven for $n>3$.

Thus we substitute

$$\xi_i(\sigma) = \xi_i^0 + \mathbf{u}_i(\sigma), \qquad i = 1, \ldots, n, \tag{26}$$

into Equation (22). The ξ_i^0 form a central configuration with the masses m_i:

$$V_{\xi_i}(\xi_1^0, \ldots, \xi_n^0) + \mu m_i \xi_i^0 = 0, \qquad i = 1, \ldots, n, \tag{27}$$

and

$$\mathbf{u}_i(\sigma) \to 0 \quad \text{as} \quad \sigma \to \infty . \tag{28}$$

Expanding about $\boldsymbol{\xi}$ yields

$$\frac{\sigma^2 + c}{2} \mathbf{u}_i'' = \mathbf{u}_i + \frac{1}{\mu m_i} \sum_{k=1}^{n} V_{\boldsymbol{\xi}_i \boldsymbol{\xi}_k}\big|_0 \mathbf{u}_k + \mathcal{O}(u^2). \tag{29}$$

By again using the components ξ_{ij}, u_{ij} of the vectors $\boldsymbol{\xi}_i$, $\mathbf{u}_i$ and defining

$$\begin{gathered} \xi_k = \xi_{ij}, \quad u_k = u_{ij} \quad \text{with} \quad k = 3(i-1) + j, \\ i = 1, \ldots, n; \quad j = 1, 2, 3; \quad k = 1, \ldots, 3n, \end{gathered} \tag{30}$$

these equations may be written as

$$\frac{\sigma^2 + c}{2} u_k'' = \sum_{l=1}^{3n} \left(\delta_{kl} + \frac{1}{\mu m_i} \frac{\partial^2 V}{\partial \xi_k \partial \xi_l}\bigg|_0 \right) u_l + \mathcal{O}(u^2), \quad k = 1, \ldots, 3n, \tag{31}$$

where i is given by

$$i = -[-k/3]$$

and δ_{kl} is the Kronecker symbol.

The method of characteristic exponents allows us now to formally determine the asymptotic behaviour of collision solutions in the n-body problem. However, only for $n \leqslant 3$ it is known that all collision solutions are obtained this way.

We seek solutions of Equation (31) of the form

$$u_k(\sigma) = c_k \sigma^\alpha + o(\sigma^\alpha), \quad \alpha < 0, \tag{32}$$

valid asymptotically for $\sigma \to \infty$, where c_k and α are constants. Substituting this into Equation (31) and equating the coefficients of the leading terms (σ^α) yields the condition

$$\frac{\alpha(\alpha - 1)}{2} c_k = \sum_{l=1}^{3n} \left(\delta_{kl} + \frac{1}{\mu} \cdot \frac{1}{m_i} \frac{\partial^2 V}{\partial \xi_k \partial \xi_l}\bigg|_0 \right) c_l \tag{33}$$

which is easily reduced to an eigenvalue problem for the matrix

$$M = \left(\frac{1}{m_{-[-k/3]}} \frac{\partial^2 V}{\partial \xi_k \partial \xi_l}\bigg|_0 \right). \tag{34}$$

If λ is an eigenvalue of M, the corresponding exponents α are, according to Equation (33), the solutions of the quadratic equation

$$\alpha^2 - \alpha - 2\left(1 + \frac{\lambda}{\mu}\right) = 0.$$

Hence

$$\alpha = \tfrac{1}{2}\left(1 \pm \sqrt{9 + 8\frac{\lambda}{\mu}}\right), \tag{35}$$

and c_k is an eigenvector of M corresponding to the eigenvalue λ.

It can be conjectured that generally (except in degenerate cases) the collison solutions may be represented as multiple Taylor series in σ^{α_i}, where the $\alpha_i<0$ are all negative numbers satisfying Equation (33).

In fact this has been proven by Siegel (1941) for $n=3$. Siegel found 3 negative exponents in the triangular case and 2 in the collinear case of the triple collision.

Since in general the exponents α_i are irrational numbers, collisions of $n(n>2)$ bodies are transcendental singularities of the solutions (branch points of infinite order). Hence, generally the analytic continuation of the solution beyond the singularity has *no* real branch, i.e. except in special cases there is no real continuation of the motion after a collision of $n>2$ bodies.

7. An Isosceles Case

The planar problem of 3 bodies with masses m_1, m_1, m_2 admits solutions such that the configuration of the three mass points remains symmetric with respect to a fixed axis (isosceles solutions).

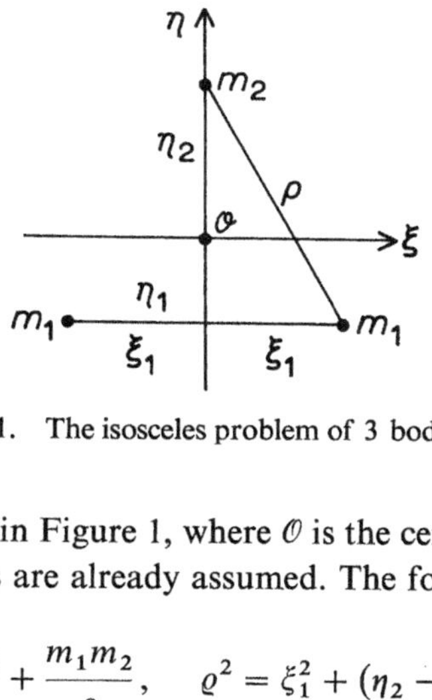

Fig. 1. The isosceles problem of 3 bodies.

Consider the situation given in Figure 1, where $\mathcal{O}$ is the center of mass, and appropriate Scheibnerian coordinates are already assumed. The force function V is

$$V=\frac{m_1^2}{2\xi_1}+\frac{m_1m_2}{\varrho}+\frac{m_1m_2}{\varrho},\qquad \varrho^2=\xi_1^2+(\eta_2-\eta_1)^2,$$

and the equations of motion (22) become

$$\tfrac{1}{2}(\sigma^2+c)\,\xi_1''=\xi_1-\frac{1}{\mu}\left(\frac{m_2\xi_1}{\varrho^3}+\frac{m_1}{4\xi_1^2}\right)$$

$$\tfrac{1}{2}(\sigma^2+c)\,\eta_1''=\eta_1+\frac{1}{\mu}\frac{m_2(\eta_2-\eta_1)}{\varrho^3}$$

$$\tfrac{1}{2}(\sigma^2+c)\,\eta_2''=\eta_2-\frac{1}{\mu}\frac{2m_1(\eta_2-\eta_1)}{\varrho^3}.$$

Subtracting the second from the third equation, substituting $\eta=\eta_2-\eta_1$, and dropping the subscript of ξ_1 yields

$$\begin{aligned} \tfrac{1}{2}(\sigma^2 + c)\,\xi'' &= \xi - \frac{1}{\mu}\left(\frac{m_2\xi}{\varrho^3} + \frac{m_1}{4\xi^2}\right) \\ \tfrac{1}{2}(\sigma^2 + c)\,\eta'' &= \eta - \frac{1}{\mu}\frac{(2m_1 + m_2)\,\eta}{\varrho^3}, \qquad \varrho^2 = \xi^2 + \eta^2 . \end{aligned} \tag{36}$$

We now consider the triangular homothetic solution given by

$$\xi = \tfrac{1}{2}, \quad \eta = \frac{\sqrt{3}}{2}, \quad \varrho = 1 . \tag{37}$$

This is an equilibrium solution of Equation (36) if μ is put equal to the total mass,

$$\mu = 2m_1 + m_2 . \tag{38}$$

In order to obtain the solutions of Equation (36) which asymptotically approach the equilibrium (37) we substitute in analogy to Equation (26)

$$\xi = \tfrac{1}{2}(1 + u), \quad \eta = \frac{\sqrt{3}}{2}(1 + v). \tag{39}$$

We now restrict ourselves to the case

$$m_1 = 28, \quad m_2 = 19, \quad \mu = 75$$

which will turn out to be one of the exceptional cases where the motion can be continued beyond a triangular triple collision. Substituting (39) and (40) into Equation (36) and expanding with respect to u and v yields

$$\begin{aligned} (\sigma^2 + c)\,u'' &= 4.86u + 1.14v - \tfrac{1659}{400}u^2 - \tfrac{57}{200}uv - \tfrac{627}{400}v^2 + \cdots \\ (\sigma^2 + c)\,v'' &= 1.5u + 4.5v - \tfrac{3}{16}u^2 - \tfrac{33}{8}uv - \tfrac{27}{16}v^2 + \cdots . \end{aligned} \tag{41}$$

The matrix M (see Equations (33) and (34)) is therefore given by

$$\frac{M}{\mu} = \begin{pmatrix} 1.43 & 0.57 \\ 0.75 & 1.25 \end{pmatrix}.$$

Its eigenvalues and eigenvectors are

$$\begin{aligned} \frac{\lambda_1}{\mu} &= 0.68 \qquad \mathbf{c}_1 = \begin{pmatrix} 0.76 \\ -1 \end{pmatrix} \\ \frac{\lambda_2}{\mu} &= 2 \qquad \mathbf{c}_2 = \begin{pmatrix} 1 \\ 1 \end{pmatrix} \end{aligned}, \tag{42}$$

and Equation (35) yields one negative exponent for each eigenvalue:

$$\alpha_1 = -1.4 \quad \alpha_2 = -2 . \tag{43}$$

A third negative exponent would be found by considering 3-dimensional motion.

As a consequence of Equation (43) and Siegel's results Equations (41) have solutions which may be expanded into a double Taylor series at $\sigma = \infty$ in the variables

$$\sigma_1 = b\sigma^{-1.4}, \quad \sigma_2 = c\sigma^{-2}, \tag{44}$$

where b and c are arbitrary constants. We assume u and v to be of the form

$$\begin{aligned} u &= 0.76 \sum_{\substack{j=1\\k=0}}^{\infty} g_{jk}\sigma_1^j\sigma_2^k \\ v &= - \sum_{\substack{j=1\\k=0}}^{\infty} h_{jk}\sigma_1^j\sigma_2^k \end{aligned} \tag{45}$$

where g_{jk}, h_{jk} are unknown coefficients with

$$g_{10} = h_{10} = 1, \tag{46}$$

since, according to (32), $\begin{pmatrix} u \\ v \end{pmatrix} \sim c_1\sigma_1$.

Writing Equations (41) as

$$\sigma^2 (1 + \sigma_2)\, u'' = P(u, v), \qquad \sigma^2 (1 + \sigma_2)\, v'' = Q(u, v)$$

and observing that

$$\sigma^2 \cdot \frac{d^2}{d\sigma^2} (\sigma_1^j\sigma_2^k) = (1.4j + 2k)(1.4j + 2k + 1)\, \sigma_1^j\sigma_2^k$$

enables us to substitute the expansions (45) into (41).

Equating coefficients yields a system of two linear equations for each pair g_{jk}, h_{jk}. Only the first one ($j=1, k=0$) has a vanishing determinant – this causes the appearance of the arbitrary constant b in Equation (44) – all the others are regular systems and can be solved recursively.

One result is

$$g_{1k} = h_{1k}; \qquad g_{1,k+1} = -\frac{(k + 0.7)(k + 1.2)}{(k + 1)(k + 2.9)}\, g_{1k}, \qquad k = 0, 1, 2, \cdots$$

which implies that the terms of the series (45) containing only first powers of σ_1 form the hypergeometric series

$$c_1\sigma_1 F(0.7, 1.2, 2.9, -\sigma_2).$$

This is, of course, a consequence of the fact that the linear terms in Equations (41) form hypergeometric differential equations.

To calculate the other coefficients up to higher orders involves a considerable amount of algebra; the results given in Table I were obtained by the Algebraic Processor SYMBAL on a CDC 6600 computer in 34 s. These coefficients define the terms of the series (45) involving powers of σ with exponents -1.4, -2.8, -3.4, -4.2, -4.8, -5.4, -5.6, -6.2, -6.8; the series contain no other exponents > -7.

TABLE I

The coefficients in the expansions Equation (45)

```
G[1,0] := 1;
H[1,0] := 1;
G[1,1] := -42/145;
H[1,1] := -42/145;
G[2,0] := -31539/37700;
H[2,0] := -2367/37700;
G[1,2] := 1309/9425;
H[1,2] := 1309/9425;
G[2,1] := 48 95219/96 13500;
H[2,1] := 3 55607/96 13500;
G[3,0] := 43 98413/84 82500;
H[3,0] := -7 59179/84 82500;
G[1,3] := -26928/3 29875;
H[1,3] := -26928/3 29875;
G[2,2] := -33 37889 39261/100 99222 08750;
H[2,2] := -2 36643 53633/100 99222 08750;
G[3,1] := -994 52113 70269/2067 93595 12500;
H[3,1] := 168 27264 01027/2067 93595 12500;
G[4,0] := -734 55468 55591/1375 09807 50000;
H[4,0] := -68 42070 30659/1375 09807 50000;
```

Since obviously all the exponents are rational numbers with denominator 5, u and v have real values also for $\sigma<0$. Therefore, in this case the motion can be continued beyond the triple collision by varying σ through infinity to negative values.

The series (45) are a means for actually calculating initial conditions which lead to a triple collision. If they are not too close to the singularity the solutions may even be continued outwards by numerical integration.

According to Equations (39), (11), (13) the inertial coordinates x, y corresponding to ξ, η are

$$x = \frac{1+u}{\sigma^2+c} \qquad y = \sqrt{3}\,\frac{1+v}{\sigma^2+c}, \tag{47}$$

and the relation between σ and time is given by Equation (12). x and y are the basis and the hight of the isosceles triangle formed by the three masses; they satisfy the differential equations

$$\frac{d^2x}{dt^2} = -\frac{19x}{r^3} - \frac{7}{x^2}; \quad \frac{d^2y}{dt^2} = -\frac{75y}{r^3}, \quad r^2 = x^2 + y^2. \tag{48}$$

Acknowledgements

The author wishes to thank the Applied Mathematics Seminar of the Swiss Federal Institute of Technology at Zürich and the Department of Aerospace Engineering of The University of Texas at Austin for supporting this research.

References

Chazy, J.: 1918, *Bull. Astron.* **35**, 321–389.
Dziobek, O.: 1900, *Astron. Nachr.* **152**, 33–46.
MacMillan, W. D. and W. Bartky: 1932, *Trans. Am. Math. Soc.* **34**, 838–875.
Moulton, F. R.: 1910, *Ann. Math.* **12**, 1–17.
Pizzetti: 1904, *Rend. Acc. Lincei* **13**, 276–283.
Scheibner, W.: 1866, *Crelle J. Reine Angew. Math.* **65**, 291.
Siegel, C. L.: 1941, *Dreierstoss. Ann. Math.* **42**, 127–168; *Gesammelte Abhandlungen* **II**, 169–210.
Sperling, H. J.: 1970, *J. Reine Angew. Math.* **245**, 15–40.
Sundman, K. F.: 1907, *Acta Soc. Sci. Fenn.* **34**, 43.
Wintner, A.: 1941, *The Analytical Foundations of Celestial Mechanics*, Princeton Univ. Press.

REGULARIZATION USING A TIME-TRANSFORMATION ONLY

D. C. HEGGIE
Institute of Astronomy, University of Cambridge, Cambridge, U.K.
and Trinity College, Cambridge, U.K.

Abstract. It is possible to regularize the perturbed Kepler problem by performing a transformation of the time only. The regularized equations are derived and interpreted, and they are shown to have properties of value in analytic and in numerical applications.

1. Derivation of the Equations

The regularization of the two-body problem by Kustaanheimo and Stiefel (Stiefel and Scheifele, 1971), which has been applied to the computational N-body problem (Peters, 1968; Aarseth, 1972; Szebehely and Bettis, 1972) and others with very happy results, achieves its end by performing transformations upon both the dependent variables, the coordinates, and the independent variable, the time. In this paper we describe a regularization that is accomplished by transformation of the time only.

We consider the perturbed Newtonian central force problem

$$\ddot{\mathbf{r}} = -\frac{k^2}{r^3}\mathbf{r} + \mathbf{F}, \tag{1}$$

where $\mathbf{r}$ is the position vector of a particle relative to the force center, k^2 is a constant, $r \equiv |\mathbf{r}|$, and $\mathbf{F}$ is the perturbing acceleration. A dot denotes differentiation with respect to ordinary time, t. As in KS-regularization, we introduce a new 'fictitious' time, s, by means of the differential equation

$$\mathrm{d}t = r\,\mathrm{d}s, \tag{2}$$

whence, denoting by a prime differentiation with respect to s, we find that

$$\frac{\mathrm{d}}{\mathrm{d}t} \equiv r^{-1}\frac{\mathrm{d}}{\mathrm{d}s} \tag{3}$$

and

$$\frac{\mathrm{d}^2}{\mathrm{d}t^2} \equiv -r^{-3}r'\frac{\mathrm{d}}{\mathrm{d}s} + r^{-2}\frac{\mathrm{d}^2}{\mathrm{d}s^2}. \tag{4}$$

From (1) and (4),

$$\mathbf{r}'' = \frac{r'}{r}\mathbf{r}' - \frac{k^2}{r}\mathbf{r} + r^2\mathbf{F}. \tag{5}$$

We define

$$P \equiv -\frac{2k^2}{r} + \frac{|\mathbf{r}'|^2}{r^2} \tag{6}$$

B. D. Tapley and V. Szebehely (eds.), Recent Advances in Dynamical Astronomy, 34–37.

and

$$\mathbf{Q} \equiv \frac{k^2}{r}\,\mathbf{r} - \frac{|\mathbf{r}'|^2}{r^2}\,\mathbf{r} + \frac{r'}{r}\,\mathbf{r}', \tag{7}$$

whence, by (3),

$$P \equiv -\frac{2k^2}{r} + |\dot{\mathbf{r}}|^2 \tag{8}$$

and

$$\mathbf{Q} \equiv \frac{k^2}{r}\,\mathbf{r} - |\dot{\mathbf{r}}|^2\,\mathbf{r} + (\mathbf{r}\cdot\dot{\mathbf{r}})\,\dot{\mathbf{r}}. \tag{9}$$

We now note that (5) may be written

$$\mathbf{r}'' = P\mathbf{r} + \mathbf{Q} + r^2\mathbf{F}. \tag{10}$$

Using (1), (8) and (9), we easily deduce expressions for $\dot{P}$ and for $\dot{\mathbf{Q}}$ whence, by (3),

$$P' = 2\mathbf{r}'\cdot\mathbf{F} \tag{11}$$

and

$$\mathbf{Q}' = -\,2\mathbf{r}\,(\mathbf{r}'\cdot\mathbf{F}) + \mathbf{r}'\,(\mathbf{r}\cdot\mathbf{F}) + \mathbf{F}\,(\mathbf{r}\cdot\mathbf{r}'). \tag{12}$$

By differentiating twice the identity $r^2 \equiv \mathbf{r}\cdot\mathbf{r}$ we obtain, by (7) and (10), the equation

$$r'' = Pr + k^2 + r\mathbf{F}\cdot\mathbf{r}. \tag{13}$$

A very similar relation is obtainable in KS-regularization. By (2) and (13) we find that

$$t''' = Pt' + k^2 + r\mathbf{F}\cdot\mathbf{r}. \tag{14}$$

Defining the quantity τ by

$$\tau \equiv t - \frac{r'}{P} \tag{15}$$

we deduce, with the aid of (2), (11) and (13), that

$$\tau' = -\frac{k^2}{P} - \frac{\mathbf{F}\cdot\mathbf{r}}{P}\,r + \frac{2\mathbf{F}\cdot\mathbf{r}'}{P^2}\,r'. \tag{16}$$

This completes the formal derivation of the equations needed in the regularization. It is not new, having already been discussed by Burdet (1967), and it was known previously to Arenstorf, Sperling and others. We proceed to discuss the interpretation of the quantities that have been introduced.

We observe from (11), (12) and (16) that P, $\mathbf{Q}$, and τ are 'elements', for their derivatives are constants in the case of unperturbed motion, i.e. when $\mathbf{F} = \mathbf{O}$. In particular, P and $\mathbf{Q}$ are then constants, and so Equation (10) is a generalization of an equation given by Szebehely (1967, p. 118) for unperturbed Keplerian motion. If h is the energy of the particle, per unit mass, then by (8) we have

$$P = 2h \tag{17}$$

in general, and if a is the osculating semi-major axis of the orbit we have

$$P = -\frac{k^2}{a}. \tag{18}$$

When $\mathbf{F} = \mathbf{O}$, (10) is the equation of motion for a displaced simple harmonic oscillator, and the position vector, $\mathbf{r}_0$, of the center of the orbit is the value of $\mathbf{r}$ at which $\mathbf{r}'' = \mathbf{O}$, whence

$$\mathbf{Q} = -P\mathbf{r}_0. \tag{19}$$

If e is the orbital eccentricity, we have

$$|\mathbf{r}_0| = e\,|a|,$$

whence, using (18) and (19),

$$|\mathbf{Q}| = k^2 e.$$

Differentiating Kepler's equation with respect to t and comparing with (2), we observe in case $P < 0$ and $\mathbf{F} = \mathbf{O}$ that

$$\omega s = E + E_0, \tag{20}$$

where E is the eccentric anomaly, E_0 is a constant and $\omega^2 \equiv -P$. By (15), τ is also linear in E.

2. Applications of the Regularization

The convenience of (10) for analytic investigations derives from two features, for in the first place this equation is linear when $\mathbf{F} = \mathbf{O}$. Then, the solution in case $P < 0$ takes the form

$$\mathbf{r} = \mathbf{A}\cos\omega s - \frac{\mathbf{Q}}{P} + \mathbf{B}\sin\omega s. \tag{21}$$

The vectors $\mathbf{A}$ and $\mathbf{B}$ which appear in (21) are constant but not arbitrary, since (6) and (7) must be satisfied. Using Equation (20) with $E_0 = 0$, and familiar expressions for Keplerian elliptic motion in terms of E, we realize that this may be achieved if we set

$$\mathbf{r} = \mathbf{a}(\cos\omega s - e) + \mathbf{b}\sin\omega s, \tag{22}$$

where $\mathbf{a}$ and $\mathbf{b}$ are taken to be vectors directed along the semi-axes of the elliptic orbit and equal to them in magnitude. The expression (22) for $\mathbf{r}$ as a finite Fourier sum in ωs also simplifies analytic work, and it is convenient that the regularization is written in terms of the physical vector $\mathbf{r}$.

Finally we briefly describe the relevance of this regularization to the computational N-body problem. Even if $\mathbf{F}$ is bounded, the right hand side of (1) is singular during a collision, when $r \to 0$. It is also an unstable equation, and for these reasons it is unsuitable for any numerical work in which relatively close approaches to the attracting centre can occur. However, if also the energy is bounded during the collision, by (17) we note that P is and so, by (6), $|\mathbf{r}'| = 0(r^{1/2})$ as $r \to 0$. Hence the righthand sides of (10) (11) and (12) are regular. Equation (10) is also stable when $\mathbf{F} = \mathbf{O}$ and $P < 0$, and so these equations are well-behaved numerically.

It remains to perform the numerical transformation of the time. Equation (16) is stable when $\mathbf{F} = \mathbf{O}$, but is singular if, as may happen in applications to the N-body problem, P passes through 0. One method by which this difficulty may be avoided is to integrate Equation (14), which is stable if $\mathbf{F} = \mathbf{O}$ and $P < 0$, and regular for $P \to 0$.

Practical aspects of the application of KS-regularization to the numerical N-body problem have been described by Aarseth (1972). The treatment of the present regularization follows similar lines, except that the equations to be integrated numerically are (10), (11), (12), and an equation for the time transformation, such as (14). The formulation is simpler than with KS-regularization, but against this two points should be considered. First, the total order of the system of equations is 11, if Equation (16) is adopted for the time transformation, whereas with KS-regularization it is 10, despite the introduction of 4-vectors. Second, the frequency ω in case $P < 0$ is twice that obtained in KS-regularization, and in practice it seems that almost twice as many steps are required per orbit as in that regularization to achieve the same accuracy.

Acknowledgements

I am grateful to my supervisor, Dr Aarseth, for introducing me to the topic of regularization, to Dr Arenstorf for a discussion on this particular method, and to the U. K. Science Research Council for a Research Studentship.

References

Aarseth, S. J.: 1972, in M. Lecar (ed.), *Gravitational N-Body Problem*, D. Reidel Publ. Co, Dordrecht, Holland, p. 373.
Burdet, C. A.: 1967, *Z.A.M.P.* **18**, 434.
Peters, C. F.: 1968, *Bull. Astron.* **3**, 167.
Stiefel, E. L. and Scheifele, G.: 1971, *Linear and Regular Celestial Mechanics*, Springer Publ. Co., Berlin.
Szebehely, V.: 1967, *Theory of Orbits*, Academic Press, New York, N.Y.
Szebehely, V. and Bettis, D. G.: 1972, in M. Lecar (ed.), *Gravitational N-Body Problem*, D. Reidel Publ. Co., Dordrecht, Holland, p. 136.

STABILIZATION OF THE DIFFERENTIAL EQUATIONS OF KEPLERIAN MOTION

J. BAUMGARTE
Mechanik-Zentrum, Lehrstuhl A, Technische Universität Braunschweig, F.R.G.

Abstract. The analytical relations of a given system of differential equations can be satisfied in a stabilized manner in order to improve the numerical accuracy of the solution by the addition of control terms to the system at hand. Examples of such relations are the energy relation in conservative systems or the analytical relations generated by outer constraints provided the LAGRANGE equations of the first kind are used. The idea to add a control term is applied to the Kepler motion.

1. Holonomic and Non-Holonomic Constraints

Our method of stabilization of a system of differential equations is based on the idea of adding a control term to the differential system at hand (Baumgarte, 1972, 1973). This idea is conveniently explained by discussing the example of a dynamical motion subjected to constraints. Later on the idea will be applied to the Kepler motion. Another method of stabilization of Kepler motion is connected with the concept of regularization of such a motion (Baumgarte, 1973; Stiefel and Scheifele, 1971). This line of approach is explained in Prof. Stiefels lectures.

Let us consider a mechanical system composed of n mass points. The equations of motion are with respect to a cartesian frame and the time t

$$m_i\ddot{x}_i - F_i = 0\,, \qquad i = 1, 2, \ldots, 3n\,, \tag{1}$$

where the F_i are the external forces. Let us now assume that the system is subjected either to a holonomic constraint

$$\tilde{N}(x_l;\, t) = 0 \tag{2}$$

or to a non-holonomic constraint

$$N(x_l,\, \dot{x}_l;\, t) = 0\,. \tag{3}$$

As an instrument for the determination of the motion of the system we adopt the principle of least curvature of Gauss.*

The principle of Gauss requires one to differentiate a non-holonomic constraint once with respect to t, whereas a holonomic constraint is differentiated twice. Thus in both cases the resulting Gaussian constraint has the form

$$f(x_l,\, \dot{x}_l,\, \ddot{x}_l;\, t) = 0\,, \tag{4a}$$

* This principle is more powerful for solving the problem under consideration than the well-known principles of Hamilton or d'Alembert. They are only able to handle the following particular cases: (1) Holonomic constraint, (2) Non-holonomic constraint, which is linear with respect to the velocities.

B. D. Tapley and V. Szebehely (eds.), Recent Advances in Dynamical Astronomy, 38–44.

where f is linear with respect to the accelerations $\ddot{x}_l$:

$$f(x_l, \dot{x}_l, \ddot{x}_l; t) = \sum_{(i)} g_i(x_l, \dot{x}_l; t)\, \ddot{x}_i + G(x_l, \dot{x}_l; t) = 0. \tag{4b}$$

Equation (4a) or (4b) is referred to as the Gaussian constraint. In case of a holonomic constraint $\tilde{N}=0$ we have for instance $f=\ddot{\tilde{N}}$.

According to Gauss' principle the curvature

$$\tfrac{1}{2}\sum_{(i)} m_i\left(\ddot{x}_i - \frac{F_i}{m_i}\right)^2 \tag{5}$$

is minimized by appropriate choice of the accelerations $\ddot{x}_i$, which are restricted by the condition (4). The appropriate mathematical technique introduces a Lagrange multiplier λ such that

$$\tfrac{1}{2}\sum_{(i)} m_i\left(\ddot{x} - \frac{F_i}{m_i}\right)^2 - \lambda f(x_l, \dot{x}_l, \ddot{x}_l; t) \tag{6}$$

is minimized unconditionally. Differentiation with respect to the accelerations yields:

$$m_i\ddot{x}_i - F_i = \lambda g_i(x_l, \dot{x}_l; t), \qquad g_i = \frac{\partial f}{\partial \ddot{x}_i}. \tag{7}$$

In order to obtain a differential system fit for numerical integration, the multiplier λ is expressed in terms of x_l, $\dot{x}_l$, t by substituting Equation (7) into Equation (4b), giving

$$\lambda = -\frac{\sum \frac{g_i F_i}{m_i} + G}{\sum \frac{g_i^2}{m_i}}, \tag{8}$$

thus the equations of motion (7) are reformulated as

$$m_i\ddot{x}_i - F_i = -\frac{\sum \frac{g_k F_k}{m_k} + G}{\sum \frac{g_k^2}{m_k}}\, g_i. \tag{9}$$

It is obvious that the principles of Hamilton and d'Alembert produce also the equations of motion (9) and, regardless of the principle under consideration, it is necessary to differentiate the original constraint $\tilde{N}=0$ twice with respect to t in order to obtain a system fit for numerical integration.

2. Example of a Holonomic Constraint

We consider a mass point on a circle with radius a in the x_1, x_2-plane. No external forces are acting. This situation implies the holonomic constraint

$$\tilde{N} = \tfrac{1}{2}(|\mathbf{x}|^2 - a^2) = 0, \qquad |\mathbf{x}|^2 = x_1^2 + x_2^2. \tag{10}$$

The Gauss-constraint is consequently

$$\ddot{\tilde{N}} = f = (\mathbf{x}, \ddot{\mathbf{x}}) + |\dot{\mathbf{x}}|^2 = 0, \tag{11}$$

$(\mathbf{x}, \ddot{\mathbf{x}}) = x_1\ddot{x}_1 + x_2\ddot{x}_2$ denotes the scalar product. Hence our vector $\mathbf{g} = (g_1, g_2)$ is

$$\mathbf{g} = \mathbf{x}. \tag{12}$$

The equations of motion (9) are in this case

$$\ddot{\mathbf{x}} = -\frac{|\dot{\mathbf{x}}|^2}{|\mathbf{x}|^2}\,\mathbf{x}. \tag{13}$$

During a numerical integration of (13) the value of $\tilde{N}$, which should be zero, may be taken as a check. By virtue of the unavoidable computational errors the computer arrives, after n steps at the instant t_0 with a value $\dot{\tilde{N}} = \varepsilon \neq 0$. Hence an erroneous value of $\tilde{N}$ remains constant and is not reduced during the subsequent integration. Later errors do not likely improve that situation. From that instant t_0 onward it follows

$$\tilde{N} = \delta + \varepsilon t; \qquad t \geqslant t_0. \tag{14}$$

Consequently the computed point spirals away from the true circle in an *unstable manner*.

3. On Numerical Integration and Stabilization

This example shows that a holonomic constraint $\tilde{N} = 0$ is violated in a linearly unstable manner by a single computational error.

In case of a non-holonomic constraint $N(x_l, \dot{x}_l; t) = 0$ (3) a similar situation cocurs.

In order to avoid instability of the originally constraint $\tilde{N}(x_l; t) = 0$ or $N(x_l, \dot{x}_l; t) = 0$ we propose the following technique.

(1) The differential equation $\ddot{\tilde{N}} = 0$ which was produced by the Gaussian rule was seen to be unstable. Thus we replace it by the stable differential equation, after choice of two constants α, β:

$$f = \ddot{\tilde{N}} + 2\alpha\dot{\tilde{N}} + \beta^2\tilde{N} = 0, \qquad \alpha > 0. \tag{15}$$

This equation is of course a consequence of the original constraint $\tilde{N} = 0$ and it reduces an erroneous value of $\tilde{N}$ exponentially. We call this technique the *stabilization of a constraint.*

Equation (15) is the new (stabilized) Gauss-constraint of the form

$$f(x_l, \dot{x}_l, \ddot{x}_l; t) = \sum_{(i)} g_i(x_l, \dot{x}_l; t)\,\ddot{x}_i + G(x_l, \dot{x}_l; t) = 0, \tag{16}$$

in which only the new quantity G differs from the old one by the aggregate $(2\alpha\dot{\tilde{N}} + \beta^2\tilde{N})$. As before this leads to the modified equations of motion

$$m_i\ddot{x}_i - F_i = -\frac{\sum \frac{g_k F_k}{m_k} + G}{\sum \frac{g_k^2}{m_k}}\, g_i. \tag{17}$$

In order to offer an example we return to the free circular motion. We had

$$\tilde{N} = \tfrac{1}{2}(|\mathbf{x}|^2 - a^2) = 0, \tag{18}$$

$$\dot{\tilde{N}} = (\mathbf{x}, \dot{\mathbf{x}}), \qquad \ddot{\tilde{N}} = (\mathbf{x}, \ddot{\mathbf{x}}) + |\dot{\mathbf{x}}|^2. \tag{19}$$

Equations (15), (16) yield

$$f = \ddot{\tilde{N}} + 2\alpha\dot{\tilde{N}} + \beta^2\tilde{N} = (\mathbf{x}, \ddot{\mathbf{x}}) + |\dot{\mathbf{x}}|^2 + 2\alpha(\mathbf{x}, \dot{\mathbf{x}}) + \tfrac{1}{2}\beta^2(|\mathbf{x}|^2 - a^2) = 0. \tag{20}$$

The result, i.e, the stabilized Equation (17), ($\mathbf{F}=0$) is

$$\ddot{\mathbf{x}} = -\frac{|\dot{\mathbf{x}}|^2 + 2\alpha(\mathbf{x}, \dot{\mathbf{x}}) + \tfrac{1}{2}\beta^2(|\mathbf{x}|^2 - a^2)}{|\mathbf{x}|^2}\,\mathbf{x}. \tag{21}$$

The new Gaussian constraint (15) and the old one $\ddot{\tilde{N}}=0$ differ by the aggregat $(2\alpha\dot{N} + \beta^2\tilde{N})$ which plays in G the role of a control term achieving the annihilation of a value $\tilde{N} \neq 0$ produced by computational errors.

(2) A non-holonomic constraint $N(x_l, \dot{x}_l; t)=0$ is modified in a similar manner. The differential relation $\dot{N}=0$ is replaced by the differential equation

$$f = \dot{N} + \gamma N = 0, \tag{22}$$

where $\gamma = \gamma(x_l, \dot{x}_l; t)$ is a positive function. Equation (22) is considered as the new Gaussian constraint. It has again the property of error reducing since by multiplication with N there results

$$\tfrac{1}{2}(N^2)^{\cdot} = -\gamma N^2, \quad \text{hence} \quad (N^2)^{\cdot} < 0,$$

thus N decreases in absolute value.

The choice $\gamma = \text{const}$ is appropriate for many problems and this reduces an erroneous value, $N \neq 0$, once more exponentially.

Remark. The coefficients $g_i(x_l, \dot{x}_l; t)$ in Equations (4b) and (16) are not influenced by our stabilizing modifications. According to Equation (7) any Gaussian constraint produces a force of reaction λg_i and therefore the direction of that force is preserved by the stabilization.

4. Stabilization of the Energy Relation

The foregoing methods of stabilization are now applied to the conservative case of motion (1) where the external forces F_i stem from a potential $U = U(x_l)$ which depends only on the position vector x_i:

$$F_i = -\frac{\partial U}{\partial x_i}. \tag{23}$$

This assumption leads to the well-known energy relation

$$N(x_l, \dot{x}_l) = \tfrac{1}{2}\sum m_i \dot{x}_i^2 + U - E_0 = 0, \tag{24}$$

(E_0 = total energy), which may be interpreted as a non-holonomic constraint. If the stabilizing technique is not put into operation, we have

$$\dot{N} = \sum m_i \dot{x}_i \ddot{x}_i + \sum \frac{\partial U}{\partial x_i} \dot{x}_i \tag{25}$$

$$g_i = m_i \dot{x}_i, \qquad G = \sum \frac{\partial U}{\partial x_i} \dot{x}_i. \tag{26}$$

Equation (8) becomes

$$\lambda = - \frac{-\sum \dot{x}_i \frac{\partial U}{\partial x_i} + \sum \frac{\partial U}{\partial x_i} \dot{x}_i}{\sum m_i \dot{x}_i^2} = 0, \tag{27}$$

and thus the stabilized Equations (9) coincide with the original Equations (1). This was to be expected since the law of energy is a consequence of the equations of motion.

In constrast to this phenomenon the stabilizing technique runs as follows. Gaussian constraint:

$$\dot{N} + \gamma N = \sum m_i \dot{x}_i \ddot{x}_i + \sum \frac{\partial U}{\partial x_i} \dot{x}_i + \gamma \left[\frac{1}{2} \sum m_i \dot{x}_i^2 + U - E_0 \right] \tag{28a}$$

$$g_i = m_i \dot{x}_i, \qquad G = \sum \frac{\partial U}{\partial x_i} \dot{x}_i + \gamma \left[\frac{1}{2} \sum m_i \dot{x}_i^2 + U - E_0 \right] \tag{28b}$$

$$\lambda = - \gamma \frac{\frac{1}{2} \sum m_i \dot{x}_i^2 + U - E_0}{\sum m_i \dot{x}_i^2}. \tag{28c}$$

The new Equations (9) of motion are

$$m_i \ddot{x}_i + \frac{\partial U}{\partial x_i} = - \frac{\gamma \left[\frac{1}{2} \sum m_k \dot{x}_k^2 + U - E_0 \right]}{\sum m_k \dot{x}_k^2} m_i \dot{x}_i \tag{29a}$$

or

$$m_i \ddot{x}_i + \frac{\partial U}{\partial x_i} = - \gamma \left[\frac{1}{2} + \frac{U - E_0}{\sum m_k \dot{x}_k^2} \right] m_i \dot{x}_i. \tag{29b}$$

The denominator is the doubled kinetic energy. The bracket in (29a) is the balance of energy. The right-hand sides of (29a) or (29b) are proportional to the momenta $m_i \dot{x}_i$, These right-hand sides may be considered to be control terms which in fact are zero during the correct motion but may not be zero during the computed motion. They should reduce computational errors.

The theory can be extended to the case where generalized coordinates are used.

5. Example: Stabilization of a Keplerian Motion

Let us consider following perturbed Keplerian motion

$$\ddot{x}_i + \frac{K^2}{r^3} x_i = \varepsilon \left[- \frac{\partial V}{\partial x_i} + P_i \right], \qquad r^2 = \sum x_k^2, \qquad i, k = 1, 2, 3, \tag{30}$$

where K^2 is the gravitational parameter of the attracting central mass and ε a small perturbation parameter. The perturbing force is split up into a force stemming from a perturbing potential $\varepsilon V(x_k, t)$ and an additional force $\varepsilon P_i(x_k, \dot{x}_k; t)$.

For stabilizing purposes the right-hand side is augmented by a control term

$$\ddot{x}_i + \frac{K^2}{r^3} x_i = \varepsilon \left[-\frac{\partial V}{\partial x_i} + P_i \right] - \gamma \frac{\frac{1}{2} \sum \dot{x}_k^2 - \frac{K^2}{r} + \varepsilon V + h}{\sum \dot{x}_k^2} \dot{x}_i, \tag{31a}$$

where

$$h = -\varepsilon \left[\frac{\partial V}{\partial t} + \sum P_k \dot{x}_k \right]. \tag{31b}$$

This procedure is motivated by the fact that in the unperturbed case $\varepsilon = 0$ Equation (31a) coincides with the Equation (29a) since the potential of the pure Kepler motion is $(-(K^2/r))$. The negative total energy is denoted by $h = -E$. According to our foregoing theoretical investigations numerical errors are reduced by the control term in case of the unperturbed motion. We expect that this is also true for a slightly perturbed motion where h is no longer constant but a *slowly* varying function.

In our experiments we used instead of t a new independent variable s defined by the relation

$$t' = \frac{\mathrm{d}t}{\mathrm{d}s} = r. \tag{32}$$

(In the sequel a prime denotes differentiation with respect to s). This transformation is motivated by the desire to construct a suitable step-length regulation during the numerical integration.

In fact a constant step-length with respect to s implies short steps with respect to t in the region where r is small and large gravitational forces and sharp bends of the trajectory occur.

The transformed Equations (31) become

$$x_i'' - \frac{\sum x_k x_k'}{r^2} x_i' + \frac{K^2}{r} x_i = \varepsilon r^2 \left[-\frac{\partial V}{\partial x_i} + P_i \right] - \\ - \frac{\gamma r^3}{\sum x_k'^2} \left[\frac{1}{2} \frac{\sum x_k'^2}{r^2} - \frac{K^2}{r} + \varepsilon V + h \right] x_i' \tag{33a}$$

$$h' = -\varepsilon \left[r \frac{\partial V}{\partial t} + \sum P_k x_k' \right]. \tag{33b}$$

In a numerical experiment we choose

$$\gamma r = \delta = \text{const} > 0, \qquad \text{hence}$$

$$x_i'' - \frac{\sum x_k x_k'}{r^2} x_i' + \frac{K^2}{r} x_i = \varepsilon r^2 \left[-\frac{\partial V}{\partial x_i} + P_i \right] - \\ - \delta \left[\frac{1}{2} - \frac{K^2 r - (\varepsilon V + h) r^2}{\sum x_k'^2} \right] x_i'. \tag{34}$$

The Equations (34), (33b), (32) describe the stabilized differential equations in the perturbed case.

The stability properties of the Equation (34) were discussed, in the unperturbed case, in great detail in the publication: 'Stability of Kepler Motion', by Schwarz (1973).

Now we give a brief account on these numerical experiments. The following two methods for computing a *unperturbed* Kepler orbit were compared:

Set N (Newton). Equation (30) (with $\varepsilon=0$),

Set St (Stabilized). Equation (34) (with $\varepsilon=0$), together with the time integration

$$t''' + 2ht' = K^2, \tag{35}$$

which is more accurate than $t'=r$ (Stiefel and Scheifele, 1971; Baumgarte, 1973).

An elliptic orbit with semi-major axis $a=1$, $K=1$, and eccentricity $c=0.95$ was computed in all experiments. Throughout all calculations we used the 4th order Runge-Kutta integration method with constant step size and calculated 2 revolutions with 100 steps per revolution. The satellite started at the pericenter.

The relative error

$$\Delta = \frac{\sqrt{(\Delta x_1)^2 + (\Delta x_2)^2}}{r}, \qquad \Delta x_i = x_{i\,(\text{exact})} - x_{i\,(\text{computed})} \tag{36}$$

after the 2 revolutions is recorded in Table I.

TABLE I

N	*St*
1.68×10^3	$\delta=0$: 7.65×10^{-2}
	$\delta=5$: 5.59×10^{-4}

Comment. The right-hand column shows that the stabilization achieves a gain of about two decimals. Probably this gain will increase when many revolutions are to be computed.

The left-hand column shows that the integration of Kepler's equation without step regulation is out of competition.

References

Baumgarte, J.: 1972, *Comp. Methods Appl. Mech. Engin.* **1**, 1–16.
Baumgarte, J.: 1973, to appear in *Celest. Mech.*
Stiefel, E. L. and Scheifele, G.: 1971, *Linear and Regular Celestial Mechanics*, Springer-Verlag, Berlin–Heidelberg–New York.
Schwarz, H. R.: 1973, to appear in *Comp. Methods Appl. Mech. Engin.*

THE PARTICULAR SOLUTIONS OF LEVI-CIVITA*

F. NAHON
Université de Paris VI, Paris, France

Abstract. To every invariant manifold (Σ) of an Hamiltonian system (H) is associated a particular solution. The trajectory is the set of points of (Σ) for which H is extremal. The time is determined by quadratures.

This is an extension of the well-known steady solution associated with an ignorable coordinate.

The Problem to be considered can be stated as follows. Consider a conservative Hamiltonian, $H(p, q)$, with N degrees of freedom, which admits the invariant relation $F(p, q)=0$. Note that

$$\frac{\mathrm{d}F}{\mathrm{d}t} = 0, \quad \text{for} \quad F = 0. \tag{1}$$

Let Σ be the manifold of the equation $F=0$. We proceed:

(a) To reduce the integration of the canonical system (H) on the manifold Σ and

(b) To find the 'particular solution' which extremizes H on Σ.

For example, if

$$H = \tfrac{1}{2}\left[p_r^2 + \frac{p_\theta^2}{r^2}\right] - U(r)$$

is the Hamiltonian for the motion of a point $M(r, \theta)$ in a plane under the action of a central force $F=\mathrm{d}U/\mathrm{d}r$, then the manifold Σ depends on the constant k, i.e.:

$$\Sigma_k \colon p_\theta = k \,.$$

The particular solution depends also on the parameter k, e.g., it is the motion

$$r = r(k) \rightarrow \frac{\mathrm{d}U}{\mathrm{d}r} + \frac{k^2}{r^3} = 0 \,.$$

This case is typical for the case of a cyclic variable:

$$q_N \text{ cyclic} \rightleftarrows \frac{\partial H}{\partial q_N} = 0 \rightarrow p_N = k \,.$$

The Levi-Cività problem is one generalization of this case.

The Method, to be used can be described as follows. Let us arbitrarily choose p_N, q_N

* Levi-Civita, 'Drei Abhandlungen über adiabatische Invarianten', *Opere Mathematiche* **4**.

B. D. Tapley and V. Szebehely (eds.), Recent Advances in Dynamical Astronomy, 45–52.

and call them $p_N = p'$, $q_N = q'$, respectively. The remaining variables are (p_i, q_i); $i = 1, \ldots, n = N-1$. Let us solve the equation $F = 0$ for p':

$$F(p, q, p', q') = 0 \rightleftarrows p' = f(p, q, q') \tag{2}$$

and let us designate by $h(p, q, q')$ the restriction of H to Σ:

$$h(p, q, q') = H(p, q, p' = f, q'). \tag{3}$$

One has the identities between the partial derivatives

$$\begin{aligned} &\frac{\partial F}{\partial \chi} + \frac{\partial F}{\partial p'}\frac{\partial f}{\partial \chi} \equiv 0 \\ &\frac{\partial H}{\partial \chi} + \frac{\partial H}{\partial p'}\frac{\partial f}{\partial \chi} \equiv \frac{\partial h}{\partial \chi} \end{aligned} \qquad (\text{where } \chi = p_i \text{ or } q_i \text{ or } q'). \tag{4}$$

Let us write the other part of the canonical equations (H) in three groups:

(A) $\quad p' = f(p, q, q')$

(B) $\quad \dfrac{dq'}{dt} = \dfrac{\partial H}{\partial p'}$

(C) $\quad \dfrac{dp_i}{dt} = -\dfrac{\partial H}{\partial q_i}$

$\quad \dfrac{dq_i}{dt} = \dfrac{\partial H}{\partial p_i}.$

By virtue of the identities of Equation (4), the group (C) can be written in the form

$$\begin{aligned} dp_i &= -\frac{\partial h}{\partial q_i}\, dt + \frac{\partial f}{\partial q_i}\, dq' \\ dq_i &= \frac{\partial h}{\partial p_i}\, dt - \frac{\partial f}{\partial q_i}\, dq' \end{aligned} \tag{5}$$

and appear as a combination of two differential systems, K_1, K_2:

System K_1 is canonical with a time variable t, $dq'/dt = 0$ and system K_2 is canonical with a time variable q', $dt/dq' = 0$.

It seems necessary, in order to integrate (C), to know the dependence $q'(t)$; we are going to see that it *is not necessary* (that this leads to a kind of separation of variables).

Using (1)

$$\frac{dF}{dt} = 0, \quad \text{for} \quad f = 0.$$

Thats i, $[H, F] = 0$ for $F = 0$ (the brackets are the Poisson brackets).

Let us introduce the dual system, canonical F, time τ

$$\left.\begin{array}{l} \dfrac{\mathrm{d}p_i}{\mathrm{d}\tau} = -\dfrac{\partial F}{\partial q_i} \\ \dfrac{\mathrm{d}q_i}{\mathrm{d}\tau} = \dfrac{\partial F}{\partial p_i} \end{array}\right\} \rightleftarrows \left\{\begin{array}{ll} (\mathrm{A}') & p' = f(p, q, q') \\ (\mathrm{B}') & \dfrac{\mathrm{d}q'}{\mathrm{d}\tau} = \dfrac{\partial F}{\partial p'} \\ (\mathrm{C}') & \dfrac{\mathrm{d}p_i}{\mathrm{d}q'} = -\dfrac{\partial f}{\partial q} \\ & \dfrac{\mathrm{d}q_i}{\mathrm{d}q'} = \dfrac{\partial f}{\partial p} \\ & (i = 1, \ldots N \end{array}\right. \tag{6}$$

where (6) follows by virtue of the identities of Equation (4). One knows that

$$[H, F] = -[F, H] = -\frac{\mathrm{d}H}{\mathrm{d}\tau}.$$

We must write that $\mathrm{d}H/\mathrm{d}\tau = 0$ for $F = 0$; that is,

$$\frac{\mathrm{d}h}{\mathrm{d}\tau} = 0 \quad \text{for system} \quad F \rightleftarrows \frac{\mathrm{d}h}{\mathrm{d}q'} = 0.$$

One has then translated the hypothesis (1) in terms of h and f through the following equations

$$\frac{\mathrm{d}h}{\mathrm{d}q'} = \frac{\partial h}{\partial q'} + [h, f] = 0 \tag{7}$$

where the brackets designate the Poisson brackets restricted to the variables, p_i, q_i.

As one has also $\mathrm{d}h/\mathrm{d}t = 0$ for the system K_1, which has a conservative Hamiltonian h, we will put forth *Theorem* 1.

Theorem 1. The differential system verified by p_i, q_i on Σ is a linear combination of two canonical systems (K_1, time variable t) (K_2, time variable q') which have in common the first integral $h(p, q, q') = \text{constant}$.

As a First Consequence consider the particular solution of Levi-Cività.

Let us consider the equations

$$\begin{array}{ll} \dfrac{\partial h}{\partial p_i} = 0 & p_i = \hat{p}_i(q') \\ & \rightleftarrows \\ \dfrac{\partial h}{\partial q_i} = 0 & q_i = \hat{q}_i(q'). \end{array} \tag{8}$$

For these values the relation (7) shows that, in general,

$$\frac{\partial h}{\partial q'} = 0.$$

That is, if

$$\frac{\partial f}{\partial p_i}, \quad \frac{\partial f}{\partial q_i}$$

are defined for those values for which h is extremal, then $h(\hat{p}_i, \hat{q}_i, q') = \hat{h}$ constant will be an extremum of h on Σ.

The relations (8) are invariant for the system K_1. They constitute, in effect, a family of solutions of equilibrium. They are also invariant for K_2: in effect h is a first integral of K_2, therefore, keeping the extremal value $\hat{h}$. There may define, then, a particular solution of H on the manifold Σ:

$$p' = f(\hat{p}, \hat{q}, q')$$
$$p = \hat{p}(q')\, q = \hat{q}(q')$$
$$\frac{dq'}{dt} = \frac{\partial H}{\partial p'} = \text{function of } q' \text{ only}.$$

We can put forth then:

Theorem 2. To any invariant relation of a canonical system (H), is associated a particular solution.

The trajectory is the loci of points on Σ where H is the extremum; the solution is determined through a quadrature.

Second Consequence: Let us forget for a moment the problem of the reduction of H on Σ, and consider the system

$$dx = X(x, t, \tau)\, dt + Y(x, t, \tau)\, dt \tag{9}$$

where x = the vector $(x_1, x_2, \ldots, x_n)$ and t, τ are two parameters pertaining to the time.

Let us give $x = x_0$ for $t = t_0$, $\tau = \tau_0$.

Is one able to calculate $x = x_1$ for $t = t_1$, $\tau = \tau_1$?

If we are given in the plane of the time variables the two points $p_0(t_0, \tau_0)$, $p_1(t_1, \tau_1)$ and an arc γ joining the points:

$$\gamma : t = t(\lambda), \qquad \tau = \tau(\lambda).$$

Equation (9) becomes an ordinary differential equation and one can calculate x_1 by virtue of Cauchy's theorem. But this solution depends on the path (γ) in general.

Definition. One says that the system (9) is integrable in the sense of Morera if the solution does not depend on the path γ, that is from choosing

$$t(\lambda), \quad \tau(\lambda).$$

For example:

$$\left.\begin{aligned} dx &= -\, y\, dt + y\, d\tau \\ dy &= x\, dt + y\, d\tau \end{aligned}\right\} \to x + iy = (x_0 + iy_0)\, e^{\tau + it}.$$

Necessary conditions are easy to find.

Let us suppose (9) is integrable in the sense of Morera:

$$x = f(x_0, t_0, \tau_0, t, \tau)$$

that is,

$$x = f(t, \tau),$$

(by neglecting the initial constants). The integrability conditions are

$$\frac{\partial^2 x}{\partial t \partial \tau} = \frac{\partial^2}{\partial \tau \partial t}$$

which yield after a simple calculation:

$$X \cdot \frac{\partial Y}{\partial x} - Y \frac{\partial X}{\partial x} = \frac{\partial X}{\partial \tau} - \frac{\partial Y}{\partial t}. \tag{10}$$

Equation (10) is a vector equation equal to n scalar equations.

A particular case is when X and Y are hamiltonian fields, that is,

$$X \left| \begin{matrix} -\dfrac{\partial H}{\partial p} \\ \dfrac{\partial H}{\partial q} \end{matrix} \right. \qquad Y \left| \begin{matrix} -\dfrac{\partial F}{\partial p} \\ \dfrac{\partial F}{\partial q} \end{matrix} \right. .$$

The conditions (10) reduce to a single condition

$$[H, F] = \frac{\partial H}{\partial \tau} - \frac{\partial F}{\partial t} \tag{11}$$

The fact that these conditions are sufficient from the theory of the integrability of a system of differential forms is known under the name of *Theorem of Frobenius.** Let us admit the theorem and apply to the system (C) the condition (11) with $H \to h$, $F \to -f$, $\tau \to q'$.

One verifies that (11) reduces to (7) and one can put forth

Theorem 3. (reduction of H on Σ). The differential system (C) is integrable in the sense of Morera.

Consequences:

(a) The two systems K_1, K_2 have in common $2n$ first integrals (which are therefore integrals of H).

(b) One can integrate (C) without knowing $q'(t)$, that is to say to calculate
$p_i = \phi_i$ (initial values, t, q')
$q_i = \psi_i$ (initial values, t, q').
One substitutes the values in (b): $dq'/dt = \partial H/\partial p'$ and it remains to resolve a differential

* Elie Cartan, *Leçons sur les Invariants Intégraux.*

equation of the first order:

$$\frac{dq'}{dt} = \text{function of } (q', t).$$

Consider the following example as an application: Interpretation of the KS theory.

Problem: Let there be in R_4 a mass point M with coordinates (u_1, u_2, u_3, u_4) moving under the action of a force derived from the force function $\Omega(u)$. The Hamiltonian of the system, with a time variable τ distinguished from the physical time variable t, is then

$$K = \tfrac{1}{2}\left(\sum_1^4 p_i^2\right) - \Omega(u), \tag{12}$$

where p_i is the conjugate of the $u_i = du_i/d\tau$, one considers the following question: determine $\Omega(u)$ in a way that K admits the invariant relation $F=0$ with

$$[F = u_4 p_1 - u_3 p_2 + u_2 p_3 - u_1 p_4]; \tag{13}$$

and (2) let Σ' be the manifold $F=0$ and find the reduced motion of K on Σ. As a Preliminary; Let us recall the properties of the KS matrix:

$$L(u) = \begin{vmatrix} u_1 & -u_2 & -u_3 & u_4 \\ u_2 & u_1 & -u_4 & -u_3 \\ u_3 & u_4 & u_1 & u_2 \\ u_4 & -u_3 & u_2 & -u_1 \end{vmatrix}$$

(a) L is orthogonal, $LL^T = (\sum_1^4 u^2)E$.
Let us put $R^2 = \sum_4^1 u_i^2$

(b) $L(u)\,du = \begin{bmatrix} dx_1 \\ dx_2 \\ dx_3 \\ \omega \end{bmatrix}$ with $\begin{aligned} x_1 &= \tfrac{1}{2}[u_1^2 - u_2^2 - u_4^2 - u_3^2] \\ x_2 &= [u_1 u_2 - u_3 u_4] \\ x_3 &= [u_1 u_3 + u_2 u_4] \\ \omega &= [u_4 du_1 - u_3 du_2 + u_2 du_3 - u_1 du_4] \end{aligned}$

ω is 'non-holonomic', that is, it is not integrable.

(c) Let us put $r^2 = \sum_1^3 x^2$, then one has $r = R^2/2$.

(d) $\sum_1^3 dx_i^2 + \omega^2 = (2r)\sum_1^4 du_i^2$.

Let us return to the problem of solving the first question. It is necessary to write that $dF/dt = 0$. In the calculation of dF/dt, the variables p disappear and there remains

$$u_4 \frac{\partial\Omega}{\partial u_1} - u_3 \frac{\partial\Omega}{\partial u_2} + u_2 \frac{\partial\Omega}{\partial u_3} - u_1 \frac{\partial\Omega}{\partial u_1} = 0 \tag{14}$$

which is a linear partial differential equation for Ω for which one knows three particular solutions, x_1, x_2, x_3 (orthogonality of $L(u)$). Therefore, Ω is an arbitrary function of x_1, x_2, x_3. The solution of the second question; i.e, the reduction of K on Σ: $F=0$.

Perform the change of variables $x_1(u)$, $x_2(u)$, $x_3(u)$ by considering them to be arbitrary independent functions of $x_4(u)$.

Perform this changing of variables by a change of canonical variables. Let p_i be conjugant to x_1.

Then one has:

$$2T = \sum_1^4 \dot{u}_1^2 = \frac{1}{2r}\left[\sum_1^3 \dot{x}_1^2 + \omega^2\right].$$

(By virtue of (d)); assume that ω is expressed as the mean of the new variables

$$\omega = \sum_1^4 \alpha_i \dot{x}_i$$

The P_i are given by

$$\frac{\partial T}{\partial \dot{x}_i}: \quad \begin{aligned} P_1 &= \frac{1}{2r}\left[\dot{x}_1 + \omega\alpha_1\right] \\ \dot{P}_4 &= \frac{1}{2r}\,\omega\alpha_4 . \end{aligned}$$

The calculation seems laborious. But on Σ, $\omega=0$, therefore $P_4=0$, and one obtains the reduced system $\bar{K}$ on Σ by simply making $K{:}P_4=0$, from which

$$P_i = \frac{1}{2r}\,\dot{x}_i \qquad (i = 1, 2, 3)$$

and

$$\bar{K} = 2r\left(\tfrac{1}{2}\sum_1^3 P_i^2\right) - \Omega\,(x_1, x_2, x_3).$$

Let us consider the change of the energy k, by making a change of the time variable

$$\mathrm{d}\tau = \frac{1}{2r}\,\mathrm{d}t .$$

The change leads to $H=1/2r(\bar{K}-k)$. That is,

$$H = \tfrac{1}{2}\sum_1^3 P_i^2 - \frac{\Omega\,(x_1, x_2, x_3) + k}{2r} \tag{16}$$

for a mass point $P(x_1, x_2, x_3)$ of R_3 under the action of the force function

$$V = \frac{\Omega\,(x) + k}{2r}. \tag{17}$$

Take Ω of the harmonic oscillaror to be $-\omega^2R^2/2=-\omega^2 r$

Then we have

$$V = -\frac{\omega^2}{2} + \frac{k}{2r}.$$

We can then put forth the following:

Theorem: The Keplerian motions of point $P(x)$ in R_3 are identical to the reduced motions of a harmonic oscillator of R_4, on the invariant manifold $F=0$.*

* For a study of particular solutions and a number of applications see Madame Losco, Doctoral Thesis (1972), Faculty of Sciences of Besançon, La Bouloie, Besançon.

EXAMPLE OF INTEGRATION OF STRONGLY OSCILLATING SYSTEMS

O. GODART

University of Louvain, Belgium

Abstract. Dynamical systems admitting solutions very near quasi periodic functions are integrated by a modification of classical difference methods.

A deferred correction can be determined by the condition of energy.

1. Introduction

Many dynamical problems can be expressed as solutions of differential equations of the second order where the first derivative is absent. Methods of integration without the evaluation of this first derivative are economical and seem more appropriate. Such are the Störmer and Cowell method. However it appears that in the long run, the calculated trajectory deviates from the real one whatsoever is the care taken in the choice among the classical multi-step formulae of integration. If first integrals exist, the errors made in their verification can be used to correct the deviations cause by accumulation of discretization and rounding off errors. It is worthwhile then to do the supplementary work necessary for their evaluation.

In a problem of $2n$ degrees of freedom with m first integrals, we can express the original variables and their first derivatives as function of $2n$ parameters including the m constants of the integrals. If from the analysis of the precision in the integration process, it is possible to choose n-m supplementary parameters insensitive to the errors of integration, the corrections due to the integrals are easily determined to the first order by a variation technique.

On the other side, difference methods basic in the classical methods are mainly suitable for approximations by polynomials. If the solutions are better fitted by another type of functions, appropriate special difference corrections are interesting.

In particular, difficulties for integrating step by step differential equations with strongly oscillating functions of the independant variable can be solved by a proper choice of coefficients of higher differences in the integrating formula. For example, see the paper of Stiefel and Bettis (1969).

The same problem may appear in some autonomous dynamical problems where periodicity is less obvious. The problem of Störmer is such an example. In the case corresponding to the motion of charged particles in the Van Allen belt, the original integrating Störmer formula fails but the use of both types of corrections, oscillation and first integral enables us to compute the trajectories very near the essential singularity.

The treatment of a more general case which could have some interest in dynamical astronomy can be done in the same way.

B. D. Tapley and V. Szebehely (eds.), Recent Advances in Dynamical Astronomy, 53–60.

2. Integration of Strongly Oscillating Functions

Let us consider a two degrees of freedom dynamical problem described by a Hamiltonian:

$$2H = X_1^2 + X_2^2 + Q^2(x_1, x_2) = \varepsilon^2 . \tag{1}$$

X_1 and X_2 are the moments conjugate to the two variables x_1, x_2, ε is a constant. In most physical cases, the equation $Q(x_1, x_2) = 0$ can be solved with a parametric angle y (see with slight difference of notations Godart, 1970).

$$x_1 = g(y) \cos y \qquad x_2 = g(y) \sin y$$

and if w is the distance to this curve of any point of the plane x_1, x_2 it could generally be expressed by

$$x_1 = g(y) \cos(y) + w \cos(y + v) \qquad x_2 = g(y) \sin y + w \sin(y + v) \tag{2}$$

the auxiliary angle v being determined from $g(y)$: $\tan v = -(\mathrm{d}g/\mathrm{d}y)/g(y)$. The expression of Q in the new variables y, w, choosen as canonical, can be expanded in power series of w

$$Q = \omega(y)\, w + Q(w^2) . \tag{3}$$

Let us call the momenta conjugate to y, w; Y and W; the Hamiltonian takes the form

$$2H \equiv W^2 + \frac{Y^2}{J^2(w, y)} + Q^2(w, y) = \varepsilon^2 , \tag{4}$$

where J is the Jacobian of the transformation. When the hamiltonian constant ε^2 is small, W, Y, and w are small and, to the first order; w is periodic of argument $q \cong \int \omega(y)\, \mathrm{d}t$ with an amplitude of the order of $\varepsilon/\omega(y)$.

On the contrary, $\mathrm{d}y/\mathrm{d}t$ is of the order of ε^2 and y will change slowly. $\omega(y)$ can be considered as the pulsation of the oscillating W and, as consequence of x_1 and x_2. Another convenient slowly varying parameter is the angle p made by the trajectory with the line $w = 0$ defined by the following relations

$$W = \varepsilon \cos p \cos q \qquad Y = \varepsilon J \sin p \tag{5}$$

and

$$Q(x_1, x_2) = \varepsilon \cos p \sin q \tag{6.1}$$

giving the expressions of the original momenta:

$$X_1 = \frac{\mathrm{d}x_1}{\mathrm{d}t} = \varepsilon [\cos p \cos q \cos(y + v) - \sin p \sin(y + v)] \tag{6.2}$$

$$X_2 = \frac{\mathrm{d}x_2}{\mathrm{d}t} = \varepsilon [\cos p \cos q \sin(y + v) + \sin p \cos(y + v)] . \tag{6.3}$$

To complete the expressions of the original variables into the three angular parameters

y, p, q; let us eliminate w between the Equations (2)

$$x_1 \sin(y+v) - x_2 \cos(y+v) = g(y) \sin v. \tag{6.4}$$

The canonical variables of the dynamical system (1) will be composed of a slowly varying function of time and an oscillating part of short period. The main constraint on the step h of numerical integration will require that $\omega(y)\,h$ is a small quantity. In fact, if we consider a simple sinusoïdal function S we get for the central difference of order $2r$, $\delta^{2r}S = (-u)^r S$ where

$$u = 4 \sin^2 \frac{\omega h}{2} \tag{7}$$

reaching 1 for $\omega h = \pi/3$.

This may impose a prohibitive small step. However it is always possible to choose h such that the secular term may be approximated by a polynomial of small degree, in such a way that higher differences let us say from δ^4 onwards will depend on the oscillating part.

The integration formula are expressed by truncated series of differences; if we write the remaining terms on the form: $\delta^4 L(\delta^2) + \delta^5 L'(\delta^2)$ where L and L' are operators expandable in series of δ^2.

In the case of one period ω^2, it will be sufficient to add to the classical integrating formula the two terms

$$L(-u)\,\delta^4 + L'(-u)\,\delta^5. \tag{8}$$

There will be two such terms for each period. Moreover, if the amplitude of one periodic term changes significatively in the interval of integration, it is possible to take it into account in considering the limiting case of two periods ω_1, ω_2, tending to a common limit. In the case of a central difference scheme, $L'(-u)=0$ and the value of $L(-u)$ is smaller (see for more details Godart, 1971).

It is then quite advantageous to proceed in two steps: a predictor without oscillating correction, a corrector utilizing central differences with the appropriate $L(u)$ correction. If ω increases along the integration, the length of the step ought to be decreased. We found that it was more expeditive without special inconvenience to divide the interval by 2 when h reaches 60°. Interpolating formula for oscillating functions have been devised in using the same idea and noticing that

$$\mu S = \cos \frac{\omega h}{2} S. \tag{9}$$

When ω decreases, economy of computing time and a decrease of total rounding of errors could be achieved by increasing the step (see Godart, 1971).

From the equations of motions of the dynamical system (1):

$$\frac{\mathrm{d}x}{\mathrm{d}t} = X \qquad \frac{\mathrm{d}X}{\mathrm{d}t} = -Q \frac{\partial Q}{\mathrm{d}x} \tag{10}$$

the elimination of X gives the second order differential equations:

$$\frac{d^2x}{dt^2} = - QQ_x$$

Q_x being an abbreviation for $\partial Q/\partial x$.

The physical problem does not require the knowledge of X except for few special points on the trajectory. However their numerical values are necessary for the condition of energy (1), which will be used as partial check and eventual correction. Finally it was found more economical to calculate the first derivatives every two steps and to use their values in the predictors formulae:

$$p_{r+1} = x_r + hX_r + h^2 [Q_{r-1} Q_{x,r-1} - 4Q_r Q_{x,r}]/6 \tag{11}$$

$$p_{r+2} = x_r + 2hX_r \pm 2h^2 [Q_r Q_{x,r} + 2Q_{r+1} Q_{x,r+1}]/3 . \tag{12}$$

The index r is related to the step and gives the value of the independent variable $t = rh$.

As a correcter formula, the following central difference equation was used

$$x_{r+1} = 2x_r - x_{r-1} - \frac{h^2}{12} [Q_{r-1} Q_{x,r+1} + 10Q_r Q_{x,r} + Q_{r+1} Q_{x,r+1}] . \tag{13}$$

As Q_{r+1}; $Q_{x,r+1}$ depend on x_{r+1}, this equation could be solved by iteration. To avoid numerous evaluations of Q and Q_x, we define the small increment such that $x_{r+1} = = p_{r+1} + \xi_{r+1}$. Neglecting the second and higher power of ξ_r and terms $h\xi$, we get

$$Q_{r+1}(x) = Q_{r+1}(p) + \xi \cdot Q_{x,r+1}(p)$$

$$Q_{x,r+1}(x) = Q_{x,r+1}(p) + \xi Q_{xx,r+1}(p) .$$

If this result is introduced in (13), the equation for the formula becomes:

$$\xi_{r+1} \left[1 + \frac{h^2}{12} (Q^2_{x,r+1}(p) + Q_{r+1}(p) Q_{x \cdot x,r+1}(p) \right] =$$

$$= 2x_r - x_{r-1} - p_{r+1} + \frac{h^2}{12} \times$$

$$\times [Q_{r+1}(p) Q_{x,r+1}(p) + 10Q_x Q_{x,r} + Q_{r-1} Q_{x,r-1}] . \tag{14}$$

Although this requires the supplementary calculation of the second partial derivatives of Q, only an approximate value of these quantities is sufficient. In our specific problem, an easy evaluation can be obtained when the parametric angle y is known from the Equation (6.4) where the predicted values of x_1 and x_2 are introduced. The pulsation $\omega(y)$ which is necessary for the oscillation correction is then obtained. Finally, we get the required approximations:

$$QQ_{x_1x_2} + Q^2_{x_1} = \omega^2(y) \cos^2(y + v)$$

$$QQ_{x_2x_2} + Q^2_{x_2} = \omega^2(y) \sin^2(y + v)$$

$$QQ_{x_1x_2} + Q_{x_1} Q_{x_2} = \omega^2(y) \sin(y + v) \cos(y + v) .$$

It will then be necessary only to compute Q and Q_x for the predicted values. The oscillation correction c_r can be expressed

$$c_{x\cdot r} = h^2 L(\omega_r h)\,\delta^4(Q_r Q_{xr}) = h^2 L(\omega_r h) \times \\ \times \left[Q_{r+2}Q_{xr+2} - 4Q_{r+1}Q_{xr+1} + 6Q_r Q_{xr} - 4Q_{r-1}Q_{xr-1} + Q_{r-2}Q_{xr-2}\right]. \tag{15}$$

In this case, we write

$$L(\omega_r h) = -\frac{l_2(\omega_r h)}{240}$$

where $l_2 = 1$ for $\omega = 0$ increasing to 1.146 for $\omega h = 60°$ (see Godart, 1971).

To compute the fourth difference p_{r+2} must be known and that correction will be determined when that predictor is evaluated. We finally get a modified value of x_r:

$$m_r = p_r + \xi_r + \varrho_r$$

Q and Q_x being known, a two step quadrature formula will be interesting for the evaluation of the next first derivative. By the Simpson rule, we get a predicted value:

$$P_{r+2} = X_r - \frac{h}{3}\left[Q_{r+2}Q_{x\cdot r+2} + 4Q_{r+1}Q_{x\cdot r+1} + Q_x Q_{xr}\right]. \tag{16}$$

A similar oscillating correction can be computed when p_{r+3} is known in this case $L(\omega_r h) = -l_1(\omega_r h)/90$; l_1 increasing to 1.141 for $\omega h = 60°$. A first value of X_{r+2} will be calculated from the predicted values p_{r+2}. The modified value will be obtained from

$$M_{r+2} - P_{r+2} = -\frac{h}{3}\left[Q^2_{x\cdot r+2}(p) + Q_{r+2}(p)\,Q_{x\cdot xr+2}(p)\right] \times \\ \times \left[\xi_{r+2} + c_{x\cdot r+2}\right] = \frac{hl_1(\omega_r h)}{270}\,\delta^4\left[Q_{r+1}Q_{x\cdot r+1}\right]. \tag{17}$$

3. Corrections due to Energy Integral

Every two steps we shall be able to verify the condition of constant energy. As the phase coordinates: x_1, x_2, X_1, X_2 can be expressed by four independent variables, the three angle parameters p, q, y and the energy coordinate ε, which is necessarily constant along a trajectory. The corrections from the predictor values may be considered as due to errors on the value of these parameters. If they are small, they will be additive and we may write

$$\delta x = \frac{\partial x}{\partial \varepsilon}\,\delta\varepsilon + \frac{\partial x}{\partial q}\,\delta q + \frac{\partial x}{\partial y}\,\delta y + \frac{\partial x}{\partial p}\,\delta p$$

with a similar relation for X.

The corrections δq, δp, δy will have no effect on the energy integral. A preliminary analysis of the propagation of errors in the process described will indicate whether in the actual problem, the weakly unstable central difference formulae will ultimately introduce important deviations and also which of the parameters are liable to be especially sensitive. To cut short rather lengthily developments, by using methods similar to Henrici (1963), it can be shown that the overal error for x and X is on the form

$$\delta x, \delta X \cong C_1 \cos(r\omega_1 h) + C_2 \sin(r\omega_2 h) + \\ + C_3 \cos(r\varepsilon\omega_2 h) + C_4 \sin(r\varepsilon\omega_2 h) \tag{18}$$

where ω_1 is a pulsation tending to $\omega(y_r)$ when h goes to zero; and ω_2 another pulsation characteristic of the motion in y. The error is of the same order of magnitude as ω_1, the factor ε indicating that the pulsation in y tends to zero with ε. The constants C_1, C_2, C_3, C_4 will not be very sensitive to the error due to δy. The oscillating correction will have, as a main effect, the correction for the phase error δq. The small correction δp will be performed also in such a way that the final evaluation of the hamiltonian integral will complete the correction in evaluating $\delta\varepsilon$.

From the modified values of the first derivatives (17) and the predicted values and their first derivative as well as the corrections, we can write the equation of energy to the first order:

$$M_1^2 + M_2^2 + Q^2(p) + 2Q(p)\, Q_x(p)\, (\xi + c) = \varepsilon^2 + 2\varepsilon\delta\varepsilon \tag{19}$$

$\delta\varepsilon$ being determined, the corresponding corrections of x and X can be evaluated. The corrected factors of the moments are

$$\frac{\partial X}{\partial \varepsilon} = \frac{X}{\varepsilon}.$$

To compute the partial variation of x_1, x_2 in ε, let us derive (2)

$$\frac{\partial x_1}{\partial \varepsilon} = \cos(y + v)\frac{\partial w}{\partial \varepsilon} \qquad \frac{\partial x_2}{\partial \varepsilon} = \sin(y + v)\frac{\partial w}{\partial \varepsilon}.$$

Deriving (6.1) we get

$$\frac{\partial Q}{\partial w}\frac{\partial w}{\partial \varepsilon} = \cos p \sin q.$$

But

$$\frac{\partial Q}{\partial w} = \frac{\partial Q}{\partial x_1}\cos(y + v) + \frac{\partial Q}{\partial x_2}\sin(y + v).$$

Then

$$\frac{\partial x_1}{\partial \varepsilon} = \frac{Q\cos(y + v)}{\varepsilon\left[\dfrac{\partial Q}{\partial x_1}\cos(y + v) + \dfrac{\partial Q}{\partial x_2}\sin(y + v)\right]}$$

$$\frac{\partial x_2}{\partial \varepsilon} = \frac{Q\sin(y + v)}{\varepsilon\left[\dfrac{\partial Q}{\partial x_1}\cos(y + v) + \dfrac{\partial Q}{\partial x_2}\sin(y + v)\right]}. \tag{20}$$

As the energy constant is not computed for the odd values of r, we have to interpolate $\delta\varepsilon$, which is zero on the step before, and generally increases monotonically to a quite small value $\delta\varepsilon$ after one step. A tentative value will be to take half the value at the intermediate step.

TABLE I

Oscillating parameters of Equations (15) and (17)

ωh(in°)	u	l_1	l_2
0	1	1	1
10	0.0303845	1.0036341	1.0037555
20	0.1206148	1.0145929	1.0152713
30	0.2679492	1.0332732	1.0344143
40	0.4679111	1.0600565	1.0621106
50	0.7144248	1.0956892	1.0990502
60	1	1.1411984	1.1461291

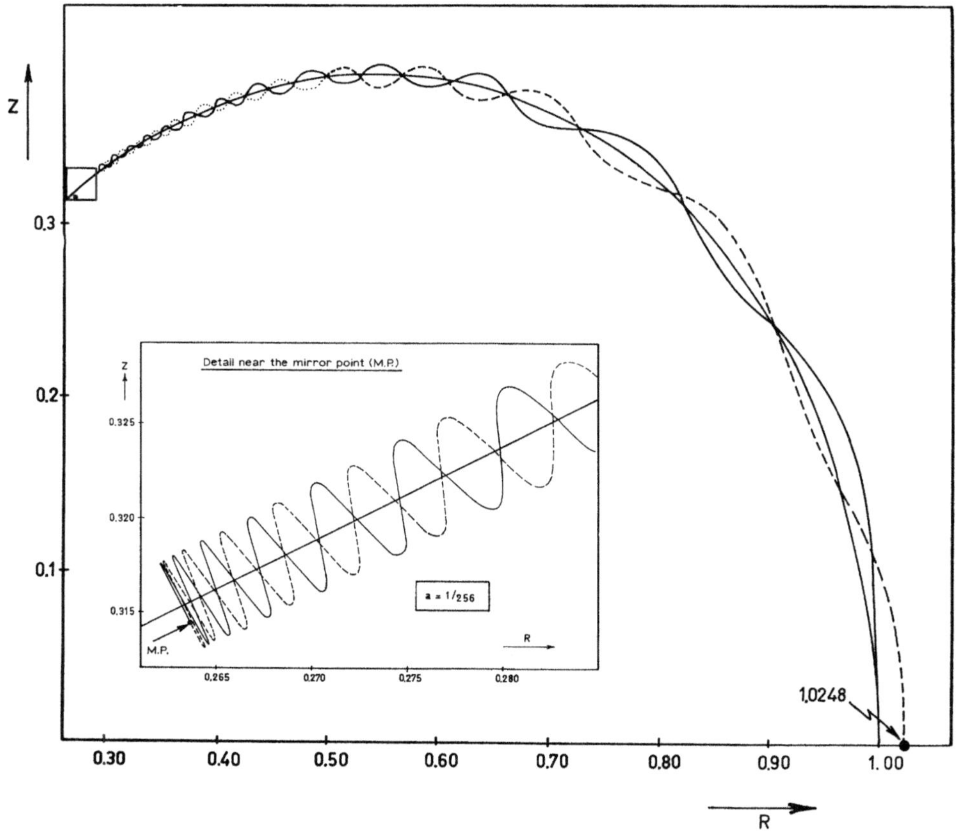

Fig. 1. An example of oscillating trajectory.

The final value of x will then be given by

$$x = p + \xi + c + \frac{\partial x}{\partial \varepsilon}\,\delta\varepsilon\,.$$

It is then advisable to compute the corrected value of $Q^*(x)$, $Q_x^*(x)$. The comparison with $Q(p)$ will indicate if the step is still suitable. Then $Q(p)$ is replaced by $Q(x)$ for the following calculations.

The operations were grouped in blocks repeating itself every two steps. The flow diagram although a bit involved is quite straight-forward following the sequence

$$p_{2r+1}\,(11)\,;\quad \xi_{2r+1}\,(14)\,;\quad c_{x\cdot 2r+1}\,(15)\,;$$

$$M_{2r-2}\,(17)\,;\quad X_{2r-2}\,(20)\,;\quad p_{2r+2}(12)\,;$$

$$\xi_{2r+2}\,(14)\,;\quad c_{x\cdot 2r}\,(15)\,;\quad P_{2r+2}\,(16)\,.$$

The numbers between parenthesis refer to the number of the formulae of this paper.

An example of the trajectory is reproduced in Figure 1. For certain trajectories we could make the following check: the value of x_1 for two successive crossings of $x_2=0$ lost only one decimal of precision.

References

Godart, O.: 1970, in Reider (ed.), *Periodic Orbits Stability and Resonances.*
Godart, O.: 1971, *Ann. Soc. Sci. Bruxelles* **85**, 83.
Henrici, P.: 1963, *Error Propagation for Difference Methods*, J. Wiley, Publ. N.Y.
Stiefel, E. and Bettis, D. G.: 1969, *Numer. Math.* **13**, 154.

THE APPLICATION OF RECURRENCE RELATIONS TO SPECIAL PERTURBATION METHODS

W. BLACK
Glasgow University, Glasgow, Scotland

Abstract. The integration of the equations of motion by explicit Taylor series using recurrence relations is compared with a classical one step method for the case of two body motion. The Taylor method is then applied in turn to the methods of Cowell and Encke in rectangular coordinates, and to a set of perturbational equations using as an example the restricted three body problem. An indication is given of the conditions under which each method is most efficient.

1. Introduction

When large scale numerical investigations are to be carried out on dynamical systems, it is of considerable importance that the numerical method chosen for the integration uses as little machine time as possible. It is also important to ensure that the integration algorithm does not cause inaccuracies to become too large giving misleading results. The work about to be described was carried out to determine which methods are appropriate to which situations.

2. Comparison of RK4 and Taylor Series Methods

Firstly let us consider the choice of numerical integration method. It has recently become very popular to use high order Taylor series expansion methods where the high order derivatives are obtained numerically by the use of recurrence relations. Indeed it has been shown that for many of the equations of Celestial Mechanics, this method outstrips classical multi- and one-step methods as far as machine time is concerned. We decided to carry out a series of tests to compare the Taylor series method with a classical Runge-Kutta fourth order method. The problem we considered was simply the two body problem with orbits of various eccentricities. The Taylor series program was written to allow us to specify the order of the expansion, n, as input data. Both programs varied the step size so that the local truncation error at each step was less than ε for each of the dependent variables.

Figure 1 shows the total machine time for 1 orbit plotted against the order of expansion, n. This is for a circular orbit and for four different accuracy criteria. Note that each graph has a value of n for which the machine time is a minimum. As n is increased the step size which may be used for a given ε obviously increases. However the extra work done as n becomes large cancels out the saving in machine time which can be gained from the larger step. We therefore would expect a value to exist which gives a minimum in machine time. The times for RK4 are plotted beside the y-axis and the numbers along the graphs refer to the number of steps taken per orbit.

B. D. Tapley and V. Szebehely (eds.), Recent Advances in Dynamical Astronomy, 61–70.

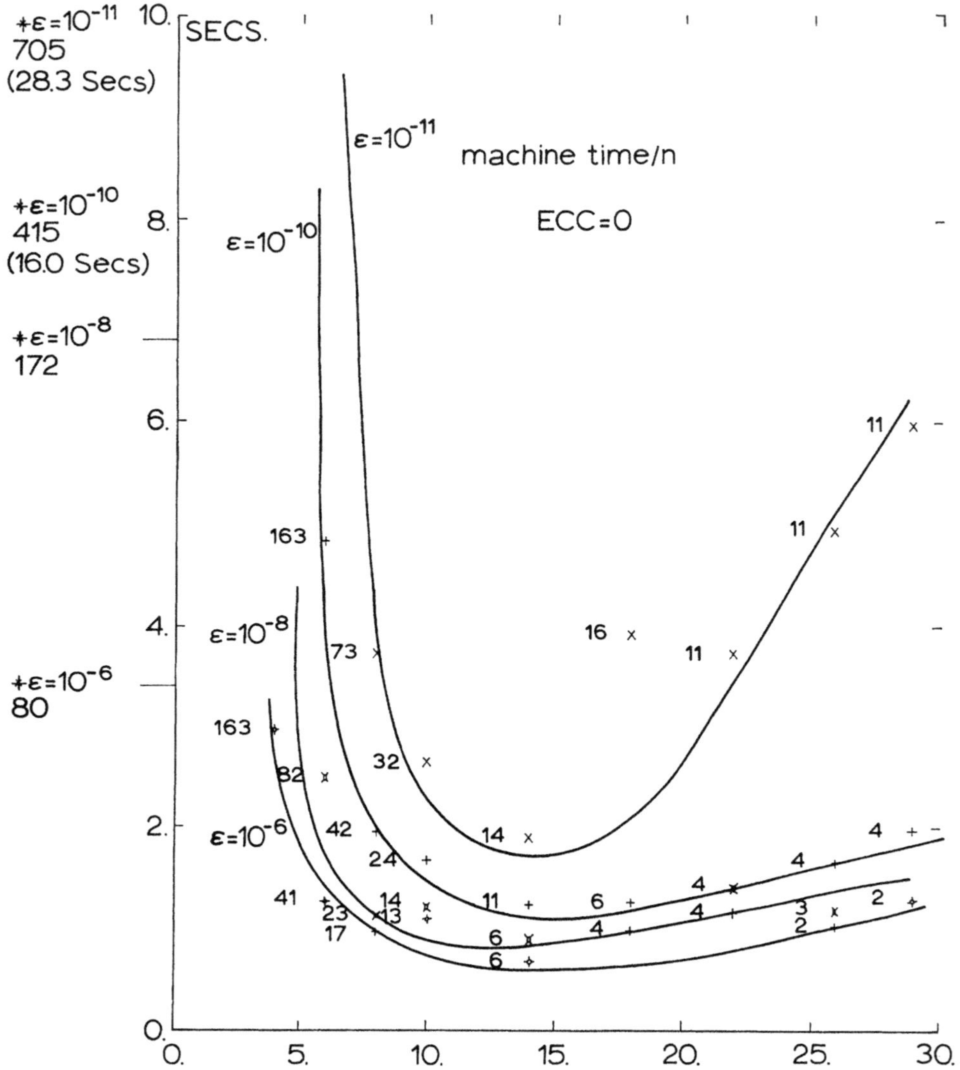

Fig. 1.

Figure 2 is similar to the previous one. Here the eccentricity is 0.95. The times for RK4 are again plotted beside the graph.

To compare the relative efficiency of the TS method to RK4, we calculate the ratio of the time taken per orbit by RK4 to the minimum time taken per orbit by the TS method i.e. the parameter is >1 when RK4 is less efficient. Table I shows the relative efficiency for a series of accuracy criteria and eccentricities. Note that the efficiency of TS increases as ε increases, but decreases as e increases. However in all cases TS is more economical in its use of machine time.

A question which naturally follows at this point is: "Can we also gain accuracy by

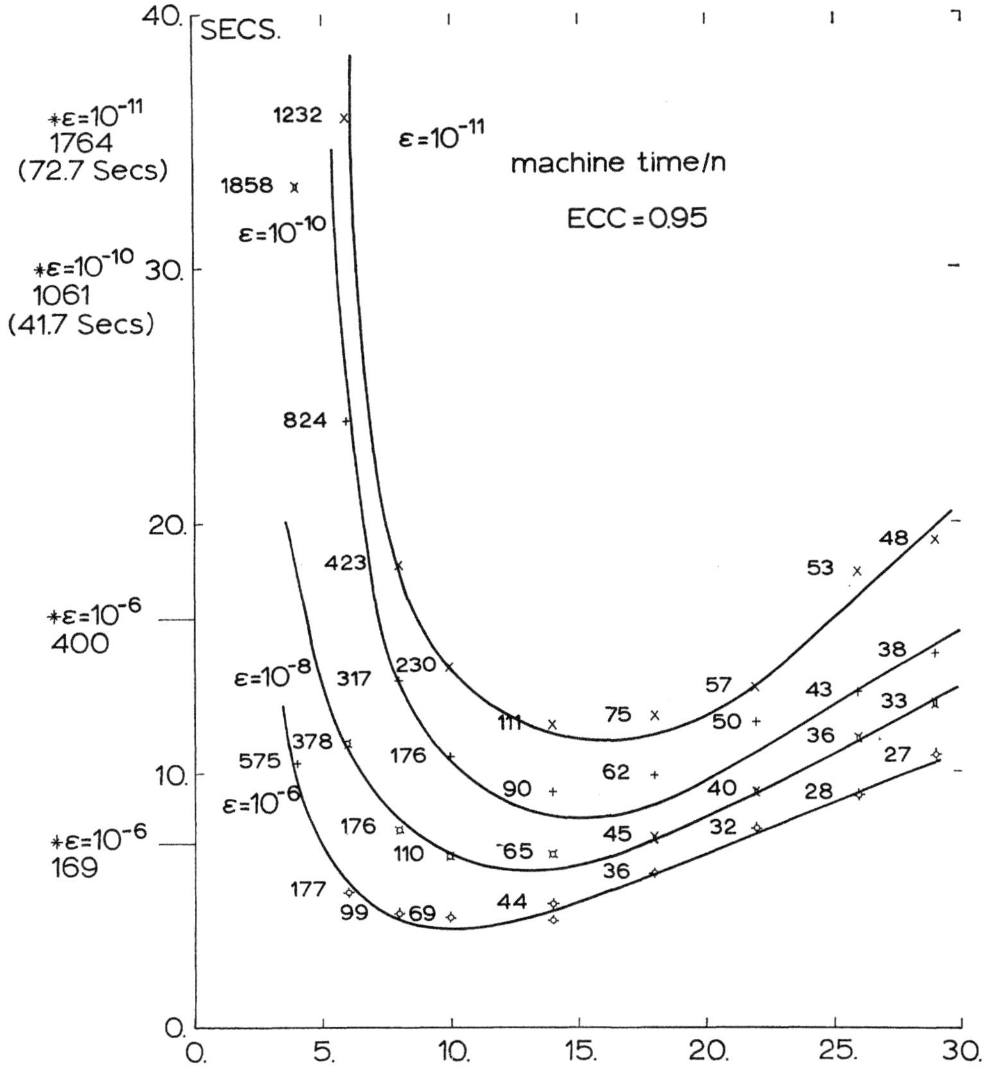

Fig. 2.

TABLE I

Relative efficiency of TS to RK4

	$\varepsilon = 10^{-6}$	$\varepsilon = 10^{-8}$	$\varepsilon = 10^{-10}$	$\varepsilon = 10^{-11}$
$e = 0$	4.9	7.4	12.8	15.0
$e = 0.25$	2.4	3.4	7.2	7.5
$e = 0.67$	1.6	2.7	5.2	6.1
$e = 0.95$	1.6	2.4	4.5	6.1

TABLE II

Errors after 10 orbits for $e = 0$ and $\varepsilon = 10^{-11}$

n	Error in non-zero comps. $\times 10^{10} (x, \dot{y}, \dot{z})$	Error in zero comps. $\times 10^{10} (\dot{x}, y, z)$	Error in semi-major axis $\times 10^{10}$	Error in angular momentum $\times 10^{10}$
4	71	467	200	54
6	2	1302	9	2
8	1	4	0	1
10	1	79	1	1
14	1	81	0	0
18	1	13	0	0
22	1	6	0	0
29	1	6	0	0
RK4	1	3501	4	1

TABLE III

Errors after 10 orbits for $e = 0.95$ and $\varepsilon = 10^{-11}$

n	Mean prop. error in initially non zero comps. $\times 10^{10} (x, \dot{y}, \dot{z})$	Mean error in initially zero comps. $\times 10^{10} (\dot{x}, y, z)$	Error in semi-major axis $\times 10^{10}$	Mean prop. error in components of angular momentum $\times 10^{10}$
4	2278	18435759	58149	96
6	6	892209	669	2
8	5	817983	980	1
10	7	1055589	349	0.5
14	9	1428088	1281	0.3
18	11	1313246	960	0.4
22	9	1178265	786	0.1
29	8	1116325	262	0.6
RK4	5	583540	1921	3

TABLE IV

Errors after 10 orbits for $e = 0$ and $\varepsilon = 10^{-6}$

n	Error in non-zero comps. $\times 10^{10} (x, \dot{y}, \dot{z})$	Error in zero comps. $\times 10^{10} (y, z, \dot{x})$	Error in semi-major axis $\times 10^{10}$	Error in angular momentum $\times 10^{10}$
4	258310	47553026	791942	198063
6	3435	618374	10375	2595
8	892	229350	6102	1526
10	4	878	14	3
14	1	40	5	1
18	0	88	1	0
22	0	61	0	0
29	16	131	1	0
RK4	3300010	18024801	392286	533647

using the TS method rather than RK4?" The answer is shown in Tables II and III. Since the solution of the 2-body problem can be found analytically we were able to calculate the errors in the position, velocity and elements of the orbit after 10 orbits. In the case of $\varepsilon = 10^{-11}$ the errors from the two methods are comparable. However in the case of $\varepsilon = 10^{-6}$ (Table IV), TS appears to be much more accurate than RK4. Remember that the minimum machine time occurred at $\sim n = 10$ in this case. It was found that the errors arose due to rounding and due to a small error in the period after which the programs stopped. By repeating the calculation using double precision arithmetic and by correcting the period, errors were almost entirely eliminated in the TS case. This demonstrates that truncation error is negligible in high order Taylor series methods.

On the basis of the above facts we decided to continue our investigation using the Taylor series method and the 3 programs about to be discussed used TS algorithms.

3. Three Special Perturbation Methods

When dynamical systems more complicated than the 2-body case are to be investigated, there are three basic procedures which may be used to obtain the numerical solutions.

(a) Cowell's method. In this method the equations of motion describing the system are integrated directly.

(b) Encke's method. Here the differential equations describing the departure of the actual motion from the motion in some reference orbit are numerically integrated. In most applications the reference orbit has been taken to be the osculating 2-body orbit at some epoch. However, as the actual motion gradually departs more and more from the rererence orbit, the old reference orbit has to be replaced by a new one.

(c) Perturbational Equations. In this method we integrate numerically the differential equations which describe the changes in quantities which are constant for undisturbed 2-body motion. e.g. Lagrange's Planetary Equations describe the changes in the standard two body orbital elements.

It was our intention to gain some insight into the differences in speed and accuracy of these 3 methods. The test case which we used was the coplanar restricted 3-body problem. Orbits of an infinitesimal mass, P, about a large mass, S, were considered. The body P was perturbed by another body J which was of mass μ times that of S and which was prescribed to move around S in a circular orbit of radius unity. We considered cases where μ was between 10^{-3} and 10^{-6}. Two types of orbit of P about S were considered.

(i) An almost circular orbit of initial semi-major axis 0.5 and eccentricity 0.1.

(ii) A highly eccentric orbit whose initial semi-major axis was 0.25 and whose initial eccentricity was 0.9. In both cases the body P was integrated for 3 orbits about S.

In the Encke method the reference orbit was taken to be the osculating two body orbit at the start of the integration. This was updated whenever the accuracy criteria supplied as data caused the program to half the step size.

The set of perturbational equations used are the differential equations for $\mathbf{h}$, $\boldsymbol{\varepsilon}$ and

λ_p. Here **h** is the osculating angular momentum vector, **ε** is a vector drawn from S towards pericentre in the osculating orbit and proportional to the osculating eccentricity known as Hamilton's integral and λ_p is the difference between the true longitude in the actual orbit and the true longitude in the osculating orbit. These equations have a very simple form and are very convenient for numerical integration compared to other sets which have been proposed e.g. Lagrange's Planetary Equations.

4. Speed Comparison

Let us begin by comparing the speeds of Encke's and Cowell's methods. All timings will be assumed from now on to refer to the value of n which gives a minimum in machine time for the problem being considered. We define the relative efficiency of Encke to Cowell as the ratio of the minimum time for Cowell to the minimum time for Encke i.e. parameter >1 implies Encke is more efficient (Tables V and VI).

TABLE V

Relative efficiency of Encke's to Cowell's method for initial eccentricity = 0.9

	$\varepsilon = 10^{-5}$	$\varepsilon = 10^{-11}$	$\varepsilon = 10^{-15}$
$\mu = 10^{-6}$	10.2	4.0	2.6
$\mu = 10^{-5}$	8.3	3.1	1.9
$\mu = 10^{-4}$	6.5	2.3	1.6
$\mu = 10^{-3}$	4.6	1.8	1.4

TABLE VI

Relative efficiency of Encke's to Cowell's method for initial ecccenctricity = 0.1

	$\varepsilon = 10^{-5}$	$\varepsilon = 10^{-11}$	$\varepsilon = 10^{-15}$
$\mu = 10^{-6}$	5.1	1.5	1.1
$\mu = 10^{-5}$	2.5	1.2	1.0
$\mu = 10^{-4}$	2.1	0.9	0.8
$\mu = 10^{-3}$	1.0	0.7	0.7

As would be expected the Encke method is much more efficient in the highly elliptic case than for the near circular case. Both tables show that as ε decreases, Encke's method becomes less efficient. Cowell's time is essentially independent of the perturbation, but as the perturbation μ increases Encke eventually becomes less efficient than Cowell. Indeed for $\mu > 10^{-3}$ Cowell is always more efficient than Encke.

In a similar fashion we may define the relative efficiency of the perturbational equations to Encke's method so that the ratio is >1 when the perturbational equations are more efficient (Tables VII and VIII). Here η is accuracy criteria to avoid confusion with vector **ε**.

For the almost circular orbit the table sho ws that the Encke method is approximately twice as fast as the perturbational equations for almost all mass ratios and accuracy requirements. The other table for $e=0.9$, however, shows that Encke is only slightly faster for moderate or high accuracy and in the case of low accuracy the perturbational equations win by a factor of 3. It should be noted here that we could expect the changes in $\mathbf{h}$, $\boldsymbol{\varepsilon}$, λ_p to be of the order of the perturbation which is in turn the same order as the imposed criterion. Thus we find ourselves in essence integrating the two body solution where $\mathbf{h}$, $\boldsymbol{\varepsilon}$, λ_p are constants.

It is important to realize why Encke's method is faster than the perturbational equations in most cases. The time taken per step for the perturbational equations is about half as much again as Encke, but the number of steps required is not reduced by a large enough factor to compensate. Since the form of our perturbational equations are simpler than any others we have seen, we would infer that the time per step and consequently the efficiency of other methods would be worse in comparison to ours.

Figure 3 summarizes the regions in which each method is most efficient with respect

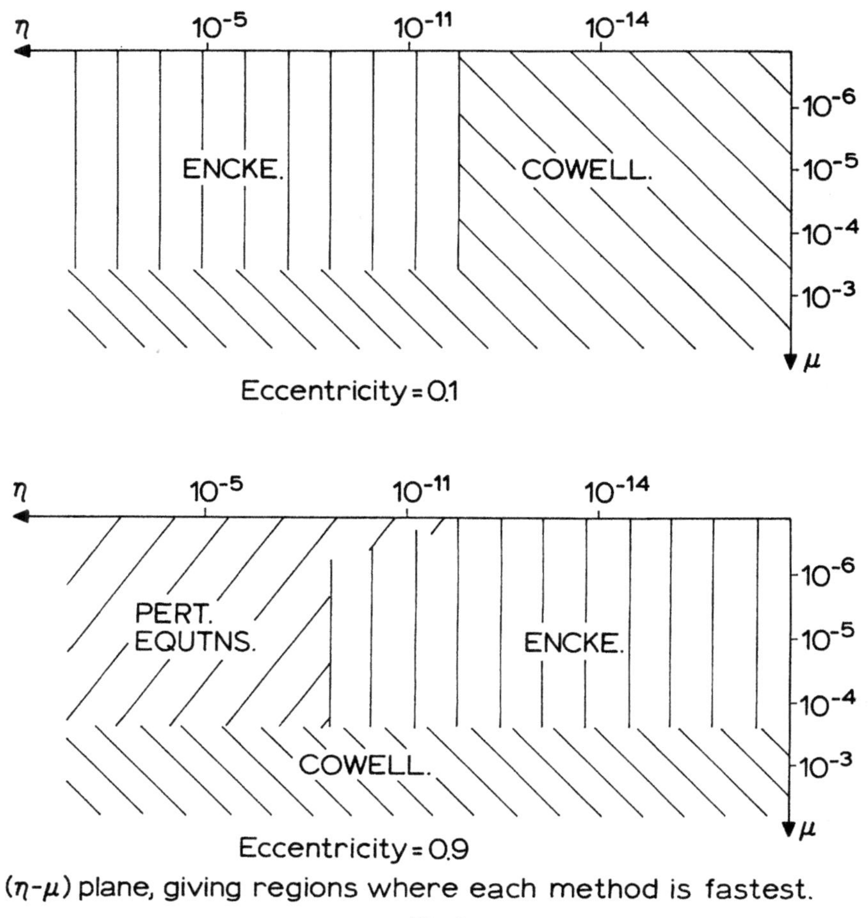

Fig. 3.

TABLE VII

The relative efficiency of the perturbational equations to Encke's method for the initial eccentricity = 0.9

	$\eta = 10^{-5}$	$\mu = 10^{-11}$	$\eta = 10^{-14}$
$\mu = 10^{-6}$	3.1	1.05	0.83
$\mu = 10^{-5}$	3.2	0.89	0.84
$\mu = 10^{-4}$	2.4	0.81	0.89
$\mu = 10^{-3}$	1.2	0.81	0.83

TABLE VIII

The relative efficiency of the perturbational equations to Encke's method for the initial eccentricity = 0.1

	$\eta = 10^{-5}$	$\eta = 10^{-11}$	$\eta = 10^{-14}$
$\mu = 10^{-6}$	1.15	0.51	0.53
$\mu = 10^{-5}$	0.91	0.64	0.56
$\mu = 10^{-4}$	0.52	0.53	0.57
$\mu = 10^{-3}$	0.54	0.55	0.52

to machine time. In all cases high perturbation implies Cowell's method is best. Only in low accuracy, highly eccentric orbits do the perturbational equations win. In all other cases Encke's method is the best.

5. Accuracy Comparison

In order to investigate the relative accuracy of the methods, the test particle was integrated for 3 orbits and then the diiection of integration reversed and the particle brought back until the time equalled zero again. Thus the differences in initial and final coordinates and components of velocities give a measure of the errors which accummulate over a fairly long period of integration.

To make the comparison between Cowell's method and Encke's method, we decided to define the relative accuracy as the ratio of the error in Cowell for n giving a minimum in machine time to the error in Encke for n giving a minimum in machine time i.e. ratio >1 implies Encke is more accurate. It should be noted howevei that accuracy can usually be gained by sacrificing machine time in both cases since higher n gives greater accuracy.

Tables IX and X show that for the very stringent accuracy criteria Encke is more accurate than Cowell but for low or moderate requirements the reverse is the case. Note that μ does not seem to affect the relative accuracy.

Similarly we define the relative accuracy of the perturbational equations to Encke's method such that the ratio is greater than 1 when the perturbational equations are more accurate. (Tables XI and XII).

We see that for very high accuracy Encke's method is far superior whereas for low

TABLE IX

The relative accuracy of Cowell's to Encke's method for initial eccentricity = 0.1

	$\varepsilon = 10^{-5}$	$\varepsilon = 10^{-11}$	$\varepsilon = 10^{-15}$
$\mu = 10^{-6}$	0.7	0.5	1.2
$\mu = 10^{-5}$	0.2	0.3	2.2
$\mu = 10^{-4}$	0.1	0.5	1.8
$\mu = 10^{-3}$	0.5	0.2	5.1

TABLE X

The relative accuracy of Cowell's to Encke's method for initial eccentricity = 0.9

	$\varepsilon = 10^{-5}$	$\varepsilon = 10^{-11}$	$\varepsilon = 10^{-15}$
$\mu = 10^{-6}$	0.5	0.2	3.6
$\mu = 10^{-5}$	0.7	0.2	5.6
$\mu = 10^{-4}$	0.3	0.4	3.4
$\mu = 10^{-3}$	0.5	1.9	6.1

TABLE XI

The relative accuracy of Encke's method to the perturbational equations for the almost circular orbit

	$\eta = 10^{-5}$	$\eta = 10^{-11}$	$\eta = 10^{-14}$
$\mu = 10^{-6}$	59.1	3.1	0.014
$\mu = 10^{-5}$	22.1	36.0	0.01
$\mu = 10^{-4}$	1.44	0.63	0.026
$\mu = 10^{-3}$	3.65	1.35	0.018

TABLE XII

The relative accuracy of Encke's method to the perturbational equations for the highly eccentric orbit

	$\eta = 10^{-5}$	$\eta = 10^{-11}$	$\eta = 10^{-14}$
$\mu = 10^{-6}$	3.25	2.70	0.079
$\mu = 10^{-5}$	0.65	1.20	0.044
$\mu = 10^{-4}$	140.4	4.09	0.092
$\mu = 10^{-3}$	637.0	0.79	0.036

or moderate accuracy the perturbational equations are better. Remember also that it was for low accuracy that the perturbational equations tended to be faster than Encke.

6. Conclusion

In summary then we have found that for high perturbations Cowell's method should be used. For perturbations in the range 10^{-3} to 10^{-6} and for high accuracy Encke is appropriate. Finally for the same range of perturbations and low accuracy the perturbational equations would appear to be best. It would appear that an Encke type approach where an improved reference orbit is used will be fruitful in a search towards greater efficiency and accuracy.

7. Acknowledgement

This is a report of work carried out by P. E. Moran, A. E. Roy and the present author at Glasgow University, Scotland and is taken from a series of 4 papers with the same title which are to be published in *Celestial Mechanics*.

NUMERICAL SOLUTION OF ORDINARY DIFFERENTIAL EQUATIONS

D. G. BETTIS

The University of Texas at Austin, Austin, Tex., U.S.A.

Abstract. Several numerical methods for solving systems of ordinary differential equations are presented, including multistep methods and single step methods. Particular emphasis is given to the application of these methods to problems in dynamical astronomy.

High order Runge-Kutta methods of the Fehlberg type for first order differential equations and also for second order equations are discussed, as well as other higher order explicit and implicit Runge-Kutta formulations. The new variable-order, variable-step multistep methods and the off-grid methods are mentioned, and the extrapolation methods outlined.

References

Fehlberg, E: 1966, 'New One-Step Integration Methods of High-Order Accuracy Applied to Some Problems in Celestial Mechanics', NASA TR R-248.

Fehlberg, E: 1968, 'Classical Fifth-, Sixth-, Seventh-, and Eight-Order Runge-Kutta Formulas with Stepsize Control', NASA TR R-287.

Fehlberg, E: 1969, 'Low-Order Classical Runge-Kutta Formulas with Stepsize Control and their Application to some Heat Transfer Problems', NASA TR R-315.

Fehlberg, E: 1970, 'Some Experimental Results Concerning the Error Propagation in Runge-Kutta Type Integration Formulas', NASA TR R-352.

Fehlberg, E: 1972, 'Classical Eight- and Lower-Order Runge-Kutta-Nystrom Formulas with Stepsize Control for Special Second-Order Differential Equations', NASA TR R-381.

Filippi, S. and Kraska, E: 1968, *Numerische Mathematik* **48.**

Gear, G. W: 1971, *Numerical Initial Value Problems in Ordinary Differential Equations*, Prentice-Hall, Inc., Englewood Cliffs, N. J.

Glasmacher, W. and Sommer D: 1970, *Implizite Runge-Kutta-Formeln*, WestdeutscherVerlag, Köln.

Lapidus, J. and Seinfeld, J. H: 1971, *Numerical Solution of Ordinary Differential Equations*, Academic Press, N.Y.

B. D. Tapley and V. Szebehely (eds.), Recent Advances in Dynamical Astronomy, 71.

PART II

THE THREE-BODY PROBLEM

RECENT ADVANCES IN THE PROBLEM OF THREE BODIES

V. SZEBEHELY

The University of Texas, Austin, Tex., U.S.A.

Abstract. These lectures discuss the persistently inherent instability dominating the dynamical behavior of three gravitationally interacting point masses. The masses of the participating bodies are of the same order of magnitude and their mutual distances are arbitrary.

The lectures are organized in the classical inverted style: first all results are presented, then the definitions and the equations of motion in various systems are described. This is followed by the development of the mathematical apparatus needed, the Lagrange-Jacobi equation, Sundman's original and modified inequalities and several dynamical escape conditions. The principal theorem regarding the stability of the system is then proved with first offering the outline, then an approximation and finally, all the details of the precise derivations. Implications of the results to questions of numerical analysis regarding the accuracy of the numerical integration of three and many-body problems are included.

1. Introduction

These lectures contain didactic material together with some new results. The subject is the general problem of three bodies with arbitrary distances between the participating bodies, the masses of which are of the same orders of magnitude. The bodies are considered to be point masses moving under the influence of their mutual gravitational attractions. The principal references are Whittaker's (1904), Chazy's (1918), Birkhoff's (1927), Wintner's (1941), Leimanis' (1958), Pollard's (1966), and Siegel's (1971) books along with the author's recently published series of papers on the subject (1967–73).

The lectures consist of three parts. In the first part the general results are described and the phenomenology of the motion explained with a simplified and approximate mathematical apparatus. The possible motions are listed, Chazy's and Birkhoff's classifications are compared with recent dynamical descriptions and the predominance of the escape-type behavior (hyperbolic-elliptic motion) is emphasized.

The second part consists of the derivation of the mathematical apparatus. It begins with the presentation of the original 18th order system of differential equations describing the problem in an inertial frame of reference. Then Jacobi's variables are introduced and the resulting 12th order system of differential equations is derived. This is followed by the presentation of the Lagrange-Jacobi second order differential equation for the moment of inertia of the system. Three forms of Sundman's inequality are derived next. Various analytical escape conditions complete the mathematical apparatus needed.

The third part utilizes the analytical results along two main lines. First the combination of Sundman's inequality with escape conditions offers a precise analytical description of the dominant mode of behavior. As the second result the limitations of the numerical integration are established and it is shown that the requirement for escape is also the principal problem for the numerical integration. The phenomenon of triple close approach, responsible for the numerical difficulties as well as for the dominant

B. D. Tapley and V. Szebehely (eds.), Recent Advances in Dynamical Astronomy, 75–106.

behavior of the system, seems to indicate, once again, how masterfully nature masks its laws.

2. Dynamic Behavior of the System

The classical work of Chazy (1918–1932) is eminently summarized by Leimanis (1958), therefore, a new system of classification is described here, given by Szebehely (1971), including not only the final but also the temporary phases. Birkhoff's (1927) classification and his remarks on Chazy's work will be mentioned at the appropriate places.

The basic parameter is the total energy of the system, which here we denote by h. Accepting the conventional ways of defining the kinetic energy by

$$T = \tfrac{1}{2} \sum_{i=1}^{3} m_i (\dot{\mathbf{r}}_i)^2 \tag{1}$$

and the potential energy by

$$F^* = - \sum_{1 \leqslant i < j \leqslant 3} \frac{G m_i m_j}{|\mathbf{r}_{ij}|} = - F, \tag{2}$$

we have $h = T - F$, where
T is the kinetic energy,
F^* is the potential energy,
F is the self-potential or Poincaré's force function,
m_i is the mass of the ith body,
$\mathbf{r}_i$ is the position vector of the ith body,
$\dot{\mathbf{r}}_i$ is the velocity vector of the ith body,
G is the gravitational constant, and
$\mathbf{r}_{ij} = \mathbf{r}_i - \mathbf{r}_j$ is the relative position of the ith body with respect to the jth body.

The case of *positive total energy* may be disposed of quickly since it leads to disruption of the system. Either all participating bodies depart on hyperbolic orbits according to $|\mathbf{r}_{ij}| \to t$, a motion which may be termed *explosion*, or two of the bodies form a binary, $|\mathbf{r}_{12}| < a$, and the third body increases its distance from this binary according to the hyperbolic law: $|\mathbf{r}_{13}|, |\mathbf{r}_{23}| \to t$. This motion is called hyperbolic-elliptic by Chazy and will be referred to here as *escape*. These unbounded motions correspond to results known from the behavior of two bodies where, with $h > 0$, the motion is unbounded (hyperbolic). This similarity between the two and three-body problems, however, is not complete since the bounded (elliptic) motion occurring in the two-body problem for $h < 0$ does not, in general, correspond to the behavior in the problem of three bodies.

The first class of motion for $h < 0$ is called *interplay*. The bodies perform repeated close approaches and $|\mathbf{r}_{ij}| < a$. Another class is termed *ejection*, when two bodies form a binary while the third body is ejected with elliptic relative velocity. Chazy's 'bounded motion', therefore, may be separated in the modern classification into two classes: interplay and ejection. As the energy of the ejected body increases it may depart on a hyperbolic orbit, leaving the binary behind. This unbounded motion occurring with $h<0$ is called (as before for $h>0$) *escape* or 'hyperbolic-elliptic' by Chazy. An important special case for applications of bounded motions is termed *revolution* when the binary formed is surrounded by the orbit of the third body. This motion occurs only with $h<0$ and its stability depends on the magnitude of the ratio ϱ/r, where ϱ is the distance between the center of mass of the binary and the third body, and r is the distance between the members of the binary. If this ratio is large, the system is stable but the original definition of the general problem of three bodies is violated since this motion restricts the distances. The character of the motion, of course, changes into interplay if $\varrho/r \sim 1$.

Another special case of bounded motion is termed *equilibrium configurations* consisting of the triangular and collinear Lagrangean solutions. It is known that these solutions are unstable when the masses are of the same order of magnitude and consequently they transit into interplay.

Finally *periodic orbits* must be mentioned which are also bounded and as far as known, unstable. The periodic orbits of the general problem do not seem to form families in the same sense we know families in the restricted problems and they are not dense.

Note that to this classification one should add the case of $h=0$ as well as the parabolic behaviors. Chazy's hyperbolic-elliptic or parabolic motion ($|\mathbf{r}_{ij}| \to t^{2/3}$) for $h=0$ are of limited significance because they call for a specific value of the energy constant. His hyperbolic-parabolic motion ($|\mathbf{r}_{12}| \to t^{2/3}$; $|\mathbf{r}_{13}|, |\mathbf{r}_{23}| \to t$) occuring with $h>0$ and parabolic-elliptic behavior ($|\mathbf{r}_{12}| < a$; $|\mathbf{r}_{13}|, |\mathbf{r}_{23}| \to t^{2/3}$) for $h<0$ are also of lower dimensionality. These classes, of course, separate the corresponding hyperbolic and elliptic cases.

Birkhoff's classification is based on the moment of inertia (I) of the three bodies, i.e. on the behavior of the function $I(t)$. In this classification *escape* for $h<0$ is associated with $I \to \infty$. *Interplay* and *periodic orbits* correspond to a uniformly bounded behavior of $I(t)$. The *equilibrium solutions* correspond to $I=$constant. With an oscillatory behavior of $I(t)$ – such that one of the three bodies recedes arbitrarily far and returns – we may associate an extreme case of *ejection*. Such oscillatory motion can occur, of course, only for $h<0$.

One of the important results of the past five years in research on the three-body problem is the discovery that the class of motions termed *escape* dominates (Agekyan, 1967; Szebehely, 1967). In other words, for arbitrary initial conditions and after a sufficiently long time the outcome of the motion is hyperbolic-elliptic. This result verifies an 'opinion' of Birkhoff according to which it is 'possible that the motions for which $I \to \infty$ as $t \to \infty$ fill up the manifold of possible motions densely'. If we add to this the conjectures that periodic orbits are neither densely distributed nor are they stable in the general problem, then the basic system of solutions of the three-body problem seems to be the escape type. Such solutions are unstable according to Laplace's definition of stability but show a remarkable persistence to changes in the initial conditions. To establish families of periodic orbits according to what is known today, requires changes in the participating masses as well as in the initial conditions; consequently they do not seem to be densely distributed (Standish, 1970; Szebehely, 1967, 1970). On the other hand the escape type orbits form continuous families as the initial conditions are changed (Szebehely, 1973). For $|h|$ small, one might expect that the manifold of motion will be filled with escape orbits since this is the case for $h \geqslant 0$. As long as no families of stable periodic orbits exist, the conjecture of densely distributed escape orbits is feasible. Numerical results seem to indicate this since the various types of motions described above all have the tendency to turn into escape orbits. Interplay is the necessary prelude leading to escape or ejection. Repeated ejections turn into escapes. Solutions near the Lagrangean solutions are unstable for the general case, as mentioned before, and turn into interplays. The known unstable periodic orbits are also surrounded by interplays as are revolutions unless restrictions are imposed in the distances.

Table I summarizes the possible motions and the various classifications.

TABLE I

Classification of possible motions

$h>0$ $I\to\infty$	hyperbolic, explosion hyperbolic-parabolic, explosion hyperbolic-elliptic, escape
$h=0$ $I\to\infty$	hyperbolic-elliptic, escape parabolic, explosion
$h<0$	bounded $I<I_0$ interplay ejection revolution equilibrium periodic orbits hyperbolic-elliptic, escape, $I\to\infty$ parabolic-elliptic, escape, $I\to\infty$ oscillating

The simplified model which explains approximately the behavior of the system consists of an already formed binary with bounding energy $E_b = -Gm_1m_2/2a$ and of another two-body problem formed by (m_1+m_2) and m_3. The energy of this system is E_e and may be obtained by

$$E_e = \frac{m_3}{2} v_3^2 + \frac{m_1 + m_2}{2} v_{12}^2 - \frac{(m_1 + m_2)\, m_3 G}{d}, \tag{3}$$

where m_1 and m_2 are the masses of the binary with semi-major axis a, m_3 is the mass of the ejected or escaping body with velocity v_3, d is the distance between the center of mass of m_1+m_2 and m_3, and v_{12} is the velocity of the center of mass of the binary.

The total energy of the system is $h=E_t=E_b+E_e$. If $E_t \geqslant 0$ then, since $E_b<0$ always, we have that $E_e>0$, and we conclude that with positive or zero total energy binary formation gives escape if $\dot{d}>0$. This means that for $E_t \geqslant 0$ there is no ejection. The above result will be shown later to be exact.

If $E_t<0$ the situation is more complicated. Since $E_e=E_t-E_b=|E_b|-|E_t|$, for escape $|E_b|>|E_t|$ is required. This can always be established with sufficiently small value of a. Therefore escape for negative total energy is associated with the formation of a close binary. If the binary formed does not have enough negative energy because its semi-major axis is large, escape does not occur but it is replaced by an ejection. The precise formulations of these results will utilize Sundman's inequality and the Lagrange-Jacobi equation, in Section 3.

3. The Equations Representing the System

A. EQUATIONS OF MOTION IN FIXED SYSTEMS AND THE INTEGRALS

The first formulation of the differential equations of motion utilizes a fixed system in which the position vectors of the three bodies are $\mathbf{r}_1, \mathbf{r}_2, \mathbf{r}_3$ as shown in Figure 1.

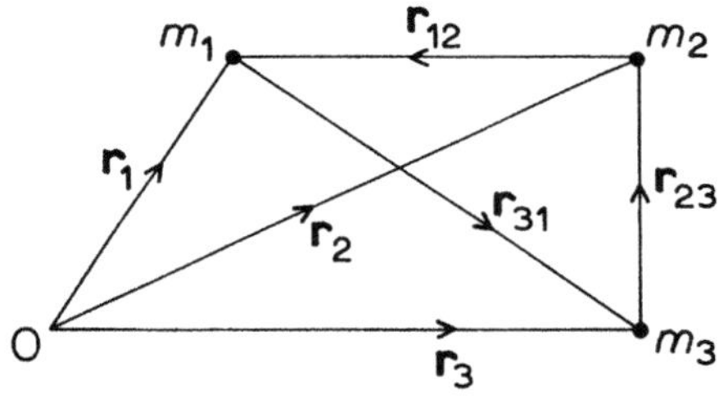

Fig. 1 The general problem of three bodies.

Introducing the vectors $\mathbf{r}_{ij}=\mathbf{r}_i-\mathbf{r}_j$ and the scalars $|\mathbf{r}_{ij}|=r_{ij}$, the equations of motion become

$$m_i\ddot{\mathbf{r}}_i = -Gm_i \sum_{i \neq j=1}^{3} \frac{m_j}{r_{ij}^3}(\mathbf{r}_i-\mathbf{r}_j), \tag{4}$$

or

$$m_i\ddot{\mathbf{r}}_i = \frac{\partial F}{\partial \mathbf{r}_i}, \tag{5}$$

where $i, j=1, 2, 3$ and F is given by Equation (2).

The six integrals of the conservation of linear momentum, also known as the integrals of the center of mass, follow from

$$\sum_{i=1}^{3} m_i\ddot{\mathbf{r}}_i = 0$$

and offer six constants of integrations $\mathbf{a}(a_1, a_2, a_3)$ and $\mathbf{b}(b_1, b_2, b_3)$ in the form

$$\frac{1}{M}\sum m_i\dot{\mathbf{r}}_i = \mathbf{a} \quad \text{and} \quad \frac{1}{M}\sum m_i\mathbf{r}_i = \mathbf{a}t + \mathbf{b},$$

where M is the total mass.

Note that the center of mass of the system is given by

$$\mathbf{r}_c = \frac{1}{M}\sum m_i\mathbf{r}_i .$$

The three integrals of angular momentum follow from

$$\sum m_i\mathbf{r}_i \times \ddot{\mathbf{r}}_i = \frac{\mathrm{d}}{\mathrm{d}t}\sum m_i\mathbf{r}_i \times \dot{\mathbf{r}}_i = 0$$

and offer three constants of integration $\mathbf{c}(c_1, c_2, c_3)$ in the form

$$\mathbf{c} = \sum m_i\mathbf{r}_i \times \dot{\mathbf{r}}_i . \tag{6}$$

Finally, the integral expressing the conservation of the total energy of the system follows from

$$\sum m_i\dot{\mathbf{r}}_i\ddot{\mathbf{r}}_i = \sum \dot{\mathbf{r}}_i \frac{\partial F}{\partial \mathbf{r}_i}$$

and offers one constant of integration

$$h = T - F$$

since

$$\sum m_i\dot{\mathbf{r}}_i\ddot{\mathbf{r}}_i = \frac{\mathrm{d}}{\mathrm{d}t}\frac{1}{2}\sum m_i\dot{\mathbf{r}}_i^2 = \frac{\mathrm{d}T}{\mathrm{d}t}$$

and

$$\sum \dot{\mathbf{r}}_i \frac{\partial F}{\partial \mathbf{r}_i} = \frac{\mathrm{d}F}{\mathrm{d}t}.$$

The original $3 \times 2 \times 3 = 18$th order system may be reduced to a sixth order system by making use of the above 10 integrals, by Jacobi's elimination of nodes, and by the elimination of the time as the independent variable. In what follows only the integrals of the center of mass will be used for a reduction to the 12th order.

B. JACOBIAN COORDINATES

The introduction of the variables proposed by Lagrange and by Jacobi is identical with utilizing the center of mass integrals for reducing the order of the system to 12 from 18. On Figure 2 the Jacobian vectors are $\mathbf{r}$ and $\boldsymbol{\varrho}$, the first connecting m_1 and m_2 and the second the center of mass of m_1 and m_2 with m_3. In order to transform the

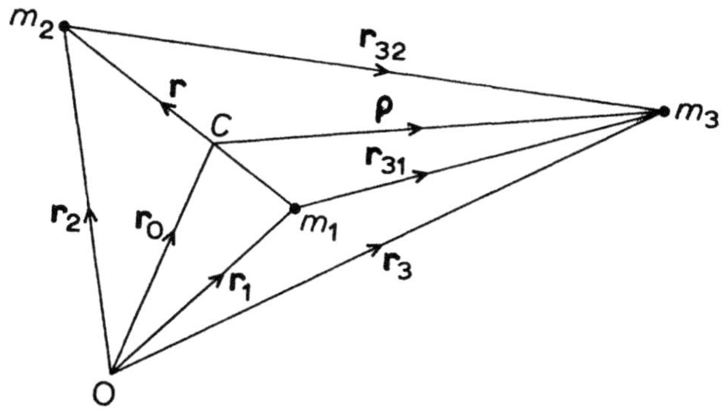

Fig. 2. Jacobian coordinates.

equations of motion into the Jacobian system we express $\mathbf{r}_{21}$, $\mathbf{r}_{32}$ and $\mathbf{r}_{31}$ with $\mathbf{r}$ and $\boldsymbol{\varrho}$. The result is

$$\begin{aligned} \mathbf{r}_{21} &= \mathbf{r}_2 - \mathbf{r}_1 = \mathbf{r}, \\ \mathbf{r}_{32} &= \mathbf{r}_3 - \mathbf{r}_2 = \boldsymbol{\varrho} - \frac{m_1}{\mu}\mathbf{r}, \\ \mathbf{r}_{31} &= \mathbf{r}_3 - \mathbf{r}_1 = \boldsymbol{\varrho} + \frac{m_2}{\mu}\mathbf{r}, \end{aligned} \tag{7}$$

since the vector pointing from m_1 to C is $(m_2/\mu)\mathbf{r}$ and from C to m_2 is $(m_1/\mu)\mathbf{r}$, with $\mu = m_1 + m_2$.

Note also that $\mathbf{r}_3 = (\mu/M)\boldsymbol{\varrho}$, since $\mathbf{r}_3 = \mathbf{r}_0 + \boldsymbol{\varrho}$ and $\mathbf{r}_0\mu + \mathbf{r}_3 m_3 = 0$ or $\mathbf{r}_0 = -\mathbf{r}_3 m_3/\mu$, having 0 at the center of mass of the system.

Now the three equations of motion are ready to be transformed. From Equation (4) we have

$$\begin{aligned} \ddot{\mathbf{r}}_1 &= G\frac{m_2}{r_{12}^3}\mathbf{r}_{21} + G\frac{m_3}{r_{31}^3}\mathbf{r}_{31}, \\ \ddot{\mathbf{r}}_2 &= G\frac{m_3}{r_{23}^3}\mathbf{r}_{32} + G\frac{m_1}{r_{12}^3}\mathbf{r}_{12}, \\ \ddot{\mathbf{r}}_3 &= G\frac{m_1}{r_{31}^3}\mathbf{r}_{13} + G\frac{m_2}{r_{23}^3}\mathbf{r}_{23}. \end{aligned} \tag{8}$$

Subtracting the first equation from the second we have

$$\ddot{\mathbf{r}} = -G\mu\frac{\mathbf{r}}{r^3} + Gm_3\left(\frac{\mathbf{r}_{32}}{r_{32}^3} - \frac{\mathbf{r}_{31}}{r_{31}^3}\right), \tag{9}$$

or

$$\ddot{\mathbf{r}} = -G\mu\frac{\mathbf{r}}{r^3} + Gm_3\left(\frac{\boldsymbol{\varrho} - \dfrac{m_1}{\mu}\mathbf{r}}{r_{32}^3} - \frac{\boldsymbol{\varrho} + \dfrac{m_2}{\mu}\mathbf{r}}{r_{31}^3}\right). \tag{10}$$

A substitution into the third of Equations (8) gives

$$\ddot{\boldsymbol{\varrho}} = -\frac{M}{\mu} G \left(\frac{m_1 \mathbf{r}_{31}}{r_{31}^3} + \frac{m_2 \mathbf{r}_{32}}{r_{32}^3} \right), \tag{11}$$

or

$$\ddot{\boldsymbol{\varrho}} = -\frac{M}{\mu} G \left(\frac{m_1 \left(\boldsymbol{\varrho} + \frac{m_2}{\mu} \mathbf{r} \right)}{r_{31}^3} + \frac{m_2 \left(\boldsymbol{\varrho} - \frac{m_1}{\mu} \mathbf{r} \right)}{r_{32}^3} \right). \tag{12}$$

Equation (10) and (12) form a 12th order system using Jacobian coordinates. Two remarks are in order. First note that these equations may also be written in the short and elegant form:

$$\ddot{\mathbf{r}} + \mu \mathbf{f}(\mathbf{r}) = (M - \mu) \left[\mathbf{f}(\boldsymbol{\varrho} - \nu \mathbf{r}) - \mathbf{f}(\boldsymbol{\varrho} + \nu^* \mathbf{r}) \right]$$

and

$$\ddot{\boldsymbol{\varrho}} = - M \left[\nu^* \mathbf{f}(\boldsymbol{\varrho} - \nu \mathbf{r}) + \nu \mathbf{f}(\boldsymbol{\varrho} + \nu^* \mathbf{r}) \right],$$

where

$$\mathbf{f}(\mathbf{x}) = G \mathbf{x} |\mathbf{x}|^{-3}, \quad \nu = \frac{m_1}{\mu} \quad \text{and} \quad \nu^* = \frac{m_2}{\mu}.$$

The second note is more interesting. Since

$$F = G \left(\frac{m_1 m_2}{r} + \frac{m_2 m_3}{r_{23}} + \frac{m_3 m_1}{r_{31}} \right) = F(\mathbf{r}, \boldsymbol{\varrho})$$

we may obtain with some diligence that

$$\ddot{\mathbf{r}} = \frac{1}{g_1} \frac{\partial F}{\partial \mathbf{r}}$$

and (13)

$$\ddot{\boldsymbol{\varrho}} = \frac{1}{g_2} \frac{\partial F}{\partial \boldsymbol{\varrho}},$$

where

$$g_1 = \frac{m_1 m_2}{\mu} \quad \text{and} \quad g_2 = \frac{m_3 \mu}{M}.$$

From Equations (13) an integral of energy follows immediately since

$$g_1 \dot{\mathbf{r}} \ddot{\mathbf{r}} + g_2 \dot{\boldsymbol{\varrho}} \ddot{\boldsymbol{\varrho}} := \frac{\partial F}{\partial \mathbf{r}} \dot{\mathbf{r}} + \frac{\partial F}{\partial \boldsymbol{\varrho}} \dot{\boldsymbol{\varrho}}$$

or

$$\tfrac{1}{2} (g_1 \dot{\mathbf{r}}^2 + g_2 \dot{\boldsymbol{\varrho}}^2) = F + h.$$

Note that in fact

$$T = \tfrac{1}{2}(g_1 \dot{\mathbf{r}}^2 + g_2 \dot{\boldsymbol{\varrho}}^2) \tag{41}$$

as may be shown by direct substitutions. Similarly, the angular momentum may be written as

$$\mathbf{c} = g_1 \mathbf{r} \times \dot{\mathbf{r}} + g_2 \boldsymbol{\varrho} \times \dot{\boldsymbol{\varrho}}. \tag{15}$$

As an exercise we may show that

$$\dot{\mathbf{c}} = g_1 \mathbf{r} \times \ddot{\mathbf{r}} + g_2 \boldsymbol{\varrho} \times \ddot{\boldsymbol{\varrho}} = 0,$$

since

$$g_1 \mathbf{r} \times \ddot{\mathbf{r}} = - g_2 \boldsymbol{\varrho} \times \ddot{\boldsymbol{\varrho}} = \frac{G}{\mu} m_1 m_2 m_3 \mathbf{r} \times \boldsymbol{\varrho} \left(\frac{1}{r_{32}^3} - \frac{1}{r_{31}^3} \right).$$

At this point we introduce the moment of inertia of the three bodies I, which will play an important role in the sequence. In general, the moment of inertia with respect to the origin of the coordinate system is defined by

$$I = \sum_{i=1}^{3} m_i \mathbf{r}_i^2. \tag{16}$$

It may be shown that the moment of inertia with respect to the center of mass is

$$\phi = \sum_{1 \leqslant i < j \leqslant 3} \frac{m_i m_j}{M} \mathbf{r}_{ij}^2, \tag{17}$$

which expression is also known as the Jacobian function. In fact according to Steiner's theorem

$$\phi = I - M \mathbf{r}_c^2, \tag{18}$$

which follows from Equation (17) by substituting

$$\mathbf{r}_{ij}^2 = \mathbf{r}_i^2 + \mathbf{r}_j^2 - 2 \mathbf{r}_i \mathbf{r}_j$$

and rearranging terms. The Jacobian function also may be written (allowing for repeated subscripts), in the form

$$\phi = \frac{1}{2} \sum_{i=1}^{3} \sum_{j=1}^{3} \frac{m_i m_j}{M} \mathbf{r}_{ij}^2$$

and therefore we have

$$\phi = \frac{1}{2M} \sum_i m_i \left(\sum_j m_j \mathbf{r}_i^2 + \sum_j m_j \mathbf{r}_j^2 - 2 \sum_j m_j \mathbf{r}_i \mathbf{r}_j \right)$$

or

$$\phi = \frac{1}{2M} \sum_i m_i \left(M\mathbf{r}_i^2 + I - 2\mathbf{r}_i \sum_j m_j \mathbf{r}_j \right).$$

This expression, on the other hand, becomes

$$\phi = \frac{1}{2M} \left[MI + MI - 2 \left(\sum_i m_i \mathbf{r}_i \right)^2 \right],$$

which is the desired result considering the previously introduced definition for the position vector of the center of mass $\mathbf{r}_c$.

In the following text the expressions for I and Φ will be used alternatively and since the origin of the coordinate system will be at the center of mass, we have $I = \Phi$.

Using the Jacobian coordinates

$$I = g_1 \mathbf{r}^2 + g_2 \boldsymbol{\varrho}^2, \tag{20}$$

which may be shown by substituting the Jacobian transformation (Equations (7)) into the Jacobian function given by Equation (17).

C. RELATION BETWEEN THE PERIMETER AND THE MOMENT OF INERTIA

It will be shown that the moment of inertia varies as the square of the perimeter, defined by

$$\sigma = r_{12} + r_{23} + r_{31}. \tag{21}$$

In fact if m' is the smallest and $\bar{m}$ and $\bar{m}'$ the two largest of the three masses, then we have

$$\frac{\sigma^2}{4 \sum m_i^{-1}} \leqslant \frac{m'}{12} \sigma^2 \leqslant I \leqslant \frac{\bar{m}\bar{m}'}{M} \sigma^2 \leqslant \bar{m}\sigma^2. \tag{22}$$

The upper and lower bounds for σ are established first in the form

$$r_1 + r_2 + r_3 \leqslant \sigma \leqslant 2(r_1 + r_2 + r_3). \tag{23}$$

Since the center of mass is inside the triangle formed by the bodies, we have, with reference to Figure 3:

$$r_j + r_k \leqslant r_{ji} + r_{ki}, \tag{24}$$

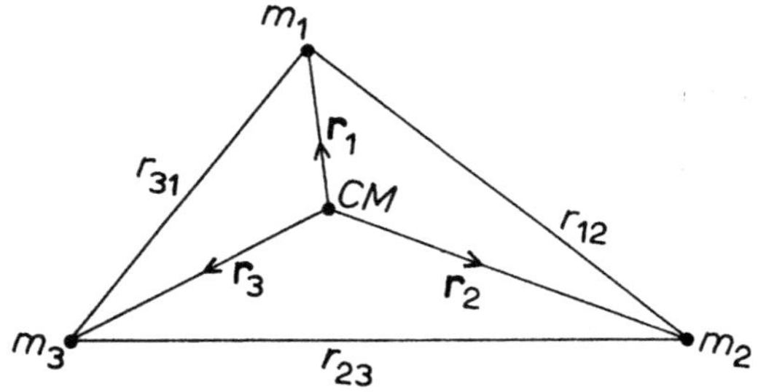

Fig. 3. Estimate for the perimeter.

where $i, j, k = 1, 2, 3$ cyclically. In this way we have

$$r_1 + r_2 \leqslant r_{13} + r_{23}, \quad r_2 + r_3 \leqslant r_{21} + r_{31} \quad \text{and} \quad r_3 + r_1 \leqslant r_{32} + r_{12}.$$

Adding these relations, one obtains

$$2(r_1 + r_2 + r_3) \leqslant 2(r_{12} + r_{23} + r_{21}) = 2\sigma$$

or

$$r_1 + r_2 + r_3 \leqslant \sigma. \tag{25}$$

Using now the triangle inequality

$$r_{jk} \leqslant r_j + r_k$$

we have

$$r_{12} \leqslant r_1 + r_2, \quad r_{23} \leqslant r_2 + r_3, \quad r_{31} \leqslant r_3 + r_1,$$

or

$$r_{12} + r_{23} + r_{31} \leqslant 2(r_1 + r_2 + r_3),$$

or

$$\sigma \leqslant 2(r_1 + r_2 + r_3). \tag{26}$$

Inequalities (25) and (26) set the bounds on σ as

$$r_1 + r_2 + r_3 \leqslant \sigma \leqslant 2(r_1 + r_2 + r_3). \tag{27}$$

The moment of inertia, on the other hand, may be estimated as

$$I = \sum m_i \mathbf{r}_i^2 \leqslant \bar{m} \sum \mathbf{r}_i^2 \leqslant \bar{m}\sigma^2, \tag{28}$$

since

$$r_1^2 + r_2^2 + r_3^2 \leqslant (r_1 + r_2 + r_3)^2 \leqslant \sigma^2.$$

A better (lower) estimate for I is obtained as follows:

$$I = \frac{1}{M} \sum m_i m_j \mathbf{r}_{ij}^2 \leqslant \frac{\bar{m}\bar{m}'}{M} \sum \mathbf{r}_{ij}^2 \leqslant \frac{\bar{m}\bar{m}'}{M} \sigma^2,$$

where $\bar{m}$ and $\bar{m}'$ are the two largest masses and where the inequality $r_{12}^2 + r_{23}^2 + r_{31}^2 \leqslant (r_{12} + r_{23} + r_{31})^2$ was utilized.

Consequently,

$$I \leqslant \frac{\bar{m}\bar{m}'}{M} \sigma^2. \tag{29}$$

The lower bound for I is obtained by means of Cauchy's inequality as follows:

$$\frac{\sigma^2}{4} \leqslant \left(\sum r_i\right)^2 = \left\{\sum \sqrt{m_i} r_i (m_i)^{-1/2}\right\}^2 \leqslant \sum m_i r_i^2 \sum m_i^{-1},$$

or

$$\frac{\sigma^2}{4} \leqslant I \sum \frac{1}{m_i} \leqslant \frac{3I}{m'},$$

and finally

$$\frac{m'\sigma^2}{12} \leqslant I \quad \text{or} \quad \frac{\sigma^2}{4\sum m_i^{-1}} \leqslant I. \tag{30}$$

A combination of inequalities (28), (29) and (30) gives the desired result announced by inequality (22). Since Cauchy's inequality shall be used again later, it may be justified to give its general form.

Let

$$A = \sum a_i^2, \quad B = \sum b_i^2 \quad \text{and} \quad C = \sum a_i b_i.$$

Then $C^2 \leqslant AB$, which states that the product of the lengths of two vectors is larger than their scalar product or

$$C = \mathbf{ab} = |\mathbf{a}|\ |\mathbf{b}| \cos(\mathbf{a}, \mathbf{b}) \leqslant |\mathbf{a}|\ |\mathbf{b}| = \sqrt{AB}.$$

D. THE LAGRANGE-JACOBI EQUATION

This equation, basic in the science of stellar dynamics, was first given by Lagrange for the problem of three bodies, see his *Oeuvres*, **IX**, 836 and VI, 260. The 14 volumes were published between 1867 and 1892. The actual work is dated 1772 but was not published until 1777. Jacobi offered this equation for the problem of n bodies in his Königsberg lectures during the winter of 1842/43, see his 'Vorlesungen über Dynamik' (actually written by Clebsch and published in 1866.)

The equation may be written as

$$\ddot{I} = 2(2T - F), \tag{31}$$

which may be transformed to various forms by $h = T - F$, such as

$$\ddot{I} = 2(F + 2h) \quad \text{and} \quad \ddot{I} = 2(T + h).$$

Equation (31) is usually proved by computing $\ddot{I}$ from Equation (16) and substituting Equations (1) and (4). After a non-trivial rearrangement Equation (31) is obtained. In these operations the upper limit of the summations may be n, in which case Jacobi's general result is obtained.

Using the Jacobian system the proof is limited to $n = 3$. From

$$I = g_1 \mathbf{r}^2 + g_2 \boldsymbol{\varrho}^2,$$

we have

$$\ddot{I} = 2(g_1 \dot{\mathbf{r}}^2 + g_2 \dot{\boldsymbol{\varrho}}^2 + g_1 \mathbf{r}\ddot{\mathbf{r}} + g_2 \boldsymbol{\varrho}\ddot{\boldsymbol{\varrho}}), \tag{32}$$

where the first two terms represent twice the kinetic energy (see Equation 14). The last two terms become

$$\mathbf{r}\frac{\partial F}{\partial \mathbf{r}} + \boldsymbol{\varrho}\frac{\partial F}{\partial \boldsymbol{\varrho}} = -F, \tag{33}$$

where the left side is obtained by substituting Equation (13) and the right side results form Euler's theorem of homogeneous functions applied to F of order -1. In this way Equation (32) may be rewritten as $\ddot{I}=2(2T-F)$ which is the desired result.

Note that $f(x_1,\ldots,x_n)$ is an mth order homogeneous function if $f(x_1\alpha, x_2\alpha,\ldots,x_n\alpha)= =\alpha^m f(x_1, x_2,\ldots,x_n)$. For such functions Euler's theorem states that

$$\sum_{i=1}^{n} \frac{\partial f}{\partial x_i} x_i = mf\,, \tag{34}$$

which may be obtained by differentiating the above definition with respect to α. This gives

$$\frac{\mathrm{d}f}{\mathrm{d}\alpha} = \frac{\partial f}{\partial(x_1\alpha)}\frac{\mathrm{d}x_1\alpha}{\mathrm{d}\alpha} + \cdots + \frac{\partial f}{\partial(x_n\alpha)}\frac{\mathrm{d}x_n\alpha}{\mathrm{d}\alpha} = m\alpha^{m-1}f$$

or

$$\frac{\partial f}{\partial(x_1\alpha)} x_1 + \cdots + \frac{\partial f}{\partial(x_n\alpha)} x_n = m\alpha^{m-1}f\,,$$

which being valid for any α, gives the desired result (Equation 34) for $\alpha=1$.

Note that

$$\sum_i \frac{\partial F}{\partial \mathbf{r}_i}\mathbf{r}_i = -F \neq 0$$

for finite $\mathbf{r}_i$, consequently $\partial F/\partial \mathbf{r}_i \neq 0$ and therefore, there are no equilibrium solutions, $\ddot{\mathbf{r}}_i=0$, of the problem of three (or n) bodies (in a fixed system of coordinates).

From the Lagrange-Jacobi equation it follows that for $h>0$, $\ddot{I}\geqslant 4h>0$ and $I\geqslant 2ht^2+bt+c$. Therefore $I\to\infty$ as $t\to\infty$ and at least one of the distances, $r_{ij}\to\infty$. The same may be shown for $h=0$. Furthermore, since $I=g_1\mathbf{r}^2+g_2\boldsymbol{\varrho}^2$, as $I\to\infty$, $\boldsymbol{\varrho}\to\infty$, if r is bounded. If a binary is formed, $|\mathbf{r}|<a$, consequently as $I\to\infty$, $\varrho\to\infty$, which corresponds to escape. If no binary is formed, $(r, \varrho)\to\infty$ as $I\to\infty$, which corresponds to explosion.

For a system with positive total energy $\ddot{I}\geqslant 2h>0$, consequently, the curve $I(t)$ is concave (from below). If a system with $h>0$ begins its motion at $t=0$, we have $I(0)>0$, $\dot{I}(0)\gtreqless 0$ and $\ddot{I}(0)>0$. The initial phase of the motion is shown in Figure 4. Note that $\dot{I}(0)=2\left[g_1\mathbf{r}(0)\dot{\mathbf{r}}(0)+g_2\boldsymbol{\varrho}(0)\dot{\boldsymbol{\varrho}}(0)\right]$ may be positive or negative and, in fact,

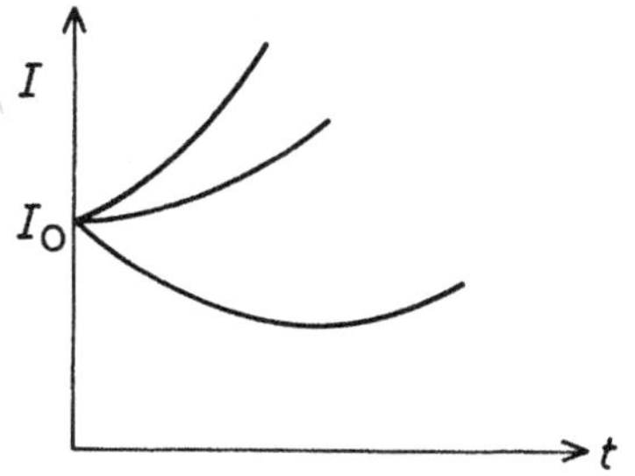

Fig. 4. Moment of inertia vs time for $h>0$.

even zero without the initial velocities being zero. (Since $h=T-F>0$, all the initial velocities cannot be zero.) The same applies for $h=0$.

If the total energy is negative, we have initially that $I(0)>0$, $\dot{I}(0)\gtreqless 0$ and $\ddot{I}(0)\gtreqless 0$. Consider first a system with zero initial velocities. In this case $\dot{I}(0)=0$, $T(0)=0$, $\ddot{I}(0)=2h<0$. The curve $I(t)$ is convex from below initially and a contraction takes place with $\dot{I}<0$. At the same time F increases (and so does T) and when the value of $F+2h=0$, or $F=+2|h|$ is reached, $\ddot{I}$ becomes zero. After this $\ddot{I}>0$ and the curve is concave from below. Now the opposite trend takes place and the contraction after reaching a value $I_{\min}$ with $\dot{I}=0$ turns into an expansion with $\dot{I}>0$ and still with $\ddot{I}>0$. As the expansion decreases the value of F, the quantity $F+2h$, becomes zero and then negative again and the curve $I(t)$ will be concave once again from below. Now let us consider once more a contraction with $\ddot{I}>0$ without restricting $\dot{I}(0)$ to 0. At the minimum of I, the value of F is large and the bodies are close together. After this time F decreases and the value of $\ddot{I}$ stays positive as long as $F>2|h|$. Therefore an explosion, when all r_{ij} increase their values to infinity, is impossible for $h<0$ since when $r_{ij}\to\infty$, $F\to 0$. In this process F will reach the value of $2|h|$ at which time $\ddot{I}$ becomes negative. The only way F can stay larger than $2|h|$ for all time is if a binary is formed.

Let the masses of the members of the binary be m_1 and m_2. Then the condition for $\ddot{I}>0$ is

$$F=G\left(\frac{m_1m_2}{r}+\frac{m_2m_3}{r_{23}}+\frac{m_3m_1}{r_{31}}\right)>2|h|.$$

With sufficiently small value of r this condition may be satisfied, no matter how large the other distances become. The condition is satisfied if, as F varies,

$$F_{\min}>2|h|.$$

Now as $\varrho\to\infty$, r_{23} and $r_{31}\to\infty$ and

$$F_{\min}=G\frac{m_1m_2}{r_{\max}},$$

where $r_{\max}=a(1+e)$ is the apogee distance of the binary in its asymptotic state. The escape condition now becomes

$$G\frac{m_1m_2}{a(1+e)}>2|h| \quad\text{or}\quad \frac{Gm_1m_2}{2|h|\,(1+e)}>a. \tag{35}$$

Note that if a circular binary orbit is formed ($e_1=0$),

$$a_1<\frac{Gm_1m_2}{2|h|}$$

and if the eccentricity is high ($e_2=1$),

$$a_2<\frac{a_1}{2},$$

therefore highly eccentric binary orbits require close binaries. Note that a special case of Equation (35) was obtained at the end of Section 2.

Figure 5 shows the initial variation of I with time for negative total energy in a few special cases.

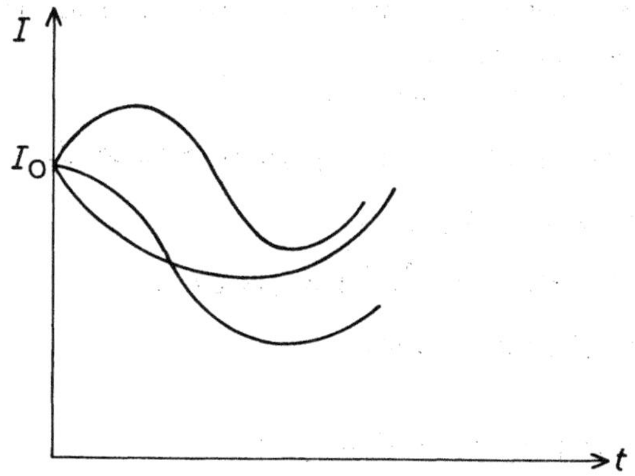

Fig. 5. Moment of inertia vs time for $h<0$.

E. SUNDMAN'S INEQUALITY

This important result connects the magnitude of the angular momentum vector $|\mathbf{c}|=c$ with the moment of inertia by the following inequality:

$$c^2 \leqslant 2IT - \tfrac{1}{4}\dot{I}^2 . \tag{36}$$

The conventional proof uses the previously mentioned triangle and Cauchy's inequalities. The magnitude of the angular momentum is

$$c = |\sum m_i \mathbf{r}_i \times v_i \leqslant \sum m_i r_i v_i |\sin \alpha_i| = \sum \sqrt{m_i} r_i \sqrt{m_i} v_i |\sin \alpha_i| ,$$

where α_i is the angle between $\mathbf{r}_i$ and $\dot{\mathbf{r}}_i = \mathbf{v}_i$.
From this, by Cauchy's inequality we have

$$c^2 \leqslant \sum m_i r_i^2 \sum m_i v_i^2 \sin^2 \alpha_i . \tag{37}$$

On the other hand

$$\dot{I} = 2 \sum m_i \mathbf{r}_i \dot{\mathbf{r}}_i = 2 \sum m_i r_i v_i \cos \alpha_i = 2 \sum \sqrt{m_i} r_i \sqrt{m_i} v_i \cos \alpha_i ,$$

from which once again by Cauchy's inequality, we have

$$\tfrac{1}{4}\dot{I}^2 < \sum m_i r_i^2 \sum m_i v_i^2 \cos^2 \alpha_i . \tag{38}$$

Addition of Equations (37) and (38) gives the desired result.

Without Cauchy's inequality the above inequality may be obtained using Jacobian coordinates as follows:

$$\mathbf{c} = g_1 \mathbf{r} \times \mathbf{v} + g_2 \boldsymbol{\varrho} \times \mathbf{V} ,$$

where $\mathbf{v}=\dot{\mathbf{r}}$, $\mathbf{V}=\dot{\boldsymbol{\varrho}}$. Let α and β be the angles between $\mathbf{r}$ and $\mathbf{v}$ on one hand and between $\boldsymbol{\varrho}$ and $\mathbf{V}$ on the other. We have

$$c^2 \leqslant g_1^2 (\mathbf{r} \times \mathbf{v})^2 + g_2^2 (\boldsymbol{\varrho} \times \mathbf{V})^2 + 2g_1 g_2 r v \varrho V \sin\alpha \sin\beta, \tag{39}$$

where the factor in the last term, representing the cosine of the angle between the vectors $\mathbf{r}\times\mathbf{v}$ and $\boldsymbol{\varrho}\times\mathbf{V}$, was dropped.

Similarly for $\dot{I} = 2(g_1 \mathbf{r}\mathbf{v} + g_2 \boldsymbol{\varrho}\mathbf{V})$ we have

$$\tfrac{1}{4}\dot{I}^2 = g_1^2 (\mathbf{r}\mathbf{v})^2 + g_2^2 (\boldsymbol{\varrho}\mathbf{V})^2 + 2g_1 g_2 r \varrho v V \cos\alpha \cos\beta. \tag{40}$$

Adding Equations (39) and (40) gives

$$c^2 + \tfrac{1}{4}\dot{I}^2 \leqslant g_1^2 r^2 v^2 + g_2^2 \varrho^2 V^2 + 2g_1 g_2 r \varrho v V, \tag{41}$$

since, in general, $(\mathbf{a}\times\mathbf{b})^2 + (\mathbf{a}\mathbf{b})^2 = a^2 b^2$.

On the other hand

$$2IT = (g_1 v^2 + g_2 V^2)(g_1 r^2 + g_2 \varrho^2)$$

or

$$2IT = g_1^2 r^2 v^2 + g_2^2 \varrho^2 V^2 + 2g_1 g_2 r \varrho v V + g_1 g_2 (rV - \varrho v)^2,$$

consequently inequality (41) may be written as

$$c^2 + \tfrac{1}{4}\dot{I}^2 \leqslant 2IT - g_1 g_2 (rV - \varrho v)^2 \leqslant 2IT, \tag{42}$$

which is equivalent to (36).

An important variation of Sundman's inequality is obtained when the kinetic energy is eliminated form Equation (36) by means of the Lagrange-Jacobi equation:

$$c^2 \leqslant (\ddot{I} - 2h) I - \tfrac{1}{4}\dot{I}^2. \tag{43}$$

Note that a weaker form of (43) is

$$c^2 \leqslant (\ddot{I} - 2h) I, \tag{44}$$

which will be used in Section 4 to study the behavior of $\ddot{I}$ with changes of I.

In the following another, rather useful form of Sundman's inequality will be derived from Equation (43). Dividing by $I>0$, rearranging and multiplying by $2/\sqrt{I}$ gives

$$0 \leqslant \left(\ddot{I} - 2h - \frac{c^2}{I} - \frac{\dot{I}^2}{4I}\right)\frac{2}{\sqrt{I}} = Z. \tag{45}$$

In this way an integrable combination is produced for the non-negative function Z. In fact it will be shown that

$$\frac{\mathrm{d}L}{\mathrm{d}t} = Z\frac{\mathrm{d}I}{\mathrm{d}t}, \tag{46}$$

where

$$L = \frac{1}{\sqrt{I}}(\dot{I}^2 + 4c^2) - 8h\sqrt{I}. \tag{47}$$

The proof only requires the computation of the time-derivative of L and then obtaining Z as $\dot{L}/\dot{I}$.

We proceed differently by integrating $Z\dot{I}$ and obtaining L directly

$$L = \int Z\dot{I}\,dt = \int \frac{2\dot{I}\ddot{I}}{\sqrt{I}}\,dt - 4h\int \frac{\dot{I}}{\sqrt{I}}\,dt - 2c^2\int \frac{\dot{I}}{I^{3/2}} - \frac{1}{2}\int \frac{\dot{I}^3}{I^{3/2}}\,dt.$$

Combination of the first and last integrals gives

$$\int \frac{2\dot{I}\ddot{I}\sqrt{I} - \frac{1}{2}\dot{I}^3 \dfrac{1}{\sqrt{I}}}{I}\,dt = \frac{\dot{I}^2}{\sqrt{I}},$$

while integration of the second and third terms presents no difficulty:

$$-4h\int \frac{dI}{\sqrt{I}} - 2c^2\int \frac{dI}{\sqrt{I^3}} = -8h\sqrt{I} + \frac{4c^2}{\sqrt{I}}.$$

Consequently, $\dot{L} = Z\dot{I}$ and the new form of Sundman's inequality expressed by Equations (45, 46 and 47) is verified. This form of the inequality says that since $Z \geqslant 0$, if I increases L does not decrease or if I decreases L does not increase.

F. ESCAPE CONDITIONS

In this section dynamical conditions for escape are established when $h<0$. We shall need only sufficient conditions for our purposes. The several sets of conditions given in the literature may all be written as follows.

If at some time, t_0

(i) $\varrho(t_0) = \varrho_0 > a$,

(ii) $\dot{\varrho}(t_0) = \dot{\varrho}_0 > 0$, and

(iii) $\dot{\varrho}_0^2 \geqslant b$,

where a and b are positive numbers, then $\varrho \to \infty$ as $t \to \infty$.

Various values for a and b are available in the literature. The sharpest estimates are given by Standish (1971):

$$a = \frac{G(m_1m_2 + m_2m_3 + m_3m_1)}{|h|}, \tag{48}$$

$$b = 2GM\left[\frac{1}{\varrho_0} + \frac{g_1}{\mu}\,\frac{a^2}{\varrho_0^2(\varrho_0 - a)}\right] \tag{49}$$

and the simplest ones by Birkhoff (1927):

$$a = \frac{2M^2G}{3|h|} \quad \text{and} \quad b = \frac{8MG}{\varrho_0}.$$

Such estimates may be obtained as follows. In order to evaluate a we must estimate the minimum inter-particle distance. Since $h = T - F$ for $h<0$, we have $|h| = F - T$.

The kinetic energy is positive, therefore $|h| \leqslant F$ or

$$|h| \leqslant G\left(\frac{m_1 m_2}{r_{12}} + \frac{m_2 m_3}{r_{23}} + \frac{m_3 m_1}{r_{31}}\right) \leqslant G\,\frac{m_1 m_2 + m_2 m_3 + m_3 m_1}{r_{\min}},$$

where in the last inequality $r_{\min}$ replaced r_{ij}. From here

$$r_{\min} \leqslant G\,\frac{m_1 m_2 + m_2 m_3 + m_3 m_1}{|h|}, \tag{50}$$

and we see that for negative total energy the minimum distance is bounded.

Note that

$$m_1 m_2 + m_2 m_3 + m_3 m_1 \leqslant \frac{M^2}{3}, \tag{51}$$

which follows from the rearrangement

$$0 \leqslant (m_1 - m_2)^2 + (m_2 - m_3)^2 + (m_3 - m_1)^2 .$$

Using (51) in (50) we have

$$r_{\min} \leqslant G\,\frac{M^2}{3\,|h|}. \tag{52}$$

This limiting value (52) of $r_{\min}$ is used for a by Tevzadze (1962) and twice of this occurs in Birkhoff's work while Standish uses Equation (50).

The estimate for the velocity follows from the integration of an estimate for the 'radial' acceleration of the escaping particle. To obtain same we first show that

$$\ddot{\varrho} \geqslant \frac{\boldsymbol{\varrho}\ddot{\boldsymbol{\varrho}}}{\varrho},$$

then compute $\boldsymbol{\varrho}\ddot{\boldsymbol{\varrho}}$ and integrate. In the first step we have from $\boldsymbol{\varrho}^2 = \varrho^2$ that

$$\boldsymbol{\varrho}\dot{\boldsymbol{\varrho}} = \varrho\dot{\varrho} \quad \text{and} \quad \boldsymbol{\varrho}\ddot{\boldsymbol{\varrho}} + \dot{\boldsymbol{\varrho}}^2 = \varrho\ddot{\varrho} + \dot{\varrho}^2 \quad \text{or} \quad \varrho\ddot{\varrho} - \boldsymbol{\varrho}\ddot{\boldsymbol{\varrho}} = \dot{\boldsymbol{\varrho}}^2 - \dot{\varrho}^2 \geqslant 0,$$

since the right side is the difference between the square of the total velocity vector and the square of one of its components.

The computation of $\boldsymbol{\varrho}\ddot{\boldsymbol{\varrho}}$ is much more complicated. Standish uses the equivalent of Equation (12) and expands $\boldsymbol{\varrho}\ddot{\boldsymbol{\varrho}}$ in Legendre polynomials. This is followed by writing a for $|\mathbf{r}|$ and 1 for all the Legendre coefficients in the process of obtaining

$$\ddot{\varrho} \geqslant -\,GM\left[\frac{1}{\varrho^2} + \frac{g_1}{\mu}\sum_{n=2}^{\infty}(n+1)\,\frac{a^n}{\varrho^{n+2}}\right]. \tag{53}$$

A more direct approach leading to a cruder estimate uses Equation (11), from which we have

$$\ddot{\varrho} \geqslant \frac{\boldsymbol{\varrho}\ddot{\boldsymbol{\varrho}}}{\varrho} = -\,\frac{MG}{\mu\varrho}\left(\frac{m_1\boldsymbol{\varrho}\mathbf{r}_{31}}{r_{31}^3} + \frac{m_2\boldsymbol{\varrho}\mathbf{r}_{32}}{r_{32}^3}\right)$$

or

$$\ddot{\varrho} \geqslant -\frac{MG}{\mu}\left(\frac{m_1 \cos\alpha}{r_{31}^2} + \frac{m_2 \cos\beta}{r_{32}^2}\right) \geqslant -\frac{MG}{\mu}\left(\frac{m_1}{r_{31}^2} + \frac{m_2}{r_{32}^2}\right).$$

Here α and β are the angles between $\boldsymbol{\varrho}$ and $\mathbf{r}_{31}$ on one hand and between $\boldsymbol{\varrho}$ and $\mathbf{r}_{32}$ on the other hand. Considering now Figure 2 we have from

$$\boldsymbol{\varrho} = \frac{m_1}{\mu}\mathbf{r} + \mathbf{r}_{32}$$

that

$$\varrho \leqslant \frac{m_1}{\mu} r + r_{32} \leqslant r + r_{32},$$

or

$$r_{32} \geqslant \varrho - r$$

and similarly

$$r_{31} \geqslant \varrho - r.$$

Substituting in the previous inequality for $\ddot{\varrho}$, we have

$$\ddot{\varrho} \geqslant -\frac{MG}{(\varrho - r)^2} \geqslant -\frac{MG}{(\varrho - a)^2}. \tag{54}$$

An even simpler value for the acceleration is obtained if for r the value of $\varrho/2$ is written. This means that $\varrho \geqslant 2r$, corresponding to Birkhoff's approach. In this case we have

$$\ddot{\varrho} \geqslant -\frac{4MG}{\varrho^2}. \tag{55}$$

The next step is to perform the integration in order to obtain an estimate for the velocity. Since $\dot{\varrho} > 0$, we have from Equation (55):

$$\dot{\varrho}\ddot{\varrho} \geqslant -\frac{4MG}{\varrho^2}\dot{\varrho}$$

or

$$\frac{\dot{\varrho}^2}{2} \geqslant +\frac{4MG}{\varrho}. \tag{56}$$

From Equation (56) we obtain $b = 8MG/\varrho_0$.

Note that an obvious correction is required in the corresponding equation given in Birkhoff's (1927) book, see p. 280.

Using Equation (54), the integration gives $MG/(\varrho - a)$ and integrating Equation (53) we obtain Equation (49).

Another solution is obtained when instead of using $r_{32}, r_{31} \geqslant \varrho - r$, the analysis is refined by employing

$$r_{32} \geqslant \varrho - \frac{m_1}{\mu} r \geqslant \varrho - \frac{m_1}{\mu} a \quad \text{and} \quad r_{31} \geqslant \varrho - \frac{m_2}{\mu} r \geqslant \varrho - \frac{m_2}{\mu} a.$$

In this way we have

$$\ddot{\varrho} \geqslant -\frac{MG}{\mu}\left(\frac{m_1}{r_{31}^2}+\frac{m_2}{r_{32}^2}\right) \geqslant -\frac{MG}{\mu}\left[\frac{m_1}{\left(\varrho-\frac{m_2}{\mu}a\right)^2}+\frac{m_2}{\left(\varrho-\frac{m_1}{\mu}a\right)^2}\right].$$

Integration gives Tevzadze's (1962) result:

$$\frac{\dot{\varrho}^2}{2} \geqslant \frac{MG}{\mu}\left[\frac{m_1}{\varrho-\frac{m_2}{\mu}a}+\frac{m_2}{\varrho-\frac{m_1}{\mu}a}\right],$$

and consequently for this case b becomes

$$b=\frac{2MG}{\mu}\left[\frac{\varrho_0-a\,\frac{m_1^2+m_2^2}{\mu^2}}{\varrho_0^2-\varrho_0 a+\frac{m_1 m_2}{\mu^2}a^2}\right].$$

The following two remarks will clarify some questions repeatedly occurring.

(1) Regarding inequality (51) an interesting generalization has been communicated to me privately by Dr U. Kirchgraber who has shown that for n bodies:

$$\sum_{1\leqslant i<j\leqslant n} m_i m_j \leqslant \frac{n-1}{2n} M^2 .$$

Therefore,

$$r_{\min} \leqslant \frac{GM^2}{|h|}\,\frac{n-1}{2n}$$

the limit of which is

$$\lim_{n\to\infty} r_{\min} \leqslant \frac{GM^2}{2|h|}.$$

Consequently, the upper bound of the minimum inter-particle distance is smaller for three than for many bodies, provided the total mass and the total energy are preserved in the comparison.

(2) In all the escape formulas we have

$$\dot{\varrho}\ddot{\varrho} \geqslant f(\varrho)\,\dot{\varrho},$$

with various functions $f(\varrho)$. By integration from t_0 to $t>t_0$ we obtain

$$\tfrac{1}{2}(\dot{\varrho}^2-\dot{\varrho}_0^2) \geqslant F(\varrho)-F(\varrho_0)$$

or

$$\tfrac{1}{2}\dot{\varrho}^2 \geqslant F(\varrho)+\left[\tfrac{1}{2}\dot{\varrho}_0^2-F(\varrho_0)\right],$$

where

$$F(\varrho)=\int f(\varrho)\,\mathrm{d}\varrho .$$

Therefore the velocity will not be zero no matter what the value of $F(\varrho)>0$ is as long as

$$\dot{\varrho}_0^2 \geqslant 2F(\varrho_0) = b\,.$$

4. Applications

A. THEORY OF ESCAPE

Consider the simplest form of Sundman's inequality,

$$c^2 \leqslant (\ddot{I} - 2h)\, I\,,$$

which was given as Equation (44). Rearrangement gives

$$\frac{c^2}{2\,|h|} - I \leqslant \frac{\ddot{I}I}{2\,|h|},$$

for $h<0$. The sign of $\ddot{I}$ is controlled by the value of I and introducing the critical value of the moment of inertia,

$$\frac{c^2}{2\,|h|} = I_c, \tag{57}$$

we have

$$I_c - I \leqslant \frac{I}{2\,|h|}\,\ddot{I}.$$

So as long as $I<I_c$, $\ddot{I}$ is positive.

If, on the other hand, $I>I_c$, $\ddot{I}$ is larger than a negative number, consequently nothing may be said regarding its sign.

Considering the complete inequality, we have

$$I_c - I + \frac{\dot{I}^2}{8\,|h|} \leqslant \frac{\ddot{I}I}{2\,|h|},$$

consequently as long as

$$I < I_c + \frac{\dot{I}^2}{2\,|h|},$$

we have $\ddot{I}\geqslant 0$. Before an inflexion point is reached, I certainly will be higher than I_c; how much higher depends on $\dot{I}$. A fast expansion with high value of $\dot{I}$ will carry I higher without inflexion than a slow expansion.

Returning now to the simplified inequality we consider that part of the curve $I(t)$ for which $I\leqslant I_c$ and have $\dot{I}=0$ at I_1. This will be a proper minimum with $\ddot{I}>0$. The curve will rise on both sides of I_1 until another point is reached, say I_2, where $\dot{I}=0$. From Sundman's modified inequality (47) we have that if $I_{\min}=I_1<I_2$, corresponding

to $\dot{I}_1 = \dot{I}_2 = 0$, then $L_1 \leqslant L_2$ and with $h<0$

$$\frac{4c^2}{\sqrt{I_1}} + 8\,|h|\,\sqrt{I_1} \leqslant \frac{4c^2}{\sqrt{I_2}} + 8\,|h|\,\sqrt{I_2}$$

or

$$4c^2\,\frac{\sqrt{I_2}-\sqrt{I_1}}{\sqrt{I_1 I_2}} \leqslant 8\,|h|\,(\sqrt{I_2}-\sqrt{I_1}).$$

Since $I_2 > I_1$ we have

$$\frac{1}{I_1}\left(\frac{c^2}{2\,|h|}\right)^2 = \frac{I_c^2}{I_1} \leqslant I_2. \tag{58}$$

Note that as $t \to \infty$, I cannot approach either zero or a constant value $I_0 \neq 0$. In the first case as I would decrease, $I \to 0$, $L \to \infty$ from Equation (47) with $c \neq 0$. But L cannot increase as I decreases, consequently I cannot approach 0.

Now attention is directed to inequality (58). For a given $I_1 = I_{\min} < I_c$ the curve $I(t)$ rises until it becomes at least as large as I_c^2/I_1. Note that $I_1 \leqslant I_c$, therefore

$$I_c \leqslant \frac{I_c^2}{I_1} \leqslant I_2,$$

so the curve rises above I_c on both sides of I_1, see Figure 6.

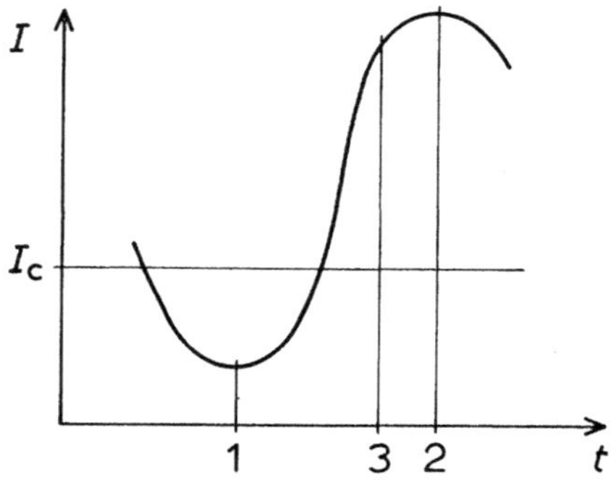

Fig. 6. Moment of inertia vs time for estimations.

Considering I_c fixed, if I_1 is sufficiently small, the curve may rise to an arbitrary high I_2 value. The value $I_c^2/I_1 = I_3$ is lower than the maximum and when it is reached, $\dot{I}_3 > 0$. We now estimate $\dot{I}$ using once again Equation (47) with $I_1 = I_{\min}$, $\dot{I}_1 = 0$, $I_2 > I > I_c$ and $\dot{I} \neq 0$. Since $L_1 \leqslant L$, we have

$$\frac{4c^2}{\sqrt{I_1}} + 8\,|h|\,\sqrt{I_1} \leqslant \frac{4c^2}{\sqrt{I}} + 8\,|h|\,\sqrt{I} + \frac{\dot{I}^2}{\sqrt{I}},$$

from which

$$\dot{I}^2 \geqslant 8\,|h|\,\sqrt{I}\,(\sqrt{I}-\sqrt{I_1})\left(\frac{I_c}{\sqrt{I_1}\sqrt{I}} - 1\right). \tag{59}$$

Consequently, for sufficiently low value of $I_1 = I_{\min}$, we have I at least as high as I_c^2/I_1 with a derivative which is at least as high as given by (59).

The next step is to return to the escape conditions established in Section 3F, and connect the requirements on ϱ_0, $\dot{\varrho}_0$ and $\dot{\varrho}_0^2$ with the above given properties of the moment of inertia.

The first escape condition is that $\varrho \geqslant a$, or

$$I = g_1 r^2 + g_2 \varrho^2 \geqslant g_1 r^2 + g_2 a^2 .$$

This condition is certainly satisfied if $I \geqslant (g_1 + g_2) a^2$, since $r = r_{\min} \leqslant a$. Therefore I must be at least as large as $(g_1 + g_2)\, a^2$. Now from Equation (58)

$$\frac{I_c^2}{I_1} = I_3 \leqslant I_2$$

and therefore

$$(g_1 + g_2)\, a^2 \leqslant \frac{I_c^2}{I_1} \quad \text{or} \quad I_1 \leqslant \frac{c^4}{(g_1 + g_2)\, 4a^2 h^2}. \tag{60}$$

Here all quantities on the right side are given in terms of the masses, and constants of integration (h and c). After substitution into (60) we have that the first escape condition requires that a minimum value of I be obtained, not larger than

$$I_{\min} \leqslant \frac{\mu M c^4}{4G^2 (\mu m_3 + m_1 m_2)^2 (M m_1 m_2 + \mu^2 m_3)}. \tag{61}$$

The second escape condition refers to $\dot{\varrho} > 0$. We know that the corresponding value $\dot{I}_3 \geqslant 0$, therefore from $\dot{I} = 2g_1 r\dot{r} + 2g_2 \varrho\dot{\varrho}$, we obtain

$$\dot{I}_3 \geqslant 2g_1 r\dot{r}$$

as the condition for $\dot{\varrho} > 0$. If $\dot{r}r < 0$, the condition is immediately satisfied, therefore we investigate the condition for

$$\dot{I}_3 \geqslant 2g_1 \left| r\dot{r} \right|, \tag{62}$$

through a bound on $|r\dot{r}|$. We shall show that $2g_1 |r\dot{r}| < A$, where A is a positive constant. Since $\dot{I}_3$ must not be smaller than $2g_1|r\dot{r}|$ and $2g_1|r\dot{r}|$ cannot be larger than A, the condition (62) is certainly satisfied if $\dot{I}_3 \geqslant A$. To find A we compute the kinetic energy:

$$T = \tfrac{1}{2} g_1 \dot{r}^2 + \tfrac{1}{2} g_2 \dot{\varrho}^2 = h + F = F - |h|,$$

from which

$$\tfrac{1}{2} g_1 \dot{r}^2 \leqslant F = G \left(\frac{m_1 m_2}{r_{12}} + \frac{m_2 m_3}{r_{23}} + \frac{m_3 m_1}{r_{31}} \right) \leqslant \frac{G \sum}{r},$$

where

$$\sum = m_1 m_2 + m_2 m_3 + m_3 m_1 .$$

Note that here $r = r_{12} = r_{\min}$. Therefore

$$4g_1^2 |r\dot{r}|^2 \leqslant 8g_1 r G \sum \leqslant 8g_1 a G \sum = A^2$$

and

$$\dot{I} \geqslant \sqrt{8ag_1G\sum} = 2G\sum\sqrt{\frac{2g_1}{|h|}}. \tag{63}$$

Now we consider the third condition, $\dot{\varrho}^2 > b$, where matters are more complicated since $b = b(\varrho)$. First ϱ and $\dot{\varrho}$ are computed from the appropriate expressions for I and $\dot{I}$ as follows:

$$\varrho = \sqrt{\frac{I - g_1r^2}{g_2}} \quad \text{and} \quad \dot{\varrho} = \frac{\dot{I} - 2g_1r\dot{r}}{2g_2\varrho}.$$

Then these results are substituted in the selected escape condition:

$$\left(\frac{\dot{I} - 2g_1r\dot{r}}{2g_2\varrho}\right)^2 \geqslant b\left[\sqrt{\frac{I - g_1r^2}{g_2}}\right]$$

or

$$(\dot{I} - 2g_1r\dot{r})^2 \geqslant 4g_2^2\,\frac{I - g_2r^2}{g_2}\,b\left[\sqrt{\frac{I - g_1r^2}{g_2}}\right].$$

Recalling now that $2g_1r\dot{r} \leqslant A$, the left side is replaced by $(\dot{I} - A)^2$ since

$$(\dot{I} - 2gr\dot{r})^2 \geqslant (\dot{I} - A)^2,$$

making the requirement on the magnitude of $\dot{I}$ more severe, or, in other words requiring a higher escape velocity.

In order to be able to be specific, Birkhoff's escape condition is selected at this point with

$$b(\varrho) = \frac{8MG}{\varrho}.$$

Then the above inequality becomes

$$(\dot{I} - A)^2 \geqslant 4g_2^2 8MG\varrho = \frac{32MGg_2^2}{\sqrt{g_2}}\sqrt{I - g_1r^2}.$$

Once again the velocity requirement is strengthened by omitting the term g_1r^2 on the right side. In this way one obtains

$$\dot{I} \geqslant A + A'I^{1/4},$$

where

$$A' = 4\sqrt{2MGg_2^{3/2}}.$$

Note that the above inequality requires a higher rate of expansion than the previous one, consequently Equation (63) may be omitted.

Since the value of $\dot{I}$ is larger than the left side of (59), the above requirement on $\dot{I}$ shall certainly be satisfied if

$$8|h|\sqrt{I}(\sqrt{I} - \sqrt{I_1})\left(\frac{I_c}{\sqrt{I_1}\sqrt{I}} - 1\right) \geqslant A + A'I^{1/4} \tag{64}$$

or if

$$\sqrt{II_1} + I_c \sqrt{\frac{I}{I_1}} - I - I_c \geqslant P,$$

where

$$P = (A + A'I^{1/4})/8\,|h| \geqslant 0. \tag{65}$$

We may now proceed by stating that the above requirement is certainly satisfied if the following is

$$I_c \sqrt{\frac{I}{I_1}} \geqslant P + I + I_c$$

or if

$$\sqrt{I_1} \leqslant \frac{I_c\sqrt{I}}{P + I + I_c} = S_1. \tag{66}$$

Note that a better (larger) lower bound may also be obtained as

$$\sqrt{I_1} \leqslant \frac{P + I + I_c - \sqrt{(P + I + I_c)^2 - 4II_c}}{2\sqrt{I}} = S_2. \tag{67}$$

Continuing with Equation (66), we compare this requirement with the previously obtained result of Equation (60) in the form

$$\sqrt{I_1} \leqslant \frac{I_c}{\sqrt{I}} = S_3,$$

where

$$\sqrt{I} = a\sqrt{g_1 + g_2}.$$

As a simple calculation shows, the requirement expressed by Equation (66) is stronger than that of Equation (60), i.e., I_1 must be smaller according to Equation (66) than according to Equation (60). With this the analysis ends since the minimum moment of inertia must be lower than the value given by Equation (66). Note that

$$S_1 \leqslant S_2 \leqslant S_3,$$

consequently the requirement on the minimum moment of inertia to be satisfied in order to obtain an escape (i.e. to satisfy all three escape requirements) is

$$I_{\min} \leqslant S_1^2,$$

where

$$S_1 = \frac{I_c\sqrt{I}}{P + I + I_c}.$$

Here

$$I = a^2(g_1 + g_2)$$
$$P = \frac{1}{8\,|h|}(A + A'I^{1/4})$$

$$I_c = \frac{c^2}{2|h|}.$$

Furthermore we recall the definitions of the quantities used in the above equations as

$$a = \frac{G\sum}{|h|},$$

$$g_1 = \frac{m_1 m_2}{\mu}, \qquad g_2 = \frac{\mu m_3}{M},$$

$$|h| = T - F,$$

$$A = \sqrt{2g_1 a G \sum},$$

$$A' = 4\sqrt{2MGg_2^{3/2}},$$

$$c^2 = \left[\sum_{i=1}^{3} m_i (\mathbf{r}_i \times \dot{\mathbf{r}}_i)\right]^2.$$

Finally the symbols in the above set are given by

$$\sum = m_1 m_2 + m_2 m_3 + m_3 m_1,$$

$$\mu = m_1 + m_2,$$

$$M = m_1 + m_2 + m_3,$$

$$T = \tfrac{1}{2} \sum m_i \dot{\mathbf{r}}_i^2,$$

$$F = G \sum_{1 \leqslant i < j \leqslant 3} \frac{m_i m_j}{r_{ij}},$$

where m_1 and m_2 are the closest pair (forming the binary), G is the constant of gravitation, $\mathbf{r}_i$ and $\dot{\mathbf{r}}_i$ are the position and velocity vectors of the ith body.

B. ASPECTS OF NUMERICAL ANALYSIS

The basic premise of this section is the sequence that (i) the dynamical system of three bodies tends to a disruption or escape which is reached sooner or later, (ii) such an escape cannot occur for negative total energy without a triple close approach, (iii) the triple collision is not continuable analytically, therefore, (iv) sufficiently close triple approaches, which must occur sooner or later, might invalidate the numerical integration unless they receive special attention. Depending on the masses and on the constants of integration the triple close approach, necessarily occurring, will show a 'closeness' of varying degree. To predict the degree of closeness is the purpose of this section. Once reliable estimates are available, the rest is left to the operator of the computer; he may wish to throw away his predictably unreliable results, he may retrace his steps and repeat his integration with a smaller stepsize or with higher precision. But he might also wish to select different initial conditions so that a critically low triple close approach does not occur prior to the dissolution of his system. And herein lies one of the many limitations of the seemingly omnipotent computer approach. The phase-space seems to be dense with escape orbits. Some initial conditions allow the reliable computation of such orbits, some render the computation meaning-less. The power of the analysis seems to be that it allows, independently of the vagaries, failures and limitations of the computer, the prediction of the outcome of the long-time behavior of the system.

A combination of the estimates for the perimeter, σ, with the estimates for the minimum moment of inertia, $I_{\min}$, is the subject of the following analysis.

The reasoning might follow these lines. With regularizable double close approaches the numerical integration of the general problem of three bodies is a routine exercise in careful programming of an appropriately high order numerical integration scheme unless a triple close approach occurs. If a triple close approach is sufficiently 'close' to satisfy the given escape conditions the numerical integration may stop before such a close approach is reached since the analytical results guarantee the escape and the 'solution' may be announced without integration. (If the details of the escape are required, then, of course, the numerical integration must 'go through' the close approach.) When a *predicted* triple close approach is not 'close enough' to produce an escape, the integration must continue until a sufficiently close triple approach is predicted. Therefore, the results of this section must be used together with the results of the previous section. The quantities to be monitored during the numerical integration are given in this section. As triple close approaches are predicted one after the other from these quantities, there are two possibilities: (i) escape will occur or (ii) escape will not occur. The ambition of the investigator at this point may be either (a) to follow the escape orbit or (b) to be satisfied that the initial conditions used result in an escape. In case (ii) the computational accuracy must be raised sufficiently so that the predicted close approach is handled reliably. If this is not done the subsequent numerical integration may be entirely without meaning. In case (i)-(b) the computation is to be stopped immediately after the prediction of the close approach, otherwise computer time is wasted.

In case (i)-(a) the accuracy must be significantly raised so that the triple close approach and the following escape are computed accurately.

The computational accuracy may be controlled by the unregularizable distance of the triple close approach. It must be assumed that the shortest distance during a close approach ($\min r_{ij} = r_{12}$, say) is regularized, consequently the critical distance governing the stepsize, or in general the accuracy, will be $\min(r_{13}, r_{23})$. But if any of these distances decreases below r_{12}, the regularization will switch to that distance. Consequently, *if the program can handle distances of the size* $r_{12} = \min r_{ij}$, *without regularization, the accuracy is called sufficient* regarding the triple close approach. The case of binary collision, when $r_{12} = 0$, must be excepted since it cannot be expected that the program handles a collision without regularization. Inasmuch as the measure of the set of initial conditions resulting in binary collisions is zero, this case may be omitted. Nevertheless, a less stringent requirement may be arrived at in the following way. The perimeter, σ, is defined as

$$\sigma = r_{12} + r_{23} + r_{31} .$$

Let $r_{12}=\min r_{ij}\leqslant r_{31}\leqslant r_{32}$, i.e. we order the distances. The smallest distance r_{12} is used in the regularized equations, consequently, it will not limit the accuracy of the computation. Attention, therefore, is directed to r_{31} which is the critical length. Note that $r_{32}<(r_{12}+r_{31})<2r_{31}$, therefore

$$\sigma \leqslant r_{12} + (r_{12} + r_{31}) + r_{31} = 2(r_{12} + r_{31}) \leqslant 4r_{31}.$$

Consequently if $I_{\min}\geqslant B$, where B is a given positive number computed at any instant as it will be shown presently, we have from Equation (22)

$$B \leqslant I_{\min} \leqslant \frac{\bar{m}\bar{m}'}{M}\sigma^2 \leqslant \frac{16\bar{m}\bar{m}'}{M} r_{31}^2.$$

Therefore, the bound for r_{31} is

$$\sqrt{\frac{BM}{16\bar{m}\bar{m}'}} \leqslant r_{31}. \tag{68}$$

A better, higher estimate may be obtained for r_{31} as follows. The Jacobian formula for the moment of inertia is

$$I = \frac{1}{M} \sum_{1\leq i<j\leq 3} m_i m_j r_{ij}^2 \leqslant \frac{1}{M} [m_1 m_2 r_{12}^2 + m_2 m_3 (r_{12} + r_{31})^2 + m_3 m_1 r_{31}^2],$$

where the same inequality $(r_{32}<r_{12}+r_{31})$ was used as before. Replacing now r_{12} by r_{31} we have

$$I \leqslant \frac{1}{M} [m_1 m_2 r_{31}^2 + 4m_2 m_3 r_{31}^2 + m_3 m_1 r_{31}^2]$$

or

$$I \leqslant \frac{m_1 m_2 + 4m_2 m_3 + m_3 m_1}{M} r_{31}^2.$$

From here

$$\sqrt{\frac{BM}{m_1 m_2 + 4m_2 m_3 + m_3 m_1}} \leqslant r_{31}. \tag{69}$$

Since

$$m_1 m_2 + 4m_2 m_3 + m_3 m_1 \leqslant 6\bar{m}\bar{m}' \leqslant 16\bar{m}\bar{m}',$$

we see that the estimate offered by Equation (69) is higher (better) than that of Equation (68).

The actual use of the above inequality is as follows. Once the estimate for $I_{\min}$ is computed in the form $B\leqslant I_{\min}\leqslant I$, we know that I cannot go below the value of B. Consequently, r_{31} cannot go below the value of $\sqrt{BM}(m_1m_2+4m_2m_3+m_3m_1)^{-1/2}$. Since $r_{32}\geqslant r_{31}$ and r_{12} is regularized, the smallest unregularized distance the computation is concerned with is given by the left side of Equation (69). In other words, all unregularized distances will be larger than the one given above.

The following pages will establish the appropriate values for $I_{\min}$.

First a set of estimates is established for negative total energy. When $I_{\min} \leqslant S_1^2$ (see Equation (66)), escape occurs and the computation is finished. Prior to this, however, several relative minima of the moment of inertia might occur and these must be handled accurately on the computer unless the rest of the computation should be rendered meaningless. Consequently, prediction of every $I_{\min}$ is essential in order to have reliable computer success.

It is recalled that the critical value of I, $I_c = c^2/2|h|$ will be above any significant minima. This value might be low (or even zero) when the angular momentum of the system is small (or zero). Therefore, the computation of this value is the first step and the computer must be able to handle values below this.

A predicted value of a minimum of I below I_c is

$$I_{\min} \geqslant \frac{I_c^2}{I_{\max}} \tag{70}$$

which result is identical with Equation (58) when $I_{\min} = I_1$ and $I_2 = I_{\max}$. Note that it is now presupposed that $t_2 < t_1$, $I(t_2) > I(t_1)$ and $\dot{I}(t_2) = \dot{I}(t_1) = 0$ as shown in Figure 7.

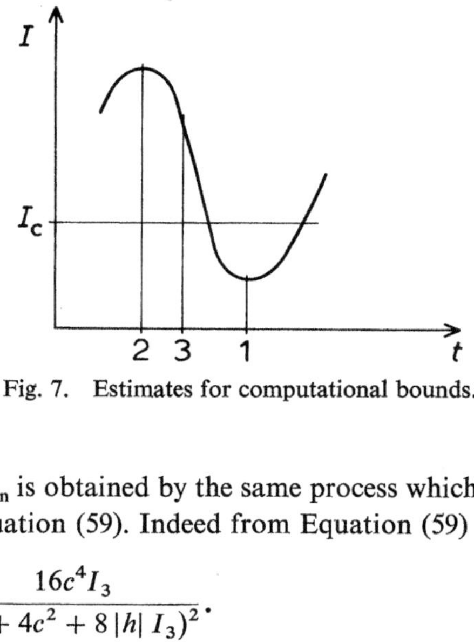

Fig. 7. Estimates for computational bounds.

Another value of $I_{\min}$ is obtained by the same process which was used to estimate $\dot{I}$ in Section 4A, see Equation (59). Indeed from Equation (59) we have

$$I_{\min} \geqslant \frac{16c^4 I_3}{(\dot{I}_3^2 + 4c^2 + 8|h|\, I_3)^2}. \tag{71}$$

This is obtained by having $t_2 < t_3 < t_1$, and $I_2 > I_3 > I_1 = I_m$ with $\dot{I}_2 = \dot{I}_1 = 0$, $\dot{I}_3 < 0$, as also shown in Figure 7. From $L_3 \geqslant L_1$ we have

$$\frac{4c^2}{\sqrt{I_3}} + 8|h|\sqrt{I_3} + \frac{\dot{I}_3^2}{\sqrt{I_3}} \geqslant \frac{4c^2}{\sqrt{I_m}} + 8|h|\sqrt{I_m} \geqslant \frac{4c^2}{\sqrt{I_m}}$$

or

$$\sqrt{I_m} \geqslant \frac{4c^2}{\dfrac{4c^2}{\sqrt{I_3}} + 8|h|\sqrt{I_3} + \dfrac{\dot{I}_3^2}{\sqrt{I_3}}},$$

from which Equation (71) immediately follows.

This result allows the prediction of the lower bound of $I_{\min}$ from any point on the curve $I(t)$ after the maximum. Note that the estimate obtained from (70) is better (higher) than the one obtained from (71) when the latter is applied at the maximum, since according to (71) we have

$$\left[\frac{16c^4 I_3}{(\dot{I}_3^2 + 4c^2 + 8|h|\,I_3)^2}\right]_{I_3 = I_{\max}} = \frac{16c^4 I_{\max}}{(4c^2 + 8|h|\,I_{\max})^2},$$

and from (70) we obtain

$$\frac{I_c^2}{I_{\max}} = \frac{c^4}{4|h|^2 I_{\max}}.$$

Consequently

$$\frac{16c^4 I_{\max}}{64|h|^2 I_{\max}^2} \geqslant \frac{16c^4 I_{\max}}{64|h|^2 I_{\max}^2 + 16c^4 + 64c^2|h|\,I_{\max}}.$$

Now we turn to the case of positive total energy, when the initial conditions completely determine the expected minimum value of the moment of inertia from $t=0$ to $t \to \infty$. If $\dot{I}(0) \geqslant 0$, the minimum value is the initial value, or $I_{\min} = I(0)$ since $\ddot{I} > 0$. If $\dot{I}(0) < 0$, we have

$$I_{\min} \geqslant \frac{16c^4 I(0)}{(\dot{I}^2(0) + 4c^2)^2}, \tag{72}$$

which is obtained from $L(0) \geqslant L_m$, since $I(0) \geqslant I_m$. Using Equation (47) we have

$$\frac{4c^2 + \dot{I}^2(0)}{\sqrt{I(0)}} \geqslant \frac{4c^2}{\sqrt{I_m}} + 8h\left(\sqrt{I(0)} - \sqrt{I_m}\right) \geqslant \frac{4c^2}{\sqrt{I_m}},$$

which when solved for $I_m = I_{\min}$ gives Equation (72).

The results pertinent to the numerical integration of three-body orbits now may be summarized in the following explicit instructions:

(1) From the initial conditions h and c^2 are to be evaluated.

(2) If $h \geqslant 0$, the values of $I(0)$ and $\dot{I}(0)$ are to be computed.

(2a) If $\dot{I}(0) \geqslant 0$, $\quad I_{\min} \geqslant I(0) = B_{2a}$.

(2b) If $\dot{I}(0) < 0$, $\quad I_{\min} \geqslant 16c^4 I(0)[\dot{I}^2(0) + 4c^2]^{-2} = B_{2b}$ as given by Equation (72).

(3) If $h < 0$, two methods are available.

(3a) Either every $I_{\max}$ is to be found and from it the following $I_{\min} \geqslant I_c^2 I_{\max}^{-1} = B_{3a}$ is to be computed according to Equation (70), with $I_c = c^2(2|h|)^{-1}$.

(3b) Or $I(t)$ is monitored and every time $I(t) \leqslant I_c$ with $\dot{I}(t) < 0$, the next $I_{\min} \geqslant 16c^4 I \times [\dot{I}^2 + 4c^2 + 8|h|I]^{-2} = B_{3b}$ is computed according to Equation (71).

(4) In all cases the smallest unregularized distance is computed from

$$\sqrt{\frac{B_i M}{m_1 m_2 + 4m_2 m_3 + m_3 m_1}} \leqslant r_{31}$$

according to Equation (69), where the shortest (regularized) distance is between m_1 and m_2 and the longest distance occurs between m_2 and m_3.

Acknowledgements

Part of these lectures were first formulated at the University of Texas at Austin during the 1971–72 academic year in my graduate class on The Theory of Orbits. I promoted my students from the restricted problem of three bodies to the general problem.

The few lectures delivered at the NATO Advanced Study Institute on this subject allowed me a short outline only. At the invitation of Prof. E. Stiefel, during my visit at ETH Zürich in October-November 1972 I have had the opportunity to work out the details. My lectures in his Seminar für Angewandte Mathematik followed this text as closely as time allowed. The tranquillity and the scientific atmosphere in his Institute in Zürich allowed the concentration necessary to complete the task which occupied my mind for ten years. My visit to Zürich was made possible by a University of Texas research leave granted by the University's Research Institute and by a National Science Foundation travel grant. NATO's contributions to these lectures were substantial since it was at the 1972 Advanced Study Institute on Dynamical Astronomy where I became aware of the general interest in this subject and I had the opportunity to discuss these matters with colleagues.

My present and former graduate students at Yale University and at the University of Texas deserve the highest praise. They helped me to establish the fact – considered by some at Cortina of some significance – that the escape is not an exception but a rule. Discussions and conferences with my associates and visitors, made possible by an ONR grant, proved invaluable.

My only hope is that the work in this field will continue and that all the untied ribbons will be neatly tied. We solved the problem of three bodies in Birkhoff's sense and we may be confident that we understand the qualitative behavior of the problem which Whittaker calls the 'most celebrated of all dynamical problems'.

References

This list of references is preceded by a few general, mostly historical remarks.

The best historical works on the general problem of three bodies are unquestionably A. Gautier's essai (1817), R. Grant's history of astronomy (1852), Prof. R. Marcolongo's treatise (1919), E. Whittaker's celebrated memoir (1900) and E. Lovett's (1911) work with many generalizations of the basic dynamical system.

The series of papers by Chazy are difficult, repetitious but essential. Siegel and his student Moser, Birkhoff and his student Lewis, Pollard and his student Saari, as well as Alekseev, Sitnikov, Merman and the Russian school, all seem to follow in the footsteps of Chazy and Sundman. Agekian's school and mine share the priviledge of finding in 1967 the phenomenon that some consider the 'solution' of the general problem. The two series of papers by these schools give the details and the numerical verification of this general result.

The list of references included below does not intend to be complete and I only hope that no major work was omitted – certainly not by design, but by negligence.

Agekian, T. A. and Anosova, Zh .P.: 1967, *Astron. Zh.* **44**, 1261.

Agekian, T. A. and Anosova, Zh. P.: 1968, *Akad. Arm. U.S.S.R., Astrophys.* **4**, 31.

Birkhoff, G. D.: 1927, *Dynamical Systems*, Am. Math. Soc. Publ., Providence, R. I.

Chazy, J.: 1918, *Bull. Astron.* **35**, 321.

Chazy, J.: 1922, *Ann. Sci. Ecole Norm.* **39**, 29.

Chazy, J.: 1927, *Bull. Soc. Math. France* **55**, 222.

Chazy, J.: 1929, *J. Math. Pure Appl.* **8**, 353.

Gautier, A.: 1817, *Essai historique sur le problème des trois corps*, Courcier Publ., Paris.

Grant, R.: 1852, *History of Physical Astronomy from the Earliest Ages to the Middle of the Nineteenth Century*, Henry G. Bohn Publ., London.

Hagihara, Y.: 1970, *Celestial Mechanics*, MIT Press, Cambridge. Mass.

Harrington, R. S.: 1968, *Astron. J.* **73**, 190.

Jacobi, C. G. J.: 1866, *Vorlesungen über Dynamik*, Reimer Publ., Berlin.

Leimanis, E. and Minorsky, N.: 1958, *Dynamics and Nonlinear Mechanics*, Wiley Publ., New York.

Lewis, D. C.: 1967, *NASA SP*-141, p. 73.

Lovett, E. O.: 1911, *Quart. J. Math.* **42**, 252.

Marcolongo, R.: 1919, *Il Problema dei Tre Corpi*, Hoepli Publ., Milano.

Merman, G. A.: 1953, *Astron. Zh.* **30**, 332.

Merman. G. A.: 1955, *Bull. Inst. Theor. Astron.* **6**, 69.

Pollard, H.: 1966, *Mathematical Introduction to Celestial Mechanics*, Prentice-Hall Publ., N.J.

Pollard, H.: 1967, *J. Math. Mech.* **17**, 601.

Pollard, H. and Saari, D.: 1968, *Arch. Rational Mech. Anal.* **30**, 263.

Pollard, H. and Saari, D.: 1970, *Celes. Mech.* **1**, 347.

Saari, D.: 1971, *Trans. Am. Math. Soc.* **162**, 267.

Siegel, C. L.: 1941, *Ann. Math.* **42**, 127.

Siegel, C. L. and Moser, J.: 1971, *Lectures on Celestial Mechanics*, Springer, Berlin.

Sitnikov, K.: 1960, *Dokl. Akad. Nauk, U.S.S.R.*, **133**, 303.

Standish, E. M.: 1970, in *Periodic Orbits and Stability*, D. Reidel Publ. Co., Dordrecht, Holland, p. 375.

Standish, E. M.: 1971, *Celes. Mech.* **4**, 44.

Stiefel, E. and Scheifele, G.: 1971, *Linear and Regular Celestial Mechanics*, Springer Publ., Berlin.

Sundman, K. F.: 1912, *Acta Math.* **36**, 105.

Szebehely, V.: 1967, *Proc. Nat. Acad. Sci., U.S.A.*, **58**, 60.

Szebehely, V.: 1970, in G. E. O. Giacaglia (ed.), *Periodic Orbits, Stability and Resonances*, D. Reidel Publ. Co., Dordrecht, Holland, p. 382.

Szebehely, V.: 1971, *Celes. Mech.* **4**, 116.

Szebehely, V.: 1972, *Astron. J.* **77**, 169.

Szebehely, V.: 1972, *Celes. Mech.* **6**, 84.

Szebehely, V.: 1972, *Proc. Nat. Acad. Sci., U.S.A.*, **69**, 1077.

Szebehely, V.: 1973, *Astron. Astrophys.* **22**, 171.

Szebehely, V.: 1973, *Proc. First European Astron. Mtg. in Athens*, 1972, Springer Press, Berlin, to appear.

Szebehely, V. and Peters, F.: 1967, *Astron. J.* **72**, 876.

Szebehely, V. and Peters, F.: 1967, *Astron. J.* **72**, 1187.

Szebehely, V. and Feagin, T.: 1973, *Celes. Mech.* **8**, to appear.

Tevzadze, G. A.: 1962, *Akad. Nauk. Arm.* U.S.S.R., **15**, 67.

Whittaker, E. T.: 1900, Rept. 69th Mtg. British Assoc., p. 121.

Whittaker. E. T.: 1904, *Analytical Dynamics*, Cambridge University Press, London and New York.

Wintner, A.: 1941, *The Analytical Foundations of Celestial Mechanics*, Princeton Univ. Press, Princeton, N.J.

PERIODIC ELLIPTIC MOTION IN THE PROBLEM OF TRHEE BODIES

R. F. ARENSTORF

Dept. of Mathematics, Vanderbilt University, Nashville, Tenn., U.S.A.

Abstract. We consider the Newtonian three-body problem in the plane for three mass points P_k with arbitrarily given masses $m_k>0$, $(k=1, 2, 3)$. Identifying the plane of motion with the ordinary complex plane we introduce the relative (complex) vectors

$$u = \mathbf{P}_2\mathbf{P}_3 = u(t), \qquad v = \mathbf{P}_1\mathbf{P}_4 = v(t),$$

where P_4 is the center of mass of P_2 and P_3. Assuming the center of mass of the three bodies to be at rest at the origin and defining the mass parameters μ and ν by

$$m_2 + m_3 = \mu(m_1 + m_2 + m_3), \qquad m_3 = \nu(m_2 + m_3)$$

the equations of motion take the form (where $\dot{} = \mathrm{d}/\mathrm{dt}$)

$$\begin{aligned} \ddot{u} + \mu F(u) &= (1-\mu)\left[F(v - \nu u) - F(v + u - \nu u)\right] \\ \ddot{v} + F(v) &= F(v) - (1-\nu)\, F(v - \nu u) - F(v + u - \nu u) \end{aligned} \tag{1}$$

with $F(u)=u|u|^{-3}$. For small $|u/v|$ this can be approximated by

$$\ddot{u} + \mu F(u) = 0, \qquad \ddot{v} + F(v) = 0; \tag{2}$$

and these two uncoupled Kepler-problems have periodic solutions of the form $u=u^*=u^*(t;\varepsilon, k, m)$, $v=v^*=v^*(t)$, in particular, where

$$\begin{aligned} u^* &= \mu^{1/3} c^2 (1 - \varepsilon \cos s)^{-1} e^{is}, \qquad v^* = e^{it}, \qquad 0 < \varepsilon < 1, \\ t &= c^3 \int_0^s (1 - \varepsilon \cos s)^{-2}\, \mathrm{d}s, \qquad c = (m/k)^{1/3} (1 - \varepsilon^2)^{1/2}, \end{aligned} \tag{3}$$

describing elliptic motion of eccentricity ε for u. Here $m>0$ and k are relatively prime integers, and the motion is periodic in the sense that $u^*(t+2\pi m)=u^*(t)$ and $v^*(t+2\pi m)=v^*(t)$ identically in t.

Our main result can now be stated: The three-body problem, as given by (1), has solutions of the form

$$u = u(t;\, \varepsilon, k, m), \qquad v = v(t;\, \varepsilon, k, m),$$

which represent periodic motions of P_1, P_2, P_3 with period $T=T(\varepsilon, k, m)$ in the sense that for the relative motion $q(t)=u(t)/v(t)$ (complex vectors)

$$q(t+T) = q(t), \quad \text{while} \quad v(t+T) = e^{ij} v(t),$$

identically in t, with real $j=j(\varepsilon, k, m)$. Here T is near $2\pi m$, the relative motion of the pair (P_2, P_3) described by the closed curve $q=q(t)$, $(0 \leqslant t \leqslant T)$ lies near the curve $q=u^*(t)/v^*(t)$, $(0 \leqslant t \leqslant 2\pi m)$, both closing after $k-m$ revolutions about $q=0$ in

B. D. Tapley and V. Szebehely (eds.), Recent Advances in Dynamical Astronomy, 107–108.

the q-plane, and v describes nearly Keplerian slowly precessing elliptic motion of small eccentricity for the pair (P_1, P_4).

These solutions exist for any preassigned masses, and any ε in $(0, 1)$ and natural number m, if only $|k|$ is large enough (above a constructable bound); i.e., if the approximating elliptic motion u^* in (3) has sufficiently small absolute magnitude $|u|^*$, (compared to $|v^*|=1$); i.e. if the ratio of distances $|P_2, P_3|/|P_1, P_4|$ is sufficiently small.

Thus identifying P_1 with the Sun, P_2 with the Earth and P_3 with the Moon, our result gains interest for the lunar theory; or for an astronautical restricted four-body problem, when adding a space vehicle to the system. Also, identifying P_1 with the planet Jupiter, P_2 with the Sun and P_3 with the planet Mercury, would give an interesting application.

A complete proof of our result can be found in Arenstorf (1968). The closest earlier known result in the three-body problem to the one outlined here is due to Moulton (1906), giving nearly *circular* motions for (P_2, P_3) and for (P_1, P_4), corresponding to $\varepsilon=0$ and arbitrary c with sufficiently small $|c|$ in (3).

In our case the approximating periodic solutions of (2) given in (3) are non-generic for the Poincaré continuation method, the degeneracy caused by resonance of the two Keplerian motions u^*, v^*. We overcome this difficulty by approximating (1), not with (2), but with the restricted three-body problem for P_3 with primaries P_1 and P_2 (moving circularly). For the latter problem the periodic solution u^* from (3) is non-degenerate relative to the periodicity condition of two successive symmetric conjunctions of P_1, P_2, P_3; as has been shown already in Arenstorf (1963). Corresponding periodic solutions for arbitrary masses m_1 and m_2 have been constructed in Arenstorf, (1966) by a generalization of the Poincaré method. The resulting periodic solutions $u=\tilde{u}(t;\, \varepsilon, k, m)$ of the restricted problem have the period $\tilde{T}=\tilde{T}(\varepsilon, k, m)$, and are non-degenerate relative to the same periodicity condition as above for the full three-body problem, since $|\tilde{T}-2\pi m|$, though very small, is different from zero. In fact, a sufficiently sharp lower estimate of this difference can be established, which, together with the fact (recognizable by careful estimations) that the restricted three-body problem gives a 'closer' approximation to the full three-body problem (1) than do the two uncoupled Kepler-problems (2), is decisive in our existence proof. This also emphasizes the importance and generic character of the restricted three-body problem (rather than of Kepler problems) for the full three-body problem, in case u/v is small; and hopefully will give rise to a generalization of the now customary techniques in astronomical perturbation theories based mostly on variables of Keplerian type.

Our result has been recently generalized to a four-body problem, which will be detailed in a future paper in Celestial Mechanics.

References

Arenstorf, R. F.: 1966, *J. Reine Angew. Math.* **221**, 113–145.
Arenstorf, R. F.: 1968, *J. Diff. Eq.* **4**. 202–256.
Arenstorf. R. F.: 1963, *Am. J. Math.* **85**, 27–35.
Moulton, F. R.: 1909, *Trans. Amer. Math. Soc.* **7**, 537–577.

ON A CONJECTURE BY POINCARÉ

G. KATSIARIS

Technical University of Athens, Athens, Greece

and

C. L. GOUDAS

University of Patras, Patras, Greece

Abstract. The conjecture of Poincaré (1892) on the 'density' of periodic orbits of the restricted 3-body problem is discussed and generalized for dynamical system with stable closed solutions. It is shown that under certain conditions the above conjecture is valid for certain domains of the phase space. Under the same conditions the 'branching' of a family of periodic orbits is shown to be 'dense'.

1. Introduction

Poincaré advanced the hypothesis that 'as close as you like' to the starting conditions of an orbit, picked at random, of the restricted problem, it is possible to find initial conditions of a closed orbit of period that can be very large, while the two paths will stay 'as close to each other as you like'. This conjecture can be expressed rigorously as follows: Let $\mathbf{x}$ be the position vector of the massless particle of the restricted problem and $\mathrm{d}\mathbf{x}/\mathrm{d}t = \mathbf{f}(\mathbf{x})$ the equation of motion. The solutions are $\mathbf{x}(\mathbf{x}_0; t)$, where $\mathbf{x}_0$ is the initial state corresponding to $t=0$, without loss of generality. If $\mathbf{x}_0$ is selected at random, then $\mathbf{x}(\mathbf{x}_0; t)$ has zero probability to be periodic. However, according to Poincaré, given $\varepsilon > 0$ there exists an x_0^*, a T^*, a $\delta(\varepsilon) > 0$ and a t^*, such that

$$\|\mathbf{x}_0 - \mathbf{x}_0^*\| < \varepsilon, \quad \|\mathbf{x}(\mathbf{x}_0; t) - \mathbf{x}(\mathbf{x}_0^*; t^*)\| < \delta(\varepsilon),$$
$$\mathbf{x}(\mathbf{x}_0^*; t) = \mathbf{x}(\mathbf{x}_0^*; t + T^*).$$

Apparently, the orbit $\mathbf{x}(\mathbf{x}_0^*; t)$ is periodic, of period T^*, while $\delta(\varepsilon)$ defines a tube, in phase space, within which the solution $\mathbf{x}(\mathbf{x}_0; t)$ is confined for all times. Attention should be given to the fact that the quantity $\|\mathbf{x}(\mathbf{x}_0; t) - \mathbf{x}(\mathbf{x}_0^*; t)\|$ need not be small since it represents the distance of the two orbits, while the argument in question is the distance of the two paths. The latter is defined as the minimum value of the quantity $\|\mathbf{x}(\mathbf{x}_0; t) - \mathbf{x}(\mathbf{x}_0^*; t_1)\|$, where t is kept fixed and t_1 receives values in the interval $[0, T^*]$.

For the particular case of the restricted problem, a large number of approximate periodic solutions have been found by numerical processes while the existence of the majority of them has been proven. Particular references of this fact are not necessary. However, the above conjecture of Poincaré has not been tested, either numerically or otherwise, although people who compute periodic orbits develop the feeling that its validity is very probable.

Before proceeding to examine the cases where the conjecture can be valid, we should point out the cases where the same conjecture is clearly not valid. And such are the cases in which the state vector $\mathbf{x}(\mathbf{x}_0; t)$ is unbounded. These cases correspond either

B. D. Tapley and V. Szebehely (eds.), Recent Advances in Dynamical Astronomy, 109–117.

to collision or to escape orbits. Poincaré apparently precluded these cases considering that they did not deserve particular mention.

The discussion from now on will cover the cases of autonomous and conservative systems with the equation of motion and variation of the forms

$$\mathrm{d}\mathbf{x}/\mathrm{d}t = \mathbf{f}(\mathbf{x}), \qquad \mathrm{d}\Delta/\mathrm{d}t = P\Delta \tag{1}$$

where

$$\Delta = (\partial x_i/\partial x_{0j}), \quad P = (\partial f_i/\partial x_j), \quad \nabla\cdot\mathbf{f} = 0.$$

The condition $\nabla\cdot\mathbf{f}=0$ safeguards the conservation of volume in phase space according to Liouville's theorem. The conservative character of the system, on the other hand, is expressed by the existence of a function $F(\mathbf{x})$, such that,

$$F(\mathbf{x}) = F(\mathbf{x}_0). \tag{2}$$

From Equation (2) we obtain the relation

$$(\nabla F)'\cdot\dot{\mathbf{x}} = 0. \tag{3}$$

For such systems Birkhoff (1927) showed that the variational matrix Δ is symplectic, i.e. the relation

$$\Delta' Q \Delta = Q, \tag{4}$$

is valid where, for the three-dimensional restricted problem, we have, $Q=(q_{ij})$, and $q_{1,2}=-q_{2,1}=-2$, $q_{i,i+3}=1$, $q_{i+3,i}=-1$, $i=1,2,3$ and $q_{i,j}=0$ for all other combinations of the suffixes.

The solutions of Equations (1) will be written as

$$\mathbf{x} = \mathbf{x}(\mathbf{x}_0; t), \qquad \Delta = \Delta(\mathbf{x}_0; t). \tag{6}$$

The conservation of volume in phase space implies that

$$\det\Delta(\mathbf{x}_0; t) = \text{const}, \tag{7}$$

and since

$$\det\Delta(\mathbf{x}_0; 0) = 1,$$

we have

$$\det\Delta(\mathbf{x}_0; t) = 1. \tag{8}$$

If we pick an $\mathbf{x}_0$ and a T at random and wish to compute an $\mathbf{x}_0^*$, not necessarily close to $\mathbf{x}_0$, such that

$$\mathbf{x}(\mathbf{x}_0^*; t+T) = \mathbf{x}(\mathbf{x}_0^*; t), \tag{9}$$

we shall have to treat relation (9) as an algebraic equation and solve it for $\mathbf{x}_0^*$. It is practical to solve instead the equation

$$\mathbf{x}(\mathbf{x}_0; T) - \mathbf{x}_0 = 0, \tag{10}$$

while $\mathbf{x}_0^*$ will be the root whose value is sought. In general for random $\mathbf{x}_0$, Equation

(10) is not satisfied, i.e. we have

$$\mathbf{x}(\mathbf{x}_0; T) - \mathbf{x}_0 \neq 0, \tag{11}$$

and an increment $\delta\mathbf{x}_0$ to $\mathbf{x}_0$ will be needed in order to satisfy it. Then we shall have

$$\mathbf{x}(\mathbf{x}_0 + \delta\mathbf{x}_0; T) - \mathbf{x}_0 - \delta\mathbf{x}_0 = 0. \tag{12}$$

A first approximation to $\delta\mathbf{x}_0$ is found by linearizing (12), whereupon the relation

$$[\Delta(\mathbf{x}_0; T) - I]\,\delta\mathbf{x}_0 = \mathbf{x}_0 - \mathbf{x}(\mathbf{x}_0; T), \tag{13}$$

is obtained. By solving the linear system (13) we obtain $\delta\mathbf{x}_0$ and repeat the above procedure from the beginning using this time $\mathbf{x}_0 + \delta\mathbf{x}_0$ in the place of $\mathbf{x}_0$. A small number of applications (five to six at most for the restricted problem) leads, usually, to $\mathbf{x}_0^*$ although the judicious choice of the initial $\mathbf{x}_0$ and the period influences substantially the effectiveness of the method.

Once $\mathbf{x}_0^*$ is found the matrix $\Delta(\mathbf{x}_0^*; T)$ acquires important properties that are useful in testing the validity of the conjecture of Poincaré. These properties are:

(i) $\Delta(\mathbf{x}_0^*; t+T) = \Delta(\mathbf{x}_0^*; t)\Delta(\mathbf{x}_0^*; T)$,
(ii) $\Delta(\mathbf{x}_0^*; T)\dot{\mathbf{x}}_0^* = \dot{\mathbf{x}}_0^*$,
(iii) $(\nabla F)'[\Delta(\mathbf{x}_0^*; T) - I] = \mathbf{0}$,
(iv) If λ eigenvalue of $\Delta(\mathbf{x}_0^*; T)$ so is λ^{-1},
(v) $\Delta'(\mathbf{x}_0^*; T)Q\Delta(\mathbf{x}_0^*; T) = Q$.

Property (i) is called the transition property, while $\Delta(\mathbf{x}_0^*; T)$ is called matrizant or transition matrix. From property (i) we can deduce the relation

$$\Delta(\mathbf{x}_0^*; nT) = \Delta^n(\mathbf{x}_0^*; T). \tag{14}$$

Property (ii) implies that $\dot{\mathbf{x}}_0^*$ is an eigenvector of $\Delta(\mathbf{x}_0^*; T)$, while the eigenvalue corresponding to it is equal to unity. Because $\partial\mathbf{f}/\partial t = 0$, any point of the solution $\mathbf{x}(\mathbf{x}_0^*; t)$ can be used as the initial state of the periodic solution. By comparing Equation (13) and property (ii) we conclude that this is precisely the meaning of the eigenvector $\dot{\mathbf{x}}_0^*$, namely, it gives the direction along which displacements of the initial conditions preserve the periodic character of the new orbit. We shall see later that all eigenvectors corresponding to unit eigenvalues have the same property. Because of property (iv) the unit eigenvalues of $\Delta(\mathbf{x}_0^*; T)$ are at least two. Thus, if

$$\operatorname{rank}[\Delta(\mathbf{x}_0^*; T) - I] = n - 2, \tag{15}$$

where n is the order of $\Delta(\mathbf{x}_0^*; T)$, then there exists a second direction, defined by the eigenvector corresponding to the second unit eigenvalue, which preserves the periodic character of the orbit as well as the period. If,

$$\operatorname{rank}[\Delta(\mathbf{x}_0^*; T) - I] = n - 1, \tag{16}$$

then there is one and only one direction preserving the periodic character and the period and this is $\dot{\mathbf{x}}_0^*$.

Property (iii) shows that ∇F is a pre-multiplying eigenvector of $\Delta(\mathbf{x}_0^*; T)$, corresponding to a unit eigenvalue. Because of Equation (3) it is deduced, without reference to property (iv), that the unit eigenvalues of $\Delta(\mathbf{x}_0^*; T)$ are, at least, two. It is remarked that although there exist always one pre-multiplying and one post-multiplying eigenvector corresponding to different unit eigenvalues, the second pre-multiplying and the second post-multiplying eigenvectors usually do not exist. They exist however, in the case when the new periodic orbit, of initial state $\mathbf{x}_0^* + \delta\mathbf{x}_0$, is of the same period as the one of initial state $\mathbf{x}_0^*$.

Property (iv) is deduced from the property

$$\Delta(\mathbf{x}_0; -t) = L\Delta(\mathbf{x}_0; t)\, L, \tag{17}$$

where, $L=(1_{ij})$, $1_{ii}=1$, $i=1, 2, \ldots, n/2$, $1_{ii}=-1$, $i=n/2+1, n/2+2, \ldots, n$, $1_{ij}=0$, $i \neq j$, and property (i). Indeed, Equation (17), for $\mathbf{x}_0=\mathbf{x}_0^*$ and $t=T$ gives

$$\Delta(\mathbf{x}_0^*; -T) = L\Delta(\mathbf{x}_0^*; T)\, L, \tag{18}$$

while property (i) becomes

$$\Delta(\mathbf{x}_0^*; -T) = \Delta^{-1}(\mathbf{x}_0^*; T). \tag{19}$$

Therefore, the matrices $L\Delta(\mathbf{x}_0^*; T)\, L$ and $\Delta^{-1}(\mathbf{x}_0^*; T)$ are equal. The matrices $\Delta(\mathbf{x}_0^*; T)$ and $L\Delta(\mathbf{x}_0^*; T)\, L$ have the same eigenvalues, while the matrices $\Delta(\mathbf{x}_0^*; T)$ and Δ^{-1} $(\mathbf{x}_0^*; T)$ have reciprocal ones. Once again we see that the unit eigenvalues appear in pairs and that at least one pair always exists.

Property (v) gives rise to $n(n-2)/2$ bilinear relations of great practical importance for the computation of orbits.

2. Search for Families

The search for periodic orbits can be conducted by analytical, analytical-numerical or purely numerical methods. In all cases the work must contain some proven criterion of existence, which can be based on analytical or numerical evidence. The numerical proof of existence should not be underestimated, since it is of the same practical value as the analytical one. One such example is given by Poincaré (1892), where use is made of the fact that if $x=x_0$, $\mu=0$, is a root of equation

$$\varphi(x, \mu) = 0 \tag{20}$$

and $(\partial\varphi/\partial x) \neq 0$ for $x=x_0$, $\mu=0$ and $\partial x/\partial\mu$ exists, then Equation (20) has a root in x for small values of μ. A numerical proof of existence of roots of (20) for small values of μ consists in computing the quantity $(\partial\varphi/\partial x)$ at $x=x_0$, $\mu=0$ and showing that this quantity is different from zero. This is the 'continuation method' developed by Poincaré. The diagram (x, μ) can be called a 'family' of roots of Equation (20).

The numerical search for periodic orbits begins with some approximation of the initial state $\mathbf{x}_0$ and of the period T. Unless the orbits sought are of some special character, such as symmetric, the condition to be fulfilled after one complete period is

$$\mathbf{x}(\mathbf{x}_0; T) - \mathbf{x}_0 = \mathbf{0}, \tag{21}$$

while the correction $\delta \mathbf{x}_0$, to be added to $\mathbf{x}_0$ when Equation (21) is not satisfied, is obtained by solving the relation

$$[\Delta(\mathbf{x}_0; T) - I]\,\delta\mathbf{x}_0 = \mathbf{x}_0 - \mathbf{x}(\mathbf{x}_0; T), \tag{22}$$

which is found by linearizing Equation (21). By successive applications of this correction process, (21) will be satisfied, whereupon $\mathbf{x}_0$ will acquire the value $\mathbf{x}_0^*$ and Equation (22) the form

$$[\Delta(\mathbf{x}_0^*; T) - I]\,\delta\mathbf{x}_0 = \mathbf{0}. \tag{23}$$

We now wish to obtain a periodic orbit of initial state $\mathbf{x}_0^{**}$ and of period T^*, where $\|\mathbf{x}_0^{**} - \mathbf{x}_0^*\| < \varepsilon$ and $|T - T^*| < \delta(\varepsilon)$. If condition (16) is satisfied, then for $T = T^*$, there exists $\mathbf{x}_0^{**}$, such that

$$\mathbf{x}_0^{**} - \mathbf{x}_0^* \propto \dot{\mathbf{x}}_0^*. \tag{24}$$

This means that $\mathbf{x}_0^{**}$ is a point on the same orbit, i.e. there exists $t = t^{**}$, such that

$$\mathbf{x}_0^{**} = \mathbf{x}(\mathbf{x}_0^*; t^{**}). \tag{25}$$

This case is of no interest except for the fact that it reminds to us that within a tube of radius $\delta(\varepsilon)$ in the n-dimensional phase space, there exists a one-dimensional manifold of points, namely the ones that define the orbit $\mathbf{x}(\mathbf{x}_0^*; t) \mid [0, T]$, that give rise to periodic orbits (in reality one periodic orbit). It is essential to note that the new initial conditions thus obtained lie in the direction of the eigenvector of $\Delta(\mathbf{x}_0^*; T)$ that corresponds to a unit eigenvalue.

If condition (15) is fulfilled, then matrix $\Delta(\mathbf{x}_0^*; T)$ has a second eigenvector, corresponding to the second unit eigenvalue. Let $\mathbf{y}_1^*(\neq \dot{\mathbf{x}}_0^*)$ be this new eigenvector. Any small change $\delta\mathbf{x}_0 \propto \mathbf{y}_1^*$ will preserve the periodic character and the period of the orbit, i.e. $\mathbf{x}(\mathbf{x}_0^* + \delta\mathbf{x}_0; t) = \mathbf{x}(\mathbf{x}_0^* + \delta\mathbf{x}_0; t + T)$. In this case increments of the type $\delta\mathbf{x}_0 = \varepsilon_0 \dot{\mathbf{x}}_0^* + \varepsilon_1 \mathbf{y}_1^*$, give rise to periodic orbits of the same period, i.e.

$$\mathbf{x}(\mathbf{x}_0^* + \varepsilon_0 \dot{\mathbf{x}}_0^* + \varepsilon_2 \mathbf{y}_1^*; t + T) = \mathbf{x}(\mathbf{x}_0^* + \varepsilon_0 \dot{\mathbf{x}}_0^* + \varepsilon_2 \mathbf{y}_1^*; t).$$

We, therefore, can conclude that within the above tube of phase space there exists a two-dimensional continuous manifold all points of which belong to periodic orbits of the same family.

We now proceed to consider the cases, where the transition matrix $\Delta(\mathbf{x}_0^*; T)$ has m unit eigenvalues (m will be even) and

$$\operatorname{rank}[\Delta(\mathbf{x}_0^*; T) - I] = n - l. \tag{26}$$

Under these conditions we can find $l(\leqslant m)$ directions, $\mathbf{y}_1^*, \mathbf{y}_2^*, \ldots, y_l^*$ preserving the periodic character and the period T of the orbit $\mathbf{x}(\mathbf{x}_0^*; t)$. Thus, for sufficiently small values, positive or negative, of $\varepsilon_0, \varepsilon_1, \varepsilon_2, \ldots, \varepsilon_l$, we have

$$\begin{aligned} \mathbf{x}(\mathbf{x}_0^* + \varepsilon_0 \dot{\mathbf{x}}_0^* + \varepsilon_1 \mathbf{y}_1^* + \varepsilon_2 \mathbf{y}_2^* + \cdots + \varepsilon_l \mathbf{y}_l^*; t + T) &= \\ = \mathbf{x}(\mathbf{x}_0^* + \varepsilon_0 \dot{\mathbf{x}}_0^* + \varepsilon_1 \mathbf{y}_1^* + \varepsilon_2 \mathbf{y}_2^* + \cdots + \varepsilon_l \mathbf{y}_l^*; t)&. \end{aligned} \tag{27}$$

As a result, within the said tube, there exists an l-dimensional continuous manifold all points of which correspond to periodic orbits of period T. Definitely Poincaré's conjecture goes beyond this conclusion.

We further assume that none of the eigenvalues $\lambda_{m+1}, \lambda_{m+2}, \ldots, \lambda_n$ is equal to unity, but that the orbit is of the stable type. This implies that

$$|\lambda_{m+k}| = 1, \qquad k = 1, 2, \ldots, n - m, \tag{28}$$

and that λ_{m+1}, etc. are complex. If λ_{m+k} is a complex eigenvalue of $\Delta(\mathbf{x}_0^*; T)$ so is $\bar{\lambda}_{m+k}$ (complex conjugate of λ_{m+k}). We can assume that $\bar{\lambda}_{m+k} = \lambda_{m+k+1}$, for $k = 1, 2, \ldots, n-m$. Let us denote by $\mathbf{y}_{m+k}$, $k = 1, 2, \ldots, n-m$ the eigenvectors of $\Delta(\mathbf{x}_0^*; T)$ corresponding to the eigenvalues of the same suffix. It is clear that increments along any of these eigenvectors will not preserve the periodic character, or the period, of the orbit $\mathbf{x}(\mathbf{x}_0^*; t)$. The elementary increments $\delta\mathbf{x}_0$ to $\mathbf{x}_0^*$ that lead to initial states of periodic orbits of period $T^* = T + \delta T$ must satisfy the condition.

$$\mathbf{x}(\mathbf{x}_0^* + \delta\mathbf{x}_0; T + \delta T) - \mathbf{x}_0^* - \delta\mathbf{x}_0 = 0, \tag{29}$$

or, after linearization

$$[\Delta(\mathbf{x}_0^*; T) - I]\, \delta\mathbf{x}_0 + \dot{\mathbf{x}}_0^*\, \delta T = 0. \tag{30}$$

Because $[\Delta(\mathbf{x}_0^*; T) - I]$ is a singular matrix, it is not possible to define $\delta\mathbf{x}_0$ by giving an arbitrary value to δT. However, we can solve Equation (30) together with the equation

$$(\nabla F)' \cdot \delta\mathbf{x}_0 = 0, \tag{31}$$

which limits the elementary displacements to those which preserve the energy of the solution $\mathbf{x}(\mathbf{x}_0^*; t)$. We thus end with the linear system

$$D(\mathbf{x}_0^*; T)\, \delta\mathbf{u}_0 = \delta\mathbf{u}_0, \tag{32}$$

where

$$D(\mathbf{x}_0^*; T) = \left[\begin{array}{c|c} \Delta(\mathbf{x}_0^*; T) & \dot{\mathbf{x}}_0^* \\ \hline (\nabla F)' & 1 \end{array}\right], \tag{33}$$

$$\delta\mathbf{u}_0 = \begin{bmatrix} \delta\mathbf{x}_0 \\ \delta T \end{bmatrix}, \tag{34}$$

and $\nabla F = \nabla F(\mathbf{x}_0^*)$.

It is important to note that the $\delta\mathbf{u}_0$'s that lead to new periodic orbits are eigenvectors of $D(\mathbf{x}_0^*; T)$. This is also true for those of the $\delta\mathbf{u}_0$'s that preserve the period, since then

$$\delta\mathbf{u}_0 = \begin{bmatrix} \delta\mathbf{x}_0 \\ 0 \end{bmatrix}. \tag{35}$$

Let us now assume that

$$\lambda_{m+k} = e^{i\varphi_k}, \tag{36}$$

where, φ_k is the kth characteristic exponent.

Therefore,

$$\varphi_{k+1} = -\varphi_k. \tag{37}$$

We further assume that for a certain value k_0 of k, it is

$$p_{k_0}\varphi_{k_0} = 2q_{k_0}\pi, \tag{38}$$

where $p_{k_0}, q_{k_0} \in I$. This implies that

$$\lambda_{m+k}^{p_{k_0}} = \lambda_{m+k+1}^{p_{k_0}} = 1. \tag{39}$$

But $\lambda_{m+k}^{p_{k_0}}$, $k=1, 2, \ldots, n-m$ are the eigenvalues of the matrix $\Delta^{p_{k_0}}(\mathbf{x}_0^*; T)$ or the matrix $\Delta(\mathbf{x}_0^*; p_{k_0}T)$. But the matrix $\Delta(\mathbf{x}_0^*; p_{k_0}T)$ is the matrizant of the periodic orbit with initial condition $\mathbf{x}_0^*$ and period $p_{k_0}T$. This orbit has $m+2$ unit eigenvalues. If

$$\operatorname{rank}[\Delta(\mathbf{x}_0^*; p_{k_0}T) - I] = n - l - 1, \tag{40}$$

then we have $l+1$ directions (eigenvectors) preserving the periodicity and the periods T and $p_{k_0}T$, while if

$$\operatorname{rank}[\Delta(\mathbf{x}_0^*; p_{k_0}T) - I] = n - l - 2, \tag{41}$$

we have $l+2$ such directions.

We now remark that for the above case

$$D(\mathbf{x}_0^*; k_0T) = \left[\begin{array}{c|c} \Delta(\mathbf{x}_0^*; k_0T) & \dot{\mathbf{x}}_0^* \\ \hline (\nabla F)' & 1 \end{array}\right] = \left[\begin{array}{c|c} \Delta^{k_0}(\mathbf{x}_0^*; T) & \dot{\mathbf{x}}_0^* \\ \hline (\nabla F)' & 1 \end{array}\right]. \tag{42}$$

This is a very important property of the matrix D.

We now assume that we can find pairs of integers p_k, q_k, such that

$$p_k\varphi_k = 2q_k\pi, \qquad k = 1, 2, \ldots, n-m. \tag{43}$$

and that p is the least common multiplier of $p_1, p_2, \ldots, p_{n-m}$. Then

$$\lambda_{m+k}^p = 1, \qquad k = 1, 2, \ldots, n-m. \tag{44}$$

Therefore, the matrix $\Delta^p(\mathbf{x}_0^*; T)$, i.e. the matrix $\Delta(\mathbf{x}_0^*; pT)$, has all its eigenvalues equal to unity. Such orbits are called 'ordinary'. Hence, if the orbit of initial state $\mathbf{x}_0^*$ is considered to be of period pT, then it is ordinary. The directions along which elementary displacements will preserve the periodicity and, perhaps, the period depend on the rank of the matrix

$$[D(\mathbf{x}_0^*; p_kT) - I]. \tag{45}$$

If, e.g. the rank of matrix (45) and the rank of $[D(\mathbf{x}_0^*; T) - I]$ differ by one or two, then one more or two, respective directions conserve the periodicity and, perhaps, the period. Further, the total number of directions along which the periodicity is preserved, but not necessarily the periods, is the rank of the matrix

$$[D(\mathbf{x}_0^*; pT) - I]. \tag{46}$$

If we assume that $p_0 = 1$ and

$$p_1 < p_2 < p_3, \ldots < p_{n-m}, \tag{47}$$

then we can compute the eigenvectors of $D(x_0^*; T)$, $D(\mathbf{x}_0^*; p_1T)$, $D(\mathbf{x}_0^*; p_2T)$, etc., that correspond to unit eigenvalues. Along these directions the period may or may not be preserved.

It is important to remark that the eigenvectors of $D(\mathbf{x}_0^*; T)$, $D(\mathbf{x}_0^*; p_1T)$, $D(\mathbf{x}_0^*; p_2T), \ldots, D(\mathbf{x}_0^*; p_kT)$, corresponding to unit eigenvalues are also eigenvectors of $D(\mathbf{x}_0^*; p_{k+1}T)$. The additional eigenvectors of $D(\mathbf{x}_0^*; p_{k+1}T)$, corresponding to unit eigenvalues, represent the new directions along which families of periods $p_{k+1}T$ 'branch' off.

After the above discussion, we can conclude that initial states $\mathbf{x}_0^{**}$ of periodic orbits 'nest' around $\mathbf{x}_0^*$, if the matrix $D(\mathbf{x}_0^*; pT)$ has n eigenvectors corresponding to its n unit eigenvalues. This, precisely, is the essence of the conjecture of Poincaré as generalized in this presentation. The stringent condition that has to be fulfilled in order that the conjecture is valid is condition (43).

3. The Case of Two Bodies

The orbits of the two body problem are 'ordinary' since all their eigenvalues are equal to unity. The matrix $D(\mathbf{x}_0; T)$, of order 7, has five eigenvectors conserving the periodicity and the period and one direction along which the period changes.

4. The Planar Restricted Problem

For stable orbits $\mathbf{x}(\mathbf{x}_0^*; t)$ we have

$$\lambda_1 = 1, \quad \lambda_2 = 1, \quad \lambda_3 = e^{i\varphi}, \quad \lambda_4 = e^{-i\varphi}. \tag{48}$$

Because $\varphi = \varphi(\mathbf{x}_0^*)$ is continuous we can assume that there always exist integers p and q, such that $p\varphi = 2q\pi$ by a appropriate selection of $\mathbf{x}_0^*$. Hence, the orbit $\mathbf{x}(\mathbf{x}_0^*; t)$, if considered of period pT, is 'ordinary'. Numerical experiments show that the matrix

$$D(\mathbf{x}_0^*; T), \tag{49}$$

has two eigenvectors corresponding to the two unit eigenvalues. One is $\dot{\mathbf{x}}_0^*$, along which the period is preserved (we practically obtain the same orbit). The other say $\mathbf{y}_1$, defines the direction along which T varies continuously.

For φ commensurable to 2π, the matrix

$$D(\mathbf{x}_0^*; pT), \tag{50}$$

besides $\dot{\mathbf{x}}_0^*$ and $\mathbf{y}_1$, which are eigenvectors of it corresponding to unit eigenvalues, can have one more eigenvector, say $\mathbf{y}_2$, corresponding to unit eigenvalues, if

$$\operatorname{rank}[D(\mathbf{x}_0^*; pT) - I] = 2, \tag{51}$$

or two eigenvectors, say $\mathbf{y}_2$ and $\mathbf{y}_3$, if

$$\text{rank}\,[D(\mathbf{x}_0^*;\, pT) - I] = 1\,. \tag{52}$$

When condition (51) holds, we have three directions along which periodicity is conserved. Along $\dot{\mathbf{x}}_0^*$ the period T is conserved while along $\mathbf{y}_1$ and $\mathbf{y}_2$ the periods T and pT vary continuously. When condition (52) holds, the period T changes continuously along $\mathbf{y}_1$, while the period pT changes continuously along $\mathbf{y}_2$ and $\mathbf{y}_3$. If condition (52) holds the conjecture of Poincaré is locally valid. Numerical experiments are necessary in order to determine which of the two conditions (51) and (52) holds.

References

Birkhoff, B.: 1927, *Dynamical Systems*, Vol. 9, American Mathematical Soc., Colloquium Publ. New York.

Poincaré, H.: 1892, *Les Méthodes Nouvelles de la Méchanique Céleste*, Vol. 1, Dover Publ., 1957.

THE THREE-DIMENSIONAL ELLIPTIC PROBLEM

G. KATSIARIS
National Technical University, Athens, Greece

Abstract. The present paper is a contribution to the study and calculation of three-dimensional periodic orbits of the elliptic restricted problem. The reference system used is the rotating-pulsating one in which the two primaries are at rest. After developing the method for calculation of three-dimensional periodic orbits and of their stability characteristics, we proceed to present numerical results which were obtained by using the computer 1906A ICL of the University of Manchester.

1. Introduction

The three-dimensional elliptic restricted three-body problem is a typical example o, non-conservative and non-integrable dynamical systems of three degrees of freedom whereas the circular problem is a characteristic example of the conservative non-integrable ones.

The importance of its study rests on the fact that in nature we encounter more frequently cases approximating the assumptions of the elliptic rather than the circular problem.

The elliptic problem has been discussed by Kopal and Lyttleton (1963), Szebehely and Giacaglia (1964), Hunter (1967) and Broucke (1969). There are also contributions which are related to the study of the stability of the motion in the equilibrium points.

Comparison between the circular and elliptic problem shows the existence of the following differences. The equations of motion of the former are autonomous while the ones of the latter are not. The elliptic problem does not admit to the energy integral. The periodic solutions of the circular problem can be grouped into families in which the period varies continuously, while the periodic orbits of the elliptic problem cannot be but of periods that are multiples of 2π. Thus, the periodic orbits of the circular case expressed in the synodical system of coordinates are not periodic with respect to the fixed frame, unless their period is commensurable to 2π, while periodic orbits of the elliptic problem are periodic in both systems.

In the elliptic problem the periodic orbits can be grouped into families on the basis of the parameters e (eccentricity of the primaries) and μ (mass of one of the primaries). Thus, if $\mathbf{x}_0$ is the initial state in phasespace of the orbit, appropriate selection can lead to a function

$$\mathbf{x}_0 = \mathbf{x}_0 (e, \mu),$$

such that $\mathbf{x}_0$ is an analytic function in both parameters. All members of a family have the same period, which in all cases must be a multiple of 2π. In the circular problem we suppose, without loss generality, that $\mathbf{x}_0 = \mathbf{x}(\mathbf{x}_0, 0)$. In the elliptic problem this is not valid. In reality this fact is used in order to distinguish between periodic orbits with initial states at perigee and periodic orbits with initial states at apogee.

B. D. Tapley and V. Szebehely (eds.), Recent Advances in Dynamical Astronomy, 118–134.

2. Equations of Motion and Variation

For a concise derivation of the elliptic problem the reader is referred to the paper by Kopal and Lyttleton (1963). These equations in a coordinate system (x, y, z), with (x, y) the plane of the two primaries, rotating with angular velocity $\boldsymbol{\omega} = \dot{v}\mathbf{k}$, with coordinates, time and masses dimensionless and with the variable distance r of the primaries as a unit of length, are as follows:

$$\begin{aligned}
\frac{d^2x}{dv^2} - 2\frac{dy}{dv} &= \frac{r}{p}\left[x - \frac{(1-\mu)(x+\mu)}{r_1^3} - \frac{\mu(x+\mu-1)}{r_2^3}\right] = \frac{r}{p}\frac{\partial\Omega}{\partial x}, \\
\frac{d^2y}{dv^2} + 2\frac{dx}{dv} &= \frac{r}{p}\left[y - \frac{(1-\mu)\,y}{r_1^3} - \frac{\mu y}{r_2^3}\right] = \frac{r}{p}\frac{\partial\Omega}{\partial y}, \\
\frac{d^2z}{dv^2} + z &= \frac{r}{p}\left[z - \frac{(1-\mu)\,z}{r_1^3} - \frac{\mu z}{r_2^3}\right] = \frac{r}{p}\frac{\partial\Omega}{\partial z}.
\end{aligned} \tag{1}$$

The parameters appearing are: v is the true anomaly of the primaries, $p = 1 - e^2$, where e is the eccentricity of the primaries,

$$r = (1 - e\cos E) = \frac{1-e^2}{1+e\cos v},$$

$$\Omega = (1-\mu)\left(\frac{1}{r_1} + \frac{r_1^2}{2}\right) + \mu\left(\frac{1}{r_2} + \frac{r_2^2}{2}\right).$$

and

$$r_1^2 = (x+\mu)^2 + y^2 + z^2, \qquad r_2^2 = (x+\mu-1)^2 + y^2 + z^2.$$

Let us introduce the notation

$$x = x_1, \qquad y = x_2, \qquad z = x_3, \qquad \frac{dx}{dv} = x_4, \qquad \frac{dy}{dv} = x_5, \qquad \frac{dz}{dv} = x_6.$$

Thus, Equations (1) become

$$\begin{aligned}
\frac{dx_1}{dv} &= x_4, \\
\frac{dx_2}{dv} &= x_5, \\
\frac{dx_3}{dv} &= x_6, \\
\frac{dx_4}{dv} &= 2x_5 + \frac{r}{p}\left[x_1 - \frac{(1-\mu)(x_1+\mu)}{r_1^3} - \frac{\mu(x_1+\mu-1)}{r_2^3}\right], \\
\frac{dx_5}{dv} &= -\,2x_4 + \frac{r}{p}\left[x_2 - \frac{(1-\mu)\,x_2}{r_1^3} - \frac{\mu x_2}{r_2^3}\right], \\
\frac{dx_6}{dv} &= -\,x_3 + \frac{r}{p}\left[x_3 - \frac{(1-\mu)\,x_3}{r_1^3} - \frac{\mu x_3}{r_2^3}\right].
\end{aligned} \tag{2}$$

If we denote by $\mathbf{x}$ the six-dimensional vector $(x_1, x_2, x_3, x_4, x_5, x_6)$ and by $\mathbf{f}$ the derivative of $\mathbf{x}$ with respect to the true anomaly υ; then the above equations become

$$\frac{\mathrm{d}\mathbf{x}}{\mathrm{d}\upsilon} = \mathbf{f}(\mathbf{x}, \upsilon). \tag{3}$$

We recall that υ is related to the time t by Kepler's second law:

$$r^2 \, \mathrm{d}\upsilon = (1 - e^2)^{1/2} \, \mathrm{d}t = p^{1/2} \, \mathrm{d}t, \tag{4}$$

Equation (3) makes obvious the non-autonomous character of the dynamical system in question due to the dependence of $\mathbf{f}$ on the independent variable υ.

The solutions of the Equation (3) are of the form

$$\mathbf{x} = \mathbf{x}(\mathbf{x}_0, \upsilon_0, \upsilon), \tag{5}$$

where

$$\mathbf{x}_0 = \mathbf{x}(\mathbf{x}_0, \upsilon_0, \upsilon_0). \tag{6}$$

The solutions (5) are analytic functions of the initial states $\mathbf{x}_0$ and υ_0. Hence, the quantities

$$\frac{\partial \mathbf{x}}{\partial x_{0j}}(\mathbf{x}_0, \upsilon_0, \upsilon), \frac{\partial \mathbf{x}}{\partial \upsilon_0} \tag{7}$$

exist and are also analytic functions of $\mathbf{x}_0$, and υ_0.

The eccentric anomaly E is related to the time t through Kepler's equation:

$$t + \varphi = E - e \sin E,$$

where φ is a phase constant.

For the starting point of the motion $t = 0$, we can take $\varphi = 0$ or $\varphi = \pi$ and consequently we can have two main categories of solutions with initial conditions

$$\mathbf{x}_0 = \mathbf{x}(\mathbf{x}_0; 0; 0), \tag{8}$$

or

$$\mathbf{x}_0 = \mathbf{x}(\mathbf{x}_0; \pi; \pi), \tag{9}$$

which correspond to the minimum and maximum separation of the two primaries.

From Equation (3) we derive the equations

$$\frac{\mathrm{d}}{\mathrm{d}\upsilon}\left(\frac{\partial \mathbf{x}}{\partial x_{0j}}\right) = \sum_{k=1}^{6} \frac{\partial \mathbf{f}}{\partial x_k} \frac{\partial x_k}{\partial x_{0j}}, \quad j = 1, 2, \ldots, 6. \tag{10}$$

Equations (10) constitute the variatonal system of the solutions with respect to their initial conditions and can be written under matrix form as follows:

$$\frac{\mathrm{d}\Delta(\mathbf{x}_0; \upsilon_0; \upsilon)}{\mathrm{d}\upsilon} = P(\mathbf{x}; \upsilon)\, \Delta(\mathbf{x}_0; \upsilon_0; \upsilon), \tag{11}$$

where

$$\Delta = \left(\frac{\partial x_i}{\partial x_{0j}}\right) \quad \text{and} \quad P = \left(\frac{\partial f_i}{\partial x_k}\right) \qquad i, j, k = 1, 2, \ldots, 6\,.$$

The analytic expressions of the elements of the matrix $P(\mathbf{x}, \upsilon)$ are:

$$\begin{aligned}
&\frac{\partial f_i}{\partial x_j} = 1\,, \quad j = i + 3\,, \quad i = 1, 2, 3\,,\\
&\frac{\partial f_i}{\partial x_j} = 0\,, \quad i = 1, 2, 3\,, \quad j = 1, 2, \ldots, 6\,, \quad j \neq i + 3\,,\\
&\frac{\partial f_4}{\partial x_5} = -\frac{\partial f_5}{\partial x_4} = 2\,,\\
&\frac{\partial f_4}{\partial x_4} = \frac{\partial f_4}{\partial x_6} = \frac{\partial f_5}{\partial x_5} = \frac{\partial f_5}{\partial x_6} = \frac{\partial f_6}{\partial x_4} = \frac{\partial f_6}{\partial x_6} = \frac{\partial f_6}{\partial x_5} = 0\,,\\
&\frac{\partial f_4}{\partial x_1} = \frac{r}{p}\left(1 - \frac{\partial^2 \Omega}{\partial x_1^2}\right), \quad \frac{\partial f_5}{\partial x_2} = \frac{r}{p}\left(1 - \frac{\partial^2 \Omega}{\partial x_2^2}\right)\\
&\frac{\partial f_4}{\partial x_2} = -\frac{r}{p}\frac{\partial^2 \Omega}{\partial x_1\,\partial x_2} = \frac{\partial f_5}{\partial x_1}, \quad \frac{\partial f_5}{\partial x_3} = -\frac{r}{p}\frac{\partial^2 \Omega}{\partial x_2\,\partial x_3} = \frac{\partial f_6}{\partial x_2}\\
&\frac{\partial f_4}{\partial x_3} = -\frac{r}{p}\frac{\partial^2 \Omega}{\partial x_1\,\partial x_3} = \frac{\partial f_6}{\partial x_1}, \quad \frac{\partial f_6}{\partial x_3} = -1 + \frac{r}{p}\left(1 - \frac{\partial^2 \Omega}{\partial x_3^2}\right).
\end{aligned} \tag{12}$$

It is easily found that

$$\frac{\partial}{\partial \upsilon} \det \Delta\,(\mathbf{x}_0;\, \upsilon_0;\, \upsilon) = \left[\det \Delta\,(\mathbf{x}_0;\, \upsilon_0;\, \upsilon)\right] \cdot \boldsymbol{\nabla} \cdot \mathbf{f}\,. \tag{13}$$

But it is

$$\boldsymbol{\nabla} \cdot \mathbf{f} = 0\,. \tag{14}$$

Therefore,

$$\frac{\partial}{\partial \upsilon} \det \Delta\,(\mathbf{x}_0;\, \upsilon_0;\, \upsilon) = 0\,,$$

or

$$\det \Delta\,(\mathbf{x}_0;\, \upsilon_0;\, \upsilon) = \text{const}\,. \tag{15}$$

Finally because $\partial x_i/\partial x_{0j} = \delta_{ij}$ for $\upsilon = \upsilon_0$, we find that

$$\det \Delta\,(\mathbf{x}_0;\, \upsilon_0;\, \upsilon) = 1\,. \tag{16}$$

This result shows that the volume in the phase space of the elliptic problem is conserved precisely as it is conserved in the circular case. This is so since the velocity-like (derivative with respect to the true anomaly) has zero divergence (see Equation (14)).

3. Symmetric Periodic Orbits

To the differential equations of the motion we apply the transformation

$$\mathbf{y} = L\mathbf{x} \qquad \upsilon_1 = -\upsilon, \tag{17}$$

where $L=(l_{ij})i, j=1, 2, \ldots, 6$ and $l_{2i-1,2i-1}=-l_{2i,2i}=1$, $i=1, 2, 3$, and $l_{ij}=0$ $i \neq j$. We thus obtain

$$\frac{\mathrm{d}(L\mathbf{y})}{\mathrm{d}(-\upsilon_1)} = \mathbf{f}(L\mathbf{y}, -\upsilon_1),$$

or

$$-L\frac{\mathrm{d}\mathbf{y}}{\mathrm{d}\upsilon_1} = -L\mathbf{f}(\mathbf{y}, \upsilon_1).$$

or finally

$$\frac{\mathrm{d}\mathbf{y}}{\mathrm{d}\upsilon_1} = \mathbf{f}(\mathbf{y}, \upsilon_1). \tag{18}$$

The differential Equations (3) and (18) are of the same form and consequently they give the same solutions for the same initial conditions. The initial conditions will be the same if, for $\upsilon=\upsilon_0$, they fulfill the relation

$$L\mathbf{x}_0 = \mathbf{x}_0. \tag{19}$$

This implies that

$$x_{02} = x_{04} = x_{06} = 0. \tag{20}$$

Due to the fact that $\upsilon=-\upsilon_1$, initial states satisfying condition (20) produce symmetric arcs with respect to the $0x_1x_3$ plane.

From the above, we can conclude that whenever the conditions $x_2=x_4=x_6=0$ are satisfied for two different points of the same solution and if the difference in the values of υ corresponding to these two points is a multiple of π, then this solution will be periodic in both rotating and non-rotating coordinate systems. Consequently, considering that for $t=0$ the conditions (20) are fulfilled, the conditions of periodicity of symmetric orbits, with respect to plane $0x_1x_3$, are the following:

$$\begin{aligned} x_2(x_{01}, x_{03}, x_{05}; \upsilon_0; e; \upsilon)_{\upsilon=\upsilon_0+k\pi} &= 0 \\ x_4(x_{01}, x_{03}, x_{05}; \upsilon_0; e; \upsilon)_{\upsilon=\upsilon_0+k\pi} &= 0 \\ x_6(x_{01}, x_{03}, x_{05}; \upsilon_0; e; \upsilon)_{\upsilon=\upsilon_0+k\pi} &= 0. \end{aligned} \tag{21}$$

4. Properties of the Jacobian Matrix

(1) The variational equation

$$\frac{\mathrm{d}\Delta(\mathbf{x}_0; \upsilon_0; \upsilon)}{\mathrm{d}\upsilon} = P(\mathbf{x}; \upsilon)\cdot\Delta(\mathbf{x}; \upsilon_0; \upsilon), \tag{22}$$

when $\mathbf{x}_0$ is the initial condition of any periodic solution, with period $T=2k\pi$ gives that

$$\frac{\mathrm{d}\Delta(\mathbf{x}_0; \upsilon_0; \upsilon+T)}{\mathrm{d}\upsilon} = P(\mathbf{x}; \upsilon + T)\cdot\Delta(x_0; \upsilon_0; \upsilon + T). \tag{23}$$

The matrix P depends on the position of the third body in phase space, which is a periodic function of υ and on the coefficient r/p which is also a periodic function in υ with period 2π. Hence, we have

$$P(\mathbf{x}; \upsilon + T) = P(\mathbf{x}; \upsilon),$$

and the Equation (23) is written:

$$\frac{\mathrm{d}\Delta(\mathbf{x}_0; \upsilon_0; \upsilon + T)}{\mathrm{d}\upsilon} = P(\mathbf{x}; \upsilon)\cdot\Delta(\mathbf{x}_0; \upsilon_0; \upsilon + T). \tag{24}$$

From Equations (22) and (24) we deduce that

$$\Delta(\mathbf{x}_0; \upsilon_0; \upsilon + T) = \Delta(\mathbf{x}_0; \upsilon_0; \upsilon)\, C. \tag{25}$$

We can have the value of the constant matrix C by putting $\upsilon=\upsilon_0$ in relation (25). For the case $\upsilon_0=0$ we receive

$$\Delta(\mathbf{x}_0; 0; \upsilon + T) = \Delta(\mathbf{x}_0; 0; \upsilon)\cdot\Delta(\mathbf{x}_0; 0; T), \tag{26}$$

while for $\upsilon=T$ the above relation becomes

$$\Delta(\mathbf{x}_0; 0; 2T) = \Delta^2(\mathbf{x}_0; 0; T).$$

After n periods we shall have

$$\Delta(\mathbf{x}_0; 0; nT) = \Delta^n(x_0; 0; T).$$

(2) It is shown (Bray-Goudas, 1967; Katsiaris-Goudas, 1970) that for the circular restricted problem the matrix $\Delta(\mathbf{x}_0; t)$ is symplectic i.e. it fulfills the relation

$$\Delta'(\mathbf{x}_0; t)\, Q\Delta(\mathbf{x}_0; t) = Q,$$

where $Q=(q_{ij})$, and $q_{1,2}=-q_{2,1}=-2$, $q_{i,i+3}=1$, $q_{i+3,i}=-1$ $i=1, 2, 3$ and $q_{i,j}=0$ for all other combinations of the suffixes.

The above property is also valid in the case of the elliptic restricted problem, i.e. the matrix $\Delta(\mathbf{x}_0; \upsilon_0; \upsilon)$ fulfillas, for every υ, the relation

$$\Delta'(x_0; \upsilon_0; \upsilon)\, Q\Delta(x_0; \upsilon_0; \upsilon) = Q. \tag{27}$$

The proof of relation (27) is based on the fact that the relation

$$P'(\mathbf{x}; \upsilon)\, Q + QP(\mathbf{x}; \upsilon) = 0,$$

is valid for every υ.

The significance of the property $\Delta' Q\Delta = Q$ is that it allows the conclusion that the matrices Δ, Δ', Δ^{-1} have the same eigenvalues. Thus if λ is an eigenvalue of Δ, so is λ^{-1}. Zero eigenvalues are excluded because the determinants of both the matrices Δ and

Δ^{-1} are equal to unity, as proved by relation (16). Unit eigenvalues also do not exist, in general, in the elliptic problem because of the non-existence of the energy integral.

(3) The transformation $\mathbf{y}=L\mathbf{x}$, $v_1=-v$, in the differential equation of motion (3) implies the transformation

$$Y = L\Delta L\,, \quad v_1 = -v\,, \tag{28}$$

where $Y=(\partial y_i/\partial y_{0j})$, $\Delta=(\partial x_i/\partial x_{0j})$, for the variational equations. Therefore, the differential Equations (3) and (11) receive the form

$$\frac{\mathrm{d}\mathbf{y}}{\mathrm{d}v_1} = \mathbf{f}(\mathbf{y}; v_1)\,, \qquad \frac{\mathrm{d}Y(\mathbf{y}_0, v_1)}{\mathrm{d}v_1} = P(\mathbf{y}, v_1)\, Y(\mathbf{y}_0, v_1)\,. \tag{29}$$

Thus, the above transformation leaves Equations (2) and (11) unchanged. This means that for the initial conditions which fulfill the relation $L\mathbf{x}_0=\mathbf{x}_0$, for $v_1=T/2$ we have

$$Y\left(\frac{T}{2}\right) = \Delta\left(-\frac{T}{2}\right). \tag{30}$$

Combining the relations (28), (30) and the relation

$$\Delta\left(\frac{T}{2}\right) = \Delta\left(-\frac{T}{2}\right)\Delta(T)\,,$$

which results from the (26) for $v=-T/2$, we obtain the relation

$$\Delta(T) = L\Delta^{-1}\left(\frac{T}{2}\right)L\Delta\left(\frac{T}{2}\right). \tag{31}$$

This relation is a characteristic property of the periodic orbits which are symmetric with respect to the $0x_1x_3$ plane. According to relation (31) the Jacobian matrix $\Delta(\mathbf{x}_0; v_0; v_0+T)$ can be computed by integration of the variational equations for only half of the period.

(4) For $v=T/2$, relation (27) becomes

$$Q^{-1}\Delta'\left(\frac{T}{2}\right)Q = \Delta^{-1}\left(\frac{T}{2}\right). \tag{32}$$

Because of (32) relation (31) becomes

$$\Delta(T) = LQ^{-1}\Delta'\left(\frac{T}{2}\right)QL\Delta\left(\frac{T}{2}\right)$$

or

$$QL\Delta(T) = \Delta'\left(\frac{T}{2}\right)QL\Delta\left(\frac{T}{2}\right). \tag{33}$$

The right side of (33) is a symmetic matrix. Indeed

$$(\Delta'QL\Delta)' = \Delta'L'Q'\Delta = \Delta'LQ'\Delta = \Delta'QL\Delta$$

because

$$LQ' = QL.$$

Thus, matrix $QL\Delta(T)$ is symmetric and by equating the corresponding elements of the $QL\Delta(T)$ we obtain 15 linear relations among the elements of the Jacobian. These relations are useful for the check of the integration program and of the accuracy of the numerical results.

5. Stability of the Solutions

As usual, the study of the stability of the solutions will be based on the eigenvalues of the Jacobian matrix $\Delta(\mathbf{x}_0, \upsilon_0, \upsilon_0+T)$. The characteristic equation

$$|\Delta(\mathbf{x}_0; \upsilon_0; \upsilon_0 + T) - \lambda I| = 0, \tag{34}$$

due to the properties of the last paragraph, will be of the form

$$\lambda^6 + \alpha_1\lambda^5 + \alpha_2\lambda^4 + a_3\lambda^3 + \alpha_2\lambda^2 + \alpha_1\lambda + 1 = 0. \tag{35}$$

Let the roots of (35) be λ_1, $1/\lambda_1$, λ_2, $1/\lambda_2$, λ_3, $1/\lambda_3$. Using the definition

$$-k_1 = \lambda_1 + \frac{1}{\lambda_1}, \qquad -k_2 = \lambda_2 + \frac{1}{\lambda_2}, \qquad -k_3 = \lambda_3 + \frac{1}{\lambda_3},$$

the characteristic roots will be found by solving the equations:

$$\lambda^2 + k_1\lambda + 1 = 0, \qquad \lambda^2 + k_2\lambda + 1 = 0, \qquad \lambda^2 + k_3\lambda + 1 = 0.$$

If we take the product of the above equations and compare it with Equation (35), we find the relations

$$\begin{aligned} k_1 + k_2 + k_3 &= \alpha_1, \\ k_1k_2 + k_2k_3 + k_3k_1 &= \alpha_2 - 3, \\ k_1k_2k_3 &= \alpha_3 - 2\alpha_1. \end{aligned} \tag{36}$$

Hence, the quantities k_1, k_2, k_3 will be roots of the cubic equation

$$z^3 - \alpha_1 z^2 + (\alpha_2 - 3)\, z - (\alpha_3 - 2\alpha_1) = 0. \tag{37}$$

Because the coefficients of the Equation (37) are real, its roots will be either three real or one real and two complex conjugates.

(a) If k_1, k_2, k_3 are real numbers, in order to have stability the relations

$$|k_1| \leqq 2, \qquad |k_2| \leqq 2, \qquad |k_3| \leqq 2, \tag{38}$$

must be valid simultaneously. In this case we shall have six characteristic roots of unit moduli.

(b) If k_1 is real and k_2, k_3 are complex conjugates we have unstable periodic orbits. Indeed, let the $\lambda^2 + k_2\lambda + 1 = 0$ have the roots $\alpha + bi$ and $1/\alpha + bi$. These roots will have moduli reciprocal numbers and if $|\alpha + bi| < 1$ then$|\ 1/\alpha + bi| > 1$.

The computation of the eigenvalues of $\Delta(\mathbf{x}_0, \upsilon_0, \upsilon)$ proceeds as follows: Let the characteristic equation of the matrix $A=(\alpha_{ij})$ of order n, be of the form

$$\lambda^n + c_1\lambda^{n-1} + c_2\lambda^{n-2} + \cdots + c_{n-1}\lambda + c_n = 0,$$

and let us denote by S_i the traces of the powers of the matrix A, i.e.

$$S_i = T_r(A^i) \quad i = 1, 2, \ldots, n.$$

The coefficients c_i of the characteristic equation and the quantities S_i are then connected by the relations

$$\mu c_\mu = - \sum_{i=1}^{\mu} c_{\mu-i} S_i, \qquad \mu = 1, 2, \ldots, n,$$

while $c_0=1$.

Thus, the computation of the coefficients of the characteristic equation of a matrix is reduced to the computation of the traces of the matrices A^i, $i=1, 2, \ldots, n$.

In the present case the computation of Δ, Δ^2, Δ^3 is enough, because the characteristic Equation (35) is a reciprocal one. But in the present work we computed all the coefficients in order to check the program and the precision of the numerical computations.

The solutions of the Equation (37) are

$$k_1 = (S_1 + S_2) + \frac{\alpha_1}{3},$$

$$k_2 = -\frac{1}{2}(S_1 + S_2) + \frac{\alpha_1}{3} + \frac{i\sqrt{3}}{2}(S_1 - S_2),$$

$$k_3 = -\tfrac{1}{2}(S_1 + S_2) + \frac{\alpha_1}{3} - \frac{i\sqrt{3}}{2}(S_1 - S_2)$$

where

$$S_1 = (r + \sqrt{q^3 + r^2})^{1/3}, \qquad S_2 = (r - \sqrt{q^3 + r^2})^{1/3} \tag{39}$$

and

$$q = \tfrac{1}{3}\alpha_2 - 1 - \tfrac{1}{9}\alpha_1^2,$$

$$r = \tfrac{1}{2}(\alpha_3 - \alpha_1) - \tfrac{1}{6}\alpha_1\alpha_2 + \tfrac{1}{27}\alpha_1^3.$$

The solutions (39) will be real and different if $q^3+r^2<0$, or real with at least two of them equal if $q^3+r^2=0$, or one real and two complex conjugates if $q^3+r^2>0$.

The stability of the periodic orbits depends on the coefficients α_1, α_2, α_3 of the characteristic equation, which are called stability coefficients. The coefficients α_1, α_2, α_3 determine a three dimensional space in which we can distinguish 12 stable-unstable regions, considering the character of the roots k_1, k_2, k_3 and their position with respect to the numbers -2 and 2, in the case they receive real values.

Geometrically the stable-unstable regions can be determined from the surfaces which the conditions

$$q^3 + r^2 = 0, \quad k_1^2 - 4 = 0, \quad k_2^2 - 4 = 0, \quad k_3^2 - 4 = 0, \tag{40}$$

represent. The first of surfaces (40) divides the space into the semi-spaces $q^3+r^2>0$ (unstable periodic orbits) and $q^3+r^2<0$ (probably stable periodic orbits) and its analytic expression is

$$4\alpha_2^3 - \alpha_1^2\alpha_2^2 - 19\alpha_1^2 + 42\alpha_1^2\alpha_2 - 8\alpha_1^4 + 108\alpha_2 - 108 - 36\alpha_2^2 + 27\alpha_3^2 - \\ - 108\alpha_1\alpha_2\alpha_3 - 54\alpha_1\alpha_3 + 4\alpha_1^3\alpha_3 = 0. \tag{41}$$

Equation (41) represent a symmetric surface with respect to the α_2 axis, and crosses the α_2 axis at the points ± 2, the α_2 axis at the point $+3$ while it does not have any common point with the α_1 axis.

The origin of the axes $(\alpha_1, \alpha_2, \alpha_3)$ belongs to the stable region.

The second of the (41) is written

$$2 + 2\alpha_2 = \pm (2\alpha_1 + \alpha_3),$$

and represents two planes, each of which is symmetric to the other with respect to the α_2 axis.

6. Computation of Periodic Solutions

The differential equations of motion and variation are integrated numerically in order to compute periodic orbits. The integration of the equations of variation is required (a) in the evaluation of the differential corrections for the convergence to a periodic orbit and of the changes along each family of periodic orbits, (b) in the stability study of the solutions and (c) in the checking of the program and of the precision of the numerical computations.

The sufficient condition for periodicity of symmetric solutions with respect to the $0x_1x_3$ plane is that the orbit should have two perpendicular crossings of the plane $0x_1x_3$ at epochs of apsides. The above criterion is a sufficient and not a necessary condition of periodicity because periodic orbits of other kinds of symmetry, or even non-symmetric, may also exist.

We consider as the initial state of the solution the position vector $\mathbf{x}_0$ which for $\upsilon = \upsilon_0$ has coordinates $(x_{01}, 0, x_{03}, 0, x_{05}, 0)$. The numerical integration is terminated when $\upsilon = \upsilon_0 + (T/2) = \upsilon_0 + k\pi$. At that moment the third body must cross the plane $0x_1x_3$ perpendicularly.

If one of the above conditions is not fulfilled the correction which we must apply to the initial conditions $\mathbf{x}_0$, in a way that the new initial conditions $\mathbf{x}_0 + \delta\mathbf{x}_0$ will correspond to a symmetric periodic orbit, can be found by solving the system:

$$\begin{aligned} x_2(x_{01} + \delta x_{01}, x_{03} + \delta x_{03}, x_{05} + \delta x_{05}; \upsilon_0 + k\pi) &= 0, \\ x_4(x_{01} + \delta x_{01}, x_{03} + \delta x_{03}, x_{05} + \delta x_{05}; \upsilon_0 + k\pi) &= 0, \\ x_6(x_{01} + \delta x_{01}, x_{03} + \delta x_{03}, x_{05} + \delta x_{05}; \upsilon_0 + k\pi) &= 0. \end{aligned} \tag{42}$$

By linearization Equations (42) become

$$\begin{aligned}
&\frac{\partial x_2}{\partial x_{01}}\,\delta x_{01} + \frac{\partial x_2}{\partial x_{03}}\,\delta x_{03} + \frac{\partial x_2}{\partial x_{05}}\,\delta x_{05} = - x_2(x_{01}, x_{03}, x_{05}),\\
&\frac{\partial x_4}{\partial x_{01}}\,\delta x_{01} + \frac{\partial x_4}{\partial x_{03}}\,\delta x_{03} + \frac{\partial x_4}{\partial x_{05}}\,\delta x_{05} = - x_4(x_{01}, x_{03}, x_{05}),\\
&\frac{\partial x_6}{\partial x_{01}}\,\delta x_{01} + \frac{\partial x_6}{\partial x_{03}}\,\delta x_{03} + \frac{\partial x_6}{\partial x_{05}}\,\delta x_{05} = - x_6(x_{01}, x_{03}, x_{05}).
\end{aligned} \tag{43}$$

The values of the coefficients and of the right hand sides of system (43), correspond to the value $v_0 + k\pi$ of the independent variable. The linear system (43) is solved for the corrections δx_{01}, δx_{03}, and δx_{05}. The entire process is then repeated using $x_{01} + \delta x_{01}$, $x_{03} + \delta x_{03}$, $x_{05} + \delta x_{05}$ as new initial conditions. A number of iterations leads to the fulfillment of the final condition,

$$(x_2^2 + x_4^2 + x_6^2)^{1/2} \leqslant \varepsilon, \tag{44}$$

where ε is suitably defined (in the present work the value 10^{-7} was employed). Condition (44) is sufficient for periodicity and symmetry of the orbit regarding the kind of initial conditions selected.

In computing periodic orbits of the elliptic problem we used as a starting point the orbits already known (Katsiaris, 1970) providing their periods are approximately $2k\pi$. For $e \neq 0$ and $\mu = \text{constant}$, the condition for periodicity and symmetry with respect to the plane $0x_1x_3$ are:

$$\begin{aligned}
&x_2(x_{01}, x_{03}, x_{05};\, e;\, v_0;\, v_0 + k\pi) = 0,\\
&x_4(x_{01}, x_{03}, x_{05};\, e;\, v_0;\, v_0 + k\pi) = 0,\\
&x_6(x_{01}, x_{03}, x_{05};\, e;\, v_0;\, v_0 + k\pi) = 0.
\end{aligned} \tag{45}$$

Orbits satisfying these conditions for e varying continuously and $k = \text{constant}$, constitute a family of periodic orbits. Once, a member of the family is found on the basis of the techniques already described, we proceed in the determination of a sufficient number of members of the same family as follows: First the system of Equations (45) is linearized in terms of the increments of the quantities x_{01}, x_{03}, x_{05} and e of the periodic orbit already known. Thus the following system is found:

$$\begin{aligned}
&\frac{\partial x_2}{\partial x_{01}}\,\delta x_{01} + \frac{\partial x_2}{\partial x_{03}}\,\delta x_{03} + \frac{\partial x_2}{\partial x_{05}}\,\delta x_{05} = - \frac{\partial x_2}{\partial e}\,\delta e,\\
&\frac{\partial x_4}{\partial x_{01}}\,\delta x_{01} + \frac{\partial x_4}{\partial x_{03}}\,\delta x_{03} + \frac{\partial x_4}{\partial x_{05}}\,\delta x_{05} = - \frac{\partial x_4}{\partial e}\,\delta e,\\
&\frac{\partial x_6}{\partial x_{01}}\,\delta x_{01} + \frac{\partial x_6}{\partial x_{03}}\,\delta x_{03} + \frac{\partial x_6}{\partial x_{05}}\,\delta x_{05} = - \frac{\partial x_6}{\partial e}\,\delta e.
\end{aligned} \tag{46}$$

Then the linear system (46) is solved for the increments δx_{01}, δx_{03}, δx_{05}, while the increment δe is given an arbitrary but small value. Keeping the new value of e fixed and using $x_{01}+\delta x_{01}$, $x_{03}+\delta x_{03}$, $x_{05}+\delta x_{05}$ as our new initial conditions we apply the technique already explained in order to obtain the member of the family corresponding to the new value of the eccentricity. The same procedure is used in order to obtain as many members of the family as are required.

It should be added here that the quantities $\partial x_i/\partial e$, required for the solution of system (46), are obtained by integrating the equation of variation of the problem with respect to the parameter e, i.e. the equation

$$\frac{\mathrm{d}\mathbf{E}}{\mathrm{d}\upsilon} = P\mathbf{E} + \mathbf{P}_1, \tag{47}$$

where

$$\mathbf{E} = \left(\frac{\partial x_i}{\partial e}\right), \quad P = \left(\frac{\partial f_i}{\partial x_j}\right),$$

and

$$\mathbf{P}_1 = \left(\frac{\partial f_i}{\partial e}\right) \quad i, j = 1, 2, \ldots, 6.$$

For $\upsilon = \upsilon_0$ we have $\mathbf{E} = \mathbf{0}$.

The families of periodic orbits can be also defined from initial conditions corresponding to the same value of the period and eccentricity while the parameter varying continuously is the mass μ of the smaller of the primaries. Such results will be given in the next paragraph. The equations used in this case for the estimation of the increments are:

$$\begin{aligned}
&\frac{\partial x_2}{\partial x_{01}}\,\delta x_{01} + \frac{\partial x_2}{\partial x_{03}}\,\delta x_{03} + \frac{\partial x_2}{\partial x_{05}}\,\delta x_{05} = -\frac{\partial x_2}{\partial \mu}\,\delta\mu,\\
&\frac{\partial x_4}{\partial x_{01}}\,\delta x_{01} + \frac{\partial x_4}{\partial x_{03}}\,\delta x_{03} + \frac{\partial x_4}{\partial x_{05}}\,\delta x_{05} = -\frac{\partial x_4}{\partial \mu}\,\delta\mu,\\
&\frac{\partial x_6}{\partial x_{01}}\,\delta x_{01} + \frac{\partial x_4}{\partial x_{03}}\,\delta x_{03} + \frac{\partial x_6}{\partial x_{05}}\,\delta x_{05} = -\frac{\partial x_6}{\partial \mu}\,\delta\mu.
\end{aligned} \tag{48}$$

In this case e is kept constant and μ is varied in steps. The equation of variation with respect to μ is

$$\frac{\mathrm{d}}{\mathrm{d}\upsilon}\left(\frac{\partial \mathbf{x}}{\partial \mu}\right) = \sum_{i=1}^{6} \frac{\partial \mathbf{f}}{\partial x_k}\frac{\partial x_k}{\partial \mu} + \frac{\partial \mathbf{f}}{\partial \mu}, \tag{49}$$

and of course $\partial \mathbf{x}/\partial\mu = \mathbf{0}$ for $\upsilon = \upsilon_0$.

7. Numerical Results

The computer program prepared for the calculations described in the previous paragraphs included variations of the initial conditions and of the parameters e and μ. One of our first results is that the continuation method of Poincaré can be applied in many cases for the parameter e. Indeed it was found that

$$\det\left(\frac{x_{2i}}{x_{0,\,2i-1}}\right) \neq 0, \qquad i = 1, 2, 3, \tag{50}$$

computed for $e=0$ and at $\upsilon=\upsilon_0+k\pi$. This means that the system (46) can be solved and that periodic orbits of the same period for $e\neq 0$ exist.

For both practical and error checking reasons the Equations (3), (11), (47) and (49) were integrated together with the equations

$$\frac{\mathrm{d}r}{\mathrm{d}\upsilon} = \frac{r^2 e \sin\upsilon}{p}, \tag{51}$$

$$\frac{\mathrm{d}t}{\mathrm{d}\upsilon} = \frac{r^2}{p^{1/2}}. \tag{52}$$

The integrals of both are known but their direct integration is easy and very useful because comparison of the results can give an estimate of the accuracy achieved.

The results listed in Table I contain a few members of a family of orbits obtained by application of the continuation method using as starting point the periodic orbit

TABLE I

	x_{01}	x_{03}	x_{05}	x_1	x_3	x_5	e
1.	−0.0771845	0.7318070	1.122952	0.0550662	0.7323378	−1.121226	0.00000
2.	−0.0767029	0.7332254	1.120385	0.0544446	0.0318542	−1.121252	0.00125
3.	−0.0762219	0.7346401	1.117834	0.0538265	0.7313682	−1.121295	0.00250
4.	−0.0757415	0.7360511	1.115298	0.0532119	0.7308679	−1.121353	0.00375
5.	−0.0752617	0.7374583	1.112779	0.0526008	0.7303652	−1.121427	0.00500
6.	−0.0743038	0.7402611	1.107787	0.0513892	0.7293408	−1.121623	0.00750
7.	−0.0733485	0.7430485	1.102859	0.0501916	0.7282910	−1.121883	0.01000
8.	−0.0723959	0.7458202	1.097993	0.0490080	0.7272161	−1.122208	0.01250
9.	−0.0714461	0.7485762	1.093190	0.0478385	0.7261162	−1.122598	0.01500
10.	−0.0704992	0.7513161	1.088450	0.0466828	0.7249915	−1.123053	0.01750
11.	−0.0695555	0.7540398	1.083771	0.0455413	0.7238420	−1.123574	0.02000
12.	−0.0686151	0.7567472	1.079155	0.0444138	0.7226680	−1.124162	0.02250
13.	−0.0676781	0.7594381	1.074600	0.0433004	0.7214695	−1.124816	0.02500
14.	−0.0667447	0.7621123	1.070107	0.0422010	0.7202469	−1.125538	0.02750
15.	−0.0658150	0.7647698	1.065675	0.0411156	0.7190003	−1.126327	0.03000
16.	−0.0639680	0.7700340	1.056993	0.0389874	0.7164357	−1.128111	0.03500
17.	−0.0621392	0.7752295	1.048553	0.0369167	0.7137773	−1.130171	0.04000
18.	−0.0603290	0.7803558	1.040352	0.0349024	0.7110269	−1.132510	0.04500
19.	−0.0585400	0.7854119	1.032389	0.0329459	0.7081863	−1.135131	0.00500
20.	−0.0567742	0.7903972	1.024661	0.0310476	0.7052572	−1.138038	0.05500
21.	−0.0550338	0.7953114	1.017166	0.0292080	0.7022415	−1.141234	0.06000
22.	−0.0533210	0.8001538	1.009903	0.0274279	0.6991414	−1.144722	0.06500
23.	−0.0516382	0.8049243	1.002871	0.0257080	0.6959587	−1.148504	0.07000
24.	−0.0499879	0.8096227	0.9960663	0.0240489	0.6926957	−1.152585	0.07500
25.	−0.0483729	0.8142487	0.9894885	0.0224515	0.6893544	−1.156967	0.08000
26.	−0.0467958	0.8188024	0.9831359	0.0209165	0.6859370	−1.161653	0.08500
27.	−0.0452595	0.8232838	0.9770071	0.0194447	0.6824457	−1.166646	0.09000

Table 1 (continued)

28.	−0.0437670	0.8276930	0.9711006	0.0180371	0.6788828	−1.171948	0.09500
29.	−0.0423213	0.8320303	0.9654152	0.0166943	0.6752506	−1.177560	0.10000
30.	−0.0409255	0.8362960	0.9599498	0.0154172	0.6715513	−1.183491	0.10500
31.	−0.0395821	0.8404905	0.9547023	0.0142059	0.6677873	−1.189737	0.11000
32.	−0.0370693	0.8486669	0.9448620	0.0119870	0.6600741	−1.203193	0.12000
33.	−0.0348085	0.8565636	0.9358862	0.0100423	0.6521295	−1.217947	0.13000
34.	−0.0328268	0.8641853	0.9277686	0.0083764	0.6439718	−1.234019	0.14000
35.	−0.0311512	0.8715373	0.9205040	0.0069927	0.6356193	−1.251425	0.15000
36.	−0.0298083	0.8786254	0.9140875	0.0058931	0.6270895	−1.270179	0.16000
37.	−0.0288239	0.8854555	0.9085144	0.0050782	0.6183999	−1.290295	0.17000
38.	−0.0282229	0.8920336	0.9037798	0.0045467	0.6095669	−1.311787	0.18000
39.	−0.0280284	0.8983658	0.8998785	0.0042953	0.6006065	−1.334664	0.19000
40.	−0.0282613	0.9044578	0.8968043	0.0043191	0.5915339	−1.358936	0.2000
41.	−0.0289406	0.9103150	0.8945499	0.0046108	0.5823634	−1.384614	0.2100
42.	−0.0300822	0.9159423	0.8931064	0.0051614	0.5731084	−1.411703	0.2200
43.	−0.0316990	0.9213441	0.8924630	0.0059595	0.5637815	−1.440211	0.2300
44.	−0.0338003	0.9265241	0.8926067	0.0069921	0.5543943	−1.470146	0.2400
45.	−0.0347549	0.9362302	0.8951904	0.0096981	0.5354812	−1.534316	0.2600
46.	−0.0471037	0.9450773	0.9006979	0.0131376	0.5164451	−1.604268	0.2800
47.	−0.0566132	0.9530728	0.9089199	0.0171453	0.4973502	−1.680072	0.3000
48.	−0.0578627	0.9602190	0.9195948	0.0215391	0.4782514	−1.761831	0.3200
49.	−0.0806455	0.9665178	0.9324108	0.0261287	0.4591970	−1.849693	0.3400
50.	−0.0946979	0.9719764	0.9470130	0.0307239	0.4402316	−1.943877	0.3600
51.	−0.1097106	0.9766131	0.9630133	0.0351437	0.4213984	−2.044684	0.3800
52.	−0.1253423	0.9804618	0.9800024	0.0392232	0.4146299	−2.152521	0.4000

TABLE II

	x_{01}	x_{03}	x_{05}	x_1	x_3	x_5	μ
1.	−0.0414174	0.8321524	0.9673756	0.0190874	0.6763086	−1.1777372	0.01
2.	−0.0500108	0.8308283	0.9489411	−0.0033826	0.6662738	−1.1763099	0.03
3.	−0.0588288	0.8290922	0.9307337	−0.0255967	0.6561579	−1.1753461	0.05
4.	−0.0679606	0.8269144	0.9128603	−0.0475581	0.6460321	−1.1747183	0.07
5.	−0.0774940	0.8242697	0.8954266	−0.0692750	0.6359726	−1.1742742	0.09
6.	−0.0875133	0.8211381	0.8785345	−0.0907587	0.6260567	−1.1738436	0.11
7.	−0.0980957	0.8175058	0.8622780	−0.1120231	0.6163591	−1.1732472	0.13
8.	−0.1093076	0.8133659	0.8467381	−0.1330832	0.6069649	−1.1723067	0.15

TABLE III

	x_{01}	x_{03}	x_{05}	x_1	x_3	x_5	μ
1.	−0.0282613	0.9044578	0.8968043	0.0043191	0.5915339	−1.3589367	0.01212
2.	−0.0284418	0.9049755	0.8823298	−0.0075434	0.5826946	−1.3680001	0.02212
3.	−0.0287568	0.9052466	0.8679784	−0.0193471	0.5739727	−1.3771336	0.03212
4.	−0.0291221	0.9052840	0.8567825	−0.0286092	0.5672105	−1.3843073	0.04000
5.	−0.0297604	0.9051068	0.8427489	−0.0403158	0.5588017	−1.3932900	0.05000
6.	−0.0306209	0.9046796	0.8289426	−0.0519720	0.5506177	−1.4020404	0.06000
7.	−0.0331053	0.9030921	0.8021321	−0.0751398	0.5350340	−1.4184684	0.08000
8.	−0.0367347	0.9005731	0.7765517	−0.0981207	0.5206212	−1.4329453	0.10000

TABLE IV

	x_{01}	x_{03}	x_{05}	x_1	x_3	x_5	e
1.	−0.0679606	0.8269144	0.9128603	−0.0475581	0.6460321	−1.1747183	0.10
2.	−0.0580853	0.8450612	0.8859946	−0.0529557	0.6267038	−1.2092522	0.12
3.	−0.0493210	0.8617560	0.8629497	−0.0571871	0.6064036	−1.2502580	0.14
4.	−0.0418846	0.8770751	0.8436138	−0.0603023	0.5854405	−129.76151	016.
5.	−0.0359869	0.8911199	0.8278304	−0.0623918	0.5641122	−1.2311081	0.18
6.	−0.0317286	0.9040061	0.8153946	−0.0635796	0.5426875	−1.4104621	0.20
7.	−0.0291413	0.9158533	0.8060567	−0.0640114	0.5213933	−1.4753856	0.22
8.	−0.0281849	0.9267761	0.7995318	−0.0638420	0.5004087	−1.5456131	0.24
9.	−0.0287596	0.9368794	0.7955124	−0.0632238	0.4798644	−1.6209402	0.26
10.	−0.0307210	0.9462551	0.7936815	−0.0622987	0.4598471	−1.7012491	0.28
11.	−0.0339844	0.9549821	0.7937228	−0.0611927	0.4404068	−1.7865230	0.30

TABLE V

	x_{01}	x_{03}	x_{05}	x_1	x_3	x_5	e
1.	−0.1093076	0.8133659	0.8467381	−0.1330832	0.6069469	−1.1723067	0.10
2.	−0.0946507	0.8322634	0.8129966	−0.1392904	0.5834762	−1.2176281	0.12
3.	−0.0814352	0.8492197	0.7836991	−0.1444150	0.5596422	−1.2695059	0.14
4.	−0.0698126	0.8644481	0.7585039	−0.1486035	0.5359139	−1.3272493	0.16
5.	−0.0598155	0.8781895	0.7369999	−0.1520068	0.5126418	−1.3902303	0.18
6.	−0.0513902	0.8906814	0.7187549	−0.1547652	0.4900598	−1.4579510	0.20
7.	−0.0444312	0.9021387	0.7033519	−0.1500069	0.4683041	−1.5300686	0.22
8.	−0.0388084	0.9127465	0.6904098	−0.1588129	0.4474384	−1.6063922	0.24
9.	−0.0344849	0.9226590	0.6795920	−0.1602580	0.4274757	−1.5311635	0.26

TABLE VI

	x_{01}	x_{03}	x_{05}	x_1	x_3	x_5	e
1.	−0.0771844	0.7318070	1.1229523	0.0550662	0.7323378	−1.1212257	0.00000
2.	−0.0776666	0.7303848	1.1255355	0.0556913	0.7328149	−1.1212147	0.00125
3.	−0.0781492	0.7289590	1.1281347	0.0563200	0.7332857	−1.1212192	0.00250
4.	−0.0786324	0.7275295	1.1307499	0.0569522	0.7337500	−1.1212394	0.00375
5.	−0.0791161	0.7260964	1.1333811	0.0575879	0.7342079	−1.1212750	0.00500
6.	−0.0800850	0.7232194	1.1386197	0.0588700	0.7351042	−1.1213925	0.00750
7.	−0.0810557	0.7203282	1.1440668	0.0601662	0.7359747	−1.1215712	0.01000
8.	−0.0820283	0.7174232	1.1495064	0.0614766	0.7368193	−1.1218105	0.01250
9.	−0.0830025	0.7145045	1.1550108	0.0628013	0.7376378	−1.1221100	0.01500
10.	−0.0839783	0.7115724	1.1605800	0.0641402	0.7384302	−1.1224691	0.01750
11.	−0.0849557	0.7086272	1.1662143	0.0654935	0.7391966	−1.1228872	0.02000
12.	−0.0859345	0.7056692	1.1719138	0.0668612	0.7399369	−1.1233639	0.02250
13.	−0.0869147	0.7026897	1.1776785	0.0682432	0.7406510	−1.1238985	0.02500

for $e=0$ and $\mu=0.01212$ with initial conditions

$$x_{01} = -0.06908557, \quad x_{02} = 0, \quad x_{03} = 0.73488119$$
$$x_{04} = 0, \quad x_{05} = 1.1106608, \quad x_{06} = 0,$$

and half-period $T/2 = 3.14142445$.

Using as starting point one member of the family given in Table I, keeping e fixed and varying μ we derive the families, members of which, are listed in Tables II and III. These families correspond to $e=0.10$ and $e=0.20$ respectively.

The Tables IV and V include periodic orbits with μ constant. Thus, for the orbits of the Table IV we have $\mu=0.07$, while for those of the Table V $\mu=0.15$. As starting points we used the orbits No. 4 and No. 8 of the Table II.

In Figure 1 we give the projections of four orbits on the plane $0x_1x_2$ while in Figure 2

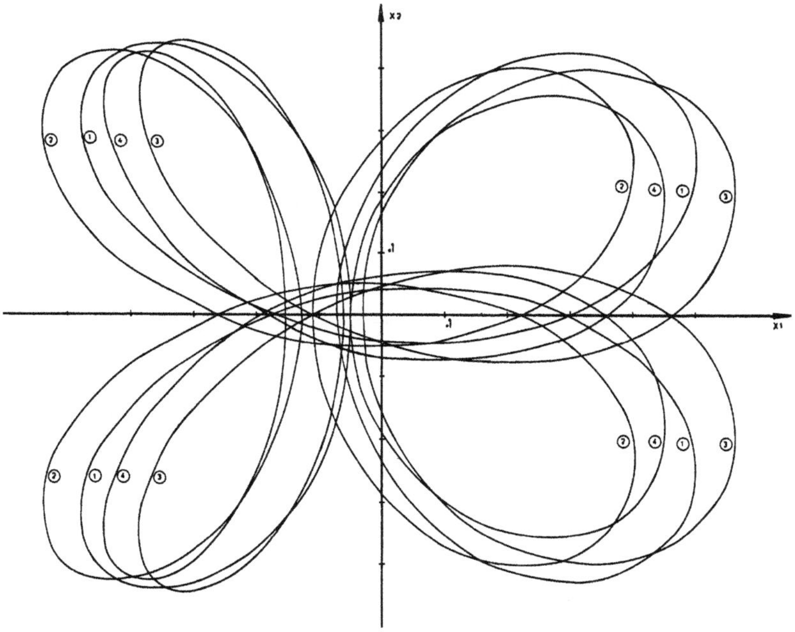

Fig. 1. The projections of four orbits on the plane $0x_1x_2$.

we give the projections of two orbits on the plane $0x_2x_3$. These orbits correspond to the following values of the parameters e and μ.:

(1) $\mu=0.07$, $e=0.10$
(2) $\mu=0.15$, $e=0.10$
(3) $\mu=0.07$, $e=0.20$
(4) $\mu=0.15$, $e=0.20$.

Finally, the Table VI includes periodic orbits with the same starting point, as the family of the Table I, and with initial states corresponding to apogee, i.e. $\mathbf{x}_0= =\mathbf{x}(\mathbf{x}_0, \pi, \pi)$.

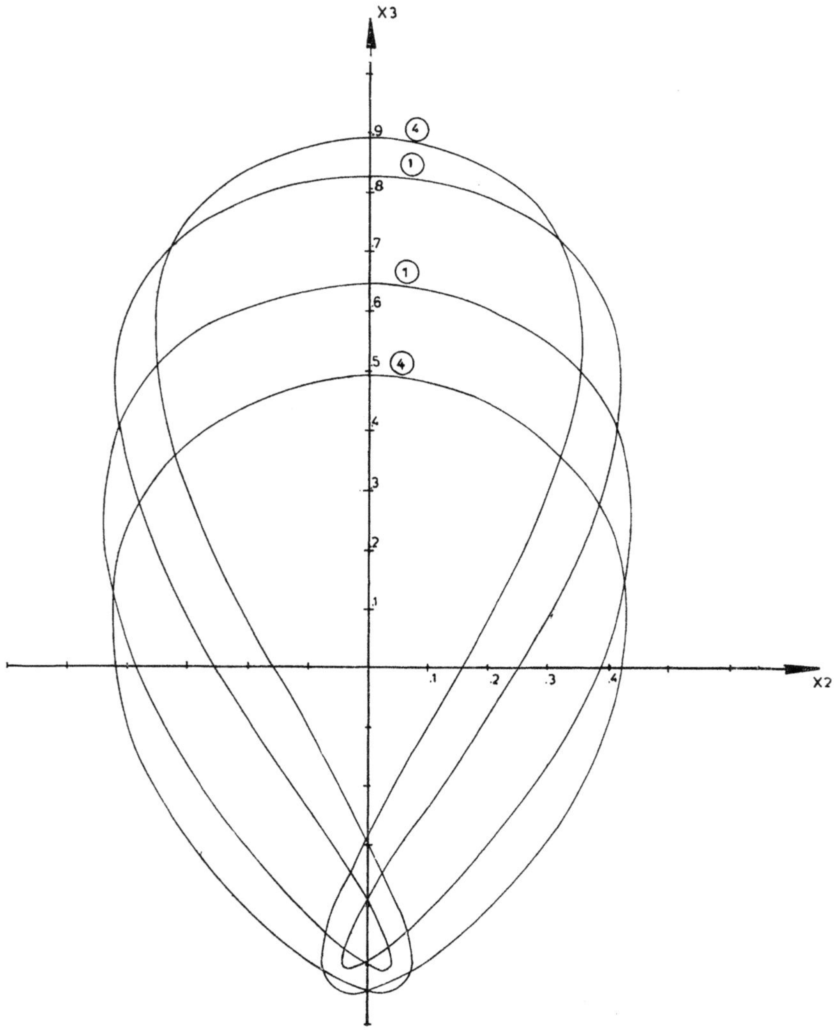

Fig. 2. The projections of two orbits on the plane $0x_2x_3$.

References

Bray, T. A. and Goudas, C. L.: 1967, *Astron. J.* **72**, 202.

Broucke, R. A.: 1969, 'Periodic Orbits in the Elliptic Restricted Three-Body Problem', Jet Propulsion Laboratory Technical Report 32–1360.

Hunter, R. B.: 1967, *Monthly Notices Roy. Astron. Soc.* **136**, 245.

Katsiaris, G.: 1970, 'Three Dimensional Periodic Orbits of Simple Symmetry in the Restricted Three Body Problem', Doctoral Dissertation.

Katsiaris, G.: 1971, *Astrophys. Space Sci.* **10**, 71.

Katsiaris, G. and Goudas, C. L.: 1970, *Astrophys. Space Sci.* **8**, 243.

Kopal, Z. and Lyttleton, L. A.: 1963, *Icarus* **1**, 455.

Szebehely, V.: 1967, *Theory of Orbits*, Academic Press, New York.

Szebehely, V. and Giacaglia, G. E. O.: 1964, Astron. **69**, 230.

SECOND AND THIRD ORDER VARIATIONS OF THE THREE DIMENSIONAL RESTRICTED PROBLEM

P. G. KAZANTZIS

University of Patras, Patras, Greece

Abstract. The numerical techniques, along with the analytic details and justification used, in computing periodic orbits and families of periodic orgits of double symmetry in the three-dimensional restricted three-body problem are described. Two distinct methods are applied in this endeavor, namely the predictor-corrector method and the method of direct expansion of the family into Taylor series. The latter is applied here for the first time.

1. Introduction

The equations of motion of the three dimensional restricted three body problem are (see e.g. Goudas, 1963)

$$\frac{\mathrm{d}x_1}{\mathrm{d}t} = x_4, \quad \frac{\mathrm{d}x_2}{\mathrm{d}t} = x_5, \quad \frac{\mathrm{d}x_3}{\mathrm{d}t} = x_6, \tag{1}$$

$$\frac{\mathrm{d}x_4}{\mathrm{d}t} = 2x_5 + x_1 - \frac{(1-\mu)(x_1+\mu)}{r_1^3} - \frac{\mu(x_1+\mu-1)}{r_2^3}, \tag{2}$$

$$\frac{\mathrm{d}x_5}{\mathrm{d}t} = -2x_4 + x_2 - \frac{(1-\mu)x_2}{r_1^3} - \frac{\mu x_2}{r_2^3}, \tag{3}$$

$$\frac{\mathrm{d}x_6}{\mathrm{d}t} = -\frac{(1-\mu)x_3}{r_1^3} - \frac{\mu x_3}{r_2^3}. \tag{4}$$

$$r_1^2 = (x_1+\mu)^2 + x_2^2 + x_3^2,$$
$$r_2^2 = (x_1+\mu-1)^2 + x_2^2 + x_3^2.$$

These admit the well-known integral

$$J = x_4^2 + x_5^2 + x_6^2 - (x_1^2 + x_2^2) - \frac{2(1-\mu)}{r_1} - \frac{2\mu}{r_2} = c \tag{5}$$

named after Jacobi.

The equations of motion can be put in the vector form

$$\frac{\mathrm{d}\mathbf{x}}{\mathrm{d}t} = \mathbf{f}(\mathbf{x}), \tag{6}$$

where

$$\mathbf{x} = (x_1, x_2, x_3, x_4, x_5, x_6)$$

and

$$\mathbf{f} = (f_1, f_2, f_3, f_4, f_5, f_6)$$

B. D. Tapley and V. Szebehely (eds.), Recent Advances in Dynamical Astronomy, 135–145.

and

$$f_1 = x_4, \quad f_2 = x_5, \quad f_3 = x_6, \tag{7}$$

$$f_4 = 2x_5 + x_1 - \frac{(1-\mu)(x_1+\mu)}{r_1^3} - \frac{\mu(x_1+\mu-1)}{r_2^3}, \tag{8}$$

$$f_5 = -2x_4 + x_2 - \frac{(1-\mu)x_2}{r_1^3} - \frac{\mu x_2}{r_2^3}, \tag{9}$$

$$f_6 = -\frac{(1-\mu)x_3}{r_1^3} - \frac{\mu x_3}{r_2^3}. \tag{10}$$

We note that

$$\frac{\partial \mathbf{f}}{\partial t} = 0$$

and

$$\nabla \cdot \mathbf{f} = 0. \tag{11}$$

As a consequence the volume in phase-space is conserved according to Liouville's lemma (see e.g. Kurth, 1957).

The solutions of (6) are of the form

$$\mathbf{x} = \mathbf{x}(\mathbf{x}_0; t), \tag{12}$$

where, $\mathbf{x}(\mathbf{x}_0; t)$ is analytic in $\mathbf{x}_0$ and t. Hence the quantities

$$\Delta = \left(\frac{\partial x_i}{\partial x_{0j}}\right) = (\mathbf{x}_{\mathbf{x}_0}) \quad \text{and} \quad \frac{\partial \mathbf{x}}{\partial t},$$

exist and are analytic in $\mathbf{x}_0$ and t.

The same is true for the quantities

$$\Delta_j = \frac{\partial \Delta}{\partial x_{0j}}, \qquad \Delta_{jk} = \frac{\partial^2 \Delta}{\partial x_{0j}\, \partial x_{0k}}. \tag{13}$$

It is evident that the matrix Δ satisfies the equation

$$\frac{\mathrm{d}\Delta}{\mathrm{d}t} = P\Delta, \tag{14}$$

where,

$$\Delta = \Delta(\mathbf{x}_0; t)$$

and

$$P = P(\mathbf{x}) = \left(\frac{\partial f_i}{\partial x_j}\right) = (P_{ij}),$$

with

$$\begin{aligned} P_{11} &= P_{12} = P_{13} = P_{15} = P_{16} = P_{21} = P_{22} = P_{23} = P_{24} = P_{26} = \\ &= P_{31} = P_{32} = P_{33} = P_{34} = P_{35} = P_{44} = P_{46} = P_{55} = P_{56} = \\ &= P_{64} = P_{65} = P_{66} = 0 \end{aligned}$$

$$P_{41} = 1 - \frac{(1-\mu)}{r_1^3} - \frac{\mu}{r_2^3} + \frac{3(1-\mu)(x_1+\mu)^2}{r_1^5} + \frac{3\mu(x_1+\mu-1)^2}{r_2^5}$$

$$P_{42} = P_{51} = \frac{3(1-\mu)(x_1+\mu)x_2}{r_1^5} + \frac{3\mu(x_1+\mu-1)x_2}{r_2^5}$$

$$P_{43} = P_{61} = \frac{3(1-\mu)(x_1+\mu)x_3}{r_1^5} + \frac{3\mu(x_1+\mu-1)x_3}{r_2^5}$$

$$P_{52} = 1 - \frac{1-\mu}{r_1^3} - \frac{\mu}{r_2^3} + \frac{3(1-\mu)x_2^2}{r_1^5} + \frac{3\mu x_2^2}{r_2^5}$$

$$P_{53} = P_{62} = \frac{3(1-\mu)x_2x_3}{r_1^5} + \frac{3\mu x_2 x_3}{r_2^5}$$

$$P_{63} = -\frac{(1-\mu)}{r_1^3} - \frac{\mu}{r_2^3} + \frac{3(1-\mu)x_3^2}{r_1^5} + \frac{3\mu x_3^2}{r_2^5}.$$

The matrix Δ_j satisfies the equation

$$\frac{d\Delta_j}{dt} = \left\{P_1 \frac{\partial x_1}{\partial x_{0j}} + P_2 \frac{\partial x_2}{\partial x_{0j}} + P_3 \frac{\partial x_3}{\partial x_{0j}}\right\} \Delta + P\Delta_j \tag{15}$$

where,

$$P_1 = \frac{\partial P}{\partial x_1}, \quad P_2 = \frac{\partial P}{\partial x_2}, \quad P_3 = \frac{\partial P}{\partial x_3}$$

and

$$\Delta_j = \left(\frac{\partial^2 x_i}{\partial x_{0k}\,\partial x_{0j}}\right),$$

for $i, k, j = 1, 2, 3, 4, 5, 6$.

Finally, the matrix Δ_{jk} satisfies the equations

$$\begin{aligned}\frac{d\Delta_{jk}}{dt} = &\left\{P_1 \frac{\partial^2 x_1}{\partial x_{0k}\,\partial x_{0j}} + P_{11} \frac{\partial x_1}{\partial x_{0k}} \frac{\partial x_1}{\partial x_{0j}} + P_{12} \frac{\partial x_2}{\partial x_{0k}} \frac{\partial x_1}{\partial x_{0j}} + \right.\\ &+ P_{13} \frac{\partial x_3}{\partial x_{0k}} \frac{\partial x_1}{\partial x_{0j}} + P_2 \frac{\partial^2 x_2}{\partial x_{0k}\,\partial x_{0j}} + P_{21} \frac{\partial x_1}{\partial x_{0k}} \frac{\partial x_2}{\partial x_{0j}} + \\ &+ P_{22} \frac{\partial x_2}{\partial x_{0k}} \frac{\partial x_2}{\partial x_{0j}} + P_{23} \frac{\partial x_3}{\partial x_{0k}} \frac{\partial x_2}{\partial x_{0j}} + P_3 \frac{\partial^2 x_3}{\partial x_{0k}\,\partial x_{0j}} + \\ &\left.+ P_{31} \frac{\partial x_1}{\partial x_{0k}} \frac{\partial x_3}{\partial x_{0j}} + P_{32} \frac{\partial x_2}{\partial x_{0k}} \frac{\partial x_3}{\partial x_{0j}} + P_{33} \frac{\partial x_3}{\partial x_{0k}} \frac{\partial x_3}{\partial x_{0j}}\right\} \Delta + \\ &+ \left\{P_1 \frac{\partial x_1}{\partial x_{0j}} + P_2 \frac{\partial x_2}{\partial x_{0j}} + P_3 \frac{\partial x_3}{\partial x_{0j}}\right\} \Delta_k + \\ &+ \left\{P_1 \frac{\partial x_1}{\partial x_{0k}} + P_2 \frac{\partial x_2}{\partial x_{0k}} + P_3 \frac{\partial x_3}{\partial x_{0k}}\right\} \Delta_j + P\Delta_{jk},\end{aligned} \tag{16}$$

where:

$$P_1 = \frac{\partial P}{\partial x_1}, \quad P_{11} = \frac{\partial^2 P}{\partial x_1^2}, \quad P_{12} = \frac{\partial^2 P}{\partial x_2\, \partial x_1}, \quad P_{13} = \frac{\partial^2 P}{\partial x_3\, \partial x_1}$$

$$P_2 = \frac{\partial P}{\partial x_2}, \quad P_{21} = \frac{\partial^2 P}{\partial x_1\, \partial x_2}, \quad P_{22} = \frac{\partial^2 P}{\partial x_2^2}, \quad P_{23} = \frac{\partial^2 P}{\partial x_3\, \partial x_2}$$

$$P_3 = \frac{\partial P}{\partial x_3}, \quad P_{31} = \frac{\partial^2 P}{\partial x_1\, \partial x_3}, \quad P_{32} = \frac{\partial^2 P}{\partial x_2\, \partial x_3}, \quad P_{33} = \frac{\partial^2 P}{\partial x_3^2}$$

$$\Delta_k = \frac{\partial \Delta}{\partial x_{0k}}, \quad \Delta_j = \frac{\partial \Delta}{\partial x_{0j}}, \quad \Delta_{jk} = \frac{\partial^2 \Delta}{\partial x_{0k}\, \partial x_{0j}}$$

The above equations together with the equations of motion have to be integrated simultaneously. The initial condition of Equations (6) will be selected according to criteria that will depend on the type of solution needed, while the initial conditions of (14), (15) and (16) at $t=0$ will be

$$\begin{aligned} \Delta(\mathbf{x}_0; 0) &= I, \\ \Delta_j(\mathbf{x}_0; 0) &= 0, \\ \Delta_{jk}(\mathbf{x}_0; 0) &= 0. \end{aligned} \tag{17}$$

2. Periodic Orbits

Solutions of the form $\mathbf{x}(\mathbf{x}_0; t)$, satisfying Equations (16), are periodic in time if

$$\mathbf{x}(\mathbf{x}_0; t + T) - \mathbf{x}(\mathbf{x}_0; t) = 0, \tag{18}$$

where T (taken as a rule different than zero) is the period. The existence of such solutions has been shown by Goudas (1961), Jefferys (1966) and Katsiaris (1969). Here we shall consider orbits of double symmetry, namely, orbits that are symmetric with respect to the planes $0x_1x_2$ and $0x_1x_3$. Such orbits have been studied by Goudas (1963) and Bray and Goudas (1966–1967).

The necessary and sufficient conditions for periodicity are

$$x_2(x_{01}, x_{05}, x_{06}; T) = 0, \tag{19}$$

$$x_4(x_{01}, x_{05}, x_{06}; T) = 0, \tag{20}$$

$$x_6(x_{01}, x_{05}, x_{06}; T) = 0, \tag{21}$$

where, $\mathbf{x}_0 = (x_{01}, 0, 0, 0, x_{05}, x_{06})$. These conditions show that it is possible, at least in principle, to express the initial conditions of such orbits in terms of the period, i.e. to obtain relations of the form

$$x_{01} = \phi_1(T), \tag{22}$$

$$x_{05} = \phi_5(T), \tag{23}$$

$$x_{06} = \phi_6(T). \tag{24}$$

This we shall do first in an analytical way and then in a numerical one using the computer.

To describe the first method, we assume that the initial state $(x_{01}, 0, 0, 0, x_{05}, x_{06})$ gives rise to a periodic orbit of period $4T$. In order that the initial state $(x_{01}^*, 0, 0, 0, x_{05}^*, x_{06}^*)$ corresponds to a periodic orbit of the same family and of period $4T^*$ we must have

$$x_2(x_{01}^*, x_{05}^*, x_{06}^*; T^*) = 0, \tag{25}$$
$$x_4(x_{01}^*, x_{05}^*, x_{06}^*; T^*) = 0, \tag{26}$$
$$x_6(x_{01}^*, x_{05}^*, x_{06}^*; T^*) = 0. \tag{27}$$

If we set

$$T^* - T = \delta T, \tag{28}$$
$$x_{01}^* - x_{01} = \delta x_{01}, \tag{29}$$
$$x_{05}^* - x_{05} = \delta x_{05}, \tag{30}$$
$$x_{06}^* - x_{06} = \delta x_{06}, \tag{31}$$

and expand conditions (25)–(27) into Taylor series, we receive the expressions

$$\sum_{n=1}^{\infty} \left(\frac{\partial}{\partial x_{01}} \delta x_{01} + \frac{\partial}{\partial x_{05}} \delta x_{05} + \frac{\partial}{\partial x_{06}} \delta x_{06} + \frac{\partial}{\partial T} \delta T \right)^n x_2 = 0, \tag{32}$$

$$\sum_{n=1}^{\infty} \left(\frac{\partial}{x_{01}} \delta x_{01} + \frac{\partial}{\partial x_{05}} \delta x_{05} + \frac{\partial}{\partial x_{06}} \delta x_{06} + \frac{\partial}{\partial T} \delta T \right)^n x_4 = 0, \tag{33}$$

$$\sum_{n=1}^{\infty} \left(\frac{\partial}{\partial x_{01}} \delta x_{01} + \frac{\partial}{\partial x_{05}} \delta x_{05} + \frac{\partial}{\partial x_{06}} \delta x_{06} + \frac{\partial}{\partial T} \delta T \right)^n x_6 = 0. \tag{34}$$

The ultimate expressions sought are the Taylor expressions of relations (22)–(24), namely the expressions

$$x_{01}^* - x_{01} = \sum_{n=1}^{\infty} A_n \, \delta T^n, \tag{35}$$

$$x_{05}^* - x_{05} = \sum_{n=1}^{\infty} B_n \, \delta T^n, \tag{36}$$

$$x_{06}^* - x_{06} = \sum_{n=1}^{\infty} C_n \, \delta T^n. \tag{37}$$

These expressions are substituted in Equations (32)–(34) and thus, by setting the coefficients of all powers of δT equal to zero we obtain the expressions of the unknown coefficients A_n, B_n, C_n, $n = 1, 2, \ldots$. We find, e.g. that

$$A_1 = \begin{vmatrix} -b_4^2 & b_2^2 & b_3^2 \\ -b_4^4 & b_2^4 & b_3^4 \\ -b_4^6 & b_2^6 & b_3^6 \end{vmatrix} \Bigg/ D \tag{38}$$

$$B_1 = \begin{vmatrix} b_1^2 & -b_4^2 & b_3^2 \\ b_1^4 & -b_4^4 & b_3^4 \\ b_1^6 & -b_4^6 & b_3^6 \end{vmatrix} \Big/ D \tag{39}$$

$$C_1 = \begin{vmatrix} b_1^2 & b_2^2 & -b_4^2 \\ b_1^4 & b_2^4 & -b_4^4 \\ b_1^6 & b_2^6 & -b_4^6 \end{vmatrix} \Big/ D \tag{40}$$

$$A_2 = \begin{vmatrix} -d_2 & b_2^2 & b_3^2 \\ -d_4 & b_2^4 & b_3^4 \\ -d_6 & b_2^6 & b_3^6 \end{vmatrix} \Big/ D \tag{41}$$

$$B_2 = \begin{vmatrix} b_1^2 & -d_2 & b_3^2 \\ b_1^4 & -d_4 & b_3^4 \\ b_1^6 & -d_6 & b_3^6 \end{vmatrix} \Big/ D \tag{42}$$

$$C_2 = \begin{vmatrix} b_1^2 & b_2^2 & -d_2 \\ b_1^4 & b_2^4 & -d_4 \\ b_1^6 & b_2^6 & -d_6 \end{vmatrix} \Big/ D \tag{43}$$

$$A_3 = \begin{vmatrix} -a_2 & b_2^2 & b_3^2 \\ -a_4 & b_2^4 & b_3^4 \\ -a_6 & b_2^6 & b_3^6 \end{vmatrix} \Big/ D \tag{44}$$

$$B_3 = \begin{vmatrix} b_1^2 & -a_2 & b_3^2 \\ b_1^4 & -a_4 & b_3^4 \\ b_1^6 & -a_6 & b_3^6 \end{vmatrix} \Big/ D \tag{45}$$

$$C_3 = \begin{vmatrix} b_1^2 & b_2^2 & -a_2 \\ b_1^4 & b_2^4 & -a_4 \\ b_1^6 & b_2^6 & -a_6 \end{vmatrix} \Big/ D, \tag{46}$$

where

$$D = \begin{vmatrix} b_1^2 & b_2^2 & b_3^2 \\ b_1^4 & b_2^4 & b_3^4 \\ b_1^6 & b_2^6 & b_3^6 \end{vmatrix}$$

and

$$\begin{aligned} d_i = {} & b_5^i A_1^2 + b_6^i B_1^2 + b_7^i C_1^2 + b_8^i + b_9^i A_1 B_1 + b_{10}^i A_1 C_1 + \\ & + b_{11}^i A_1 + b_{12}^i B_1 C_1 + b_{13}^i B_1 + b_{14}^i C_1 \qquad i = 2, 4, 6 \end{aligned}$$

$$\begin{aligned} a_i = {} & 2b_5^i A_1 A_2 + 2b_6^i B_1 B_2 + 2b_7^i C_1 C_2 + b_9^i (A_1 B_2 + A_2 B_1) + \\ & + b_{10}^i (A_1 C_2 + A_2 C_1) + b_{11}^i A_2 + b_{12}^i (B_1 C_2 + B_2 C_1) + \\ & + b_{13}^i B_2 + b_{14}^i C_2 + b_{15}^i A_1^3 + b_{16}^i B_1^3 + b_{17}^i C_1^3 + b_{18}^i + b_{19} A_1 B_1^2 + \\ & + b_{20}^i A_1 C_1^2 + b_{21}^i A_1 + b_{22}^i B_1 A_1^2 + b_{23}^i B_1 C_1^2 + b_{24}^i B_1 + \\ & + b_{25}^i C_1 A_1^2 + b_{26}^i C_1 B_1^2 + b_{27}^i C_1 + b_{28}^i A_1^2 + b_{29}^i B_1^2 + b_{30}^i C_1^2 + \\ & + b_{31}^i A_1 B_1 C_1 + b_{32}^i A_1 B_1 + b_{33}^i A_1 C_1 + b_{34}^i B_1 C_1 \qquad i = 2, 4, 6 \end{aligned}$$

$$\frac{\partial x_i}{\partial x_{01}} = b_1^i, \quad \frac{\partial x_i}{\partial x_{05}} = b_2^i, \quad \frac{\partial x_i}{\partial x_{06}} = b_3^i,$$

$$\frac{\partial x_i}{\partial T} = b_4^i, \quad \frac{\partial^2 x_i}{\partial x_{01}^2} = 2b_5^i, \quad \frac{\partial^2 x_i}{\partial x_{05}^2} = 2b_6^i,$$

$$\frac{\partial^2 x_i}{\partial x_{06}^2} = 2b_7^i, \quad \frac{\partial^2 x_i}{\partial T^2} = 2b_8^i, \quad \frac{\partial^2 x_i}{\partial x_{01}\,\partial x_{05}} = b_9^i,$$

$$\frac{\partial^2 x_i}{\partial x_{01}\,\partial x_{06}} = b_{10}^i, \quad \frac{\partial^2 x_i}{\partial x_{01}\,\partial T} = b_{11}^i, \quad \frac{\partial^2 x_i}{\partial x_{05}\,\partial x_{06}} = b_{12}^i,$$

$$\frac{\partial^2 x_i}{\partial x_{05}\,\partial T} = b_{13}^i, \quad \frac{\partial^2 x_i}{\partial x_{06}\,\partial T} = b_{14}^i, \quad \frac{\partial^3 x_i}{\partial x_{01}^3} = 6b_{15}^i,$$

$$\frac{\partial^3 x_i}{\partial x_{05}^3} = 6b_{16}^i, \quad \frac{\partial^3 x_i}{\partial x_{06}^3} = 6b_{17}^i, \quad \frac{\partial^3 x_i}{\partial T^3} = 6b_{18}^i,$$

$$\frac{\partial^3 x_i}{\partial x_{01}\,\partial x_{05}^2} = 2b_{19}^i, \quad \frac{\partial^3 x_i}{\partial x_{01}\,\partial x_{06}^2} = 2b_{20}^i, \quad \frac{\partial^3 x_i}{\partial x_{01}\,\partial T^2} = 2b_{21}^i,$$

$$\frac{\partial^3 x_i}{\partial x_{05}\,\partial x_{01}^2} = 2b_{22}^i, \quad \frac{\partial^3 x_i}{\partial x_{05}\,\partial x_{06}^2} = 2b_{23}^i, \quad \frac{\partial^3 x_i}{\partial x_{05}\,\partial T^2} = 2b_{24}^i,$$

$$\frac{\partial^3 x_i}{\partial x_{06}\,\partial x_{01}^2} = 2b_{25}^i, \quad \frac{\partial^3 x_i}{\partial x_{06}\,\partial x_{05}^2} = 2b_{26}^i, \quad \frac{\partial^3 x_i}{\partial x_{06}\,\partial T^2} = 2b_{27}^i,$$

$$\frac{\partial^3 x_i}{\partial T\,\partial x_{01}^2} = 2b_{28}^i, \quad \frac{\partial^3 x_i}{\partial T\,\partial x_{05}^2} = 2b_{29}^i, \quad \frac{\partial^3 x_i}{\partial T\,\partial x_{06}^2} = 2b_{30}^i,$$

$$\frac{\partial^3 x_i}{\partial x_{01}\,\partial x_{05}\,\partial x_{06}} = b_{31}^i, \quad \frac{\partial^3 x_i}{\partial x_{01}\,\partial x_{05}\,\partial T} = b_{32}^i, \quad \frac{\partial^3 x_i}{\partial x_{01}\,\partial x_{06}\,\partial T} = b_{33}^i,$$

$$\frac{\partial^3 x_i}{\partial x_{05}\,\partial x_{06}\,\partial T} = b_{34}^i \qquad i = 2, 4, 6.$$

The above expressions of the coefficients of the series (35)–(37) permit the direct expression of the family as a function of the period, so long as the periods are selected within the range of convergence of the corresponding series. The said coefficients are computed at the end of the quarter of the period ($t = T$) of the periodic orbit with initial state corresponding to the known exact periodic orbit. Obviously, the present method can be applied if one member of the family and the corresponding period are known and if the partial derivatives

$$\frac{\partial^n x_i}{\partial x_{01}^{n_1}\,\partial x_{05}^{n_2}\,\partial x_{06}^{n_3}\,\partial T^{n_4}}$$

with $n = n_1 + n_2 + n_3 + n_4$, $i = 2, 4, 6$, are known at the end of the first quarter of the period.

The numerical work done in this connection was extended to cover cases with $n = 3$. This implied the integration of $6 + 36 + 216 + 1248$ differential equations. Numerical results will be presented in a future publication. The computation of the initial state of an exact periodic orbit, when no analytical information is available about such orbit, is done in the way described below which is also the second method for computing families of periodic orbits. This method is based on Poincaré's 'continuity' method and is usually called the 'predictor-corrector' method. The predictor-corrector method, for unknown families of double symmetry, starts with the 'corrector' part, unless one periodic orbit is known. Otherwise the conditions of periodicity (19)–(21) are used in order to compute corrections. In this paper we shall develop the 'corrector' part (this we shall later repeat for the 'predictor') using second order variations. Indeed, equations (32)–(34), after elimination of all terms of order higher than the second, become:

$$\begin{aligned}
&\frac{\partial x_2}{\partial x_{01}}\,\delta x_{01} + \frac{\partial x_2}{\partial x_{05}}\,\delta x_{05} + \frac{\partial x_2}{\partial x_{06}}\,\delta x_{06} + \frac{\partial x_2}{\partial T}\,\delta T + \frac{1}{2}\frac{\partial^2 x_2}{\partial x_{01}^2}\,\delta x_{01}^2 + \\
&+ \frac{1}{2}\frac{\partial^2 x_2}{\partial x_{05}^2}\,\delta x_{05}^2 + \frac{1}{2}\frac{\partial^2 x_2}{\partial x_{06}^2}\,\delta x_{06}^2 + \frac{1}{2}\frac{\partial^2 x_2}{\partial T^2}\,\delta T^2 + \frac{\partial^2 x_2}{\partial x_{01}\,\partial x_{05}}\,\delta x_{01}\,\delta x_{05} + \\
&+ \frac{\partial^2 x_2}{\partial x_{01}\,\partial x_{06}}\,\delta x_{01}\,\delta x_{06} + \frac{\partial^2 x_2}{\partial x_{01}\,\partial T}\,\delta x_{01}\,\delta T + \frac{\partial^2 x_2}{\partial x_{05}\,\partial x_{06}}\,\delta x_{05}\,\delta x_{06} + \\
&+ \frac{\partial^2 x_2}{\partial x_{05}\,\partial T}\,\delta x_{05}\,\delta T + \frac{\partial^2 x_2}{\partial x_{06}\,\partial T}\,\delta x_{06}\,\delta T = 0
\end{aligned} \tag{47}$$

$$\begin{aligned}
&\frac{\partial x_4}{\partial x_{01}}\,\delta x_{01} + \frac{\partial x_4}{\partial x_{05}}\,\delta x_{05} + \frac{\partial x_4}{\partial x_{06}}\,\delta x_{06} + \frac{\partial x_4}{\partial T}\,\delta T + \frac{1}{2}\frac{\partial^2 x_4}{\partial x_{01}^2}\,\delta x_{01}^2 + \\
&+ \frac{1}{2}\frac{\partial^2 x_4}{\partial x_{05}^2}\,\delta x_{05}^2 + \frac{1}{2}\frac{\partial^2 x_4}{\partial x_{06}^2}\,\delta x_{06}^2 + \frac{1}{2}\frac{\partial^2 x_4}{\partial T^2}\,\delta T^2 + \frac{\partial^2 x_4}{\partial x_{01}\,\partial x_{05}}\,\delta x_{01}\,\delta x_{05} + \\
&+ \frac{\partial^2 x_4}{\partial x_{01}\,\partial x_{06}}\,\delta x_{01}\,\delta x_{06} + \frac{\partial^2 x_4}{\partial x_{01}\,\partial T}\,\delta x_{01}\,\delta T + \frac{\partial^2 x_4}{\partial x_{05}\,\partial x_{06}}\,\delta x_{05}\,\delta x_{06} + \\
&+ \frac{\partial^2 x_4}{\partial x_{05}\,\partial T}\,\delta x_{05}\,\delta T + \frac{\partial^2 x_4}{\partial x_{06}\,\partial T}\,\delta x_{06}\,\delta T = -\,x_4
\end{aligned} \tag{48}$$

$$\begin{aligned}
&\frac{\partial x_6}{\partial x_{01}}\,\delta x_{01} + \frac{\partial x_6}{\partial x_{05}}\,\delta x_{05} + \frac{\partial x_6}{\partial x_{06}}\,\delta x_{06} + \frac{\partial x_6}{\partial T}\,\delta T + \frac{1}{2}\frac{\partial^2 x_6}{\partial x_{01}^2}\,\delta x_{01}^2 + \\
&+ \frac{1}{2}\frac{\partial^2 x_6}{\partial x_{05}^2}\,\delta x_{05}^2 + \frac{1}{2}\frac{\partial^2 x_6}{\partial x_{06}^2}\,\delta x_{06}^2 + \frac{1}{2}\frac{\partial^2 x_6}{\partial T^2}\,\delta T^2 + \frac{\partial^2 x_6}{\partial x_{01}\,\partial x_{05}}\,\delta x_{01}\,\delta x_{05} + \\
&+ \frac{\partial^2 x_6}{\partial x_{01}\,\partial x_{06}}\,\delta x_{01}\,\delta x_{06} + \frac{\partial^2 x_6}{\partial x_{01}\,\partial T}\,\delta x_{01}\,\delta T + \frac{\partial^2 x_6}{\partial x_{05}\,\partial x_{06}}\,\delta x_{05}\,\delta x_{06} + \\
&+ \frac{\partial^2 x_6}{\partial x_{05}\,\partial T}\,\delta x_{05}\,\delta T + \frac{\partial^2 x_6}{\partial x_{06}\,\partial T}\,\delta x_{06}\,\delta T = -\,x_6\,.
\end{aligned} \tag{49}$$

We remark that $\mathbf{x}_0$ has been selected at random, but as near to a periodic orbit as possible, so that x_4 and x_6 at the crossing of the plane $0x_1x_3$ are close to zero. Relations (47)–(49) do not suffice for the unique determination of the corrections δx_{01}, δx_{05},

δx_{06} and δT. For this reason a fourth relation providing the 'steepest descent' to the family of periodic orbits will be used. Such relation is obtained in the following way: Since the correction

$$\delta \mathbf{x}_0 = (\delta x_{01}, \delta x_{05}, \delta x_{06}; \delta T)$$

must be normal to the vectors ∇x_2, ∇x_4, ∇x_6, where:

$$\nabla x_i = \left(\frac{\partial x_i}{\partial x_{01}}, \frac{\partial x_i}{\partial x_{05}}, \frac{\partial x_i}{\partial x_{06}}, \frac{\partial x_i}{\partial T}\right), \qquad i = 2, 4, 6 \tag{50}$$

the relation needed is

$$(\nabla x_2 \otimes \nabla x_4 \otimes \nabla x_6) \cdot \delta \mathbf{x}_0 = 0. \tag{51}$$

The expression within the brackets is defined as follows

$$\nabla x_2 \otimes \nabla x_4 \otimes \nabla x_6 = \begin{vmatrix} \mathbf{i}_1 & \mathbf{i}_2 & \mathbf{i}_3 & \mathbf{i}_4 \\ \dfrac{\partial x_2}{\partial x_{01}} & \dfrac{\partial x_2}{\partial x_{05}} & \dfrac{\partial x_2}{\partial x_{06}} & \dfrac{\partial x_2}{\partial T} \\ \dfrac{\partial x_4}{\partial x_{01}} & \dfrac{\partial x_4}{\partial x_{05}} & \dfrac{\partial x_4}{\partial x_{06}} & \dfrac{\partial x_4}{\partial T} \\ \dfrac{\partial x_6}{\partial x_{01}} & \dfrac{\partial x_6}{\partial x_{05}} & \dfrac{\partial x_6}{\partial x_{06}} & \dfrac{\partial x_6}{\partial T} \end{vmatrix} \tag{52}$$

while the relations

$$\mathbf{i}_v \cdot \mathbf{i}_\varrho = \delta_{v\varrho}, \qquad v, \varrho = 1, 2, 3, 4.$$

(δ_{vp} are the Kronecker deltas) are postulated. Thus the final form of the additional relation is

$$\begin{vmatrix} \delta x_{01} & \delta x_{05} & \delta x_{06} & \partial T \\ \dfrac{\partial x_2}{\partial x_{01}} & \dfrac{\partial x_2}{\partial x_{05}} & \dfrac{\partial x_2}{\partial x_{06}} & \dfrac{\partial x_2}{\partial T} \\ \dfrac{\partial x_4}{\partial x_{01}} & \dfrac{\partial x_4}{\partial x_{05}} & \dfrac{\partial x_4}{\partial x_{06}} & \dfrac{\partial x_4}{\partial T} \\ \dfrac{\partial x_6}{\partial x_{01}} & \dfrac{\partial x_6}{\partial x_{05}} & \dfrac{\partial x_6}{\partial x_{06}} & \dfrac{\partial x_6}{\partial T} \end{vmatrix} = 0. \tag{53}$$

Equations (47)–(49) and (53) suffice for the unique determination of the correction. Repetitive application of such corrections lead to the orbit sought.

A special problem rises with the selection of the suitable root of the system (47)–(49) and (53) since this system has eight different roots. The selection is made automatically by the computer which solves the above system by the Newton-Raphson method using as initial estimate of the root the solution of the linear system derived from Equations (47)–(49) and (53) after elimination of second order terms. With this we conclude the description of the 'corrector' part.

Once the first periodic orbit is found we make use of the 'predictor' method in order to obtain an approximation of a closeby periodic solution of the same family. The 'predictor' part is similar with the 'predictor' part employed usually in the first order variational method with the difference that second order terms in the 'predictor' conditions are included. These conditions are Equations (47)–(49) in which the rhs are equal to zero because we are on a periodic orbit. The problem of selecting the suitable root rises here also, and is faced in the same way, as above.

3. Numerical Work

The integration of the equations of motion and variation of the first, second and third order, wherever the latter were necessary, was done by the Runge-Kutta-Gill method since the purpose of the present work is independent of the integration technique. The routine used controls automatically the truncation error by doubling or halving the step size.

As an independent gross error control the integrals of motion and variations were calculated at each step of integration. The integral of energy is already given and the integrals of variation are the following:

$$\sum_{i=1}^{6} \frac{\partial J}{\partial x_i} \frac{\partial x_i}{\partial x_{0j}} = \frac{\partial J(\mathbf{x}_0)}{\partial x_{0j}}, \qquad j = 1, 2, 3, 4, 5, 6,$$

$$\sum_{l=1}^{6} \sum_{i=1}^{6} \frac{\partial^2 J}{\partial x_i \, \partial x_l} \frac{\partial x_i}{\partial x_{0j}} \frac{\partial x_l}{\partial x_{0k}} + \sum_{l=1}^{6} \sum_{i=1}^{6} \frac{\partial J}{\partial x_i} \frac{\partial^2 x_i}{\partial x_{0j} \, \partial x_l} \frac{\partial x_l}{\partial x_{0k}} =$$
$$= \frac{\partial^2 J(\mathbf{x}_0)}{\partial x_{0j} \, \partial x_{0k}}, \qquad j, k = 1, 2, 3, 4, 5, 6$$

and

$$\sum_{m=1}^{6} \sum_{l=1}^{6} \sum_{i=1}^{6} \frac{\partial^3 J}{\partial x_i \, \partial x_l \, \partial x_m} \frac{\partial x_i}{\partial x_{0j}} \frac{\partial x_l}{\partial x_{0k}} \frac{\partial x_m}{\partial x_{0\varrho}} +$$
$$+ \sum_{m=1}^{6} \sum_{l=1}^{6} \sum_{i=1}^{6} \frac{\partial^2 J}{\partial x_i \, \partial x_l} \frac{\partial^2 x_i}{\partial x_{0j} \, \partial x_m} \frac{\partial x_l}{\partial x_{0k}} \frac{\partial x_m}{\partial x_{0\varrho}} +$$
$$+ \sum_{m=1}^{6} \sum_{l=1}^{6} \sum_{i=1}^{6} \frac{\partial^2 J}{\partial x_i \, \partial x_l} \frac{\partial x_i}{\partial x_{0j}} \frac{\partial^2 x_l}{\partial x_{0k} \, \partial x_m} \frac{\partial x_m}{\partial x_{0\varrho}} +$$
$$+ \sum_{m=1}^{6} \sum_{l=1}^{6} \sum_{i=1}^{6} \frac{\partial^2 J}{\partial x_i \, \partial x_m} \frac{\partial^2 x_i}{\partial x_{0j} \, \partial x_l} \frac{\partial x_l}{\partial x_{0k}} \frac{\partial x_m}{\partial x_{0\varrho}} +$$

$$+\sum_{m=1}^{6}\sum_{l=1}^{6}\sum_{i=1}^{6}\frac{\partial J}{\partial x_i}\frac{\partial^3 x_i}{\partial x_{0j}\,\partial x_l\,\partial x_m}\frac{\partial x_l}{\partial x_{0k}}\frac{\partial x_m}{\partial x_{0\varrho}}+$$

$$+\sum_{m=1}^{6}\sum_{l=1}^{6}\sum_{i=1}^{6}\frac{\partial J}{\partial x_i}\frac{\partial^2 x_i}{\partial x_{0j}\,\partial x_l}\frac{\partial^2 x_l}{\partial x_{0k}\,\partial x_m}\frac{\partial x_m}{\partial x_{0\varrho}}=$$

$$=\frac{\partial^3 J(\mathbf{x}_0)}{\partial x_{0j}\,\partial x_{0k}\,\partial x_{0\varrho}},\qquad j,k,\varrho=1,2,3,4,5,6.$$

Acknowledgement

I express herewith my gratitude to Prof. C. L. Goudas for his advice and incouragement in carrying out the present work.

References

Goudas, C. L.: 1961, *Bull. Soc. Math. Grèce* (N. Serie) **2**.
Goudas, C. L.: 1963, *Icarus* **2**, 1.
Goudas, C. L. and Bray, T. A.: 1967, *Astron. J.* **72**, 202.
Goudas, C. L. and Bray, T. A.: 1967, in Z. Kopal (ed.), *Advances Astron. Astrophys*, **5**, Academic Press, New York.
Jefferys, W. H.: 1966, *Astron. J.* **71**, 7.
Katsiaris, G.: 1969, 'Simply symmetric Periodic Orbits in the Restricted Problem of Three Bodies', Doctoral Dissertation, Technical University of Athens.
Kurth, R.: 1957, *Introduction to the Mechanics of Stellar Systems*, Pergamon Press, London.

PLANAR PERIODIC ORBITS USING SECOND AND THIRD VARIATIONS

C. G. ZAGOURAS

Dept. of Mechanics, University of Patras, Patras, Greece

Abstract. The techniques applied in the calculation of closed solutions of the planar restricted problem of three bodies are presented. Use of second and third variational equations is made by means of which both orbits and entire families of orbits can be computed by straight forward expansion of the sufficient conditions for periodicity into Taylor series.

1. Equations of Motion

With the use of a rotating system of coordinates with origin at their center of mass and a suitable choice of the units of length, mass and time the equation of motion of a massless particle C, in the restricted problem of three bodies takes the form

$$\dot{\mathbf{x}} = f(\mathbf{x}), \tag{1}$$

where

$$\mathbf{x} = (x_1, x_2, x_3, x_4), \qquad f = (f_1, f_2, f_3, f_4),$$

and

$$f_1 = x_3, \qquad f_2 = x_4,$$

$$f_3 = 2x_4 + x_1 - \frac{(1-\mu)(x_1+\mu)}{r_1^3} - \frac{(x_1+\mu-1)}{r_2^3}.,$$

$$f_4 = -2x_3 + x_2 - \frac{(1-\mu)\,x_2}{r_1^3} - \frac{\mu x_2}{r_2^3}.$$

Equations (1), admit the well-known integral

$$J = \tfrac{1}{2}(x_3^2 + x_4^2) - \tfrac{1}{2}(x_1^2 + x_2^2) - \frac{(1-\mu)}{r_1} - \frac{\mu}{r_2} = \text{const}, \tag{2}$$

which is named after Jacobi.

2. Equations of Variations of the First Order

Because $\partial f/\partial t = 0$, the system is autonomous and its solutions are analytic functions of the initial state $\mathbf{x}_0$ and the time t, where

$$\mathbf{x}_0 = (x_{01}, x_{02}, x_{03}, x_{04}), \quad \text{for} \quad t = 0.$$

Thus, we can write

$$\mathbf{x} = \mathbf{x}(\mathbf{x}_0; t), \qquad \mathbf{x}_0 = \mathbf{x}(\mathbf{x}_0; 0).$$

Since the solution $\mathbf{x} = \mathbf{x}(\mathbf{x}_0; t)$ is an analytic function of the initial conditions, the

B. D. Tapley and V. Szebehely (eds.), Recent Advances in Dynamical Astronomy, 146–155.

following quantities exist:

$$\mathbf{e}_j = \frac{\partial \mathbf{x}}{\partial x_{0j}} = e_j(\mathbf{x}_0; t) \tag{3}$$

and are analytic functions of $\mathbf{x}_0$. They satisfy the equations

$$\frac{\mathrm{d}}{\mathrm{d}t}\left(\frac{\partial \mathbf{x}}{\partial x_{0j}}\right) = \sum_{l=1}^{4} \frac{\partial f}{\partial x_l}\frac{\partial x_l}{\partial x_{0j}}, \quad j = 1, 2, 3, 4$$

which can be written in matrix form,

$$\frac{\mathrm{d}\Delta}{\mathrm{d}t} = P\Delta, \tag{4}$$

where,

$$\Delta = \left(\frac{\partial x_i}{\partial x_{0j}}\right) \quad \text{and} \quad P = \left(\frac{\partial f_i}{\partial x_j}\right), \quad i, j = 1, 2, 3, 4.$$

In detail the matrix P is

$$P = \begin{bmatrix} 0 & 0 & 1 & 0 \\ 0 & 0 & 0 & 1 \\ \frac{\partial f_3}{\partial x_1} & \frac{\partial f_3}{\partial x_2} & 0 & 2 \\ \frac{\partial f_4}{\partial x_1} & \frac{\partial f_4}{\partial x_2} & -2 & 0 \end{bmatrix}, \tag{5}$$

where,

$$\begin{aligned} \frac{\partial f_3}{\partial x_1} &= 1 - \frac{1-\mu}{r_1^3} - \frac{\mu}{r_2^3} + \frac{3(1-\mu)(x_1+\mu)^2}{r_1^5} + \frac{3\mu(x_1+\mu-1)^2}{r_2^5}, \\ \frac{\partial f_3}{\partial x_2} &= \frac{3(1-\mu)(x_1+\mu)x_2}{r_1^5} + \frac{3\mu(x_1+\mu-1)x_2}{r_2^5}, \\ \frac{\partial f_4}{\partial x_1} &= \frac{\partial f_3}{\partial x_2}, \\ \frac{\partial f_4}{\partial x_2} &= 1 - \frac{1-\mu}{r_1^3} - \frac{\mu}{r_2^3} + \frac{3(1-\mu)x_2^2}{r_1^5} + \frac{3\mu x_2^2}{r_2^3}. \end{aligned} \tag{6}$$

The elements of Δ can be obtained by numerical integration of Equation (4).

3. Equations of Variations of Second Order

The differential equations of variations of second order are obtained by differentiating with respect to the initial conditions the Equation (4) of variation of first order.

Thus we obtain the expressions

$$\frac{\partial}{\partial x_{0j}}\left(\frac{\mathrm{d}\Delta}{\mathrm{d}t}\right) = \frac{\partial}{\partial x_{0j}}(P\cdot\Delta), \quad j = 1, 2, 3, 4 \tag{7}$$

or

$$\frac{\mathrm{d}\Delta_j}{\mathrm{d}t} = P_1 \frac{\partial x_1}{\partial x_{0j}} \Delta + P_2 \frac{\partial x_2}{\partial x_{0j}} \Delta + P \cdot \Delta_j, \quad j = 1, 2, 3, 4 \tag{8}$$

where,

$$\Delta_j = \frac{\partial \Delta}{\partial x_{0j}}, \quad P_1 = \frac{\partial P}{\partial x_1}, \quad P_2 = \frac{\partial P}{\partial x_2} \quad j = 1, 2, 3, 4,$$

and

$$\Delta_j = \begin{bmatrix} \frac{\partial^2 x_1}{\partial x_{01} \partial x_{0j}} & \frac{\partial^2 x_1}{\partial x_{02} \partial x_{0j}} & \frac{\partial^2 x_1}{\partial x_{03} \partial x_{0j}} & \frac{\partial^2 x_1}{\partial x_{04} \partial x_{0j}} \\ \frac{\partial^2 x_2}{\partial x_{01} \partial x_{0j}} & \frac{\partial^2 x_2}{\partial x_{02} \partial x_{0j}} & \frac{\partial^2 x_2}{\partial x_{03} \partial x_{0j}} & \frac{\partial^2 x_1}{\partial x_{04} \partial x_{0j}} \\ \frac{\partial^2 x_3}{\partial x_{01} \partial x_{0j}} & \frac{\partial^2 x_3}{\partial x_{02} \partial x_{0j}} & \frac{\partial^2 x_3}{\partial x_{03} \partial x_{0j}} & \frac{\partial^2 x_3}{\partial x_{04} \partial x_{0j}} \\ \frac{\partial^2 x_4}{\partial x_{01} \partial x_{0j}} & \frac{\partial^2 x_4}{\partial x_{02} \partial x_{0j}} & \frac{\partial^2 x_4}{\partial x_{03} \partial x_{0j}} & \frac{\partial^2 x_4}{\partial x_{04} \partial x_{0j}} \end{bmatrix} = \left(\frac{\partial^2 x_i}{\partial x_{0k} \partial x_{0j}} \right) \tag{9}$$

for $i, j, k = 1, 2, 3, 4$,

$$P_1 = \begin{bmatrix} 0 & 0 & 0 & 0 \\ 0 & 0 & 0 & 0 \\ \frac{\partial^2 f_3}{\partial x_1^2} & \frac{\partial^2 f_3}{\partial x_2 \partial x_1} & 0 & 0 \\ \frac{\partial^2 f_4}{\partial x_1^2} & \frac{\partial^2 f_4}{\partial x_2 \partial x_1} & 0 & 0 \end{bmatrix}, \tag{10}$$

$$P_2 = \begin{bmatrix} & 0 & 0 & 0 \\ 0 & 0 & 0 & 0 \\ \frac{\partial^2 f_3}{\partial x_1 \partial x_2} & \frac{\partial^2 f_3}{\partial x_2^2} & 0 & 0 \\ \frac{\partial^2 f_4}{\partial x_1 \partial x_2} & \frac{\partial^2 f_4}{\partial x_2^2} & 0 & 0 \end{bmatrix}. \tag{11}$$

The elements of the above matrices are:

$$\frac{\partial^2 f_3}{\partial x_1^2} = \frac{9(1-\mu)(x_1+\mu)}{r_1^5} + \frac{9\mu(x_1+\mu-1)}{r_2^5} - \\ - \frac{15(1-\mu)(x_1+\mu)^3}{r_1^7} - \frac{15\mu(x_1+\mu-1)^3}{r_2^7},$$

$$\frac{\partial^2 f_3}{\partial x_1 \partial x_2} = \frac{3(1-\mu)x_2}{r_2^5} + \frac{3\mu x_2}{r_2^5} - \frac{15(1-\mu)(x_1+\mu)^2 x_2}{r_1^7} - \\ - \frac{15\mu(x_1+\mu-1)^2 x_2}{r_2^7},$$

$$\frac{\partial^2 f_4}{\partial x_1^2} = \frac{\partial^2 f_3}{\partial x_1 \partial x_2} = \frac{\partial^2 f_3}{\partial x_2 \partial x_1},$$

$$\frac{\partial^2 f_4}{\partial x_1 \partial x_2} = \frac{3(1-\mu)(x_1+\mu)}{r_1^5} + \frac{3\mu(x_1+\mu-1)}{r_2^5} - \\ - \frac{15(1-\mu)(x_1+\mu)x_2^2}{r_1^7} - \frac{15\mu(x_1+\mu-1)x_2^2}{r_2^7} \tag{12}$$

$$\frac{\partial^2 f_3}{\partial x_2^2} = \frac{\partial^2 f_4}{\partial x_1 \partial x_2} = \frac{\partial^2 f_4}{\partial x_1 \partial x_2}$$

and

$$\frac{\partial^2 f_4}{\partial x_2^2} = \frac{9(1-\mu)x_2}{r_1^5} + \frac{9\mu x_2}{r_2^5} - \frac{15(1-\mu)x_2^3}{r_1^7} - \frac{15\mu x_2^3}{r_2^7}.$$

Equations (8) represent the equations of second variations.

4. Planar Symmetric Periodic Orbits

A solution $\mathbf{x}=\mathbf{x}(\mathbf{x}_0; t)$ of the system of differential Equation (1) is periodic of period $2T$, when it satisfies the relation

$$\mathbf{x}(\mathbf{x}_0; t+2T) = \mathbf{x}(\mathbf{x}_0; t), \quad T \neq 0. \tag{13}$$

In this paper, we restrict our attention to periodic orbits that are symmetrical with respect to the x-axis and re-enter at their second crossing with this axis.

It is known (e.g. Goudas, 1963, 1966) that, two symmetric arcs belong to the same solution when there is a point on each arc at which the velocity is perpendicular to the x-axis and in addition, these two points coalesce at the same point on this axis. Thus the conditions of periodicity are:

$$x_{02} = 0, \quad x_{03} = 0 \tag{14a}$$

or

$$\mathbf{x}_0 = (x_{01}, 0, 0, x_{04}),$$

and

$$x_2(x_{01}, x_{04}; T) = 0, \quad x_3(x_{01}, x_{04}; T) = 0. \tag{14b}$$

In general, for any set $(x_{01}, 0, 0, x_{04})$ of initial conditions, we have

$$\begin{aligned} x_2(x_{01}, x_{04}; T) &\neq 0, \\ x_3(x_{01}, x_{04}; T) &\neq 0. \end{aligned} \tag{15}$$

In order to determine the initial state of a periodic orbit we need to find suitable corrections, δx_{01}, δx_{04}, δT, such that while the relations $x_{02}=x_{03}=0$ will be preserved, the conditions

$$\begin{aligned} x_2(x_{01}+\delta x_{01}, x_{04}+\delta x_{04}, T+\delta T) &= 0 \\ x_3(x_{01}+\delta x_{01}, x_{04}+\delta x_{04}, T+\delta T) &= 0 \end{aligned} \tag{16}$$

will be satisfied.

This system has an infinity of solutions and we use the method of steepest descent to establish a third equation which will complete system (16) and hopefully provide us with the corrections required for a satisfactory convergence to the periodic orbit seeked.

For given values of x_{01}, x_{04} there exists a value of T for which

$$\begin{aligned} x_2(x_{01}, x_{04}; T) &= 0, \\ x_3(x_{01}, x_{04}; T) &= C \neq 0. \end{aligned} \tag{17}$$

Consider now Equations (17) as surfaces in the three-dimensional space (x_{01}, x_{04}, T) and the three dimensional curve (ε) which is their section. Every point on this curve corresponds to a set of parameters (x_{01}, x_{04}, T) satisfying Equations (17). In the same space the section of the surfaces

$$\begin{aligned} x_2(x_{01}, x_{04}; T) &= 0, \\ x_3(x_{01}, x_{04}; T) &= 0, \end{aligned}$$

is a curve (ε') which, for small c, can be considered as 'parallel' to (ε).

We wish to take a small displacement along the tangent of (ε) in the direction $\boldsymbol{\delta}^* = (\delta x_{01}^*, \delta x_{04}^*, \delta T^*)$. Therefore, we must have

$$\begin{aligned} x_2(x_{01} + \delta x_{01}^*, x_{04} + \delta x_{04}^*, T + \delta T^*) &= 0 \\ x_3(x_{01} + \delta x_{01}^*, x_{04} + \delta x_{04}^*, T + \delta T^*) &= c. \end{aligned} \tag{18}$$

Expanding Equations (18) into Taylor series and retaining only terms of up to and including the first order, we obtain the linear system (we set $\delta T^* = 1$ because we need the direction of $\boldsymbol{\delta}^*$)

$$\begin{aligned} \frac{\partial x_2}{\partial x_{01}} \delta x_{01}^* + \frac{\partial x_2}{\partial x_{04}} \delta x_{04}^* + \frac{\partial x_2}{\partial T} &= 0, \\ \frac{\partial x_3}{\partial x_{01}} \delta x_{01}^* + \frac{\partial x_3}{\partial x_{04}} \delta x_{04}^* + \frac{\partial x_3}{\partial T} &= 0. \end{aligned} \tag{19}$$

Solving this system we find that the corrections are:

$$\begin{aligned} \delta x_{01}^* &= \left\{ \frac{\partial x_3}{\partial T} \frac{\partial x_2}{\partial x_{04}} - \frac{\partial x_2}{\partial T} \frac{\partial x_3}{\partial x_{04}} \right\} \Bigg/ \left\{ \frac{\partial x_2}{\partial x_{01}} \frac{\partial x_3}{\partial x_{04}} - \frac{\partial x_2}{\partial x_{04}} \frac{\partial x_3}{\partial x_{01}} \right\}, \\ \delta x_{04}^* &= \left\{ \frac{\partial x_2}{\partial T} \frac{\partial x_3}{\partial x_{01}} - \frac{\partial x_3}{\partial T} \frac{\partial x_2}{\partial x_{01}} \right\} \Bigg/ \left\{ \frac{\partial x_2}{\partial x_{01}} \frac{\partial x_3}{\partial x_{04}} - \frac{\partial x_2}{\partial x_{04}} \frac{\partial x_3}{\partial x_{01}} \right\}, \end{aligned} \tag{20}$$

According to the method of steepest descent we have to move perpendicularly from curve (ε) to curve (ε'), i.e. in a direction $\boldsymbol{\delta} = (\delta x_{01}, \delta x_{04}, \delta T)$, such that

$$\boldsymbol{\delta}^* \cdot \boldsymbol{\delta} = 0, \tag{21}$$

or

$$\delta x_{01}^* \, \delta x_{01} + \delta x_{04}^* \, \delta x_{04} + \delta T = 0. \tag{22}$$

Equation (22) together with Equations (16) provide a system which, in general, we can solve for the unknown corrections. Expanding Equations (16) and ignoring terms of the third and higher order we obtain the system

$$l_1^{(2)}\,\delta x_{01} + l_2^{(2)}\,\delta x_{04} + l_3^{(2)}\,\delta x_{01}^2 + l_4^{(2)}\,\delta x_{04}^2 + l_5^{(2)}\,\delta x_{01}\,\delta x_{04} = 0,$$
$$l_1^{(3)}\,\delta x_{01} + l_2^{(3)}\,\delta x_{04} + l_3^{(3)}\,\delta x_{01}^2 + l_4^{(3)}\,\delta x_{04}^2 + l_5^{(5)}\,\delta x_{01}\,\delta x_{04} + x_3 = 0, \tag{23}$$

where,

$$l_1^{(i)} = \frac{\partial x_i}{\partial x_{01}} - a\,\frac{\partial x_i}{\partial T}/d, \quad l_2^{(i)} = \frac{\partial x_i}{\partial x_{04}} - b\,\frac{\partial x_i}{\partial T}/d,$$
$$l_3^{(i)} = \frac{1}{2}\frac{\partial^2 x_i}{\partial x_{01}^2} + \frac{1}{2}a^2\,\frac{\partial^2 x_i}{\partial T^2}/d^2 - a\,\frac{\partial^2 x_i}{\partial x_{01}\,\partial T}/d,$$
$$l_4^{(i)} = \frac{1}{2}\frac{\partial^2 x_i}{\partial x_{04}^2} + \frac{1}{2}b^2\,\frac{\partial^2 x_i}{\partial T^2}/d^2 - b\,\frac{\partial^2 x_i}{\partial x_{04}\,\partial T}/d,$$
$$l_5^{(i)} = ab\,\frac{\partial^2 x_i}{\partial T^2}/d^2 + \frac{\partial^2 x_i}{\partial x_{01}\,\partial x_{04}} - b\,\frac{\partial^2 x_i}{\partial x_{01}\,\partial T}/d - a\,\frac{\partial^2 x_i}{\partial x_{04}\,\partial T}/d, \qquad i = 2, 3$$

and

$$a = \frac{\partial x_3}{\partial T}\frac{\partial x_2}{\partial x_{04}} - \frac{\partial x_2}{\partial T}\frac{\partial x_3}{\partial x_{04}}, \qquad b = \frac{\partial x_2}{\partial T}\frac{\partial x_3}{\partial x_{01}} - \frac{\partial x_2}{\partial x_{01}}\frac{\partial x_3}{\partial T}$$
$$d = \frac{\partial x_2}{\partial x_{01}}\frac{\partial x_3}{\partial x_{04}} - \frac{\partial x_2}{\partial x_{04}}\frac{\partial x_3}{\partial x_{01}}.$$

5. Calculation of Periodic Orbits and Families

Results from previous work on this problem can be used in the selection of first approximations of periodic orbits. By means of the procedure described in the previous paragraphs, these approximations are improved and the initial state of a member of a family is determined. Our goal is to compute the entire family of solutions to which the above is a member and this can be done in two different methods.

The first method can be described as follows: To the initial state of the first orbit of the family an increment is added along either the tangent (first-order correction), or the parabola of common tangent with the family (second-order correction). In our numerical work both types of correction have been used. The initial conditions of the exact periodic orbit plus the above increments are used as starting conditions in the computation of the new orbit of the family, by means of the same technique of correction as described before. The same process is applied repeatedly in order to compute a number of members of the family that will describe sufficiently well the geometricall and dynamical characteristics of the entire family.

The second method involves expression of the family by a Taylor expansion. The necessary conditions for periodicity are

$$x_2(x_{01}, x_{04}; T) = 0,$$
$$x_3(x_{01}, x_{04}; T) = 0. \tag{24}$$

Once the first member of the family is obtained (say by the method described in the first part of this paragraph), the initial conditions of orbits of the same family, if put in the form $x_{01}+\delta x_{01}$, $x_{04}+\delta x_{04}$, where x_{01}, x_{04}, are the initial conditions of the known periodic orbit, will fulfill Equations (24), or their Taylor expansions,

$$\begin{aligned}
&\left(\frac{\partial}{\partial x_{01}}\,\delta x_{01}+\frac{\partial}{\partial x_{04}}\,\delta x_{04}+\frac{\partial}{\partial T}\,\delta T\right)x_2+\\
&\quad+\frac{1}{2!}\left(\frac{\partial}{\partial x_{01}}\,\delta x_{01}+\frac{\partial}{\partial x_{04}}\,\delta x_{04}+\frac{\partial}{\partial T}\,\delta T\right)^2 x_2+\\
&\quad+\frac{1}{3!}\left(\frac{\partial}{\partial x_{01}}\,\delta x_{01}+\frac{\partial}{\partial x_{04}}\,\delta x_{04}+\frac{\partial}{\partial T}\,\delta T\right)^3 x_2+\cdots=0,\\
&\left(\frac{\partial}{\partial x_{01}}\,\delta x_{01}+\frac{\partial}{\partial x_{04}}\,\delta x_{04}+\frac{\partial}{\partial T}\,\delta T\right)x_3\\
&\quad+\frac{1}{2!}\left(\frac{\partial}{\partial x_{01}}\,\delta x_{01}+\frac{\partial}{\partial x_{04}}\,\delta x_{04}+\frac{\partial}{\partial T}\,\delta T\right)^2 x_3+\\
&\quad+\frac{1}{3!}\left(\frac{\partial}{\partial x_{01}}\,\delta x_{01}+\frac{\partial}{\partial x_{04}}\,\delta x_{04}+\frac{\partial}{\partial T}\,\delta T\right)^3 x_3+\cdots=0.
\end{aligned}\tag{25}$$

System (25), truncated at the nth order term, will provide a set of two relations, which can be written in the form

$$\begin{aligned}
x_{01}^* &= x_{01}+p_1\,\delta T+p_2\,\delta T^2+\cdots+p_n\,\delta T^n\\
x_{04}^* &= x_{04}+q_1\,\delta T+q_2\,\delta T^2+\cdots+q_n\,\delta T^n
\end{aligned}\tag{26}$$

where

$$x_{01}^*=x_{01}+\delta x_{01}\,,\qquad x_{04}^*=x_{04}+\delta x_{04}\,.$$

The expression of the family in this way entails the computation of quantities such as

$$\frac{\partial x_2^{n_1+n_2+n_3}}{\partial x_{01}^{n_1}\,\partial x_{04}^{n_2}\,\partial T^{n_3}},\quad \frac{\partial x_3^{n_1+n_2+n_3}}{\partial x_{01}^{n_1}\,\partial x_{04}^{n_2}\,\partial T^{n_3}},$$

where $n_1+n_2+n_3\leqslant n$, and therefore integration of the equations of variation of the corresponding order.

In this paper we shall limit ourselves to terms up to the third order. We give below closed expressions for p_1, q_1, p_2, q_2, p_3, q_3 as well as the differential equations that the computation of these equations will involve:

$$\begin{aligned}
p_i &= (a_2b_3^{(i)}-a_3^{(i)}b_2)/(a_1b_2-a_2b_1)\\
q_i &= (a_3^{(i)}b_1-a_1b_3^{(i)})/(a_1b_2-a_2b_1)
\end{aligned},\quad i=1,2,3,$$

where

$$a_1=\frac{\partial x_2}{\partial x_{01}},\qquad a_2=\frac{\partial x_2}{\partial x_{04}},\qquad a_3^{(1)}=\frac{\partial x_2}{\partial T},$$

$$b_1=\frac{\partial x_3}{\partial x_{01}},\qquad b_2=\frac{\partial x_3}{\partial x_{04}},\qquad b_3^{(1)}=\frac{\partial x_3}{\partial T},$$

$$a_3^{(2)}=a_{42}p_1^2+a_{52}q_1^2+a_{62}+a_{72}p_1q_1+a_{82}p_1+a_{92}q_1\,,$$

$$b_3^{(2)} = a_{43}p_1^2 + a_{53}q_1^2 + a_{63} + a_{73}p_1q_1 + a_{83}p_1 + a_{93}q_1 ,$$

$$a_{4j} = \frac{1}{2}\frac{\partial^2 x_j}{\partial x_{01}^2}, \quad a_{5j} = \frac{1}{2}\frac{\partial^2 x_j}{\partial x_{04}^2}, \quad a_{6j} = \frac{1}{2}\frac{\partial^2 x_j}{\partial T^2}$$

$$a_{7j} = \frac{\partial^2 x_j}{\partial x_{01}\,\partial x_{04}}, \quad a_{8j} = \frac{\partial^2 x_j}{\partial x_{01}\,\partial T}, \quad a_{9j} = \frac{\partial^2 x_j}{\partial x_{04}\,\partial T}$$

$$a_{10j} = \frac{1}{6}\frac{\partial^3 x_j}{\partial x_{01}^3}, \quad a_{11j} = \frac{1}{6}\frac{\partial^3 x_j}{\partial x_{04}^3}, \quad a_{12j} = \frac{1}{6}\frac{\partial^3 x_j}{\partial T^3}$$

$$a_{13j} = \frac{1}{2}\frac{\partial^3 x_j}{\partial x_{01}^2\,\partial x_{04}}, \quad a_{14j} = \frac{1}{2}\frac{\partial^3 x_j}{\partial x_{01}\,\partial x_{04}^2},$$

$$a_{15j} = \frac{1}{2}\frac{\partial^3 x_j}{\partial x_{01}^2\,\partial T}, \quad a_{16j} = \frac{1}{2}\frac{\partial^3 x_j}{\partial x_{01}\,\partial T^2},$$

$$a_{17j} = \frac{1}{2}\frac{\partial^3 x_j}{\partial x_{04}^2\,\partial T}, \quad a_{18j} = \frac{1}{2}\frac{\partial^3 x_j}{\partial x_{04}\,\partial T^2}$$

$$a_{19j} = \frac{\partial^3 x_j}{\partial x_{01}\,\partial x_{04}\,\partial T}, \quad j = 2, 3,$$

and

$$\begin{aligned} a_3^{(3)} = {} & 2a_{42}p_1p_2 + 2a_{52}q_1q_2 + a_{72}p_2p_1 + a_{72}p_1q_2 + a_{82}p_2 + a_{92}q_2 + \\ & + a_{102}p_1^3 + a_{112}q_1^3 + a_{122} + a_{132}p_1^2q_1 + a_{142}q_1^2p_1 + a_{152}p_1^2 + \\ & + a_{162}p_1 + a_{172}q_2^2 + a_{182}q_1 + a_{192}p_1q_1 , \\ b_3^{(3)} = {} & 2a_{43}p_1p_2 + 2a_{53}q_1q_2 + a_{73}p_2p_1 + a_{73}p_1q_2 + a_{83}p_2 + a_{93}q_2 + \\ & + a_{103}p_1^3 + a_{113}q_1^3 + a_{123} + a_{133}p_1^2q_1 + a_{143}q_1^2p_1 + a_{153}p_1^2 + \\ & + a_{163}p_1 + a_{173}q_2^2 + a_{183}q_1 + a_{193}p_1q_1 . \end{aligned}$$

Thus it becomes evident that the coefficients of the expansion (26) are determined uniquely because they are, in all cases, the solution of simultaneous equations.

The differential equations of first and second variations have been already given and thus we shall give below the equations of third variations. These are:

$$\begin{aligned} \frac{\mathrm{d}\Delta_{jh}}{\mathrm{d}t} = {} & \left\{P_{11}\frac{\partial x_1}{\partial x_{0h}}\frac{\partial x_1}{\partial x_{0j}} + P_{12}\frac{\partial x_2}{\partial x_{0h}}\frac{\partial x_1}{\partial x_{0j}} + P_{21}\frac{\partial x_1}{\partial x_{0h}}\frac{\partial x_2}{\partial x_{0j}} + \right. \\ & \left. + P_{22}\frac{\partial x_2}{\partial x_{0h}}\frac{\partial x_2}{\partial x_{0j}} + P_1\frac{\partial^2 x_1}{\partial x_{0h}\,\partial x_{0j}} + P_2\frac{\partial^2 x_2}{\partial x_{0h}\,\partial x_{0j}}\right\}\Delta + \\ & + \left\{P_1\frac{\partial x_1}{\partial x_{0j}} + P_2\frac{\partial x_2}{\partial x_{0j}}\right\}\Delta_h + \left\{P_1\frac{\partial x_1}{\partial x_{0h}} + \frac{\partial x_2}{\partial x_{0h}}\right\}\Delta_j + \\ & + P\Delta_{jh}, \quad j, h = 1, 2, 3, 4, \end{aligned}$$

where

$$\Delta_{jh} = \frac{\partial \Delta_j}{\partial x_{0h}}, \quad \Delta_h = \frac{\partial \Delta}{\partial x_{0h}} \quad j, h = 1, 2, 3, 4,$$

$$P_{11} = \frac{\partial P_1}{\partial x_1}, \quad P_{12} = \frac{\partial P_1}{\partial x_2}, \quad P_{21} = \frac{\partial P_2}{\partial x_1}, \quad P_{22} = \frac{\partial P_2}{\partial x_2}.$$

No numerical results will be presented here. However, a description of the procedure followed will be given. The integration method employed is that of Runge-Kutta-Gill 4th order. By this method the 340 equations (of motion, first, second and third variations) were integrated. Faster methods developed in the past ten years (e.g. use of Steffensen's series) could also be used but the present work and its goals are independent of the integration method. The version of the Runge-Kutta-Gill employed halves or doubles the step-length so that an optimum step-size maintaining the prescribed accuracy will be used. The accuracy of the integration is also checked by means of the integrals of motion first, second and third variations, as well as the symplectic property and Schwartz's identities.

The integral of motion is given by expression (2). The integrals of the equations of first, second and third variations are:

$$\sum_{i=1}^{4} \frac{\partial J}{\partial x_i} \frac{\partial x_i}{\partial x_{0j}} = \frac{\partial J(\mathbf{x}_0)}{\partial x_{0j}}, \qquad j = 1, 2, 3, 4,$$

$$\sum_{l=1}^{4} \sum_{i=1}^{4} \frac{\partial^2 J}{\partial x_i \, \partial x_l} \frac{\partial x_i}{\partial x_{0j}} \frac{\partial x_l}{\partial x_{0k}} + \sum_{l=1}^{4} \sum_{i=1}^{4} \frac{\partial J}{\partial x_i} \frac{\partial^2 x_i}{\partial x_{0j} \, \partial x_l} \frac{\partial x_l}{\partial x_{0k}} = \frac{\partial^2 J(\mathbf{x}_0)}{\partial x_{0j} \, \partial x_{0k}}, \qquad j, k = 1, 2, 3, 4.$$

and

$$\begin{aligned} &\sum_m \sum_l \sum_i \frac{\partial^3 J}{\partial x_i \, \partial x_l \, \partial x_m} \frac{\partial x_i}{\partial x_{0j}} \frac{\partial x_l}{\partial x_{0k}} \frac{\partial x_m}{\partial x_{0\varrho}} + \\ &+ \sum_m \sum_l \sum_i \frac{\partial^2 J}{\partial x_i \, \partial x_l} \frac{\partial^2 x_i}{\partial x_{0j} \, \partial x_m} \frac{\partial x_l}{\partial x_{0k}} \frac{\partial x_m}{\partial x_{0\varrho}} + \\ &+ \sum_m \sum_l \sum_i \frac{\partial^2 J}{\partial x_i \, \partial x_l} \frac{\partial x_i}{\partial x_{0j}} \frac{\partial^2 x_l}{\partial x_{0k} \, \partial x_m} \frac{\partial x_m}{\partial x_{0\varrho}} + \\ &+ \sum_m \sum_l \sum_i \frac{\partial^2 J}{\partial x_i \, \partial x_m} \frac{\partial^2 x_i}{\partial x_{0j} \, \partial x_l} \frac{\partial x_l}{\partial x_{0k}} \frac{\partial x_m}{\partial x_{0\varrho}} + \\ &+ \sum_m \sum_l \sum_i \frac{\partial J}{\partial x_i} \frac{\partial^3 x_i}{\partial x_{0j} \, \partial x_l \, \partial x_m} \frac{\partial x_l}{\partial x_{0k}} \frac{\partial x_m}{\partial x_{0\varrho}} + \\ &+ \sum_m \sum_l \sum_i \frac{\partial J}{\partial x_i} \frac{\partial^2 x_i}{\partial x_{0j} \, \partial x_l} \frac{\partial^2 x_l}{\partial x_{0k} \, \partial x_m} \frac{\partial x_m}{\partial x_{0\varrho}} = \frac{\partial^3 J(\mathbf{x}_0)}{\partial x_{0j} \, \partial x_{0k} \, \partial x_{0\varrho}}, \\ &\qquad j, k, \varrho = 1, 2, 3, 4. \end{aligned}$$

The corrections to the initial state are obtaining by solving Equations (23) by means of the Newton-Raphson method.

Acknowledgements

This work has been done at the University of Patras under the supervision of Prof. C. L. Goudas to whom I am grateful for his valuable advise.

References

Darwin, G. H.: 1911, *Periodic orbits*, Acta Math., New York.
Goudas, C. L.: 1963, 'Three dimensional periodic orbits and their stability', Astronomical contributions from the Univ. of Manchester.
Goudas, C. L. and Bray, T.: 1966, 'Doubly symmetric orbits about the collinear Lagrangian points'.
Katsiaris, G.: 1969, 'Simply symmetric periodic orbits in the restricted problem of three bodies', Doctoral Dissertation, Technical University of Athens.
Szebehely, V.: 1967, *Theory of orbits*, Academic Press.

ELLIPTIC RESTRICTED PROBLEM: FOURTH-ORDER STABILITY ANALYSIS OF THE TRIANGULAR POINTS

E. RABE

University of Cincinnati Observatory, Cincinnati, Ohio, U.S.A.

Abstract. The expansions considered in a preceding investigation of small free oscillations about the triangular points of the elliptic restricted problem (Rabe, 1970) have been carried to the fourth powers of the eccentricity e and the mass parameter κ. For the appropriate fixed values of the frequency parameter Z, the resulting expressions for the determinant equation $D=0$ reduce directly to the desired relations between those e- and κ-values which determine the transition curves between stability and instability in the κ-e plane. For small and moderate values of e, these analytical results are in good agreement with those of earlier numerical studies, as well as with previous analytical ones of lower order.

The stability of the triangular points L_4, L_5 in the elliptic restricted problem has been studied numerically by Danby (1964), on the basis of properly selected integrations of the first-order variation equations. Subsequently Alfriend and Rand (1969), using the two variable expansion method, obtained analytical results of the form $\kappa(e, e^2)$ for the transition curves bounding regions of stability in the κ-e plane. κ denotes the mass parameter.

The present author has proposed a method (Rabe, 1970) for determining the small free oscillations about L_4 and L_5 in the form of exponential series proceeding in powers of e, where the applicable frequency Z has to satisfy the determinant equation $\Delta=0$. Here Δ is the determinant of the coefficients of an infinite number of homogeneous linear equations, resulting from the substitution of the assumed solution into the linearized differential equations. These differential equations have periodic coefficients depending on the basic frequency N of the relative motion of the two primaries, so that the solution involves terms with arguments or exponents depending on all possible integral multiples of N, in addition to their dependence on Z. Only terms up to the order of e^2 had been considered in the original short exposition of the method and of its results concerning the stability of L_4 and L_5 in the elliptic problem, and the determinant Δ had been approximated by that of the principal 6×6 linear equations. It had been noted, though, that the consideration of an expanded 10×10 set of coefficients, from a set of 10 equations, would enable one to carry the expansion of Δ to the inclusion of the power e^4, and thus to obtain Z to a correspondingly higher degree of approximation.

It will be convenient to denote by D the infinite determinant which results from Δ if each linear equation is first divided by N^2. This is in line with the introduction of the normalized frequency Z by Equation (30) of the earlier paper (Rabe, 1970), which in effect changed the unit of time so that N is now represented by $Z=+1$. For further

B. D. Tapley and V. Szebehely (eds.), Recent Advances in Dynamical Astronomy, 156–160.

simplification of the numerical coefficients of D, it will be useful to introduce

$$\kappa = \frac{27}{4}\nu \tag{1}$$

as the new mass parameter, instead of the earlier $\nu = \mu/N^4$.

In terms of Z, κ, and e, the expansion of the 10×10 approximation of D takes the form

$$\begin{aligned} D = {} & [(576Z^4 - 2800Z^6 + 5596Z^8 - 5925Z^{10} + 3581Z^{12} - \\ & - 1250Z^{14} + 246Z^{16} - 25Z^{18} + Z^{20}) + \kappa(-576Z^2 + 880Z^4 - \\ & - 28Z^6 - 295Z^8 + 24Z^{10} + 10Z^{12} - 20Z^{14} + 5Z^{16}) + \\ & + \kappa^2(1200Z^2 - 1860Z^4 + 230Z^6 + 330Z^8 + 90Z^{10} + 10Z^{12}) + \\ & + \kappa^3(144 - 20Z^2 + 590Z^4 + 140Z^6 + 10Z^8) + \kappa^4(24 + 55Z^2 + \\ & + 5Z^4) + \kappa^5] + e^2[(13824Z^4 - 58992Z^6 + 102936Z^8 - \\ & - 95319Z^{10} + 51168Z^{12} - 16386Z^{14} + 3060Z^{16} - 303Z^{18} + \\ & + 12Z^{20}) + \kappa(-15840Z^2 + 14704Z^4 + 3058Z^6 + 1156Z^8 - \\ & - 4456Z^{10} + 1856Z^{12} - 562Z^{14} + 84Z^{16}) + \kappa^2(33208Z^2 - \\ & - 27678Z^4 + 2124Z^6 + 4238Z^8 + 888Z^{10} + 180Z^{12}) + \\ & + \kappa^3(3912 - 1370Z^2 + 7112Z^4 + 1674Z^6 + 156Z^8) + \kappa^4(326 + \\ & + 527Z^2 + 48Z^4)] + e^4\left[\left(71136Z^4 - 336122Z^6 + \frac{1249017}{2}Z^8 - \right.\right. \\ & - \frac{1188315}{2}Z^{10} + \frac{2551517}{8}Z^{12} - \frac{202743}{2}Z^{14} + \frac{75465}{4}Z^{16} - \\ & \left. - 1874Z^{18} + \frac{597}{8}Z^{20}\right) + \kappa\left(-153720Z^2 + 118811Z^4 + \right. \\ & + \frac{166205}{4}Z^6 + \frac{272839}{8}Z^8 - \frac{234923}{4}Z^{10} + 22918Z^{12} - \\ & \left. - \frac{11251}{2}Z^{14} + \frac{5529}{8}Z^{16}\right) + \kappa^2\left(31392 + 285673Z^2 - \right. \\ & - 208906Z^4 + \frac{133927}{8}Z^6 + \frac{248663}{8}Z^8 + \frac{37293}{8}Z^{10} + \\ & \left. + \frac{11709}{8}Z^{12}\right) + \kappa^3\left(33952 + 5564Z^2 + \frac{361051}{8}Z^4 + \frac{39175}{4}Z^6 + \right. \\ & \left.\left. + \frac{7275}{8}Z^8\right) + \kappa^4\left(\frac{1589}{4} + \frac{5645}{8}Z^2 + \frac{249}{4}Z^4\right)\right] + \cdots. \end{aligned} \tag{2}$$

For any prescribed values of e and κ, the oscillation frequency Z has to be determined from

$$D = 0, \tag{3}$$

and the points L_4 and L_5 will be stable only if Z is real. Once Z has been obtained from (3), the coefficients u_r, v_r of the solution u, v can be determined from the 10 linear equations. One of the complex coefficients, say u_0, remains arbitrary. In general, Equation (3) has four different roots Z between 0 and $+1$, but two of these were found to be redundant in the construction of the complete solution u, v (Rabe, 1971). It will be seen that the $\kappa - e$ boundaries of stability are associated with two particular and fixed double-roots of (3). It may be noted that for any root Z the resulting u, v is a particular solution in terms of a pulsating coordinate frame, so that the actual linear displacement from L_4 or L_5 will be given by $\xi = \varrho u$, $\eta = \varrho v$, where ϱ is the variable distance between the primaries.

For $e=0$, or for the circular restricted problem, the expression (2) for D reduces to its first square bracket, which involves all powers of κ up to κ^5 and all even powers of Z up to Z^{20}. It is easily verified that the two 'critical' values of the mass parameter (Danby, 1964) are $\kappa_1 = \frac{1}{4}$ and $\kappa_2 = \frac{3}{16}$, in terms of the κ used in this study. If the e-independent part of D is denoted by D_0, it will be found that

$$D_0 = 0 \tag{4}$$

for $\kappa_1 = \frac{1}{4}$, $Z_1 = 1/\sqrt{2}$, as well as for $\kappa_2 = \frac{3}{16}$, $Z_2 = \frac{1}{2}$. Moreover, the partial derivative

$$\begin{aligned}\partial D_0/\partial (Z^2) = [&(1152 Z^2 - 8400 Z^4 + 22384 Z^6 - 29625 Z^8 + \\ &+ 21486 Z^{10} - 8750 Z^{12} + 1968 Z^{14} - 225 Z^{16} + 10 Z^{18}) + \\ &+ \kappa(-576 + 1760 Z^2 - 84 Z^4 - 1180 Z^6 + 120 Z^8 + \\ &+ 60 Z^{10} - 140 Z^{12} + 40 Z^{14}) + \kappa^2 (1200 - 3720 Z^2 + \\ &+ 690 Z^4 + 1320 Z^6 + 450 Z^8 + 60 Z^{10}) + \kappa^3 (-20 + \\ &+ 1180 Z^2 + 420 Z^4 + 40 Z^6) + \kappa^4 (55 + 10 Z^2)]\end{aligned} \tag{5}$$

also vanishes for κ_1, Z_1 and κ_2, Z_2, so that

$$\partial D_0/\partial Z = 2Z\, \partial D_0/\partial (Z^2) = 0 \tag{6}$$

is likewise true for these two sets of particular values κ_j, Z_j. Finally, since from

$$\begin{aligned}\partial^2 D_0/\partial (Z^2)^2 = [&(1152 - 16800 Z^2 + 67152 Z^4 - 118500 Z^6 + \\ &+ 107430 Z^8 - 52500 Z^{10} + 13776 Z^{12} - 1800 Z^{14} + \\ &+ 90 Z^{16}) + \kappa (1760 - 168 Z^2 - 3540 Z^4 + 480 Z^6 + \\ &+ 300 Z^8 - 840 Z^{10} + 280 Z^{12}) + \kappa^2 (-3720 + \\ &+ 1380 Z^2 + 3960 Z^4 + 1800 Z^6 + 300 Z^8) + \\ &+ \kappa^3 (1180 + 840 Z^2 + 120 Z^4) + 10 \kappa^4]\end{aligned} \tag{7}$$

and

$$\partial^2 D_0/\partial Z^2 = 4Z^2\, \partial^2 D_0/\partial (Z^2)^2 + 2\, \partial D_0/\partial (Z^2) \tag{8}$$

one obtains $\partial^2 D_0/\partial Z^2 = +256$ for $j=1$, but $\partial^2 D_0/\partial Z^2 = -297/2$ for $j=2$, it follows that $D_0(\kappa_1, Z)$ has a minimum for $Z = Z_1 = 1/\sqrt{2}$, while $D_0(\kappa_2, Z)$ has a maximum for $Z = Z_2 = \frac{1}{2}$. Consequently, even in the restricted problem with $e=0$, Z_1 and Z_2 are double-roots of $D_0 = 0$. For $Z_2 = \frac{1}{2}$, this has been noted before (Rabe, 1970).

In the elliptic problem, with $e \neq 0$, Z has to satisfy the condition $D=0$, instead of $D_0=0$. If κ_j, Z_j denote the critical κ_1, Z_1 *or* κ_2, Z_2 of the circular problem with $e=0$, the complete D of Equation (2) may be expanded as a three-variable Taylor series in powers of the quantities

$$\Delta\kappa = \kappa - \kappa_j, \qquad \Delta Z = Z - Z_j, \qquad \Delta e = e. \tag{9}$$

Because of $\partial D_0/\partial Z = 0$ in both cases, the lowest power of ΔZ involved in these series will be $(\Delta Z)^2$. Considering only the dominant terms, the resulting equations $D=0$ are approximated by

$$\begin{aligned} &(1) \qquad 64(\Delta\kappa) - 32e^2 + 128(\Delta Z)^2 = 0 \quad [\text{for } j=1] \\ &(2) \qquad 297(\Delta\kappa)^2 - \frac{9801}{256}e^2 - \frac{297}{4}(\Delta Z)^2 = 0 \quad [\text{for } j=2]. \end{aligned} \tag{10}$$

If arbitrary values of $(\Delta\kappa)_j$ and e are admitted, the resulting

$$(\Delta Z)_1^2 = \frac{1}{4}e^2 - \frac{1}{2}(\Delta\kappa)_1, \qquad (\Delta Z)_2^2 = 4(\Delta\kappa)_2^2 - \frac{33}{64}e^2 \tag{11}$$

may be positive or negative, or zero. For $(\Delta Z)^2 > 0$, one gets two real and unequal roots $Z = Z_j \pm (\Delta Z)_j$ and stability, while for $(\Delta Z)^2 < 0$ the resulting imaginary $(\Delta Z)_j = \pm iR_j$ lead to conjugate complex roots Z and to instability. Therefore, the transitional real *double*-roots of (3) are possible only if $(\Delta Z)_j^2 = 0$, or if the critical frequencies $Z_1 = 1/\sqrt{2}$ and $Z_2 = \frac{1}{2}$ of the circular problem remain fixed even in the elliptic problem. The $\kappa - e$ boundaries between stability and instability are curves *defined* by these two fixed roots Z_1 and Z_2. While higher-order terms have been omitted in Equations (10) and (11), their inclusion would merely modify the simple results $\pm(\Delta Z)_j$ into absolutely slightly different $+(\Delta Z)_{j+}$, $-(\Delta Z)_{j-}$, with similar modifications for imaginary $(\Delta Z)_j$, without changing the fact that only the assumption $(\Delta Z)_j \equiv 0$ permits the continued existence of the basic double-roots Z_j when $e \neq 0$.

Because of this requirement $(\Delta Z)_j = 0$, the contemplated three-variable Taylor expansion of D is reduced to one in powers of $\Delta\kappa$ and e, for the purpose of finding the $\Delta\kappa(e)$ representing the boundaries of stability. On the basis of Equation (2) for D, all the terms factored by e^2 and e^4 can be carried up to the order of $(\Delta\kappa)^4$, but the following results are more than sufficient to obtain $\Delta\kappa$ to the order of e^4, or at least to that of e^3.

(1) For $\kappa_1 = \frac{1}{4}$, $Z_1 = 1/\sqrt{2}$:

$$\begin{aligned} D = {} & [64\Delta\kappa + 432(\Delta\kappa)^2 + 353(\Delta\kappa)^3 + \cdots] + e^2[-32 + 2236\Delta\kappa + \\ & + 14389(\Delta\kappa)^2 + \cdots] + e^4\left[-\frac{4727}{4} + \frac{993403}{32}\Delta\kappa + \cdots\right] = 0, \end{aligned} \tag{12}$$

or

$$\Delta\kappa = \frac{1}{2}e^2 - \frac{177}{256}e^4 + \cdots,$$

(2) For $\kappa_2 = \frac{3}{16}$, $Z_2 = \frac{1}{2}$:

$$\begin{aligned} D = {} & [297\,(\Delta\kappa)^2 + 207\,(\Delta\kappa)^3 + 39\,(\Delta\kappa)^4 + \cdots] + \\ & + e^2 \left[-\frac{9801}{256} - \frac{629505}{8192}\,\Delta\kappa + \frac{1151091}{128}\,(\Delta\kappa)^2 + \cdots \right] \\ & + e^4 \left[-\frac{18492705}{16384} + \frac{497152863}{65536}\,\Delta\kappa + \cdots \right] = 0\,, \end{aligned} \tag{13}$$

or

$$\Delta\kappa = \pm \frac{\sqrt{33}}{16}\,e + 0.04165\,e^2 \mp 0.14926\,e^3 + \cdots.$$

The coefficient of e^4 in the second $\Delta\kappa$ remains undetermined, because the D in (13) has no term involving $\Delta\kappa$ linearly.

The analytical results in Equations (12) and (13) are in good agreement, for small and moderate values of e, with the second-order ones by Alfriend and Rand (1969), and with the numerical findings by Danby (1964). In particular, the important point denoted by 'D' in Danby's Figure 1 for the transition curves may be obtained from the present Equations (12) and (13) by requiring that

$$\kappa_1 + (\Delta\kappa)_1 = \kappa_2 + (\Delta\kappa)_2\,. \tag{14}$$

Considering the terms up to and including e^3 in both $(\Delta\kappa)_j$, it is found that the two boundaries come together for $\kappa = 0.296$ and $e = 0.302$. Danby's comparable result, properly converted with respect to the present mass parameter, was $\kappa = 0.302$, $e = 0.314$.

References

Alfriend, K. T. and Rand, R. H.: 1969, *AIAA J.* **7**, 1024.

Danby, J. M. A.: 1964, *Astron. J.* **69**, 165.

Rabe, E.: 1970, in G. E. O. Giacaglia (ed.), *Periodic Orbits, Stability and Resonances*, Reidel Publ. Co., Dordrecht, Holland, p. 33.

Rabe, E.: 1971, *Bull. AAS* **3**, 270.

A LINEAR DESCRIPTION OF THE SECOND SPECIES SOLUTIONS

P. GUILLAUME
Institut d'Astrophysique, Liège, Belgium

Abstract. From Breakwell-Perko's Matching Theory, we deduce I/O Equations relating the 'histories' of a second species solution before and after its passage near the Moon. We give some applications of these equations.

1. Introduction

A second species solution of the restricted problem is, roughly speaking, a solution which passes too close to the Moon to be considered as a weakly perturbed keplerian orbit around the Earth. Near the Moon, it suffers from it an important deflexion.

The concept of second species solution was introduced by Poincaré (1899), which, however, did not give a complete analysis of them. Kevorkian (1962, 1963, 1964, 1968) gave an approximate analytical description of these solutions.

But we shall restrict our attention to the very efficient study by Breakwell and Perko (1965, 1966). After a brief summary of their theory, we shall deduce from it approximate equations relating the 'histories' of a second species solution before and after its passage near the Moon. Lastly, we shall give some applications of these equations.

2. The Matching Theory (Breakwell-Perko, 1965, 1966)

The following definitions are used: $\mathbf{r}$ is the position vector **ES** from the Earth to the Satellite. Let us consider the function $\mathbf{r}(t, \mu)$ which is defined on the set $]t_a, t_b[\times]0, \mu_0[$ $(\mu_0 \ll 1)$ and which is a solution of the differential equations of the planar circular restricted problem which can be written in the form (see Figure 1):

$$\ddot{\mathbf{r}} = -\frac{\mathbf{r}}{r^3} + \mu \mathbf{g}(\mathbf{r}, t).$$

$\mathbf{r}(t, \mu)$ is, by definition, a Breakwell-Perko second species solution or simply an SSS, if

(1) $\varrho(\mu) \equiv \inf_{t_a < t < t_b} \varrho(t, \mu) = 0(\mu)$ where $\boldsymbol{\varrho}(t, \mu)$ is the position vector **MS** from the Moon to the Satellite

$$\boldsymbol{\varrho}(t, \mu) = \mathbf{r}(t, \mu) - \mathbf{R}(t)$$

and

$$\varrho(t, \mu) = |\boldsymbol{\varrho}(t, \mu)|.$$

(2) $r(t)$ being the limit of $\mathbf{r}(t, \mu)$ as μ tends to 0, there exists a t-value t_1 such that

$$\varrho(t_1) \equiv r(t_1) - \mathbf{R}(t_1) = 0$$

B. D. Tapley and V. Szebehely (eds.), Recent Advances in Dynamical Astronomy, 161–174.

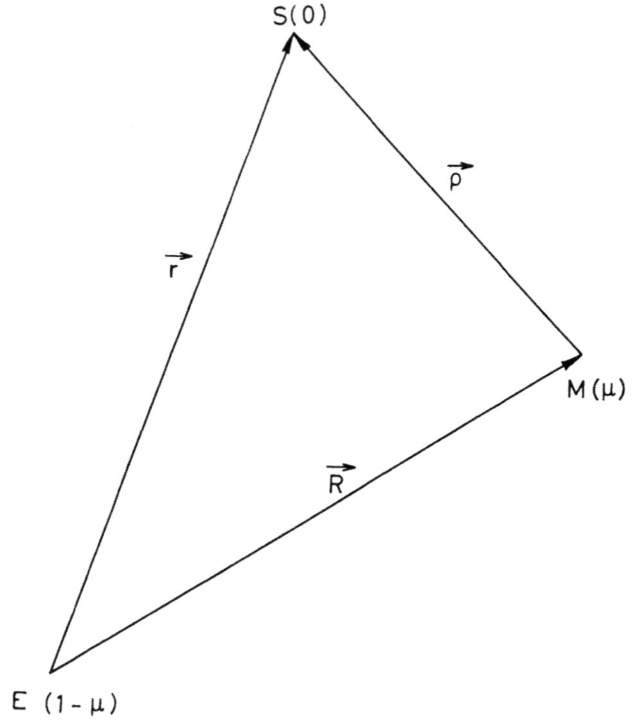

Fig. 1.

and an interval $[t_0, t_0']$ including t_1 and in which $\boldsymbol{\varrho}(t)$ is different from zero for all t-values different from t_1.

$r(t)$ is called the basic orbit of $\mathbf{r}(t, \mu)$: considered in the interval $[t_0, t_0']$, it is the union of two different conics respectively defined on $[t_0, t_1]$ and $[t_1, t_0']$ (Figure 2).

(3) δt_0 and $\delta t_0'$ being chosen $\leqslant 0(\mu)$ and in such a way that

$$t_1 \in [t_0 + \delta t_0, t_0' + \delta t_0'] \subset]t_a, t_b[,$$

the position and velocity vectors must satisfy the inequalities

$$\delta \mathbf{r}_0' \equiv \mathbf{r}(t_0' + \delta t_0', \mu) - r(t_0') \leqslant 0(\mu);$$
$$\delta \mathbf{v}_0' \equiv \dot{\mathbf{r}}(t_0' + \delta t_0', \mu) - r(t_0') \leqslant 0(\mu).$$

We shall restrict our attention to the SSS in the interval $[t_0 + \delta t_0, t_0' + \delta t_0']$. Breakwell and Perko found two different approximate expansions to $0(\mu^2)$ before its closest approach to the Moon:

– The first one is valid outside a circle centered at M and of radius $\varrho_1 = 0(\mu^{1/2})$; the first approximation of this 'outer expansion' is, of course, a conic which we take as the osculating conic to $\mathbf{r}(t, \mu)$ at $t_0 + \delta t_0$ and which we designate by $\mathbf{r}^{(0)}(t)$

$$\mathbf{r}^{(0)}(t_0 + \delta t_0) = \mathbf{r}(t_0 + \delta t_0, \mu)$$
$$\dot{\mathbf{r}}^{(0)}(t_0 + \delta t_0) = \dot{\mathbf{r}}(t_0 + \delta t_0, \mu).$$

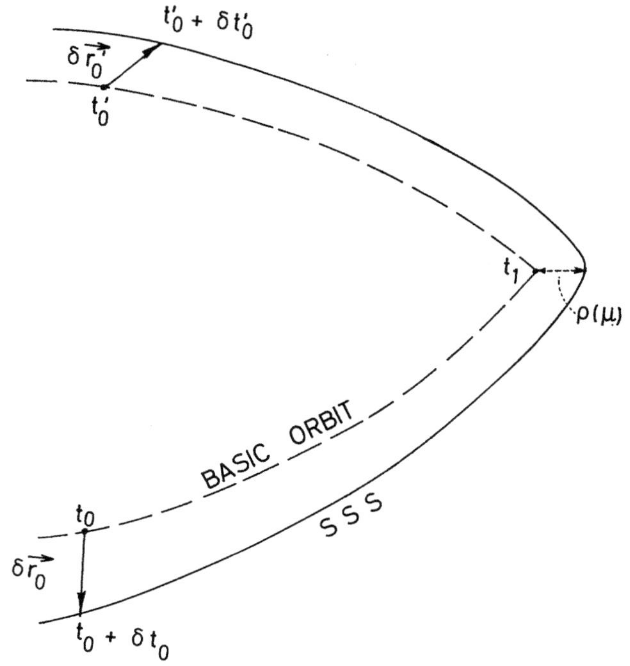
t'0 + δt'0
δr'0
t'0
t1
ρ(μ)
BASIC ORBIT
SSS
t0
δr0
t0 + δt0

Fig. 2.

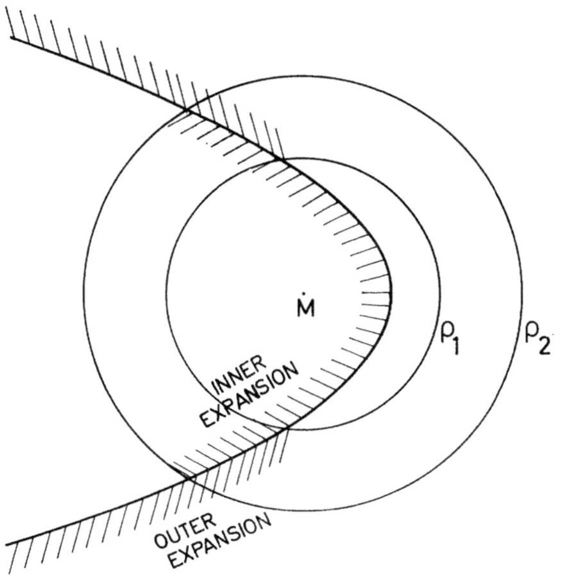
M
ρ1
ρ2
INNER EXPANSION
OUTER EXPANSION

Fig. 3.

– The second one is valid in a circle centered also at the Moon and of radius $\varrho_2 = 0(\mu^{1/2}) > \varrho_1$; this is the 'inner expansion' whose first approximation $\boldsymbol{\varrho}^{(0)}(t, \mu)$ is an hyperbola which is chosen as the osculating hyperbola to $\boldsymbol{\varrho}(t, \mu)$ at the time t_p of the closest approach to M (See Figure 3):

$$\dot{\varrho}(t_p, \mu) = 0$$

$$\boldsymbol{\varrho}^{(0)}(t_p, \mu) = \boldsymbol{\varrho}(t_p, \mu); \quad \dot{\boldsymbol{\varrho}}^{(0)}(t_p, \mu) = \dot{\boldsymbol{\varrho}}(t_p, \mu).$$

Since $\varrho_2 > \varrho_1$, inner and outer expansions can be matched in the transition ring $\varrho_1 \leqslant \varrho \leqslant \varrho_2$. This leads to the Matching Equations which can be written as follows:

$$\boldsymbol{\Delta} = (t_p - t_1)\,\mathbf{V} + \delta\mathbf{r}^{(0)}(t_1) + \mu\mathbf{r}_1^{(b)}(t_1, t_0) - \frac{\mu}{V^3} \ln \frac{\mu}{2V^3 \left|\sin \frac{\beta^{(0)}}{2}\right|}\,\mathbf{V} + 0(\mu^2) \tag{1}$$

where

$$\mathbf{v}_\infty = \mathbf{V} + \delta\mathbf{v}^{(0)}(t_1) + \mu\mathbf{v}_1^{(b)}(t_1, t_0) + 0(\mu^2) \tag{2}$$

– $\boldsymbol{\Delta}$ is the position vector of the 'incoming' asymptote along which S approximately arrives near M;

– $\mathbf{v}_\infty$ is the velocity at infinity on this asymptote (similar definitions hold for $\boldsymbol{\Delta}'$ and $\mathbf{v}'_\infty$ which are related to the 'outgoing' asymptote, see Figure 4);

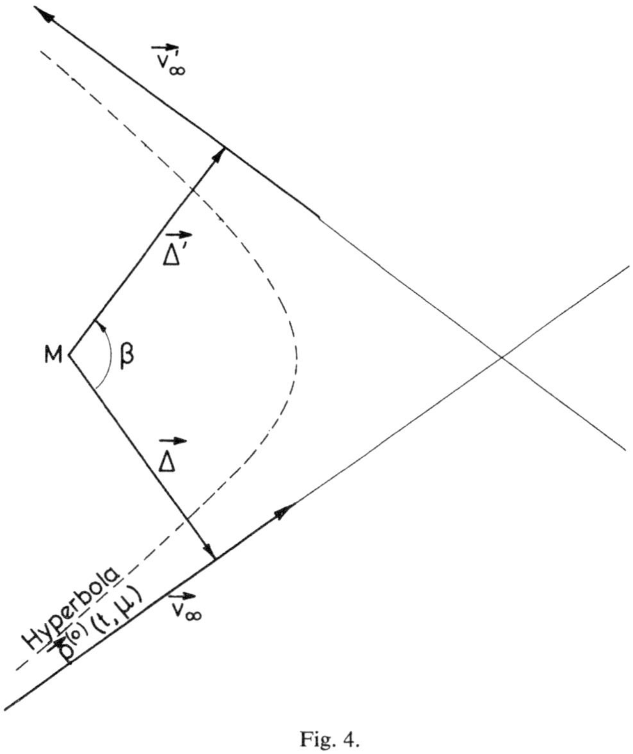

Fig. 4.

– $\mathbf{V}$ is the velocity relative to M on $\mathbf{r}(t)$ at the time t_1 of the collision

$$\mathbf{V} = \dot{\boldsymbol{\varrho}}(t_1) = \dot{r}(t_1) - \dot{\mathbf{R}}(t_1)\,; \tag{3}$$

in the planar circular restricted problem, $\dot{\mathbf{R}}(t_1) = \mathbf{e}_3 \wedge \dot{\mathbf{R}}(t_1)$; furthermore, at t_1, $\mathbf{R}(t_1) = r(t_1)$. So, we have also

$$\mathbf{V} = \dot{r}(t_1) - \mathbf{e}_3 \wedge r(t_1).$$

In other words, $\mathbf{V}$ is the synodical velocity at t_1 on the basic orbit.

$$\begin{aligned} -\delta\mathbf{r}^{(0)}(t) &= \mathbf{r}^{(0)}(t) - r(t) \\ \delta\mathbf{v}^{(0)}(t) &= \dot{r}^{(0)}(t) - \dot{r}(t). \end{aligned} \tag{4}$$

– $\mu\mathbf{r}_1^{(b)}(t_1, t_0)$ is the bounded part of the first order perturbation of M on the motion of S. More precisely, let the classical first order perturbation of M on S be

$$\mu\mathbf{r}_1(t, t_0) = \mu \int_{t_0}^{t} \frac{\partial r(t)}{\partial r(t')}\, \mathbf{g}(r(t'), t')\, \mathrm{d}t'.$$

This quantity behaves as $-(\mu/V^3)\ln(t_1 - t)\,\mathbf{V}$ in the sense that, as t tends to t_1, they both become infinite, but their difference remains bounded. It is this difference that we designate by $\mu\mathbf{r}_1^{(b)}(t, t_0)$

$$\begin{aligned} \mu\mathbf{r}_1^{(b)}(t, t_0) &= \mu\mathbf{r}_1(t, t_0) + \frac{\mu}{V^3}\ln(t_1 - t)\,\mathbf{V}\,; \\ \mu\mathbf{v}_1^{(b)}(t, t_0) &= \frac{\mathrm{d}}{\mathrm{d}t}\left[\mu\mathbf{r}_1^{(b)}(t, t_0)\right]. \end{aligned} \tag{5}$$

– $\beta^{(0)}$ is the angle $(\mathbf{V}, \mathbf{V}')$, where $\mathbf{V}'$ is the synodical velocity on the basic orbit at the departure from M (see Figure 5).

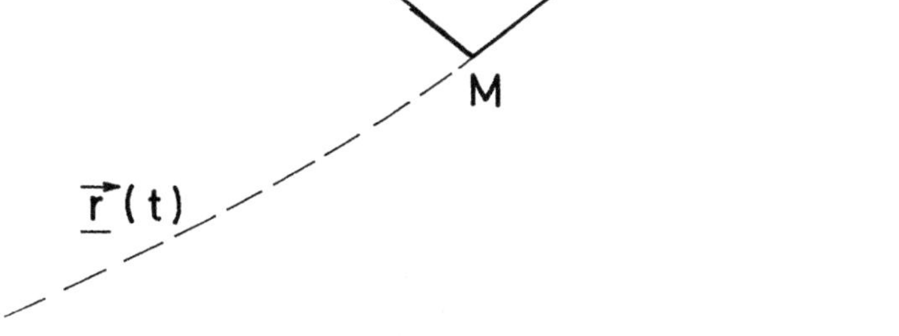

Fig. 5

Similar equations hold for the portion of the SSS after its closest approach to the Moon

$$\boldsymbol{\Delta}' = (t_p - t_1)\,\mathbf{V}' + \delta\mathbf{r}'^{(0)}(t_1) + \mu\mathbf{r}_1'^{(b)}(t_1, t_0') + \\ + \frac{\mu}{V^3}\ln\frac{\mu}{2V^3\left|\sin\dfrac{\beta^{(0)}}{2}\right|}\,\mathbf{V}' + 0(\mu^2) \qquad (1')$$

$$\mathbf{v}_\infty' = \mathbf{V}' + \delta\mathbf{v}'^{(0)}(t_1) + \mu\mathbf{v}_1'^{(b)}(t_1, t_0') + 0(\mu^2). \qquad (2')$$

Remarks

(1) Taking the scalar product of Equation (1) with $V^{-2}\cdot\mathbf{V}$, one deduces the difference $t_p - t_1$ in terms of $\delta\mathbf{r}_0$ and $\delta\mathbf{v}_0$ implicitly contained in $\delta\mathbf{r}^{(0)}(t_1)$

$$t_p - t_1 = -\frac{\mathbf{V}}{V^2}\cdot\left(\delta\mathbf{r}^{(0)}(t_1) + \mu\mathbf{r}_1^{(b)}(t_1, t_0)\right) + \frac{\mu}{V^3}\ln\frac{\mu}{2V^3\left|\sin\dfrac{\beta^{(0)}}{2}\right|} + 0(\mu^2). \qquad (6)$$

(2) The first order approximation of Equations (2) and (2′) is:

$$\mathbf{V} = \mathbf{v}_\infty + 0(\mu); \qquad \mathbf{V}' = \mathbf{v}_\infty' + 0(\mu).$$

These equations imply

(a) since $v_\infty = v_\infty'$ for all values of μ, that

$$V = V'. \qquad (7)$$

(b) if β is the angle $(\mathbf{v}_\infty, \mathbf{v}_\infty')$, that

$$\beta = \beta^{(0)} + 0(\mu). \qquad (8)$$

So, we can call $\beta^{(0)}$ zero order deflexion angle.

3. The Input Output Equations

$\boldsymbol{\Delta}'$ and $\mathbf{v}_\infty'$ are obtained from $\boldsymbol{\Delta}$ and $\mathbf{v}_\infty$ through a rotation of angle β:

$$\boldsymbol{\Delta}' = R(\beta)\cdot\boldsymbol{\Delta}; \qquad \mathbf{v}_\infty' = R(\beta)\cdot\mathbf{v}_\infty.$$

That gives, using Equations (1), (2), (1′) and (2′):

$$(t_p - t_1)\,\mathbf{V}' + \delta\mathbf{r}'^{(0)}(t_1) + \mu\mathbf{r}_1'^{(b)}(t_1, t_0') + \frac{\mu}{V^3}\ln\frac{\mu}{2V^3\left|\sin\dfrac{\beta^{(0)}}{2}\right|}\,\mathbf{V}' = \\ = R(\beta)\cdot\left[(t_p - t_1)\,\mathbf{V} + \delta\mathbf{r}^{(0)}(t_1) + \mu\mathbf{r}_1^{(b)}(t_1, t_0) - \frac{\mu}{V^3}\ln\frac{\mu}{2V^3\left|\sin\dfrac{\beta^{(0)}}{2}\right|}\,\mathbf{V}\right] + 0(\mu^2) \qquad (9)$$

$$\mathbf{V}' + \delta\mathbf{v}'^{(0)}(t_1) + \mu\mathbf{v}_1'^{(b)}(t_1, t_0') = \\ = R(\beta)\cdot\left[\mathbf{V} + \delta\mathbf{v}^{(0)}(t_1) + \mu\mathbf{v}_1^{(b)}(t_1, t_0)\right] + 0(\mu^2).$$

β is expressed in terms of $\mathbf{v}_\infty$ and of the angular momentum h of the inner hyperbola by the classical formula of the hyperbolic motion

$$\cot \frac{\beta}{2} = \frac{v_\infty h}{\mu}.$$

Furthermore, $h = (\mathbf{\Delta} \wedge \mathbf{v}_\infty) \cdot \mathbf{e}_3$ can be written in terms of $\delta \mathbf{r}^{(0)}(t_1)$ and $\delta \mathbf{v}^{(0)}(t_1)$ by means of Equations (1) and (2):

$$h = [(\delta \mathbf{r}^{(0)}(t_1) + \mu \mathbf{r}_1^{(b)}(t_1, t_0)) \wedge \mathbf{V}] \cdot \mathbf{e}_3 + 0(\mu^2).$$

And eventually, since $v_\infty = V + 0(\mu)$, we get for:

$$\cot \frac{\beta}{2} = \frac{-V[\delta \mathbf{r}^{(0)}(t_1) + \mu \mathbf{r}_1^{(b)}(t_1, t_0)] \cdot (\mathbf{e}_3 \wedge \mathbf{V})}{\mu} + 0(\mu). \tag{10}$$

Equations (9) and (10) can be interpreted as follows:
Knowing the conditions at t_0 on the SSS, and thus also $\delta \mathbf{r}^{(0)}(t_1)$ and $\delta \mathbf{v}^{(0)}(t_1)$, Equations (9) and (10) give $\delta \mathbf{r}'^{(0)}(t_1)$ and $\delta \mathbf{v}'^{(0)}(t_1)$ and eventually the conditions at t_0' on the SSS in question. All that can be done without considering explicitly what happens near the Moon. In other words, given the conditions at t_0 on the SSS, conditions which we call the 'input variables', Equations (9) and (10) act as a black box that furnishes directly the 'output variables' i.e. the conditions at t_0'. That is the reason why we have called Equations (9) the Input-Output Equations or more simply the I/O Equations.

Now, it can be shown that, because of the existence of the Jacobian integral, they are not independent. For a given basic orbit, we have shown that they reduce in fact to the following three simple scalar independent equations:

$$V(\mathbf{e}_3 \wedge \mathbf{V}) \cdot [\delta \mathbf{r}^{(0)}(t_1) + \mu \mathbf{r}_1^{(b)}(t_1, t_0)] = -\mu \cot \frac{\beta^{(0)}}{2} + 0(\mu^2) \tag{11}$$

$$V(\mathbf{e}_3 \wedge \mathbf{V}') \cdot [\delta \mathbf{r}'^{(0)}(t_1) + \mu \mathbf{r}_1'^{(b)}(t_1, t_0')] = -\mu \cot \frac{\beta^{(0)}}{2} + 0(\mu^2) \tag{12}$$

$$\mathbf{V}' \cdot \delta \mathbf{r}'^{(0)}(t_1) - \mathbf{V} \cdot \delta \mathbf{r}^{(0)}(t_1) + \mathbf{V}' \cdot \mu \mathbf{r}_1'^{(b)}(t_1, t_0') - \mathbf{V} \cdot \mu \mathbf{r}_1^{(b)}(t_1, t_0) - \\ - \frac{2\mu}{V} \ln \frac{\mu}{2V^3 \left| \sin \frac{\beta^{(0)}}{2} \right|} = 0 + 0(\mu^2). \tag{13}$$

Let us note, at this point, that Equations (1), (1'), (11), (12) and (13) are only valid if $\beta^{(0)}$ is different from zero. We shall discuss it later.

Because of the third point of the definition of an SSS, $\delta \mathbf{r}^{(0)}(t_1)$ can be replaced by its differential with respect to δt_0 and to the difference between the conditions in the synodical frame at $t_0 + \delta t_0$ and t_0 respectively on the SSS and on its basic orbit

$$\delta x_0 = x(t_0 + \delta t_0, \mu) - x(t_0), \qquad \delta y_0 = y(t_0 + \delta t_0, \mu) - y(t_0)$$
$$\delta \dot{x}_0 = \dot{x}(t_0 + \delta t_0, \mu) - \dot{x}(t_0), \qquad \delta \dot{y}_0 = \dot{y}(t_0 + \delta t_0, \mu) - \dot{y}(t_0),$$

similarly

$$\delta x_0' = x(t_0' + \delta t_0', \mu) - x(t_0'), \qquad \delta y_0' = y(t_0' + \delta t_0', \mu) - y(t_0')$$
$$\delta \dot{x}_0' = \dot{x}(t_0' + \delta t_0', \mu) - \dot{x}(t_0'), \qquad \delta \dot{y}_0' = \dot{y}(t_0' + \delta t_0', \mu) - \dot{y}(t_0').$$

Furthermore, since the Jacobian constant plays an important role in the restricted problem, it is useful to replace $\delta\dot{y}_0$ by its differential in terms of δx_0, μ and the difference δC between the Jacobian constants $C+\delta C$ and C respectively on the SSS and on its basic orbit. Thus we substitute to $\delta\mathbf{r}^{(0)}(t_1)$ and $\delta\mathbf{r}'^{(0)}(t_1)$ respectively

$$\mathbf{r}_U(t_1, t_0)\cdot\delta U_0 - \mathbf{V}\,\delta t_0 \quad \text{and} \quad \mathbf{r}'_{U'}(t_1, t'_0)\cdot\delta U'_0 - \mathbf{V}'\,\delta t'_0$$

where

$$\delta U_0 = (\delta x_0, \delta y_0, \delta\dot{x}_0, \delta C), \qquad \delta U'_0 = (\delta x'_0, \delta y'_0, \delta\dot{x}'_0, \delta C)$$

and $\mathbf{r}_U(t_1, t_0)$ and $\mathbf{r}'_{U'}(t_1, t'_0)$ correspond to the usual linear operator applied to δU_0 and $\delta U'_0$.

Eventually, the *I*/O Equations take the form:

$$Y_U(t_1, t_0)\cdot\delta U_0 + \mu Y_1^{(b)}(t_1, t_0) = -\frac{\mu}{V^2}\cot\frac{\beta^{(0)}}{2} + 0(\mu^2) \tag{14}$$

$$Y'_{U'}(t_1, t'_0)\cdot\delta U'_0 + \mu Y_1'^{(b)}(t_1, t'_0) = -\frac{\mu}{V^2}\cot\frac{\beta^{(0)}}{2} + 0(\mu^2) \tag{15}$$

$$\begin{aligned}\delta t'_0 - \delta t_0 = {} & X'_{U'}(t_1, t'_0)\cdot\delta U'_0 - X_U(t_1, t_0)\cdot\delta U_0 + \mu X_1'^{(b)}(t_1, t'_0) - \\ & - \mu X_1^{(b)}(t_1, t_0) - \frac{2\mu}{V^2}\ln\frac{\mu}{2V^3\left|\sin\dfrac{\beta^{(0)}}{2}\right|} + 0(\mu^2).\end{aligned} \tag{16}$$

X and Y are the components of $\mathbf{r}$ in a reference system whose axis are directly parallel respectively to $\mathbf{V}$ and $\mathbf{e}_3\wedge\mathbf{V}$. Similar definitions hold for X' and Y' relative to $\mathbf{V}'$ and $\mathbf{e}_3\wedge\mathbf{V}'$. (See Figure 6.)

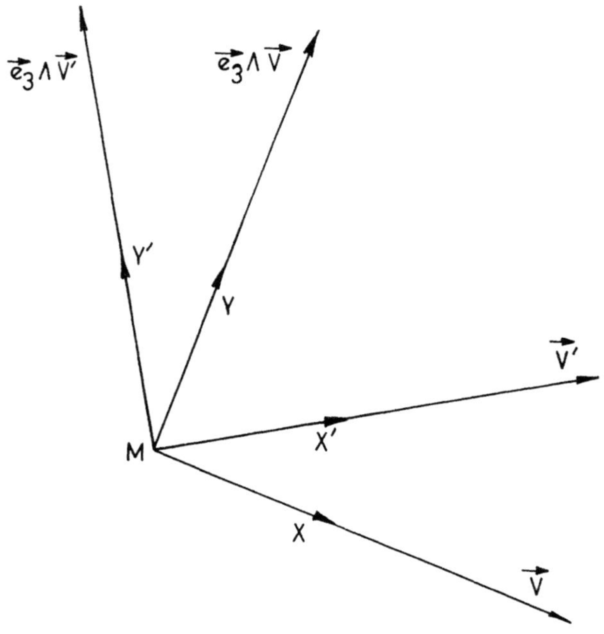

Fig. 6.

4. Applications

(A) In his numerical investigations of symmetric periodic orbits of the restricted problem (i.e. orbits which cross twice orthogonally the axis EM), Broucke (1962, 1968) obtained a family represented by the curves B_1 and B_2 in the (x_0, C)-plane (x_0: abscissa of the point of orthogonal crossing; C: Jacobian constant) (Figure 7).

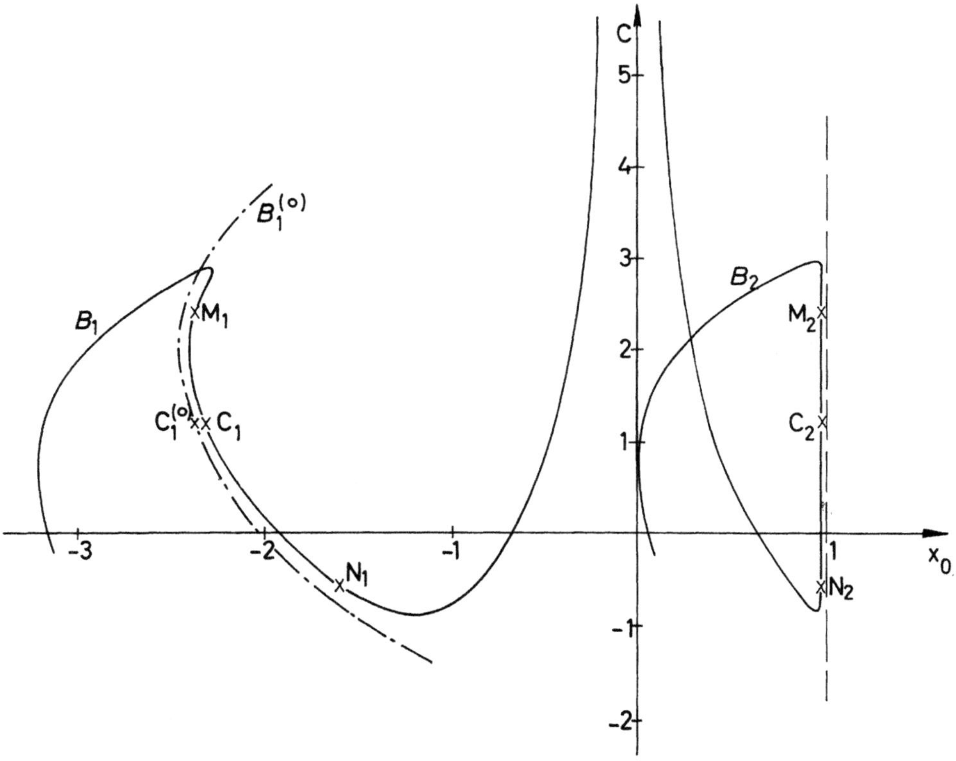

Fig. 7.

In this diagram, a symmetric periodic orbit is represented by two points of same ordinate C (Figure 8).

Let us consider the portions of the curves respectively comprised between M_1 and N_1 on the one hand and M_2 and N_2 on the other. The arc M_2N_2 is close to the straight line of equation $x_0 = 1$, which has been drawn in dotted lines. The points C_1 and C_2 of same ordinate correspond to a solution which crosses orthogonally EM once far from $M(\leftrightarrow C_1)$ and once close to $M(\leftrightarrow C_2)$.

This symmetric periodic orbit can be considered as an SSS.

But, given a symmetric periodic SSS $\mathbf{r}(t, \mu)$, what can we say of its basic orbit $r(t)$?

(1) This basic orbit encounters twice M;

(2) It is symmetric with respect to EM, and in particular

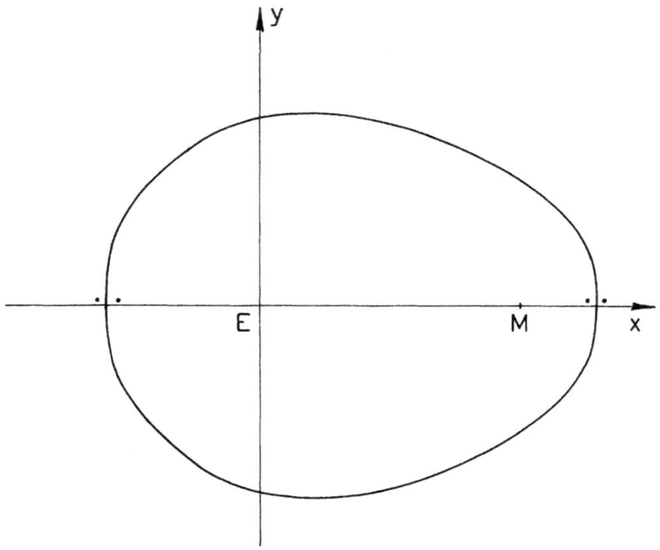

Fig. 8.

– it crosses orthogonally EM (at $X_1^{(0)}$ on Figure 9)

– its velocities $\mathbf{V}$ and $\mathbf{V}'$ at the collision are symmetric to each other with respect to the y-axis

$$2\alpha + \beta^{(0)} = 0 \quad (\text{mod } \pi). \tag{17}$$

Such an orbit is called by Henon (1968) arc of a conic 'S' with consecutive collisions. These conics S are distributed into families which are represented by curves in the (x_0, C)-plane (x_0: abscissa of the point $X_1^{(0)}$ of orthogonal crossing of the synodical axis by the conic S). One of these curves is drawn in dashed dotted line on Figure 7 and is designated by $B_1^{(0)}$.

Now the problem is the following: given values of x_0 and C corresponding to a point $C_1^{(0)}$ of $B_1^{(0)}$, let us consider the conic S with consecutive collisions which corresponds to these values. Let us choose as initial time the time of symmetric conjunction (at $X_1^{(0)}$) on this conic, which we shall designate by $\mathbf{r}(t)$

$$x(0) = x_0, \quad y(0) = 0, \quad \dot{x}(0) = 0, \quad \dot{y}(0) = \dot{y}_0.$$

$\mathbf{r}(t)$ encounters M at the times τ and $-\tau$. Let us now search an SSS of basic orbit $\mathbf{r}(t)$ and which crosses orthogonally EM at the initial time at X_1 near $X_1^{(0)}$

$$x(0, \mu) = x_0 + \delta x_0, \quad y(0, \mu) = 0, \quad \dot{x}(0, \mu) = 0, \quad \dot{y}(0, \mu) = \dot{y}_0 + \delta \dot{y}_0$$
$$\delta x_0, \delta \dot{y}_0 \leqslant 0(\mu).$$

Let $C + \delta C$ be its Jacobian constant. The initial conditions of this SSS, which has a

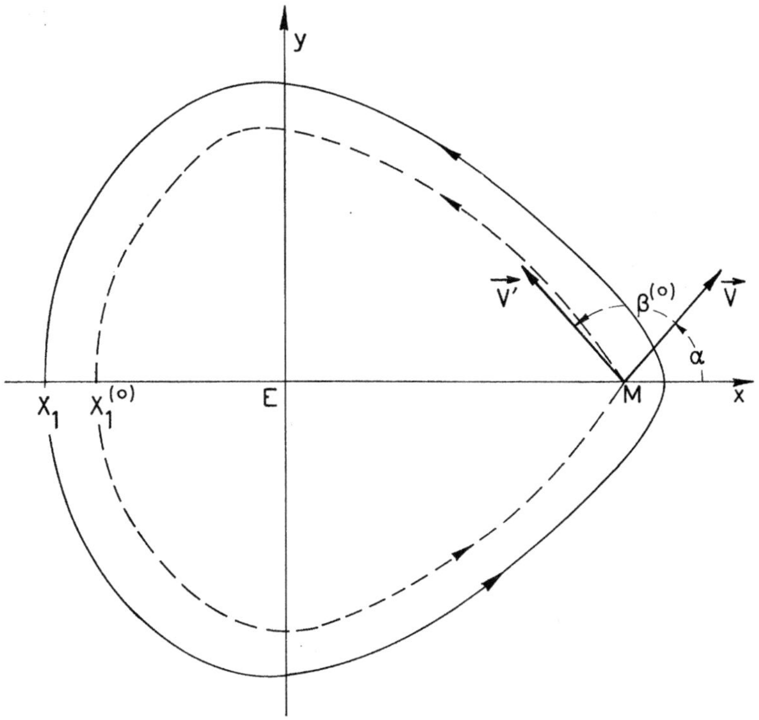

Fig. 9.

given zero order deflexion angle $\beta^{(0)}$, must verify the I/O Equation (14) which becomes here

$$Y_x(\tau, 0)\,\delta x_0 + Y_C(\tau, 0)\,\delta C = -\frac{\mu}{V^2}\cot\frac{\beta^{(0)}}{2} - \mu Y_1^{(b)}(\tau, 0) + 0(\mu^2). \qquad (18)$$

In order to take the symmetry conjunction near M into account, we have simply, at this order, to replace $\cot(\beta_1^{(0)}/2)$ by $-\tan\alpha$ (Equation (17)).

What does Equation (18) mean?

(1) If we keep δC equal to 0, we get a value for δx_0 in terms of μ, x_0 and C. This gives to $0(\mu^2)$ the position of C_1 on the curve B_1 (see Figure 7).

(2) If we equate to 0 the first member of Equation (18), we get the equation of the tangent to $B_1^{(0)}$ at $C_1^{(0)}$. In other words, B_1 is approximately obtained in the neighbourhood of $C_1^{(0)}$ by a parallel displacement of $B_1^{(0)}$ proportional to μ.

As for the half period, $\tau + \delta\tau$, it is simply obtained from Equation (6), by replacing $t_p - t_1$ by $\delta\tau$ in this equation:

$$\delta\tau = -V^{-1}\left(X_x(\tau, 0)\,\delta x_0 + X_C(\tau, 0)\,\delta C + \mu X_1^{(b)}(\tau, 0)\right) + \\ + \frac{\mu}{V^3}\ln\frac{\mu}{2V^3\left|\sin\frac{\beta^{(0)}}{2}\right|} + 0(\mu^2).$$

(B) One can deduce an other result from the previous considerations.

But, first of all, let us prove that

$$Y_1^{(b)}(-\tau, 0) = Y_1^{(b)}(\tau, 0). \tag{19}$$

The periodic SSS which corresponds to C_1 (see Figure 7) is symmetric. So, if it crosses orthogonally the axis EM at $t = \tau + \delta\tau$, a new symmetric conjunction must occur at $t = -(\tau + \delta\tau)$. Thus, from the I/O Equation (15), we get

$$Y_x(-\tau, 0)\,\delta x_0 + Y_C(-\tau, 0)\,\delta C = \frac{\mu}{V^2}\tan\alpha - \mu Y_1^{(b)}(-\tau, 0) + 0(\mu^2). \tag{20}$$

(The primes have been omitted because all quantities are to be computed along the same conic as for Equation (18).) But, since the coefficients of δx_0 and δC in Equations (18) and (20) can be proved to be equal because of the symmetrical character of the basic conic $\mathbf{r}(t)$ with respect to $t = 0$:

$$Y_x(-\tau, 0) = Y_x(\tau, 0); \quad Y_C(-\tau, 0) = Y_C(\tau, 0), \tag{21}$$

the difference of Equations (18) and (20) furnishes the desired Equation (19).

Now let us search for an SSS $\mathbf{r}(t, \mu)$ which is periodic but not necessarily symmetric and which has $r(t)$ as basic orbit. We choose still the initial time at the crossing of the synodical axis near X_1^0 (see Figure 9)

$$x(0, \mu) = x_0 + \delta x_0, \quad y(0, \mu) = 0, \quad \dot{x}(0, \mu) = \delta\dot{x}_0, \quad \dot{y}(0, \mu) = \dot{y}_0 + \delta\dot{y}_0.$$

These initial conditions must verify the I/O Equation (14):

$$Y_x(\tau, 0)\,\delta x_0 + Y_{\dot{x}}(\tau, 0)\,\delta\dot{x}_0 + Y_C(\tau, 0)\,\delta C = \\ = -\mu\cot\frac{\beta^{(0)}}{2} - \mu Y_1^{(b)}(\tau, 0) + 0(\mu^2). \tag{22}$$

Then, $2\tau + \delta\tau$ being the time of the second crossing of EM by the SSS, let us write as follows the conditions on $\mathbf{r}(t, \mu)$ at $2\tau + \delta\tau$:

$$x(2\tau + \delta\tau, \mu) = x_0 + \delta x_\tau, \quad y(2\tau + \delta\tau, \mu) = 0,$$
$$\dot{x}(2\tau + \delta\tau, \mu) = \delta\dot{x}_\tau, \quad \dot{y}(2\tau + \delta\tau, \mu) = \dot{y}_0 + \delta\dot{y}_\tau.$$

These quantities must satisfy Equation (15), which becomes:

$$Y_x(\tau, 2\tau)\,\delta x_\tau + Y_{\dot{x}}(\tau, 2\tau)\,\delta\dot{x}_\tau + Y_C(\tau, 2\tau)\,\delta C = \\ = -\mu\cot\frac{\beta^{(0)}}{2} - \mu Y_1^{(b)}(\tau, 2\tau) + 0(\mu^2).$$

The periodicity conditions allow us to replace δx_τ, $\delta\dot{x}_\tau$ respectively by δx_0, $\delta\dot{x}_0$;

furthermore we can substitute $(-\tau, 0)$ to the arguments $(\tau, 2\tau)$ because of the autonomous character of the differential equations of the restricted problem in the synodical axis. Hence, we get:

$$Y_x(-\tau, 0)\,\delta x_0 + Y_{\dot{x}}(-\tau, 0)\,\delta \dot{x}_0 + Y_C(-\tau, 0)\,\delta C = \\ = -\mu \cot \frac{\beta^{(0)}}{2} - \mu Y_1^{(b)}(-\tau, 0) + 0(\mu^2). \qquad (23)$$

Then, taking into account Equations (19) and (21), subtracting Equations (22) and (23) and computing the coefficients of $\delta\dot{x}_0$, we obtain:

$$V \cos\alpha \cdot x_0 \cdot \dot{y}_0 \cdot \delta\dot{x}_0 = 0(\mu^2).$$

Thus, because $\alpha \neq \pi/2$ (in order that $\beta^{(0)} \neq 0$: see Equation (17)), and if the basic orbit starts neither at the Earth nor on the zero velocity curve, we must have

$$\delta\dot{x}_0 = 0(\mu^2). \qquad (24)$$

We obtain thus the following result, correct to $0(\mu^2)$: an SSS which is periodic and has a conic S with consecutive collisions as a basic orbit is symmetric periodic. This result is important. For it can be proved that the conics S are the only basic conics of the SSS defined at the beginning of this paper.

5. Conclusion

The Matching Theory gives a description of the SSS valid to $0(\mu^2)$. But the I/O Equation (14) relating the initial conditions depends on the history of the SSS after its passage close to the Moon only through the zero order deflexion angle. For instance, whether or not the SSS crosses orthogonally the axis EM after the passage near the Moon has no infiuence on this equation. This influence has disappeared in the terms $0(\mu^2)$. So it can be interesting to compute these terms. They have been computed by Breakwell and Perko. But their results must still be checked because of the complicated character of the computations. Furthermore, terms of larger order in μ would be useful in practical applications, for example in the computation of a trajectory of a satellite passing close to Mars or Venus.

We have found that the expressions to be manipulated in the Matching Theory can be put into the form of Poisson series and handled by analytical compilers as MAO or ESP (Rom, 1970, 1971). The results will appear in a next paper.

The Matching Theory is not yet applicable to a SSS which passes through the Earth or the Moon. But it seems possible to extend it to such cases by using regularizing variables as Burdet's (1968).

We have constructed a new Matching Theory which remains applicable when $\beta^{(0)} = 0$. That will be presented in a next paper. As a conclusion, we can say that it seems that the Matching Theory is a very powerfull tool in the study of every kind of Second Species Solutions.

References

Breakwell, J. M. and Perko, L. M.: 1966, in R. L. Duncombe and V. G. Szebehely (eds.), *Progress in Astronautics and Aeronautics*, Vol. **17**, Academic Press, New York, London.
Broucke, R.: 1962, 'Recherche d'orbites périodiques dans le problème restreint plan (Système Terre-Lune)', Doctoral Dissertation.
Broucke, R.: 1968, JPL Techn. Rep., 32–1168.
Burdet, C. A.: 1968, 'Theory of Kepler Motion: The General Perturbed Two Body Problem', Doctoral Dissertation.
Guillaume, P.: 1971, 'Solutions périodiques symétriques du problème restreint des trois corps pour de faibles valeurs du rapport des masses', Doctoral Dissertation.
Henon, M.: 1968, *Bull. Astron.* **3**, 377–402.
Kevorkian, J.: 1962, *Astron. J.* **67**, 204–211.
Kevorkian, J. and Lagerstrom, P. A.: 1963, *J. Mécanique* **2**, 189–218.
Kevorkian, J. and Lagerstrom, P. A., 1963, *J. Mécanique* **2**, 493–504.
Kevorkian, J. and Lagerstrom ,P. A.: 1966, in R. L. Duncombe and V. G. Szebehely (eds.), *Progress in Astronautics and Aeronautics*', Vol. **14**. Academic Press, New York, London.
Kevorkian, J. and Lancaster, J. E.: 1968, *Astron. J.* **73**, 791–806.
Perko, L. M.: 1965, 'Asymptotic Matching in the Restricted Three-Body Problem', Ph. D. Dissertation.
Poincaré, H.: 1893, *Les méthodes nouvelles de la mécanique céleste*, Vol. **3**, Gauthier-Villars, Paris.
Rom, A.: 1970, *Celes. Mech.* **1**, 301–319.
Rom, A.: 1971, *Celes. Mech.* **3**, 331–345.

PART III

THE *N*-BODY PROBLEM AND STELLAR DYNAMICS

PROBLEMS OF STELLAR DYNAMICS

G. CONTOPOULOS
University of Thessaloniki, Greece

1. Introduction

A few years ago we had an IAU Symposium on the Theory of Orbits in Thessaloniki, Greece, that brought together for the first time people working in Celestial Mechanics and Stellar Dynamics.

Although the interaction between the two groups was not as extensive as we would have liked, still in the subsequent years there has been some useful exchange of ideas between Celestial Mechanicians and Stellar Dynamicists. The interaction centered on two subjects:

(a) The theory of Integrals of Motion, and

(b) The N-body problem.

Stellar Dynamics has established, at the same time, contacts with Plasma Physics.

Plasma Dynamics has several problems in common with Stellar Dynamics and Celestial Mechanics. At the background there is a common N-body problem with inverse square forces. One branch of Plasma Dynamics deals with Orbit Theory, and several other problems have their analogues in the Galaxy, or in other stellar systems. It is of interest to see how Plasma Physicists, unaware of the work done in Stellar Dynamics and Celestial Mechanics since Poincaré, not only attacked similar problems and found similar results, but in many cases went much further. Their advances in recent years have been quite impressive. Thus we have found it necessary to learn their methods and apply them to our own problems.

I think that a natural extention of Celestial Mechanics is in the direction of Stellar Dynamics and Plasma Physics. Otherwise, whatever the interest of accurate theory of motion of various bodies in the Solar System it cannot provide life at infinitum to Celestial Mechanics.

I have thought of a few problems of Stellar Dynamics that would be of interest to Celestial Mechanicians.

The first class of problems refers to the general behavior of orbits and integrals of motion in a *time independent field*. This is divided into two-dimensional and many dimensional problems.

2. Motions in two Dimensions

Typical cases of two dimensional problems are the orbits in a rotating meridian plane of an axisymmetric galaxy, and in the plane of symmetry of a spiral galaxy.

The motions in time-independent two dimensional fields are now well understood.

A first step in this study is to find the periodic orbits.

B. D. Tapley and V. Szebehely (eds.), Recent Advances in Dynamical Astronomy, 177–191.

As proved by Arnold and Moser periodic orbits of the stable type are surrounded, in general, by tori containing quasi periodic orbits, with two frequencies ω_1, ω_2. The intersections of these tori by a surface of section are invariant curves. On each invariant curve a rotation number is defined, which is equal to the ratio of the two frequencies ω_1/ω_2.

In integrable cases the whole surface of section is filled with closed invariant curves. Curves along which ω_1/ω_2 is rational (n/m) are composed of periodic orbits. However, in non-integrable systems which are near integrable ones, we have only a small number of periodic orbits with a given rational rotation number n/m, usually one stable and one unstable. The invariant curves near such stable periodic points* form a set of islands, while the asymptotic curves emanating from the unstable periodic points cross each other in an intricate way forming zones of instability around the original periodic point (Poincaré, 1899; Birkhoff, 1927, 1935).

The area covered by each set of islands and zone of instability, is of the order of $\varepsilon^{(m+n)/2}$ (Contopoulos, 1965) where ε is a measure of deviation of the non-integrable system from the integrable one.

Thus, although there is an infinity of zones of instability around the original periodic point, their measure is not only finite, but small, if ε is small. As found in specific cases by Deprit and Henrard (1966) it is quite possible that the non-integrable system behaves like an integrable one within all the accuracy provided by double precision in the computer. Namely no islands or zones of instability are discernible, and the characteristic exponents of the periodic orbits of the n/m type are zero to the limits of the accuracy of the computer.

However, as the perturbation increases, the islands and zones of instability increase in size, and at the same time the range of values of ω_1/ω_2 also increases, thus the number of important zones of instability (where n and m are small) increases. When the closed invariant curves separating various zones of instability disappear, the asymptotic curves of various types cross each other. The intersection of asymptotic curves emanating from the same periodic orbit are homoclinic points, while the intersections of asymptotic curves of different periodic orbits heteroclinic points. Bartlett and Wagner (1970) have given examples of intersections of such zones of instability and heteroclinic orbits in a particular dynamical system (Figure 1).

We have presented evidence (Contopoulos, 1971) that the appearance of zones of instability is always connected with the appearance of heteroclinic orbits. In fact if we draw a 'rotation curve' (Figure 2) giving the rotation number as a function of the distance from the initial periodic point, we see that whenever there is a 'dissolution of invariant curves' (which corresponds to the complicated behavior of the asymptotic curves, and nearby open invariant curves, in the zones of instability), there is an abrupt change of the rotation number. In the zones of instability we have indications that, besides the main periodic orbits of type n/m, there is an infinity of periodic orbits corresponding to nearby rational numbers. As no closed invariant curve separates the

* We use the expression 'periodic point' to indicate the intersection of the surface of section by the periodic orbit.

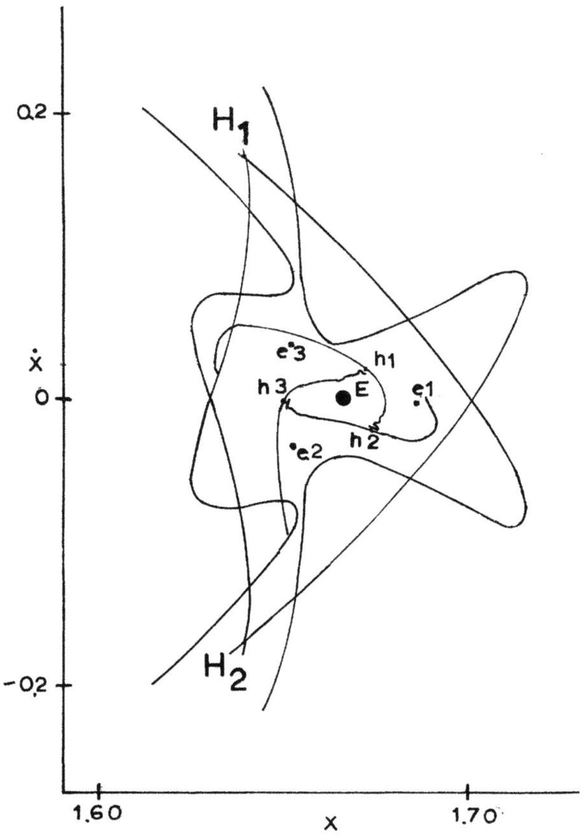

Fig. 1. Intersections of asymptotic curves from the unstable periodic points (H_1, H_2) and (h_1, h_2, h_3). Around the stable periodic point E are marked 3 stable periodic points (adapted from Bartlett and Wagner, 1970).

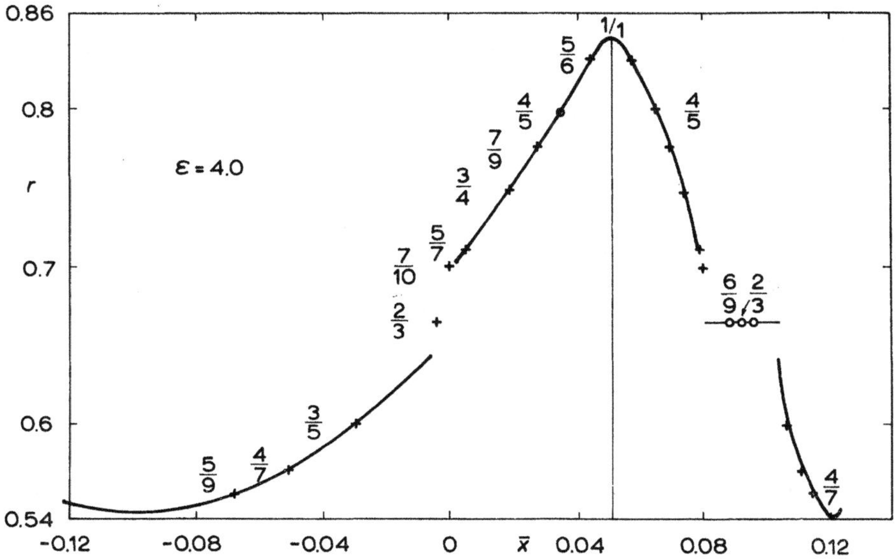

Fig. 2. The rotation number r versus the $\bar{x}$-coordinate in a particular dynamical problem.

asymptotic curves of these periodic orbits we have probably intersections corresponding to heteroclinic orbits.

This is, in our opinion, the actual meaning of the 'overlapping' or 'interaction of resonances', which is now accepted as an explanation of the 'dissolution' of invariant curves (Chirikov, 1967; Contopoulos, 1967; Filonenko *et al.*, 1967; Rosenbluth *et al.*, 1966; Walker and Ford, 1969). It means that orbits starting in the neighborhood of an unstable periodic orbit approach orbitrarily closely other periodic orbits of a different type.

In many zones of instability we have found also a new type of periodic orbits, which we call 'irregular' (Contopoulos, 1970a). Usually periodic orbits are generated from the unperturbed problem by gradually changing the perturbation ε. Namely, as ε varies from zero, a few periodic orbits of the unperturbed problem (in many cases only two) are continued and generate branches of resonant periodic orbits, which produce islands and zones of instability. At the same time, however, there appear new families of periodic orbits, unrelated to the unperturbed problem. Their characteristic curves seem to come from infinity and return to infinity. In many cases such periodic orbits do not have a definite rotation number. This also shows the complexity of the orbits in the zones of instability.

Usually the 'region of stability' around a periodic point, i.e. the region where we see closed invariant curves and insignificant zones of instability, is defined by the asymptotic curves of the nearest 'important' resonant periodic orbits (where n, m are small). As the perturbation increases the region of stability usually decreases, and finally disappears, while the central periodic point becomes unstable.

This behavior has been studied recently in the case of the restricted three-body problem by Mullins and Bartlett (1971). A measure of the area covered by the intermingled asymptotic curves is provided by the area of the first loop between two asymptotic curves emanating from adjacent invariant points of the same periodic orbit (Figure 1) as compared to the whole area of the cell formed by extending the inner asymptotic curves up to their first intersections.

According to Mullins and Bartlett, if the ratio of these areas is smaller than 10^{-5} practically the whole cell is stable, while if this ratio is larger than 0.02 there is practically no stable region around the central periodic point. However Hénon's (1966) results show that a certain stable region is left even beyond the case where one would expect the asymptotic curves to fill the whole inner space. In fact while Mullins and Bartlett find that we should have only 'infinitesimal stability' or instability, for i-type orbits if the Jacobi constant is smaller than 3.875, Hénon finds appreciable regions of stability for $C \geqslant 3.75$ and even for small ranges of C smaller than that. It seems that in certain cases the zones of instability carefully avoid certain regions. Thus we may have stable periodic orbits for very large perturbations.

The question now arises whether stable periodic orbits appear for arbitrarily large values of the perturbation, or whether all orbits become eventually unstable.

Chirikov (1971) mentions Sinai's conjecture that stable periodic orbits are dense in every ergodic region. Of course the islands around such stable periodic orbits are so small that their total measure is very small. A rough explanation of the format:on of such stable islands is as follows. Let us consider the successive intersections of all orbits by a surface of section as defining a mapping T. The equations of motion require that this mapping is area preserving. Thus at each point the mapping has a direction of dilatation and one of contraction, e.g., in the case of an unstable periodic orbit we have dilatation along one asymptotic curve and contraction along the other. The neighborhood of the unstable point is divided into four sectors, two of which are dilatating, and two contracting. If now we have interaction of many closely located unstable points, it may be that at certain points the dilatational and contractional effects counteract each other and thus small stable regions are formed.

Small stable islands in an ergodic 'sea' have been found by several authors (Hénon, 1966; Contopoulos, 1971; Chirikov, 1971). Such islands become extremely small for large perturbations, and their practical significance becomes small.

On the other hand there is a possibility that at least all the regular families of periodic orbits become unstable for large perturbations.

In a diagram giving the trace of the monodromy matric versus the perturbation ε, a family of periodic orbits is represented by a curve (Figure 3). If the trace is between 0 and 4 we have stability. A family leaving the stability region generates, according to rules due originally to Poincaré (for details see Hénon, 1965, and Contopoulos, 1970b) new families, or branches of families, one of which is stable (unless the stability curve has a maximum there). Thus the stability lost by the original family is inherited by some successor.

However our calculations (Contopoulos, 1970b) show that the angle formed by the stability curve inside the stability region with the ε-axis tends to 90°, as ε increases. Thus according to Ford (private communication) it may be that an infinite number of successive stable regular families are congested in a region near a certain ε_{max} and for $\varepsilon > \varepsilon_{max}$ we do not have any stable (regular) periodic orbits.

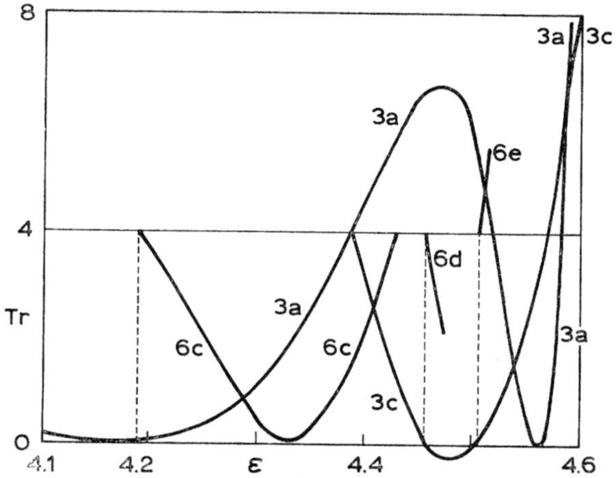

Fig. 3. The stability index (trace) versus the perturbation ε in a particular dynamical problem.

However, even if this is true, we may have, for larger perturbations, islands due to irregular periodic families. In fact near the minima of the characteristic curves of irregular families we have always small stable regions. Thus if, as it looks probable, new irregular families appear for ever increasing perturbations, there are always islands of stability in the ergodic regions, even for very large perturbations.

We must stress here that in cases where we have empirical evidence of ergodicity (zones of instability, etc), different orbits may behave quite differently, even if both could be called ergodic in a certain region.

A systematic study of seemingly ergodic orbits is very important in view of possible applications to systems of more dimensions.

3. Motions in many Dimensions

Studies of orbits in more than two dimensions have started only relatively recently. In such cases Arnold and Moser's theorems, proving quasi-periodicity of orbits near a periodic orbit of the stable type, do not secure stability. In fact the integral surfaces in a case of $2n$ dimensions are of n dimensions and they do not isolate a neighborhood of the periodic orbit that cannot be reached by orbits starting far away. On the contrary there is a conjecture, due to Arnold and Avez (1967) that in general all orbits are unstable, i.e. orbits starting arbitrarily near a given orbit may go arbitrarily near every point in phase space. This is called 'Arnold instability' or 'Arnold diffusion' by Chirikov (1971).

Thus we may have the following curious situation. On one hand most of the available phase space is covered with good integral surfaces, i.e. the great majority of orbits are not ergodic. However an arbitrarily small perturbation can make every orbit ergodic.

Of course we have also a fast diffusion whenever we have interaction of two or more important (low order) resonances. Such a diffusion is called by Chirikov 'streamer diffusion'.

Chirikov made several numerical experiments to find the effects of Arnold instability. Although his results are not conclusive it may be that this instability plays an important role in some cases.

In applying the theory to the solar system he found some cases, where Arnold's diffusion may be important in times smaller than the life-time of the solar system, namely:

(a) The large eccentricity and inclination of the orbit of Pluto.

(b) The breaking of a planet between Mars and Jupiter, due to a close approach to Jupiter, and

(c) The gaps in the distribution of asteroids.

A more careful study of this instability is therefore quite desirable.

Experiments with many-dimensional dynamical systems are quite time consuming. Thus most of the experiments done until now are made with mappings, which have the same general characteristics as dynamical systems.

Froeschlé has studied recently (Froeschlé, 1971; Froeschlé and Scheidecker, 1972) such systems, namely a four dimensional mapping T:

$$\begin{aligned} x_1 &= x_0 + a_1 \sin(x_0 + y_0) + b \sin(x_0 + y_0 + z_0 + t_0) \\ y_1 &= x_0 + y_0 \\ z_1 &= z_0 + a_2 \sin(z_0 + t_0) + b \sin(x_0 + y_0 + z_0 + t_0) \\ t_1 &= z_0 + t_0 . \end{aligned}$$

This study showed the following:

(1) If for $b=0$ both the $x-y$ and $z-t$ partial mappings are isolating, then T is also isolating for small b. This proves that no Arnold diffusion is apparent in this case.

(2) If one, or both partial mappings are semi-ergodic in a certain region then T is also semi-ergodic.

(3) A measure of ergodicity is given by the eigenvalues λ_1^n, λ_2^n of the tangential mapping T^{n*} of T^n.

In practically isolating cases the eigenvalues are almost equal to 1, while in semi-ergodic cases they grow exponentially with n; in fact $\log_{10}|\lambda_1^n|$ and $\log_{10}|\lambda_2^n|$ grow almost linearly with n.

However different semi-ergodic systems have quite different eigenvalues. Deviations from the original orbit are faster as b is larger. This again shows that the ergodic behavior is different in different 'ergodic' cases.

It is obviously of great importance to extend these results to more dimensions. Of special importance is to check the existence and possible role of Arnold's diffusion. Such a study would provide better understanding of the ergodic problem and of the foundations of statistical mechanics. It would give some insight in problems of many degrees of freedom, like the n-body problem, where the non-existence of isolating integrals of motion does not secure similarity between systems with different initial conditions.

4. Violent Changes of Potential

Lynden-Bell (1967) considered systems where the change of the potential is so violent that it may be considered stochastic. He found that such systems satisfy a new form of statistics, similar to Fermi-Dirac's. Many numerical experiments have been made recently to check the applicability of Lynden-Bell's statistics.

The first numerical experiments indicated that one dimensional many body systems, which are initially far from equilibrium, relax violently and their coarse grained distribution function approach Lynden-Bell's. However, recent experiments by Lecar and his associates (Cuperman *et al.*, 1969; Lecar and Cohen, 1971) cast doubt on the close association between violence of relaxation and degree of applicability of the new statistics.

The main question is what is meant by 'violent' relaxation. The basic assumption leading to Lynden-Bell's statistics is that every phase element spreads out evenly throughout phase space, so that every non infinitesimal cube in the 6-dimensional phase space contains samples of every initial phase element. The only restriction is that the *total* energy and angular momentum of the N-particle system is conserved.

If Lynden-Bell's assumption were generally valid, we would have a kind of H-theorem for Collisionless Dynamics, i.e. a tendency towards a 'final' state, similar to the state of maximum entropy. However it seems that there is no general rule like this and there are many equilibria, different from Lynden-Bell's.

The problem that we face now is two-fold:

(1) To find the possible equilibria and how far their region of influence extends, i.e. how far can a system be initially in order to remain 'close' to this equilibrium for ever (without collisions).

(2) In many 'violent' cases only part of the system (that with small energies) satisfies Lynden-Bell's statistics. Is there any practical way of deciding which systems or what parts of a system are 'violent' enough?

The first problem is connected with the general problem of 'small' oscillations around an equilibrium. We know from specific examples that 'small' may actually be fairly large in practice. The problem is to see how far the non-linear problem behaves like the linear one.

The second problem is connected with the behavior of the integrals of motion of the N-body problem, besides the classical 10 integrals. The extreme cases of isolating and ergodic integrals are relatively easy. Isolating integrals must be included in Lynden-Bell's formulas, ergodic integrals can be safely ignored. However there are many cases in between that make the situation practically hopeless. There seem to be quasi-isolating integrals that do not allow particles to go into specific areas of the $6N$-dimensional phase space. Such integrals are well conserved in certain parts of phase space, while they vary widely in other parts. Some hopes have been based in the projection effects of the $6N$-dimensional phase space on the 6-dimensional phase space. It is expected that although a system may not be ergodic in the $(6N-10)$-dimensional phase-space, its projection may well be so. This subject is worth of further study.

Our experience with integrals of motion in many dimensions is extremely limited. At the present moment I would expect information from the opposite end, namely from numerical experiments. I would consider any numerical experiments on the applicability of Lynden-Bell's statistics as providing also evidence about the degree of ergodicity of the integrals of motion of the particular N-body problem treated.

5. Resonances in the Dipole Field

As an example of an interesting problem from Plasma Physics I will mention the problem of motion of charged particles in the van Allen belts. Dragt (1965) has studied the motion of particles in the meridian plane of a magnetic dipole field, which is characterized by the (reduced) Hamiltonian

$$H = \tfrac{1}{2}(p_r^2 + p_z^2) + V$$

where

$$V = \tfrac{1}{2}\left[\frac{1}{r} - \frac{r}{(r^2 + z^2)^{3/2}}\right].$$

In this field the particle gyrates about the guiding field line

$$r = \cos^2 \theta$$

The particle is trapped provided its total (reduced) energy H is less than 1/32.

Dragt compared the results of numerical integration of orbits with Alfvén's theory of adiabatic invariants and a more accurate canonical theory. The orbits are represented by their invariant curves on the surface of section $(r, \dot{r})$. These invariants are, in general, closed around the central point, which represents the orbit through the origin of the dipole. The comparison between theory and numerical results is satisfactory except in resonance cases where the invariant curves are islands.

Similar results were found by Braun (1970), who applied Moser's theory to the dipole field. He proved rigorously the existence of quasi-periodic orbits and of good invariant curves when the ratio of the frequencies is not very near a resonance.

We have started in Thessaloniki a study of resonances in the dipole field using the methods developed in the case of the 'third integral'.

After a number of transformations of variables we bring the Hamiltonian in the form

$$\begin{aligned} H = {} & \tfrac{1}{2}(a^2 + p_a^2 + \omega^2\beta^2 + p_\beta^2) + \\ & + 9(a^2 + p_a^2)^2 + \tfrac{9}{16}\left[\frac{5}{2} - \frac{3}{\omega^2} + \frac{5}{4(1-4\omega^2)}\right](\omega^2\beta^2 + p_\beta^2)^2 + \\ & + \tfrac{9}{16}\left[11 - \frac{2}{\omega^2} + \frac{2\omega^2(2\omega^2 - 5)}{(1-4\omega^2)^2}\right](a^2 + p_a^2)(\omega^2\beta^2 + p_\beta^2) + H_6 + \cdots, \end{aligned}$$

where H_6 contains terms of 6th order, and $\omega^2 = 18h$, where h is the value of the Hamiltonian.

Thus the system is made separable up to the terms of 5th degree.

If ω is a simple rational number, the theory of the adiabatic invariants fails. This is a fact known already to the first authors that applied adiabatic invariants (see e.g., Sommerfeld, 1951), but no effort apparently has been made to find other constants of motion in resonance cases until recently (Jaeger and Lichtenberg, 1972).

Our present work aims at constructing such constants of motion, near each particular resonance.

If we write

$$\Phi'_{10} = \tfrac{1}{2}(a^2 + p_a^2), \qquad \Phi'_{20} = \tfrac{1}{2}(\omega^2\beta^2 + p_\beta^2),$$

we find the rotation number (Contopoulos and Hadjidemetriou, 1967)

$$r = -\frac{(\partial H/\partial \Phi'_{10})}{\omega(\partial H/\partial \Phi'_{20})} \pmod 1 = -$$

$$-\frac{1}{\omega}\left\{1+\tfrac{9}{4}\left[21+\frac{2}{\omega^2}-\frac{2\omega^2(28\omega^2-5)}{(1-4\omega^2)^2}\right]\Phi'_{10}+\right.$$
$$\left.+\tfrac{9}{2}\left[3+\frac{2}{\omega^2}+\frac{112\omega^4-5}{4(1-4\omega^2)^2}\right]\Phi'_{20}\right\}\ (\mathrm{mod}\ 1)$$

to this approximation. We have a resonance whenever r is rational. It is obvious that this happens only if ω is near this rational.

We have a computer program that gives the form of a 'third' integral, when ω is rational, if the higher order terms of the Hamiltonian are known. This can be used to find invariant curves, and other characteristics of orbits near resonances. It can also give the extent of each resonance region, and, finally, an estimate of the conditions necessary for the overlapping of resonances and the breakdown of the 'third' integral. The usefulness of such calculations for mirror machines and for the van Allen belts are obvious.

6. Orbits in Spiral Galaxies

As an example of a problem of current interest in Stellar Dynamics, where methods similar to those of Celestial Mechanics can be applied, is the problem of orbits in a spiral galaxy.

We consider, in particular, orbits on the plane of symmetry of a galaxy where the potential is of the form

$$V = V_0(r) + A(r)\cos\left[\Phi(r) - 2\theta'\right] \tag{1}$$

where

$$\theta' = \theta - \Omega_s t \tag{2}$$

is the azimuth in a frame of reference rotating with the angular velocity Ω_s of the spiral pattern. The minima of potential are along two spirals

$$\Phi(r) - 2\theta' = \pm\pi$$

symmetric with respect to the origin. The maxima of density are very near the minima of potential in general.

Assuming that the velocities of individual stars do not differ very much from the circular and that the amplitude A of the wave is small we find that most orbits fill distorted rings around the origin (Figure 4). Such orbits have been studied numerically by Barbanis (1968). Barbanis studied, in particular, the invariant curves on a plane $(x, \dot{x})$, by plotting the coordinates x, and the corresponding components of velocity $\dot{x}$, of the successive intersections of the orbits by the x-axis, for a given value of the Hamiltonian H (Jacobi integral)

$$H = \tfrac{1}{2}(\dot{x}^2 + \dot{y}^2) - \tfrac{1}{2}\Omega_s^2 r^2 + V_0(r) + A(r)\cos\left[\Phi(r) - 2\theta'\right] \tag{4}$$

If we assume $y=0$ we have $\theta'=0$, $r=x$, and the invariant curves are inside the limiting curve

$$\dot{x}^2 = 2H + \Omega_s^2 x^2 - 2V_0(x) - 2A(x)\cos\left[\Phi(x)\right]. \tag{5}$$

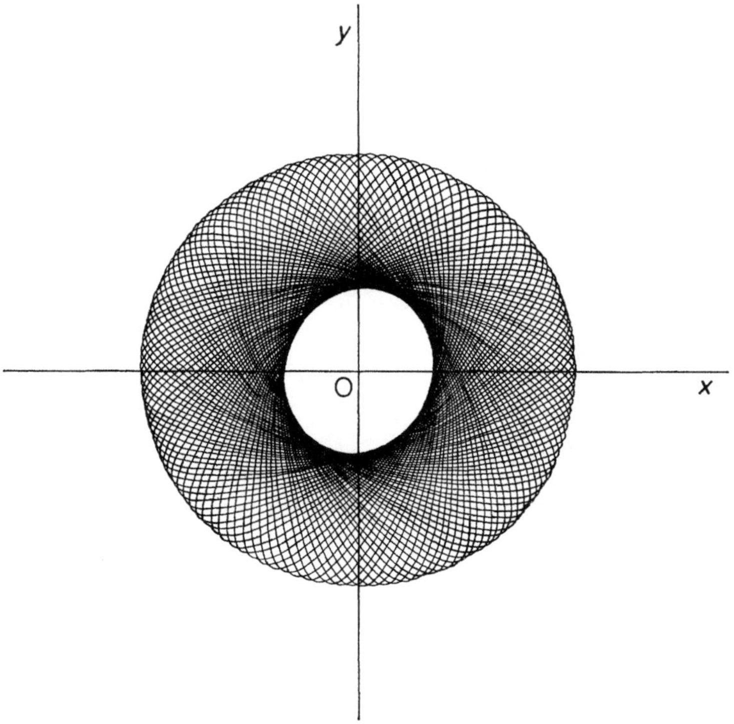

Fig. 4. An epicyclic orbit in a spiral galaxy.

If we consider the spiral potential as a perturbation, the unperturbed invariant curves (for $A=0$) are found if we use the second integral of motion, namely the angular momentum

$$r^2\dot{\theta} = J_0, \tag{6}$$

which in the rotating reference system is written

$$\Omega_s r^2 + \dot{y}x - \dot{x}y = J_0. \tag{7}$$

If we solve this equation for $\dot{y}$ and assume $y=0$ we have

$$\dot{y} = \frac{(J_0 - \Omega_s x^2)}{x} \tag{8}$$

and if we insert this value in Equation (4) we find

$$2H = \dot{x}^2 + \frac{J_0^2}{x^2} + 2V_0(x) - 2J_0\Omega_s. \tag{9}$$

In the special case of a (circular) periodic orbit we have $\dot{x}_c=0$, and for $y_c=0$,

$$J_0 = J_c = r_c^2\Omega_c = [x_c^3 V_0'(x_c)]^{1/2}, \tag{10}$$

where Ω_c is the circular angular velocity at $r=r_c=x_c$, and

$$2H = \frac{J_c^2}{x_c^2} + 2V_0(x_c) - 2J_c\Omega_s. \tag{11}$$

An invariant curve of an orbit with the same Jacobi constant H and a value of J_0 near the above, $J_0=J_c+\Delta J_0$ is

$$\dot{x}^2 + \frac{J_0^2}{x^2} - 2J_0\Omega_s + 2V_0(x) = \frac{J_c^2}{x_c^2} - 2J_c\Omega_s + 2V_0(x_c),$$

or

$$\dot{x}^2 + \frac{2J_c\Delta J_0}{x_c^2} + \left[\frac{3J_c}{x_c^4} + V_0''(x_c)\right](\Delta x)^2 - 2\Delta J_0\Omega_s + \cdots = 0,$$

or

$$\dot{x}^2 + \kappa_c^2(\Delta x)^2 \simeq 2\Delta J_0(\Omega_s - \Omega_c), \tag{12}$$

where κ_c is the epicyclic frequency. Thus the invariant curve is approximately an ellipse. If the circular orbit is inside the corotation distance, then $\Omega_c > \Omega_s$, therefore we must have $\Delta J_0 < 0$.

If an invariant starts at a point $x_{\min}$ of the x-axis ($\dot{x}=0$) with

$$\dot{y} = \dot{y}_1 = 2H - 2V_0(x_{\min}) + \Omega_s^2 x_{\min}^2 \tag{13}$$

we can find J_0 from Equation (7), written in the form

$$J_0 = \Omega_s x_{\min}^2 + \dot{y}_1 x_{\min}. \tag{14}$$

Thus

$$\frac{\mathrm{d}J_0}{\mathrm{d}x_{\min}} = \frac{x_{\min}}{\dot{y}_1}\left[\frac{J_0^2}{x_{\min}^3} - V_0'(x_{\min})\right] \tag{15}$$

and this is positive, because at the minimum of the orbit the centrifugal force is larger than the attraction. Thus if J_0 decreases from J_c the corresponding $x_{\min}$ decreases continuously from x_c.

After J_0 is fixed the value of $x_{\max}$ is found from Equation (9) if we set $\dot{x}=0$. Then

$$\frac{\mathrm{d}J_0}{\mathrm{d}x_{\max}} = \frac{x_{\max}}{\dot{y}_2}\left[\frac{J_0^2}{x_{\max}^3} - V_0'(x_{\max})\right], \tag{16}$$

where

$$\dot{y}_2 = \frac{J_0}{x_{\max}} - \Omega_s x_{\max}. \tag{17}$$

The last factor in Equation (16) is negative, thus $\mathrm{d}J_0/\mathrm{d}x_{\max}$ is negative as long as $J_0 > \Omega_s x_{\max}^2$. Therefore, as J_0 decreases, $x_{\max}$ increases, until we reach the case $J_0 = \Omega_s x_{\max}^2$. In this case $\dot{y}_2=0$ and the invariant curve is tangent to the limiting curve at the point ($x_{\max}$, $\dot{x}=0$).

For even smaller J_0 we have $J_0 < \Omega_s x_{\max}^2$, i.e. $\mathrm{d}J_0/\mathrm{d}x_{\max} > 0$; thus as J_0 decreases $x_{\max}$ decreases also. In these cases there is a value of $x=x_m$ between $x_{\min}$ and $x_{\max}$ for which $J_0 = \Omega_s x_m^2$. At this point the invariant curve reaches the limiting curve and is tangent to it. In fact the solution $\dot{x}_I^2$ of Equation (9) is smaller than the solution $\dot{x}_H^2$ of Equation (5) for $A=0$, except if $J_0 = \Omega_s x_m^2$. In this point we have also

$$\left(\frac{\mathrm{d}\dot{x}_I^2}{\mathrm{d}x}\right)_m = \left(\frac{\mathrm{d}\dot{x}_H^2}{\mathrm{d}x}\right)_m, \tag{18}$$

hence $(\mathrm{d}\dot{x}_I/\mathrm{d}x)_m = (\mathrm{d}\dot{x}_H/\mathrm{d}x)_m$.

We notice that for $x<x_m$ we have $\dot{y}>0$ while for $x>x_m$ we have $\dot{y}<0$; thus the invariant curves, which are tangent to the limiting curve are composed of an arc with $\dot{y}>0$ (left of x_m) and an arc with $\dot{y}<0$ (right of x_m). As J_0 becomes smaller and smaller the value of $x_{\max}$ may become even smaller than x_c.

If we increase now H the circular orbit moves outwards, because

$$\frac{\mathrm{d}H}{\mathrm{d}x_c} = \left(\frac{J_c}{x_c^2} - \Omega_s\right)\frac{\mathrm{d}J_c}{\mathrm{d}x_c} = \frac{(\Omega_c - \Omega_s)}{2\Omega_c} x_c \kappa_c^2 > 0\,. \tag{19}$$

In the perturbed case, $A \neq 0$, the invariant curves become distorted, and all the usual phenomena of non-integrable systems appear (resonant periodic orbits, islands, and partial dissolution of invariant curves).

Of particular interest are two resonance cases

(a) $2(\Omega_0 - \Omega_s) = \kappa_0$ (Inner Lindblad resonance) and

(b) $\Omega_0 = \Omega_s$ (Particle Resonance).

In the first case the topology of the invariant curves is quite different from the non-resonant case. Instead of one basic set of closed invariant curves (one inside the other), we have two such sets, as in the resonance 2/1 of the galactic potential (Contopoulos and Moutsoulas, 1966). As this case has been described in detail elsewhere (Contopoulos, 1970c) we will not discuss it further.

On the other hand at the particle resonance we have four equilibrium points, two of them unstable, at the minima of potential (L_1, L_2) and two stable, at the maxima of potential (L_4, L_5). The situation is quite similar to the case of the Lagrangian points of the restricted three-body problem. There are short and long period orbits around L_4, L_5 and most non-periodic orbits are quasi periodic with one short and one long period.

The orbits are either rings, around L_4, or L_5 or they fill elongated regions, that look like bananas (Figure 5), thus we call them banana orbits. The theory of such orbits has been developed in a recent paper (Contopoulos, 1973).

It is of interest to summarize our results concerning the invariant curves near L_4, or L_5.

If the Hamiltonian H has a value above a certain minimum the limiting curve (5) is open. However many orbits remain trapped near L_4, or L_5. Their invariant curves are in general two closed curves near L_4, or L_5. In many cases these invariant curves intersect each other (Figure 6).

In some cases the two curves combine into one figure eight curve, and the corre-

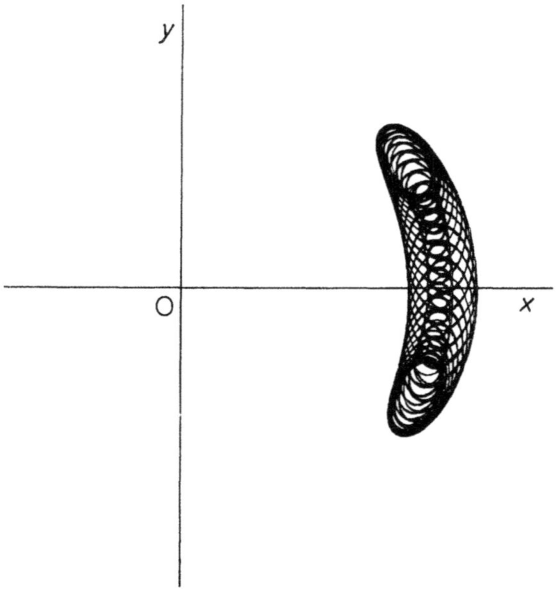

Fig. 5. A banana-type orbit in a spiral galaxy.

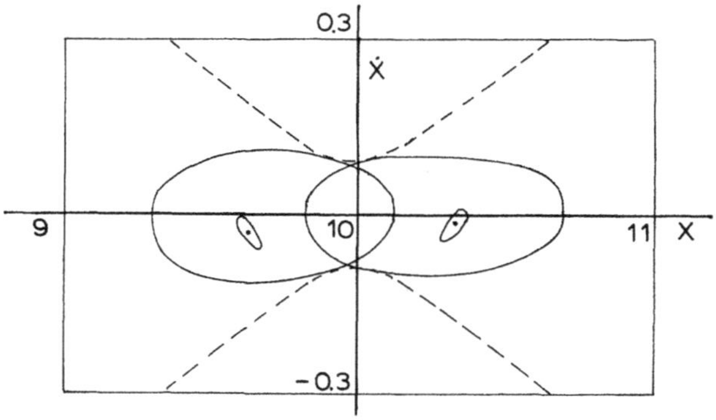

Fig. 6. Invariant curves of trapped orbits near the corotation point ($x = 10$). The dashed curve represents the limiting curve (5).

sponding orbits are quite asymmetric with respect to L_4. Finally there are periodic and non-periodic orbits that leave L_4 completely outside.

Such theoretical and numerical calculations have special interest in the theory of spiral structure. It must be remarked here that the necessary estimates do not require great accuracy. Simple order of magnitude theoretical results are sufficient for many applications. This does not mean that we can relax accuracy in our orbit calculations, because without sufficient accuracy we may find the integrals of motion, or the orbits, to show some effects of evolution in time, which are not real. However after we know what effects to expect we do not need many detailed calculations to estimate their influence.

This is one example where an interesting physical effect can be estimated with the help of a low order theory.

On the other hand there are problems that need a higher order theory to be understood. In the present case in order to distinguish, theoretically, trapped from untrapped orbits near L_4 a higher order theory is necessary.

I hope that these examples give an idea of the wealth of problems in Stellar Dynamics that await further research, which can be done easily by people with a background in Celestial Mechanics. I believe that an interaction between Celestial Mechanics and Stellar Dynamics is still useful and would be useful in the future also.

References

Arnold, V. I. and Avez, A.: 1967, *Problèmes ergodiques de la Mécanique classique,* Gauthier-Villars, Paris.

Barbanis, B.: 1969, *Astron. J.* **73**, 784.

Bartlett, J. H. and Wagner, C. A.: 1970, *Celes. Mech.* **2**, 228.

Birkhoff, G. D.: 1927, *Dynamical Systems,* American Mathematical Society, Providence, R.I.

Birkhoff, G. D.: 1935, *Mem. Pont. Acad. Sci. Novi Lyncaei,* **1**, 85.

Braun, M.: 1970, *J. Diff. Equ.* **8**, 294.

Chirikov, B. V.: 1967, *Soviet Phys. Dokl.* **12**, 612.

Chirikov, B. V.: 1971, *Nucl. Phys. Inst., Siberian Section USSR Acad. Sci.,* Rep. No. 267.

Contopoulos, G.: 1965, *Astron. L.* **70**, 526.

Contopoulos, G.: 1967, *Bull. Astron.* (3) **2**, 223.

Contopoulos, G.: 1970a, *Astron. J.* **75**, 96.

Contopoulos, G.: 1970b, *Astron. J.* **75**, 508.

Contopoulos, G.: 1970c, *Astrophys. J.* **160**, 113.

Contopoulos, G.: 1971, *Astron. J.* **76**, 147.

Contopoulos, G.: 1973, *Astrophys. J.* (in press).

Contopoulos, G. and Hadjidemetriou, J. D.: 1968, *Astron. J.* **73**, 86.

Contopoulos, G. and Moutsoulas, M.: 1966, *Astron. J.* **71**, 687.

Cuperman, S., Goldstein, S., and Lecar, M.: 1969, *Monthly Notices Roy. Astron. Soc.* **146**, 161.

Deprit, A. and Henrard, J.: 1969, *Boeing Res. Lab. Math. Note,* No. **629**.

Dragt, A. J.: 1965, *Rev. Geophys.* **3**, 255.

Filonenko, N. N., Sagdeev, R. Z., and Zaslavsky, G. M.: 1967, *Nucl. Fusion* **7**, 253.

Froeschlé, C.: 1971, *Astrophys. Space Sci.***14**, 40.

Froeschlé, C. and Scheidecker, J. P.: 1972, *Astron. Astrophys.* (to be published).

Hénon, M.: 1965, *Ann. Astrophys.* **28**, 992.

Hénon, M.: 1966, *Bull. Astron.* (3) **1**, 57.

Jaeger, E. F. and Lichtenberg, A. J.: 1972, *Ann. Phys.* **71**, 319.

Lecar, M. and Cohen, L.: 1971, *Astrophys. Space Sci.* **13**, 397.

Lynden-Bell, D.: 1967, *Monthly Notices Rey. Astr. Soc.* **136**, 101.

Mullins, L. D. and Bartlett, J. H.: 1972, Preprint, University of Alabama.

Poincaré, H.: 1899, *Les méthodes nouvelles de la mécanique céleste III*, Gauthier-Villars, Paris.

Rosenbluth, M. N., Sagdeev, R. Z., and Taylor, J. B.: 1966, *Nucl. Fusion* **6**, 297.

Sommerfeld, A.: 1951, *Atombau und Spektrallinien,* F. Vieweg und Sohn, **1**, pp. 370, 698.

Walker, G. H. and Ford, J.: 1969, *Phys. Rev.* **188**, 416.

INVARIANT MANIFOLDS IN CELESTIAL MECHANICS

W. T. KYNER

Dept. of Mathematics, University of New Mexico, Albuquerque, N. M., U.S.A.

Abstract. Several examples of invariant manifold problems in celestial mechanics are discussed including the determination of the topological character of the Kepler energy manifold and the derivation of a priori bounds for the artificial satellite problem.

The purpose of this lecture is to discuss several applications of invariant manifold theory to celestial mechanics. Some of these are recent, others very old. Since all have been published elsewhere, many of the details will be omitted.

The most widely used invariant manifold in conservative problems is the energy manifold. Let us first consider the Kepler problem, i.e., the reduced form of the two body problem. It is obvious that if the total energy is negative, then the motion is bounded. However, the determination of the topological nature of the manifold and the characterization of the flow on the manifold is not trivial. This problem is closely related to regularization and the corresponding linear theory of celestial mechanics which was developed by Stiefel and his colleagues (1972). Its solution (due to Moser, 1970) can be stated as follows: If the total energy is negative, the energy manifold of the Kepler problem is topologically equivalent to the unit tangent bundle of the n-dimensional punctured sphere $\hat{S}^n$. Furthermore, the flow defined by the Kepler problem is equivalent to the geodesic flow on $\hat{S}^n$. To prove this, Moser extended the stereographic projection of $\hat{S}^n$ onto R^n to a mapping of the corresponding tangent bundles, i.e., $T(\hat{S}^n)$ onto $T(R^n)$, and derived the induced flow. It is important to note that the image of $\hat{S}^n$ is the momentum space of the Kepler problem rather than the configuration space. One reason for this is that the hodographs of the Kepler problem are circles which are parameterized by arc length, a regularized variable corresponding to the eccentric anomaly. Another point of view is that of abstract dynamical systems (Abraham, 1967) where it is shown that the momentum space, the cotangent bundle of the configuration space, has a standard symplectic structure. This result is basic to the abstract theory. Its relevance here is highlighted by the fact that Folk (1935) applied a stereographic projection to the momentum variables in his 1935 analysis of the hydrogen atom, the quantum mechanical analogue of the Kepler problem.

Another application of the invariant manifolds is in the derivation of error bounds for the main problem of artificial satellite theory, i.e., the analysis of the bounded orbits in an axisymmetric gravitational field. The deviation from the Newtonian field can be considered as a perturbation. Brouwer (1959) and Kozai (1959) contributed most to the solution of the satellite problem. Brouwer used the von Zeipel method-Kozai, a variant of the method of averaging. Both methods are formally equivalent and can be shown to be rigorously equivalent on finite time intervals if a priori bounds on the variables are available. Such bounds follow (see Kyner, 1965) from the results of MacMillan (1910). He employed the two known constants of the motion (i.e., in-

B. D. Tapley and V. Szebehely (eds.), Recent Advances in Dynamical Astronomy, 192–196.

variant manifolds), the energy and the polar component of the angular momentum, and showed that if the product of these two constants is strictly negative and if the perturbation is sufficiently small, then the motion is confined to a region in space whose bounding surface is homeomorphic to a torus. Analogous results for the restricted problem of three bodies are obtained with the aid of the Jacobi integral (see Chapter 4 of Szebehely, 1967).

MacMillan's hypothesis of strict negativity excluded polar orbits. We therefore pose the following problem: prove that polar orbits with negative energy cannot spiral into the attracting center. We note that boundedness is an immediate consequence of the negative energy assumption. We have chosen this problem, not because of its intrinsic importance, but rather as an example of one which can evidently not be solved by the classical methods of dynamics. However, its solution is an immediate consequence of the fundamental results of Kolmogorov, Arnold, and Moser (The KAM theory) (Arnold and Avez, 1968; Siegel and Moser, 1971).

The Hamiltonian of the problem (see Chapter 17 of Brouwer and Clemence (1967)) can be written

$$F = \frac{\mu^4}{2L^2} - \frac{\mu^4}{4L^6} J_2 R^2 \left(\frac{a}{r}\right)^3 [1 + 3\cos 2(f+g)]. \tag{1}$$

We take the longitude of the ascending node equal to zero thus freeing the letter h which will be used to denote the total energy. The set of orbits with the same total energy satisfy the constraint, $F+h=0$. This constraint will permit us to transform the problem into the framework of the KAM theory. In terms of the Poincaré variables,

$$\begin{aligned} &L = \sqrt{\mu a}\,, \quad U = l + g\,, \\ &x = \sqrt{2(L-G)}\cos g\,, \quad y = -\sqrt{2(L-G)}\sin g\,, \end{aligned} \tag{2}$$

the governing differential equations are

$$\begin{aligned} \frac{\mathrm{d}L}{\mathrm{d}t} &= \frac{\partial F}{\partial U}, \quad & \frac{\mathrm{d}U}{\mathrm{d}t} &= -\frac{\partial F}{\partial L}, \\ \frac{\mathrm{d}x}{\mathrm{d}t} &= \frac{\partial F}{\partial y}, \quad & \frac{\mathrm{d}y}{\mathrm{d}t} &= -\frac{\partial F}{\partial x}. \end{aligned} \tag{3}$$

Since we are considering only those trajectories with the same total energy, the flow is restricted to a three dimensional manifold of the four dimensional phase space. Analytically, we accomplish this by setting $h=-F$ and solving for L. If the perturbing term of the potential is small, this can be done. We obtain

$$\begin{aligned} L &= K(x, y, U, h, J_2) = \\ &= L_0 \left\{1 - \frac{\mu^2}{4L_0^4} J_2 R^2 \left(\frac{a}{r}\right)^3 [1 + 3\cos 2(f+g)]\right\} + 0(J_2^2), \end{aligned} \tag{4}$$

where $L_0 = \mu/(-2h)^{1/2}$. We note (see Kyner (1968) for a detailed discussion of this type of isoenergetic reduction) that if U, the mean latitude, is taken as the independent variable, then K is the Hamiltonian of the reduced system of differential equations which determine the flow on the energy manifold, i.e.,

$$\frac{\mathrm{d}x}{\mathrm{d}U} = -\frac{\partial K}{\partial y}, \quad \frac{\mathrm{d}y}{\mathrm{d}U} = \frac{\partial K}{\partial x}. \tag{5}$$

The requirement that the perturbing term be small deserves explanation since it seems to assume that the variable r is bounded from below, the desired conclusion of our study.

Let $\varrho = x^2 + y^2 = 2(L-G)$. In a domain in phase space defined by

$$0 \leqslant \varrho < \bar{\varrho}, \quad \underline{L} < L < \bar{L}, \quad 0 \leqslant U \leqslant 2\pi,$$

we have that a/r is bounded and the representation (4) is valid. If we can prove that the orbits with initial values $(U=0)$ in the domain are contained in the domain for all U (mod 2π), then the orbits are bounded away from the attracting center. In fact, we shall show that most of the orbits are quasi-periodic and lie on invariant tori. The orbits in the gaps between two tori cannot intersect them and therefore always remain between them.

Let us take x, y, U as cartesian coordinates in a domain of phase space defined by $0 \leqslant \varrho < \bar{\varrho}$, all U. This cylinder can be considered as the interior of a torus if we reduce $U \bmod 2\pi$. If $J_2 = 0$, the center of the torus is a circular orbit. It generates $(J_2 \neq 0)$ a nearly circular orbit with period 2π in U (see MacMillan, 1910). We shall show that this periodic orbit is the center of a family of invariant manifolds, each of which is homeomorphic to a torus. This will be done with the aid of the Moser Twist Mapping Theorem (Siegel and Moser, 1971): Let

$$\theta_1 = \theta + \varepsilon\gamma(\varrho) + \varepsilon^2 Q(\varrho, \theta, \varepsilon), \quad \varrho_1 = \varrho + \varepsilon^2 P(\varrho, \theta, \varepsilon)\; 0 < \varepsilon \leqslant 1,$$

denote an analytic area preserving mapping of the annulus $0 < a_1 < \varrho < a_2$ into itself with $\gamma' \neq 0$. Then for sufficiently small ε there exists an invariant curve

$$C: \theta = \Phi + \varepsilon v(\Phi, \varepsilon), \quad \varrho = \varrho_0 + \varepsilon u(\Phi, \varepsilon),$$

u, v real analytic with period 2π in Φ and such that on C the mapping is defined by

$$\Phi_1 = \Phi + \varepsilon\gamma(\varrho_0),$$

where $a_1 < \varrho_0 < a_2$. The twist, $\omega = \varepsilon\gamma(\varrho_0)$, satisfies the inequalities

$$\left|\frac{\omega}{2\pi} - \frac{p}{q}\right| \geqslant \sigma(\varepsilon)\, q^{-5/2} \quad \text{for all integers} \quad p, q \text{ with } q > 0.$$

The parameter $\sigma = 0(\varepsilon)$.

The analytic Hamiltonian system (5) induces an area preserving mapping from the cross-section $U=0$ to the cross-section $U=2\pi$ by following an orbit with initial conditions $(x_0, y_0, 0)$ to the point $(x_1, y_1, 2\pi)$. Analytically, we have

$$
\begin{aligned}
x_1 &= x_0 - \int_0^{2\pi} \frac{\partial K}{\partial y}\, \mathrm{d}U = x_0 + 2\pi \frac{\partial \bar{K}}{\partial y} + 0(J_2^2), \\
y_1 &= y_0 + \int_0^{2\pi} \frac{\partial K}{\partial x}\, \mathrm{d}U = y_1 + 2\pi \frac{\partial \bar{K}}{\partial x} + 0(J_2^2),
\end{aligned}
\tag{6}
$$

where (see Kyner (1968) for a corresponding analysis)

$$
\begin{aligned}
\bar{K} &= L_0 - \frac{\mu^2}{4G_0^3} J_2 R^2, \\
\varrho_0 &= x_0^2 + y_0^2 = 2(L_0 - G_0).
\end{aligned}
$$

Let

$$
x = \sqrt{\varrho}\cos\theta, \qquad y = -\sqrt{\varrho}\sin\theta, \qquad \gamma(\varrho_0) = \frac{3\mu^2}{4G_0^4}\frac{J_2 R^2}{2\pi}
$$

(compare with (2)), then it follows that

$$
\begin{aligned}
\varrho_1 &= \varrho_0 + 0(J_2^2), \\
\theta_1 &= \theta_0 + \gamma(\varrho_0) + 0(J_2^2).
\end{aligned}
\tag{7}
$$

Note that $\gamma(\varrho_0)$ is a monotonic increasing function.

We can now conclude that there exists a one parameter family $C(\omega)$ of invariant curves which are dense in themselves and almost fill the annulus $0 \leqslant \varrho < \bar{\varrho}$ (see Kyner (1968) for a more detailed discussion). In the x, y, U (mod 2π) space, the set of orbits with initial values on $C(\omega)$ generate a torus. Since the rotation number (the average slope) is irrational, the orbits on the torus form a dense subset. The flow is said to be ergodic on the torus. The atypical orbits which do not lie on invariant tori lie in the gaps between nearby tori. Since the invariant tori depended continuously on ϱ and h, we say that the typical orbit is toroidal stable in that it lies on an invariant torus and under a perturbation of initial conditions, it either is on a nearby torus or is trapped between two nearby tori. Toroidal stability is a generalization of orbital stability wherein a curve under perturbation must stay in a neighborhood of its original position. Clearly, the nearly circular periodic orbit generated by the center of the torus is orbitally stable.

References

Abraham, R.: 1967, *Foundations of Mechanics*, W. A. Benjamin, New York-Amsterdam.

Arnold, V. I. and Avez, A.: 1968, *Ergodic Problems of Classical Mechanics*, W. A. Benjamin, New York-Amsterdam.

Brouwer, D.: 1959, *Astron. J.* **64**, 378–397.

Brouwer, D., Clemence, G. M.: 1967, *Methods of Celestial Mechanics*, Academic Press, New York-London.

Fock, V.: 1935, *Z. Phys.* **98**, 145–154.

Kozai, Y.: 1959, *Astron. J.* **64**, 367–377.

Kyner, W. T.: 1965, *J. Soc. Indust. Appl. Math.* **13**, 136–171.
Kyner, W. T.: 1968, *Mem. Am. Math. Soc.* **81**, 27 pp.
MacMillan, W. D.: 1910, *Trans. Am. Math. Soc.* **11**, 55–120.
Moser, J.: 1970, *Comm. Pure Applied. Math.* **23**, 609–635.
Siegel, C. L., Moser, J.: 1971, *Lectures on Celestial Mechanics,* Springer, Berlin-Heidelberg, New York.
Stiefel, E. L. and Scheifle, G.: 1972, *Linear and Regular Celestial Mechanics,* Springer, Berlin-Heidelberg-New York.
Szebehely, V.: 1967, *Theory of Orbits,* Academic Press, New York-London.

THEORETICAL ASPECTS OF N-BODY COMPUTER SIMULATIONS

S. J. AARSETH
Institute of Astronomy, Cambridge, England

Abstract. A general discussion is given of some basic dynamical phenomena and their relevance to computer simulations of small N-body systems. The main topics include particle relaxation, mass segregation, escape mechanisms and binary evolution. Most features of the dynamical evolution can be understood by simple theoretical considerations, but concepts of statistical mechanics do not always apply. Numerical results of calculations with up to 500 particles show that one dominant binary is always formed near the center. Subsequent close encounters with other members tend to produce energetic escapers and the binary evolution therefore provides an energy sink mechanism which has important dynamical consequences.

1. Introduction

The main purpose of N-body calculations is to understand better the long-term behaviour of small self-gravitating systems. Consequently, the objectives of numerical work in stellar dynamics differ in many respects from celestial mechanics, where the emphasis is on accuracy and where the solutions are usually much better behaved. Although the motivation for experimental stellar dynamics is based on the existence of star clusters, most computer simulations are concerned with isolated systems. In the first instance, the numerical results may be used to test theoretical models which are based on simplifications and hence may not be dynamically consistent. Secondly, the parameter dependence of dynamical processes may be studied in considerable detail, leading to a clarification of theoretical concepts.

The direct approach is free from assumptions and provide complete information for a given set of initial conditions. However, numerical methods cannot yield unique solutions for sufficiently long time intervals because of the strong non-linearity of gravitational interactions. In order to be meaningful, such results must therefore be statistically significant as well as dynamically consistent. The former requirement implies that richer systems can be studied with greater confidence, albeit with increased expenditure of computing time. Alternatively, repeated integrations of many smaller systems can readily be made in order to distinguish between statistical uncertainties and systematic effects. The two procedures are not equivalent, however, because of the particle number dependence and it is therefore important to evaluate the general validity of single examples.

Recent advances in computer technology together with the introduction of more powerful numerical methods make it possible to study systems with several hundred interacting particles. It is very encouraging that already the dynamics of typical open star clusters and clusters of galaxies may be simulated by direct calculations. At first the problem seems prohibitive because of the infinity of initial conditions, but the actual choice may be guided by observational and theoretical considerations. We adopt

B. D. Tapley and V. Szebehely (eds.), Recent Advances in Dynamical Astronomy, 197–207.

the working hypothesis that individual positions and velocities at the beginning are relatively unimportant for the long-term evolution. Furthermore, the assumption of spherical symmetry can be justified by observations but is not absolutely essential. However, in a study of general dynamical phenomena it is important to include both heavy and light particles; for example the relative distribution may be selected from actual mass functions. It is mainly this requirement which complicates the theoretical treatment and makes the direct approach so attractive in spite of the limitations.

In the present paper we discuss several aspects of dynamical evolution and emphasize the connection between experimental results and theoretical considerations. Only some of the most common phenomena are included for simplicity, but the subsequent discussion should nevertheless serve to increase the confidence in the numerical results.

2. Relaxation Effects

Stellar systems can be completely specified by a distribution function $f(m, \mathbf{r}, \dot{\mathbf{r}}, t)$ which gives the probability density of a particle with mass m being at the position $\mathbf{r}$ with velocity $\dot{\mathbf{r}}$ at time t. Given N such mass points with arbitrary initial conditions, their mutual interactions modify the function f. In the absence of close encounters the distribution function satisfies the Liouville equation and the general solution depends on at most six isolating integrals of the motion for each particle. Many attempts have been made to obtain steady-state solutions which are self-consistent, but the forms of these integrals are not known for general configurations.

The regularity of rich stellar systems justifies the introduction of simplifying assumptions, using either spherical or rotational symmetry. In addition, most observed systems are sufficiently old to ensure approximate dynamical equilibrium. Two kinds of isolating integrals are known explicitly for such systems; i.e. binding energy and angular momentum, but only one component of the latter vector is conserved if there is rotational symmetry. These concepts, which play a fundamental role in theoretical discussions, are also extremely useful for describing the dynamics of simulated systems where even approximate conservation may not apply. When comparing the behaviour of particles of different masses, it is more convenient to work with the binding energy per unit mass, which for the i'th particle is defined by

$$E_i = -G \sum_{\substack{j=1 \\ j \neq i}}^{N} \frac{m_j}{|\mathbf{r}_i - \mathbf{r}_j|} + \tfrac{1}{2}\dot{\mathbf{r}}_i^2, \tag{1}$$

where G is the gravitational constant. We note that $E_i<0$ implies bound motion, whereas the particle will escape to infinity if $E_i>0$ for all time; the condition $E_i=0$ therefore defines the local escape velocity. Secondly, we introduce the corresponding angular momentum per unit mass,

$$\mathbf{J}_i = \mathbf{r}_i \times \dot{\mathbf{r}}_i. \tag{2}$$

Although the function $f(m, E, \mathbf{J}, t)$ does not specify the computed systems uniquely, the quantities E_i and $\mathbf{J}_i$ may nevertheless be used to characterize individual orbits. However, the analogy with the two-body description using an equivalent semi-major axis and eccentricity is not very useful because only a few bound orbits are approximately closed.

Except for the well-known conservation properties of isolated systems, no precise information about the general N-body problem is available. One of the main tasks in stellar dynamics is to calculate the rate of change of the fundamental quantities (1) and (2) by including the contributions from all other particles according to the equation of motion

$$\ddot{\mathbf{r}}_i = -G \sum_{\substack{j=1 \\ j \neq i}}^{N} \frac{m_j (\mathbf{r}_i - \mathbf{r}_j)}{|\mathbf{r}_i - \mathbf{r}_j|^3}. \tag{3}$$

The total attraction (3) may be divided into a regular force field which is mainly due to distant members and an irregular component arising from neighbouring particles. The latter component modifies the orbital parameters which would otherwise remain approximately constant. Consequently, the actual orbits deviate gradually from the collisionless solution. The concept of relaxation time has been introduced in order to estimate the time-scale for significant changes of the orbital parameters. Assuming a steady-state system with constant density and equal masses, the relaxation time for a typical particle is given by (Chandrasekhar, 1942)

$$t_E = 0.136 \left(\frac{N \mathscr{R}^3}{G \bar{m}} \right)^{1/2} \frac{1}{\ln (N/2^{3/2})}, \tag{4}$$

where $\mathscr{R}$ is a scale factor to be associated with the total radius and $\bar{m}$ is the mean particle mass.

Relaxation times based on numerical calculations may be estimated by considering the energy change ΔE_i during an interval Δt. However, this procedure does not lead to a unique answer since consecutive energy changes are usually uncorrelated. Comparisons with theory are therefore subject to some uncertainty because of the choice of time interval as well as differences in the density distribution. Further problems arise when particles of different masses are combined in order to obtain a meaningful average for the whole system. Nevertheless, simple comparisons of numerical results and theoretical predictions give satisfactory agreements.

Although the direct calculations include all encounters consistently, the effect of neighbouring members cannot readily be separated from the total energy change. Further insight into the relative contributions of near and distant particles may be achieved by differentiating Equation (1) with respect to time. Replacing $\ddot{\mathbf{r}}_i$ by the explicit force summation of Equation (3) gives the simplified expression

$$\dot{E}_i = -G \sum_{\substack{j=1 \\ j \neq i}}^{N} \frac{m_j \dot{\mathbf{r}}_j \cdot (\mathbf{r}_i - \mathbf{r}_j)}{|\mathbf{r}_i - \mathbf{r}_j|^3}. \tag{5}$$

We note that although Equation (5) does not contain the velocity of particle i, the motion of the other particles is modified by the masspoint m_i. Two limiting cases of energy conservation may be distinguished. First, the binding energy E_i is a constant of the motion for any system if all the other members are constrained to be at rest. This artificial situation is of theoretical interest since it permits comparison with the predicted angular momentum change due to orbital deflections, and hence the evaluation of an equivalent relaxation time (Chandrasekhar, 1942). Such numerical calculations have the advantage of being very fast and the results for many homogeneous systems with up to 2500 equal-mass particles are in very good agreement with the theoretical expressions (Standish and Aksnes, 1969).

Equation (5) shows that contributions to the binding energy change fall off rapidly with distance. On the other hand, this inverse square effect is to some extent compensated by the greater number of distant particles. Nevertheless, the contributions to halo orbits arising from a central spherical mass distribution tend to cancel, giving an arbitrarily small binding energy change as the distance approaches infinity. Hence the approximate collisionless nature of bound halo orbits justifies the use of longer integration intervals when the total force changes are relatively smooth. Conversely, particles near the center are not significantly affected by a spherical distribution of distant halo members and escaping stars may therefore be excluded from the calculations after a while. Furthermore, statistical fluctuations also decrease with distance because of the smaller density and rms velocity. This implies that for systems near equilibrium the main relaxation effects on central particles are due to the closest neighbours.

From total energy conservation and the form of Equation (5) it follows that the net effect of individual interactions is to produce energy changes of opposite sign. Using Equation (1) we have

$$\sum_{i=1}^{N} m_i E_i = \Phi + E_0 \,, \tag{6}$$

where Φ and E_0 is the total potential energy and binding energy, respectively. Differentiation with respect to time then gives

$$\sum_{i=1}^{N} m_i \dot{E}_i = \frac{\mathrm{d}\Phi}{\mathrm{d}t} \,. \tag{7}$$

Collective mass motions may therefore change the binding energies significantly, unless the requirement for dynamical equilibrium is approximately satisfied; i.e. $2T + \Phi \simeq 0$, where T is the total kinetic energy. This well-known virial theorem condition is based on time averaging, but the calculations show that the tendency for unstable systems to oscillate is rapidly damped out. In addition to overall mass motions, close encounters between two particles m_k and m_l contribute to the summation (7) by an amount

$$m_k \dot{E}_k + m_l \dot{E}_l = - m_k m_l \frac{\mathbf{R} \cdot \dot{\mathbf{R}}}{R^3} \,, \tag{8}$$

where $\mathbf{R} = \mathbf{r}_k - \mathbf{r}_l$, etc. However, such interactions are usually of short duration and may be regarded as isolated events which do not affect the dynamics of more distant particles.

A careful determination of the particle relaxation time must ensure that departures from approximate equilibrium do not invalidate the results. The time-scale for such contributions is closely associated with the mean crossing time, defined by

$$t_{\mathrm{cr}} = \frac{2\sqrt{2}\mathscr{R}^{3/2}}{(GN\bar{m})^{1/2}}. \tag{9}$$

Hence the two effects can only be separated if the relaxation time is somewhat greater than the crossing time. The relevant ratio is given by

$$t_E/t_{\mathrm{cr}} = 0.05\,N/\ln\,(N/2^{3/2}), \tag{10}$$

and the two time-scales are approximately equal for $N = 60$. With the present upper limit of $N = 500$ (Aarseth, 1972), $t_E \simeq 5t_{\mathrm{cr}}$. However, central relaxation times are much smaller than the typical value given by Equation (4) and should still be meaningful for a few hundred particles if the calculations extend over many crossing times.

One simple approach to the relaxation time problem is to calculate an instantaneous value for each particle, defined by

$$\tau_E = \frac{\dot{\mathbf{r}}_i^2}{2\,|\dot{E}_i|}. \tag{11}$$

Representative estimates may then be obtained by averaging over different volumes or by combining results for similar masses. Unperturbed close binaries may provide spurious contributions (cf. Equation (8)), hence only the center of mass effect should be included if the two-body motion is bound. Dividing the system with $N = 500$ into concentric shells, we find that the average relaxation time ranges from $\langle\tau_E\rangle \lesssim 0.2t_{\mathrm{cr}}$ near the center to $\langle\tau_E\rangle \simeq 80t_{\mathrm{cr}}$ at a distance containing about 80% of all members. The corresponding range of mean, hence $\langle\tau_E\rangle$ density exceeds four powers of ten hence $\langle\tau_E\rangle$ is consistent with the theoretical inverse square root dependence on mass density. The results cannot be compared directly with Equation (4), but the predicted value is well within the range of the experimental results where different masses are combined.

To conclude this part, we emphasize that the evolution rate of isolated stellar systems is determined by particle relaxation effects. Hence it is essential to examine many different types of interactions before the long-term behaviour can be understood theoretically.

3. Mass Segregation and Escape

Theoretical discussions of particle interactions usually assume the two-body approximation where the individual contributions are independent of each other. Although the general principles of two-body encounters between particles of different masses are well understood, it is extremely difficult to give a satisfactory treatment of the dynamical consequences. However, the general mass dependence can be discussed with some confidence.

Consider a close hyperbolic encounter between a heavy and a light particle. On the average, the latter will gain kinetic energy and will therefore become less bound with respect to the other members. Consequently, light particles tend to increase their apocenter distances, whereas the heavy particles become more centrally concentrated and attain higher velocities in order to balance the greater attraction. The initial velocity excess is reversed during this phase if the outward motions are bound, but the light particles again move faster when returning towards the inner regions. Thus the principle of equipartition of kinetic energy is only consistent with mass segregation when the comparison is made at the same distance, rather than using time averaged values. This simple discussion shows that concepts in statistical mechanics cannot be applied to stellar systems without considering the dynamical implications.

Once a particle becomes less strongly bound, its mean free path between close encounters increases and further energy changes take place on a longer time-scale. However, bound particles ejected from the center in elongated orbits subsequently return for further interactions which may speed up their motions if the velocities are still below the local equipartition value. If the condition of strict equipartition between two particles of mass m_1 and m_2 is reached, the velocity of the latter is given by

$$v_2^2 = \frac{m_1}{m_2} v_1^2 . \tag{12}$$

Let v_1 be characteristic of the central rms velocity which exceeds the overall rms velocity $\langle v \rangle$ by a small factor δ depending on the actual configuration. Now the condition for bound motion requires that $E_2 < 0$, corresponding to $v_2 < v_e$, where v_e is the local escape velocity. A dynamically relaxed system may be described approximately by a polytrope of index five (Wielen, 1967), giving $v_e \simeq 2.6 \langle v \rangle$ and $\delta \simeq 1.3$. It follows that the equipartition velocity may exceed the central escape velocity for mass ratios $m_1/m_2 \gtrsim 4$. However, the equipartition process cannot be used to infer that the probability of escape by two-body encounters should increase in direct propertion to the mass ratio.

An improved description of the escape phenomenon can be obtained from more detailed considerations of the two-body mechanism. Following the classical treatment (Chandrasekhar, 1942), the final velocity of particle m_2 interacting with particle m_1 is given by

$$v_2'^2 = V_g^2 + \frac{2m_1}{m_1 + m_2} V_g V \cos \varphi' + \left(\frac{m_1}{m_1 + m_2} \right)^2 V^2 , \tag{13}$$

where

$$V_g = \frac{|m_1 \mathbf{v}_1 + m_2 \mathbf{v}_2|}{m_1 + m_2} , \tag{14}$$

$$V = |\mathbf{v}_2 - \mathbf{v}_1| , \tag{15}$$

and φ' is the angle between $\mathbf{V}_g$ and the relative velocity at the end of the encounter.

Let us first evaluate Equation (13) for the case $m_2 \ll m_1$. If $v_1 \leqslant v_e$ and $v_2 \leqslant v_e$, the limiting velocity of the light particle is $3v_e$ which is achieved for $\varphi' = 2\pi$. To illustrate the mass dependence of the terminal velocity, this limit reduces to $v_2' = 2v_e$ if $m_1 = 3m_2$ for the same relative motion. However, the maximum escape velocity does not occur for head-on collisions if $m_2 > 0$. Thus collisions with $m_2 = m_1$ simply lead to exchange of velocity, whereas the upper limit $v_2' = \sqrt{2}\, v_e$ occurs if $\mathbf{v}_2$ is perpendicular to $\mathbf{v}_1$.

The considerations above indicate how the presence of heavy particles should lead to escape of light members. It also follows from Equation (13) that the outcome of such encounters is essentially independent of the mass ratio, provided that the latter exceeds unity by a moderate factor. Furthermore, as the masses become more equal, the general equipartition condition (12) only appears to be a consequence of the gravitational sling process (13) if the angle dependence of the maximum velocity is neglected, whereas escape is still possible if the masses are equal. We also remark that the two-body escape process operates without the extreme conditions adopted above; i.e. $v_1 = v_2 = \langle v \rangle$ implies that $v_2' = 3\langle v \rangle$ for $\varphi' = 2\pi$ and $m_2 = 0$. Finally, the idealized description does not take into account the acceleration of the local center of mass as well as more complicated types of interactions which would increase the escape rate further.

Close encounters are much less important outside the central region because of the smaller density. To some extent, the longer relaxation time of more distant light members is compensated by the smaller energy change required to satisfy the escape condition $E_i > 0$. It is rather difficult to ascertain theoretically which of these two effects are most important. The numerical results obtained so far indicate that the escape rate of light particles is approximately independent of mass, but this conclusion does not necessarily hold for richer systems or very long time intervals.

The net effect of the dynamical evolution is to produce a more centrally concentrated structure together with an extended halo distribution. Most bound halo members are first ejected from the center in elongated orbits from which they return for further interactions. Numerical experiments have confirmed the discrete nature of the escape process emphasized in advance of the first calculations (Hénon, 1960), whereas earlier treatments assumed a diffusion process in energy space, arising from small random contributions. Furthermore, direct solutions provide detailed information about the escape mechanism itself. Five different processes have been classified (van Albada, 1968):

(1) Two-body, triple and multiple encounters.

(2) Encounters between binaries and field particles.

(3) Ejection of one member from a bound triple system.

(4) Exchange of one binary component with an incoming particle.

(5) Weakly bound halo members may escape during further central passages.

These considerations are based on calculations with small particle numbers, i.e. $N \leqslant 24$, but all processes also take place in the larger systems which have been explored by direct methods.

The wide variety of significant escape phenomena studied numerically provide a better framework for describing the long-term behaviour of stellar systems. In contrast, current theories only employ the simple assumption of two-body encounters and theoretical discussions are therefore likely to underestimate the actual escape rate. However, the basic principles of stellar dynamics may still be used to outline the general features of the evolution. It can readily be demonstrated that two-body encounters provide the dominant relaxation mechanism for steady-state systems in the limit of large particle numbers. Hence theoretical treatments should become more reliable for richer systems which cannot be investigated by present numerical techniques.

4. Binary Evolution

The virial theorem condition implies that two-body encounters are usually hyperbolic and of short duration. Although the energy change with respect to other members may be significant, the subsequent relative motion of such encounters remain hyperbolic if the external perturbation is small. On the other hand, suitable three-body encounters can give rise to binary formation with the third particle absorbing the hyperbolic excess energy. The probability of such events is small in homogeneous systems, but conditions are more favourable in the high-density core which develops during the evolution. Secondly, the ejection of strongly bound members from the shrinking core leads naturally to the formation of one close central binary. The latter process is inevitable and occurs on a much shorter time-scale than is required for the majority of particles to escape. Although the total formation rate is small, the numerical calculations show that even one close binary may affect the whole evolution significantly, at least for $N \lesssim 500$.

When other particles are present, it is no longer adequate to define binaries by the usual expression

$$\frac{1}{a} = \frac{2}{R} - \frac{\dot{\mathbf{R}}^2}{G(m_k + m_l)}, \tag{16}$$

where $a > 0$ for elliptic motion. Instead the stability of binary orbits depends on the size of the semi-major axis compared to the local inter-particle separation. The relative strength of the external field may be estimated by the invariant perturbation

$$\gamma = \frac{|\mathbf{F}|\, R^2}{m_k + m_l}, \tag{17}$$

where $\mathbf{F}$ is the tidal acceleration of the relative motion. Thus $\gamma \ll 1$ at the apocentre usually ensures stability during the subsequent revolution, whereas $\gamma \simeq 1$ may lead to rapid disintegration or the exchange of companions.

Once formed, a wide binary may either increase its binding energy or be destroyed by further encounters. Suitable conditions for binary disruption by an incoming third particle may be obtained from the reversal of all velocities at the time of formation.

The two processes are not entirely symmetrical, however, since the velocities of field particles are normally less than the local escape velocity, whereas the latter may easily be exceeded after a favourable interaction. This imbalance therefore provides a mechanism for increasing the fraction of the total energy absorbed by the binary, once the threshold energy for dissociation has been reached. Accordingly, we introduce an energy parameter λ defined by

$$\frac{Gm_k m_l}{2a} = -\lambda E, \tag{18}$$

where E denotes the total energy of all bound members; i.e. the quantity $E_0 - E$ represents the excess kinetic energy carried away by escapers. Binaries may be considered close in a dynamical sense if the binding energy exceeds the mean kinetic energy of field particles; this criterion corresponds to $\lambda > 1/N$. On the other hand, the upper limit $\lambda = 1$ would be reached if all other members were ejected in escape orbits.

Results of binary evolution for several systems with $N=250$ and one system with $N=500$ have been discussed elsewhere (Aarseth, 1972). The property of increasing λ-values is common to all these examples. In the systems with different particle masses, an evolution measure $\lambda \simeq 0.5$ was reached after 6–18 mean crossing times. Furthermore, the time-scales are well correlated with the maximum mass ratios which fall in the range 80:1 to 16:1. The final binaries tend to be formed by some of the most massive members after successive exchanges of companions, hence the evolution is in the direction of more energetically favourable configurations. Although the cross-section for exchange is small, close binaries are usually found near the center and therefore benefit from the overall central focusing as well as the mass segregation effect.

The dynamics of interactions with binaries may be illustrated by considering in isolation a bound system consisting of one binary with mass components m_1 and m_2, together with an approaching third particle of mass m_3. Unless exchange takes place, the total energy before and after the encounter is approximated well by the expression

$$E_0 = m_3 E_3 + \tfrac{1}{2}(m_1 + m_2)\,\dot{\mathbf{r}}_{\mathrm{cm}}^2 - \frac{Gm_1 m_2}{2a}, \tag{19}$$

where $\dot{\mathbf{r}}_{\mathrm{cm}}$ denotes the binary velocity with respect to the combined center of mass. If the third particle escapes, $E_3 > 0$ and the new binding energy of the binary exceeds the total energy in absolute value. Similarly, linear momentum conservations implies a binary velocity

$$\dot{\mathbf{r}}_{\mathrm{cm}} = -\frac{m_3}{m_1 + m_2}\,\dot{\mathbf{r}}_3 . \tag{20}$$

Thus an escaping particle may impart a significant recoil velocity to the binary, resulting in a greater kinetic energy of the center of mass motion. This process is also efficient when the three-body system is subject to the attraction of an overall mass distribution. The binary would then lose kinetic energy by the equipartition process during wider encounters when the relative motion is approximately conserved. Kinetic energy produced by energetic interactions involving three particles is therefore transferred to other central members via the center of mass motion.

Most particles interacting strongly with massive binaries satisfy the condition $m_3 \ll m_1 + m_2$, since the number of heavy members is relatively small for realistic mass distributions. The numerical calculations show that such close encounters may lead to energetic escapers with final velocities exceeding three or four times the rms value. These large escape velocities can also be understood qualitatively; i.e. it is the binary relative motion rather than the overall rms velocity which is relevant. In order to illustrate this idea explicitly, we introduce the invariant ratios

$$\alpha = \frac{\dot{\mathbf{r}}_3^2}{\langle \dot{\mathbf{R}}^2 \rangle}, \tag{21}$$

and

$$\beta = \frac{\langle \dot{\mathbf{R}}^2 \rangle}{\langle v^2 \rangle}. \tag{22}$$

Combining Equations (16) and (18) together with the equilibrium condition $|E| = T = 0.5\, N\bar{m}\langle v^2 \rangle$, the second parameter reduces to the general expression

$$\beta = \frac{\lambda N \bar{m}\,(m_k + m_l)}{m_k m_l}. \tag{23}$$

Hence the condition $\lambda \gg 1/N$ also implies that $\beta \gg 1$ by nearly the same amount. On the other hand, favourable interactions with binaries may produce third particle velocities reaching $\alpha \simeq 1$. This condition then guarantees a very large escape velocity, i.e. $\alpha\beta \gg 1$, since $v_e \lesssim 3\langle v \rangle$ for most theoretical models.

The energy sink behaviour of close binaries implies that the rest of the system expands with time. Actual calculations also show that the corresponding evolution rate begins to decrease when λ becomes significant. This behaviour may be understood by considering a new system comprising all single bound particles as well as binary center of mass configurations. If we neglect interactions between escaping particles and adopt the center of mass approximation for binaries, the energy required to disperse this system to infinity is given by

$$E_f = E_0 - E_\infty - E_b, \tag{24}$$

where E_∞ and E_b is the total binding energy of all escapers and binaries, respectively. Both these quantities are usually zero at the beginning of the calculations, hence the total energy of the free particles, E_f, increases towards zero as a result of the evolution. Assuming approximate equilibrium, the crossing time of the new system then becomes

$$t_{\mathrm{cr}} = \frac{\sqrt{2}\, M_f^{5/2}}{4\,|E_f|^{3/2}}, \tag{25}$$

where M_f is the total mass of all bound members. The dynamical time-scale is closely associated with the evolutionary crossing time defined here. Finally, we emphasize that for $N \lesssim 500$ the critical stage $\lambda \simeq 0.5$ is reached when only about 10 percent of all particles have escaped. Thus Equation (25) predicts much longer time-scales for the subsequent evolution.

References

Aarseth, S. J.: 1972, in M. Lecar (ed.), *Gravitational N-Body Problem,* D. Reidel Publ. Co., Dordrecht, Holland p. 88.
Albada, T. S. van: 1968, *Bull. Astron. Inst. Neth.* **19**, 479.
Chandrasekhar, S.: 1942, *Principles of Stellar Dynamics,* University of Chicago, Chicago.
Hénon, M.: 1960, *Ann. Astrophys.* **23**, 668.
Standish, E. M. and Aksnes, K.: 1969, *Astrophys. J.* **158**, 519.
Wielen, R.: 1967, *Veroeffentl. Astron. Rechen-Inst. Heidelberg,* No. 19.

INTEGRATION METHODS OF THE *N*-BODY PROBLEM

S. J. AARSETH
Institute of Astronomy, Cambridge, England

Abstract. Considerable efforts have been made over the past decade to develop numerical methods for studying the gravitational *N*-body problem. Different approaches are required, depending on whether the emphasis is on accuracy or efficiency. Simple methods may provide accurate solutions if the computer word length is sufficiently large, but such investigations are only feasible for relatively small systems or short time intervals. However, the computing time requirement rapidly becomes excessive because the number of mutual interactions increases with the square of the particle number. More economical methods must therefore be employed in order to study richer systems over significant time intervals.

Integrations of large systems favour the use of high-order force differences since the past information can readily be remembered. Practical experience indicates that four or five orders are adequate, but the optimum order is machine dependent. We discuss an integration scheme which is based on a fourth-order polynomial with an additional corrector. Each particle is advanced in time by a variable interval which is governed by the convergence rate of the corresponding force polynomial. Dynamical consistency is ensured by using a low-order prediction of the co-ordinates of all other particles at the time of a force calculation. The individual time-step method is simple to use and works efficiently for a variety of particle numbers and density distributions.

Calculations of close encounters are rather time-consuming and subject to loss of accuracy from rounding errors. Binaries with short periods are especially troublesome, necessitating the introduction of special techniques. Fortunately, dominant two-body encounters may be integrated with improved accuracy using the Kustaanheimo-Stiefel regularization treatment which transforms the equations of motion into a more well-behaved form. Critical encounters may now be studied with relative ease and the method is also powerful for large perturbations. In addition, this new technique provides a more natural description of the two-body motion than is offered by direct calculations.

The regularization technique can be extended to include an arbitrary number of simultaneous two-body encounters and may readily be used in conjunction with the basic integration method. We discuss procedures for combining the two methods efficiently, together with a scheme for automatic decision-making. Finally, some dynamically consistent approximations are considered which speed up the calculations without undue loss of accuracy. A detailed description of both these integration methods has been given elsewhere (Aarseth (1971), *Astrophys. Space Sci.* **14**, 118).

B. D. Tapley and V. Szebehely (eds.), Recent Advances in Dynamical Astronomy, 208.

NUMERICAL STUDIES OF THE GRAVITATIONAL PROBLEM OF N-BODIES

R. WIELEN

Astronomisches Rechen-Institut, Heidelberg, Germany

Abstract. The dynamical evolution of stellar systems containing a few hundred stars can be studied by numerically integrating the equations of motion of all the stars as an N-body problem. From a methodic point of view, these numerical N-body experiments belong to two branches of dynamical astronomy: celestial mechanics and stellar dynamics. As in celestial mechanics, the basic procedure is the computation of individual orbits of mass points in the exact gravitational field of all the other bodies. However, these individual orbits are not the information at which we aim in stellar dynamics. Hence, we finally deduce from all the individual orbits the desired statistical information about the evolution of the system as a whole ('macroscopic properties'). Of course, the individual orbits also give valuable insight into the 'microscopic' behaviour of a stellar system.

What can we hope to learn from N-body experiments? There are two basic applications of the results: First, we can test the statistical theories of stellar dynamics. For that purpose, the physical situation should be chosen as simple as possible in order to test the basic assumptions and predictions of the theories. For example, we should study isolated spherical clusters of stars of equal mass. Only if the theories can successfully describe the results obtained from such simple experiments, should we introduce additional complications like different masses of the stars, external fields, etc. Second, we can simulate the dynamical evolution of real stellar systems. At present, we are able to handle systems of up to $N=500$ stars in the numerical experiments. This number N is typical for many open star clusters and clusters of galaxies. Hence, we may compare the N-body experiments directly with these types of astronomical objects without any further theory, provided that the models include all the physically relevant effects. For example, for open cluster we have to include the actual spectrum of stellar masses, perhaps a mass loss due to the internal evolution of the stars, the tidal field of the Galaxy, gravitational shocks by passing HI-clouds, etc..

The N-body experiments have the advantage that they are, as far as possible, free from mathematical assumptions which are not physically inherent in the problem. In contrast, all the presently available statistical theories of stellar dynamics have to introduce such additional assumptions. However, there also exists an hitherto unsolved fundamental problem in the interpretation of N-body experiments: The solutions of the general N-body problem are in most cases highly unstable. Due to this basic physical instability, we cannot trace the true individual orbits over long periods of time by any numerical technique. However, although the individual orbits of the stars are not fully reliable, it is generally argued, and supported by the comparison of different experimental results, that the derived statistical properties of the dynamical

B. D. Tapley and V. Szebehely (eds.), Recent Advances in Dynamical Astronomy, 209–210.

evolution of the whole stellar system are not biased by this microscopic instability.

In numerical experiments for clusters containing up to 500 stars, the following general results have been found: (1) The central density in the cluster increases with time. (2) A large halo of stars is formed in the outermost regions of a cluster. (3) The rate of dynamical evolution is strongly increased by unequal masses of the stars. (4) The most massive stars segregate towards the center of the cluster. (5) In the core, one or a few binaries are formed which absorb most of the binding energy of the cluster. (6) During close encounters, some stars gain enough energy for escape. The escapers are produced suddenly, preferentially in the core of the cluster, and not by a slow diffusion process. The rate of escape depends only weakly on the mass of a star. (7) The mean velocity of the stars in the cluster decreases with the distance from the center. (8) In the core of a cluster, the velocity distribution is isotropic. (9) In the outer parts of the cluster, the motion of the stars is primarily in the radial direction and hence strongly non-isotropic. (10) No equipartion of the kinetic energy takes place among stars of different mass. (11) The virial theorem is, on the average, nicely fulfilled for bound clusters. (12) A slow rotation does not significantly affect the dynamical evolution as outlined above. (13) The rate of escape is drastically increased by the tidal field of the Galaxy. For a typical open cluster ($N = 500$ stars, total mass of 250 solar masses, median radius in projection of about 1 parsec), the numerical experiments predict a total lifetime of about 5×10^8 yrs. (14) A mass loss of massive stars has only a minor effect on the dynamical evolution of open clusters.

For a detailed discussion of the concepts and of the results of numerical experiments on the gravitational problem of N-bodies, we refer to the general references quoted below.

General References

Proceedings of Colloquia

IAU Symposium No. 25: 'The Theory of Orbits in the Solar System and in Stellar Systems' (ed. by G. Contopoulos), Academic Press, London-New York, 1966.
'Colloque sur les méthodes nouvelles de la dynamique stellaire', *Bull. Astron.* **2** (3), 1–285 (1967).
'Symposium on Computer Simulation of Plasma and Many-Body Problems', NASA SP-153, 1967.
'IAU Colloquium on the Gravitational Problem of N-bodies', *Bull. Astron.* **3** (3), 1–311 (1968).
IAU Colloquium No. 10, 'Gravitational N-Body Problem' (ed. by M. Lecar), D. Reidel Publ. Comp., Dordrecht-Holland, 1972. Also partly reprinted in *Astrophys. Space Sci.* **13**, 279–495 (1971) and **14**, 3–178 (1971).

Forthcoming Review Papers

Aarseth, S. J.: 1973, *Vistas in Astronomy,* Pergamon Press, London (to be published).
Wielen, R.: *Proceedings of the First European Astronomical Meeting, held in Athens* 1972, Springer-Verlag Berlin-Heidelberg-New York (to be published).

STAR ENCOUNTER SIMULATIONS

J. R. DORMAND

Mathematics Department, Teesside Polytechnic, Middlesbrough, Teesside, England

Abstract. A short description of the 'limacoid' technique for the simulation of disruptive star encounters is given, together with some results connected with the capture theory of the origin of the solar system.

1. Introduction

The study of encounters between stars is of particular importance to the capture theory of the origin of the solar system. This theory, suggested by Woolfson (1964), predicts the capture of material by the Sun from a light, diffuse, pre-main sequence star during a close encounter. Such an encounter might take place in a dense young star cluster in which both stars were formed.

The problem of simulating numerically an encounter was first tackled by Woolfson (1964), who devised a two dimensional star model which consisted of a number of mass points filling the outer regions with a single more massive point at the center. The star was placed on a parabolic orbit with respect to the Sun, which was taken to be a point mass, and the equations of motion of each mass point were solved simultaneously using a Runge-Kutta method. Some of the mass points from the outer regions of the star were captured by the Sun and it was suggested that these points could represent the protoplanetary material.

Although this model demonstrated the qualitative aspects of the capture theory it has some serious defects from a physical point of view. For example:

(1) it was two dimensional – a real protostar would be three dimensional;

(2) a discrete system of mass points is used to model a continuous distribution of matter. The first defect is easy to overcome – at the cost of extra computations; the second, however, is rather more serious. In such a system all mass points interact gravitationally with all the others; the model can therefore suffer from localised point effects in which singularities in the equations of motion can occur. It was found for example that under gravitational forces alone the model took on a fragmented appearance very early in the simulation. This unnatural pooling of points had to be overcome by introducing repulsion forces between points, corresponding to the physical reallity of gaseous matter resisting the tendency to build up regions of high density.

2. The Limacoid Method

When Prof. Woolfson and the present author re-examined the star encounter problem a few years later it was decided to pay special attention to the above points. It was thought that a mass point star model was still a useful system from the point of view of numerical analysis, but it was felt desirable to compute the gravitational field of the

B. D. Tapley and V. Szebehely (eds.), Recent Advances in Dynamical Astronomy, 211–215.

configuration of mass points *as a whole* rather than using individual pairs of points. It is clear that such a procedure would avoid the singularity problem.

In order to be able to do this we must first consider the shape of a protostar during its encounter with the sun. According to Jeans (1929) the tidally distorted protostar would fill an equipotential surface which is rather pearshaped. The profile of such a surface can be reasonably approximated by part of a limacon curve whose polar equation is

$$r = \frac{a(k\cos\theta - 1)}{k - 1}, \qquad |\theta| \leqslant \arccos(1/k),$$

where a is the length of the axis and k is a measure of the 'eccentricity' (see Figure 1). When k becomes large the curve approaches a circle.

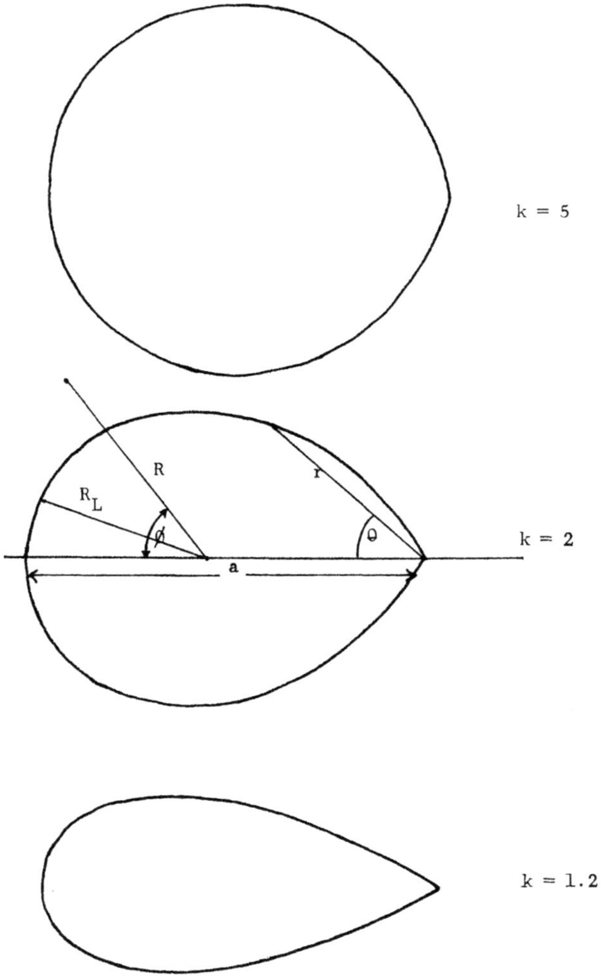

Fig. 1. The limacon: $r = a(k\cos\theta - 1)/(k - 1)$ for three values of k.

Accordingly we took the shape of the distorted protostar to be the figure of revolution formed by rotating the limacon about its axis of symmetry; the figure bounded by this surface will be called a 'limacoid'. It was hoped that limacoids could be fitted to configurations of mass points representing a protostar and that the gravitational acceleration of each mass point would be given by the limacoidal field.

Unfortunately the gravitational field of the limacoid is not amenable to analytical treatment except at the tip where the attraction is given by

$$F = \pi \varrho G a (2k + 1)(k - 1)/3k^2,$$

where ϱ is the density (assumed to be uniform). However the symmetry of the limacoid can be used to compute its field numerically and this has been done for many values (1782) of R, ϕ, and k, where R and ϕ are coordinates with respect to the center of mass of the limacoid (see Figure 1). At each (R, ϕ, k) the center of action of the force, as a displacement from the center of mass along the symmetry axis, and the ratio of the force to that of a sphere placed at the center of action were computed. A polynomial surface was then fitted to the data for each of the two gravitational parameters; the maximum degrees of the three variables R, ϕ and k were 9, 4 and 4 respectively giving 250 polynomial coefficients for each surface. An orthogonal polynomial fitting method was used.

We have not yet discussed the three dimensional point distribution to which the limacoid must be fitted before our field polynomials can be used. Since we are chiefly interested in the outer regions of the protostar, which are likely to suffer disruption, most of its mass (75%) was concentrated at a single point, while the remainder was distributed on two concentric spherical shells of points. To achieve a fairly uniform distribution of points they were placed on the normals to the faces and the vertices of regular polyhedra. A hexoctahedron (321) was used for the outer shell and an icositetrahedron (311) formed the inner shell. The direction ratios of the normals to the 48 faces of the hexoctahedron are obtained by permuting the number triple (321). The vertices were also used to give an additional 26 points and in a similar way the inner shell received 50 points. For the purposes of computation we need only consider a hemisphere since the portion of the protostar above the orbital plane will always remain a mirror image of that below the plane.

During the encounter simulation each shell of points is taken to represent the boundary of a limacoid of constant density. The equations of motion of each mass point are integrated simultaneously using a fourth order Runge-Kutta method and at the beginning of each step a limacoid is fitted to each of the shells of points. The appropriate limacoid is determined by the two parameters k and a. The former is found from the ratio $(\bar{R}_L)^2/\bar{R}_L^2$, where R_L is the distance from the center of mass to the boundary of the limacoid and the bar denotes mean value. This ratio has been computed for a range of k values and will clearly approach unity as k increases and the limacoid approaches a sphere. Once k has been determined a is easily found from the actual $\bar{R}_L$ computed earlier for $a = 1$. The position of the limacoid axis is fixed by fitting a least squares line to the point configuration, neglecting the Z-coordinates since the XY-plane contains the star orbit.

3. Results

Figure 2 shows the result of an encounter simulation using the limacoid method. The profile of the protostar in the early stages of the encounter seems to be limacoidal to high accuracy but, following perihelion, the fit is less good and the limacoidal treatment is discontinued shortly afterwards. Mass points in the escaping filament with distances greater than 1.4 rad from the center of the protostar are not included in the limacoid fitting.

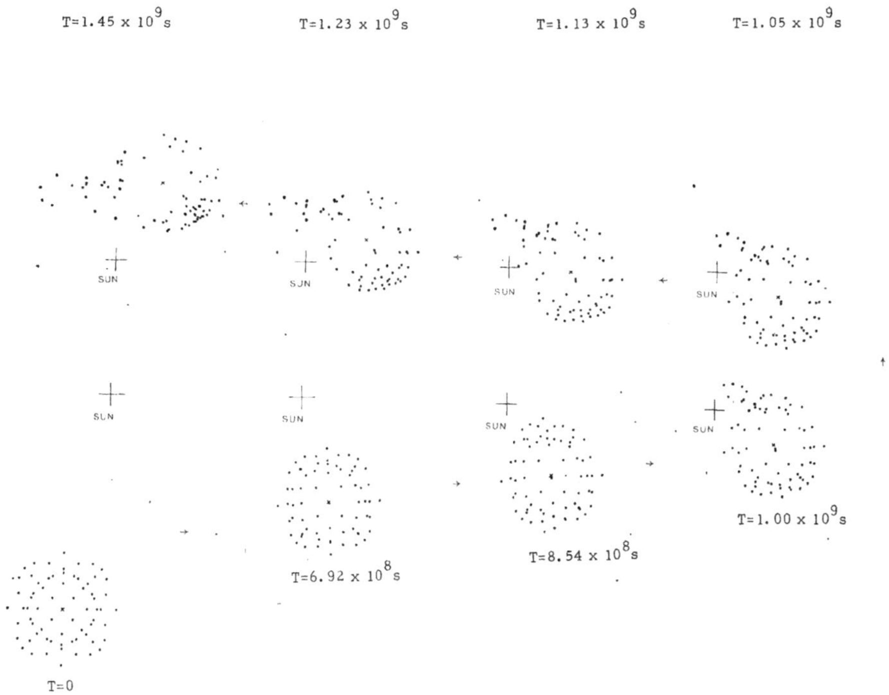

Fig. 2. Encounter between Sun and protostar. Only x and y coordinates of points above orbital plane are plotted. Cross marks position of central point of protostar.

The protostar in the example shown had a mass of $0.25\,M_{\odot}$ and a radius of 16 AU giving a density of about 10^{-11} g cm^{-3}. Its orbit was hyperbolic with eccentricity 1.5 and center to center perihelion was 20 AU. In this case 21/125 mass points were captured by the Sun (4% of the star mass) into orbits with perihelia in the range 0.2 to 10 AU and eccentricities from 0.73 to 0.99. Results of other encounter simulations are given by Dormand and Woolfson (1971).

Initial tests of encounters have always been made using a simple three body model and it has been found that material can be captured by the Sun from the protostar under a wide range of initial conditions. With the above star model capture can take place if the star orbit has eccentricity less than about 3. Different mass ratios have also been tried showing that this parameter is not critical.

The above model star was constrained to collapse at a small constant rate (in isolation) but some protostars in free-fall collapse have also been examined in which solar capture took place.

References

Dormand, J. R. and Woolfson, M. M.: 1971, *Monthly Notices Roy. Astron. Soc.* **151**, 307.
Jeans, J. H.: 1929, *Astronomy and Cosmogony,* Cambridge University Press.
Woolfson, M. M.: 1964, *Proc. Roy. Soc.* **A282**, 485.

YUKAWA POTENTIAL AND STELLAR ENCOUNTER

O. GODART
University of Louvain, Louvain, Belgium

Abstract. If a cosmological constant is accepted, at great distances, newtonian is modified in a Yukawa potential. There will then be a limit for stable circular orbits and then orbits of elliptic types. Moreover, the integrals to evoluate global effects of stellar encounters converge.

1. Introduction

In the classical book of Chandrasekhar, *Principles of Stellar Dynamics*, the evaluation of motion's deviation due to accumulative stallar encounters meets with the mathematical difficulty of a logarithmic divergent integral. This was solved in a somewhat empirical way by restricting the impact distance of binary encounters to the mean distance of the stars in the stellar system considered. In any way, the final result is not very sensitive to the choice of this upper limit. Improvements have been brought by statistical considerations for example by a work of Hénon 1958. In fact, we have to take into account of multiple encounters leading to very complex effects. Similar problems exist in the theory of plasma where ions are surrounded by clouds of electrons forming a screen. When they are supposed to be in thermodynamic equilibrium, the potential of the ions is given by $e^{-r/r_0}/r$ where r_0 is the Debye screening length equal to $r_0 = \sqrt{(kT/4\pi ne^2)}$ where T is the temperature and n is the equilibrium number density of the electrons. At the limit, T tending to infinity or n to zero, we find the Newtonian potential. The same type of potential is met in the theory of mesons and was called potential of Yukawa. However, in the theory of gravitation, there are masses of only one sign and it seems that the transposition of plasma considerations is unacceptable. But when we are dealing with stellar systems, the dimensions are such that relativistic corrections to the Newtonian potential could be envisaged, chiefly because the Chandrasekhar integrals extend to the infinity where a simple Newtonian potential will introduce difficulties in relation to Mach's principle. A first work done, in computing Chandrasekhar integrals with a potential as $1/r^{1+\varepsilon}$, where ε is a strictly positive constant, has shown their convergence. In particular, taking $\varepsilon = 1$ the integrals can be made exact and give the time of relaxation slightly bigger than those of Chandrasekhar.

If one accepts in the theory of general relativity, the existence of cosmological constant Λ, a closer similitude to plasma theory is reached. In fact, that constant could be associated to a repulsive force proportional to the distance counter-balancing the Newtonian attraction in $1/r$. Another interpretation would consist in considering it as a fictitious negative constant density of mass somewhat similar to the smoothed cloud of electrons:

$$\varrho_\Lambda = \frac{c^2 \Lambda}{8\pi G}.$$

B. D. Tapley and V. Szebehely (eds.), Recent Advances in Dynamical Astronomy, 216–221.

Taking into account the Λ term, the potential would take the form

$$\frac{e^{-\alpha\sqrt{\Lambda} r}}{r}$$

where α is a dimensionless constant of the order of 1, Λ is a small quantity such that $1/\Lambda \simeq 10^{55}\ \mathrm{cm}^2$.

The screen distance r_0 is quite big, but its effect is not inappreciable in the galactic scale. Even, if the cosmological constant is taken nil, a Yukawa potential will be introduced if we consider an indefinite homogeneous medium containing localized regions where the density differs from the mean ϱ_0. If this mean is subtracted from local density, we get perturbed regions with positive or negative gravitational masses. These quasi-particles move with respect to the homogeneous substratum. A theory of such quasi-particles gives a potential proportional to

$$\frac{e^{-r/r_0}}{r} \quad \text{(see Saslaw, 1968)},$$

where r_0 is very near the Jeans' length of the stellar system divided by 2π. Before studying the central problem of the stellar encounter with such a potential, it is worth while to examine the two body problem.

2. Analysis of the Solutions

As in the general case of central forces, the relative motion of two masses m_1 and m_2 will be in an invariant plane and the polar coordinates r, θ wil satisfy the integral of constant moment of momentum

$$r^2 \frac{\mathrm{d}\theta}{\mathrm{d}t} = h \tag{1}$$

and of constant energy

$$\frac{1}{2}\left[\left(\frac{\mathrm{d}r}{\mathrm{d}t}\right)^2 + r^2\left(\frac{\mathrm{d}\theta}{\mathrm{d}t}\right)^2\right] - G(m_1 + m_2)\frac{e^{-r/r_0}}{r} = E. \tag{2}$$

Let dimensionless variables

$$r' = \frac{r}{r_0} \qquad t' = \frac{h}{r_0^2}. \tag{3}$$

We shall drop here after the prime and write for (1) and (2)

$$r^2 \frac{\mathrm{d}\theta}{\mathrm{d}t} = 1 \tag{4}$$

$$\frac{1}{2}\left[\left(\frac{\mathrm{d}r}{\mathrm{d}t}\right)^2 + r^2\left(\frac{\mathrm{d}\theta}{\mathrm{d}t}\right)^2\right] - \mu\frac{e^{-r}}{r} = \varepsilon. \tag{5}$$

We can, eliminating t between both expressions, obtain the differential equation of the trajectories by using the independent variable $u=1/r$. We get then

$$\frac{1}{2}\left(\frac{\mathrm{d}u}{\mathrm{d}\theta}\right)^2 + u^2 - \mu u e^{-1/u} = \varepsilon .$$

If θ is considered as a time, the function $V(u, \mu)=u^2/2-\mu u e^{-1/u}$ can be assimilated to a force function.

For any particular value of $u=u_\varepsilon$, the acceleration is nil for $(\delta V/\delta u)=0$ giving

$$\mu = \mu_c = \frac{u_c^2 \, e^{1/u_c}}{1 + u_c} .$$

The inverse is not true because $\mu(u_c)$ reaches a minimum for

$$u_l = \frac{\sqrt{5} - 1}{2} \quad \text{equals to} \quad \frac{e^{1+\sqrt{5}/2}}{2 + \sqrt{5}} = \mu_l .$$

Then for any value of $\mu<\mu_l$, $V(u, \mu)$ will be a constantly increasing function and as a consequence for any value of $\varepsilon>0$ there will exist orbits starting from infinity $(u=0)$ and reaching a minimum distance $V(u, \mu)=\varepsilon$ and then returning to infinity.

There are orbits of the hyperbolic type.

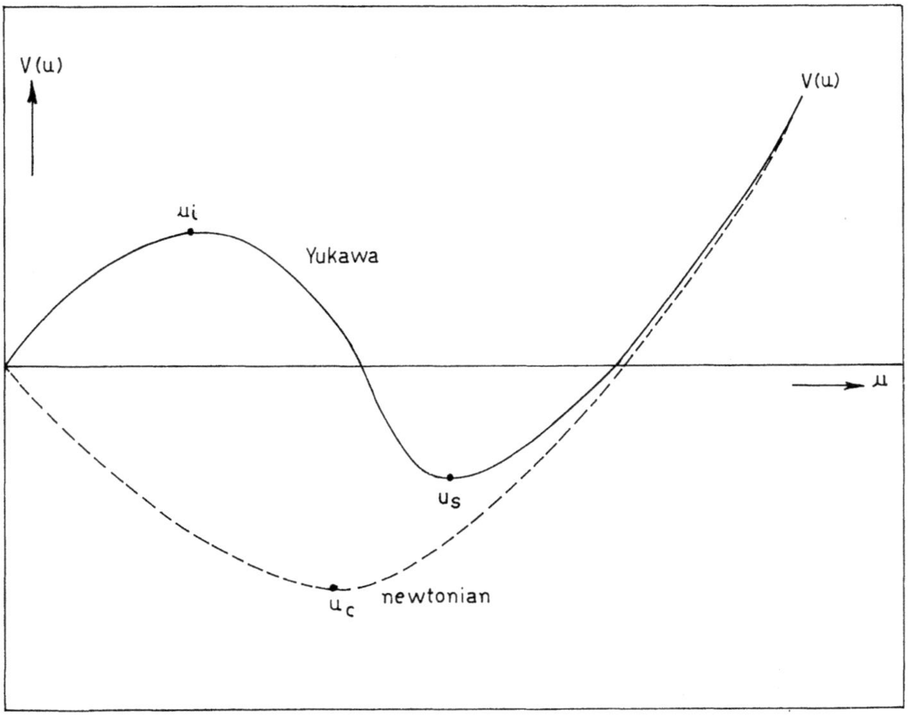

Fig. 1. Potential when $\mu > \mu_l$

For $\mu > \mu_e$, $V(u, \mu)$ does not increase monotonically from 0 to infinity, but reaches a maximum for $u = u_{ci}$ and a minimum for $u = u/_{ls}$. (Figure 1.)

If the constant of energy of a particular orbit is $\varepsilon_i = V(u_{ci}, \mu)$ or $\varepsilon_s = V(u_{cs}, \mu)$ they are circular orbits

$$\varepsilon = \frac{u_c^2(1 - u_c)}{2(1 + u_c)}.$$

For $\varepsilon > \varepsilon_i$ we have again hyperbolic orbits. But for $\varepsilon_s < \varepsilon < \varepsilon_i$, we shall get orbits oscillating between u_1 and u_2. We shall call them orbits of an elliptic type. Moreover if we call ε_0 the maximum of 0 and ε_s between $\varepsilon_0 < \varepsilon < \varepsilon_i$, there will exist also orbits coming from infinity to a finite distance that we shall call hyperbolic of the second kind.

To distinguish the two circular orbits existing for $\mu = \mu_e$, let us write the variation equation where $u = u_c + \delta u$, we get

$$\frac{\mathrm{d}^2 \delta u}{\mathrm{d}\theta^2} = \left(\frac{\partial^2 V}{\partial u^2}\right)_{u=u_c} \delta u = \frac{1 - u_c - u_c^2}{u_c(1 + u_c)} \delta u$$

$\delta^2 V/\delta u^2$ is positive for $u < u_e$ and then the orbit is unstable. It is negative for $u > u_e$ and the orbit is then stable (see Figure 2). ε will also reach a maximum for

$$u = u_l\,; \qquad \varepsilon_l = \frac{5\sqrt{5} - 11}{4} = 0{,}045\,.$$

For $\varepsilon > \varepsilon_l$ there will be only hyperbolic orbits.

Also for $\varepsilon = \varepsilon_i$, $\mu = \mu_i$ there will be three orbits, one circular, one asymptotic, the limit of the hyperbolic orbit of the second kind and another asymptotic limit of the elliptic orbits.

It is interesting to make a comparison with the newtonian case

$$V(u) \equiv \frac{u^2}{2} - \mu u\,.$$

$V(u)$ instead of increasing for u small decreases to reach a minimum for a circular orbit where $\mu = \mu_c$ and $\varepsilon_i = -u_c^2/2$.

Then for $\varepsilon < 0$ we shall get elliptic orbits and $\varepsilon > 0$ hyperbolic ones. However, it could be noticed that the complexity of orbits in the Yukawa potential is not cumbersome for the specific problem of encounters. If we introduce the impact distance D, the speed at infinity V_∞.

We have $V_\infty D = h$ then

$$\varepsilon = \frac{1}{2}\left(\frac{r_0}{D}\right)^2.$$

As r_0 is great, for all nearby impact distances, considered $\varepsilon > \varepsilon_1$

$$D < \tfrac{10}{3} r_0\,.$$

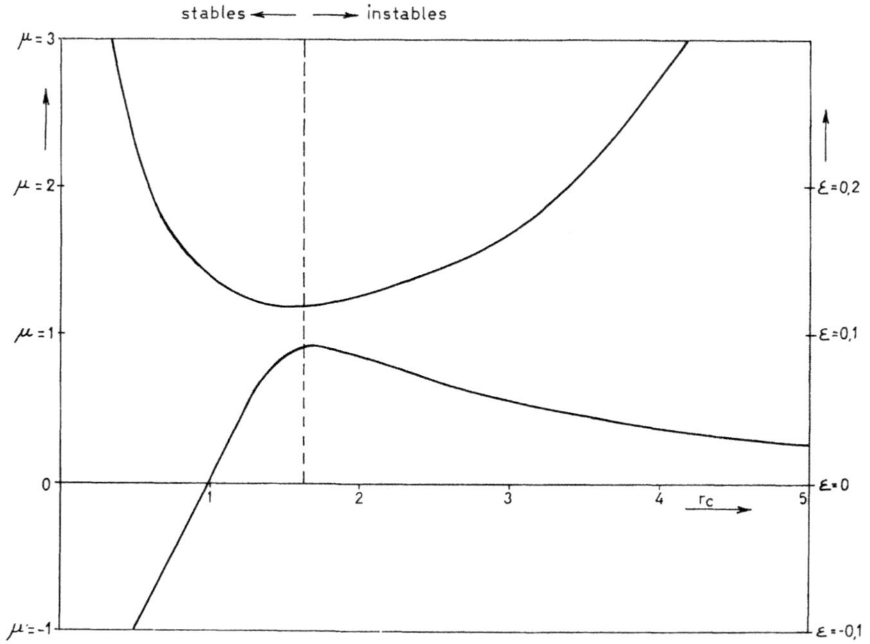
stables
instables
μ = 3
μ = 2
μ = 1
0
μ = -1
ε = 0,2
ε = 0,1
ε = 0
ε = -0,1
1
2
3
4
5
r_c

Fig. 2.

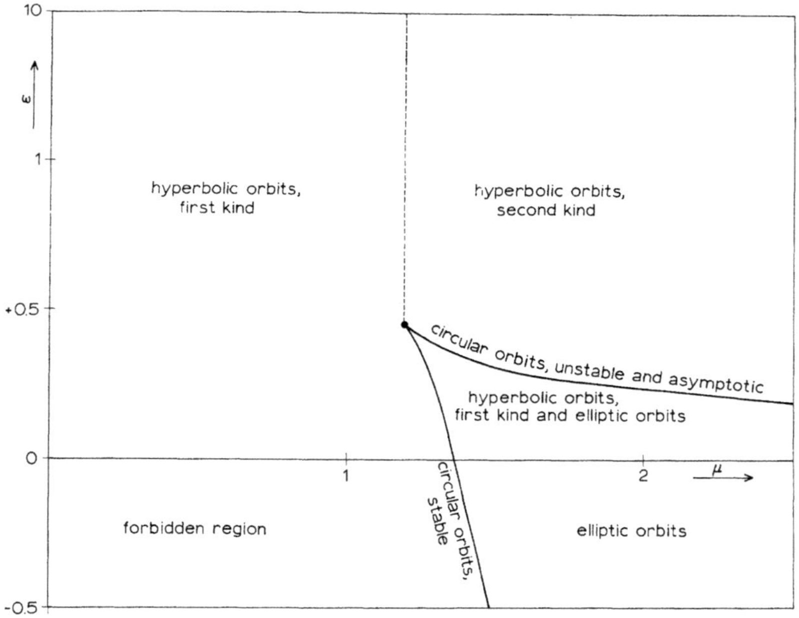
10
1
+0.5
0
-0.5
ε
1
2
μ
hyperbolic orbits, first kind
hyperbolic orbits, second kind
circular orbits, unstable and asymptotic
hyperbolic orbits, first kind and elliptic orbits
forbidden region
circular orbits, stable
elliptic orbits

Fig. 3.

Another representation can be made in a diagram ε, μ; the loci of zero speed at a certain distance u_1 is a straight line.

When two such straight lines cross, for that ε, μ; there will exist an elliptic orbit; circular orbits are the limits when the crossing angle tends to zero; and the envelope of such a straight line gives the curve $\varepsilon_1\mu$ for circular orbits (Figure 3).

References

Henon, H.: 1958, *Ann. Astrophys.* **21**, 186–216.
Saslaw, W.C.: 1968, *Monthly Notices Roy. Astron. Soc.* **141**, 71–108.

AN INVESTIGATION OF THE RÔLE OF TURBULENCE IN THE FORMATION OF STAR CLUSTERS BY A POINT-MASS SIMULATION

C. AUST
University of York, Heslington, York, England

Abstract. An investigation of the role of turbulence in the fragmentation of a collapsing interstellar cloud is described. The cloud is represented by a set of discrete point masses, whose motion under the action of suitably designed inter-point forces is followed numerically.

Numerical studies of the evolution of star clusters usually assume rather arbitrary starting conditions, based on the empirical result that the final state of the cluster (the core-halo type of structure, evaporation of the more energetic stars, etc.) seems to be independent of the chosen initial conditions. However there are aspects of the properties and early evolution of star clusters which can only be explained by investigating the actual formation process of the cluster. These include close binary formation, the initial mass function of cluster members, whereabouts and in what order stars of different masses appear in the cluster, and the interaction of protostars with uncondensed cloud material. We at the University of York are also interested in the probabilities of encounters between stars in various states of evolution with regard to the likelihood of close encounters which satisfy the requirements of Woolfson's capture theory of the origin of the solar system (Woolfson, 1964; Dormand and Woolfson, 1971).

Current ideas on the subject of cluster formation are that, one way or another, an interstellar gas cloud of about one thousand solar masses becomes unstable towards gravitational collapse. As the cloud material becomes more dense, it undergoes fragmentation, a process whereby parts of the cloud collapse to very high densities while the mean density of the cloud remains relatively low. It is supposed that stars form in these higher density regions, and that dynamical interactions between the stars eventually halt the overall collapse of the protocluster. Any remaining low density material will either be accreted by the stars, or be pushed out of the cluster by radiation pressure from hot O or B type stars. The evolution of the cluster would then proceed in the usual manner.

The details of the fragmentation process have never been satisfactorily explained. Grzędzielski (1966) investigated the possibility of partial fragmentation of a pregalaxy in the early phase of contraction. He proposed a mechanism whereby unstable configurations of high density were formed due to the cumulative effect of random shock waves, generated by large velocity differences supposed to exist initially in the pregalaxy. The fact that the interstellar medium is highly turbulent has led us to suppose that a similar fragmentation process could occur in collapsing proto-clusters.

The way in which we have chosen to study this problem is to represent a collapsing

B. D. Tapley and V. Szebehely (eds.), Recent Advances in Dynamical Astronomy, 222–228.

spherical cloud by a set of discrete point-masses distributed throughout the cloud in such a way as to reasonably represent any chosen radial density distribution. An inter-point force-law has been designed to simulate the action of both gravitational and thermal forces. For the time being, we are considering only isothermal collapse, in the absence of magnetic fields and angular momentum.

The equations of motion of the points are being integrated by a suitably modified version of the fourth-order polynomial method with individual time-steps developed by Aarseth (1971). The integration routine is initialized by three cycles of a fourth-order Runge-Kutta procedure. The presence of repulsive thermal forces will ensure that, in general, very close encounters between points do not occur, so maintaining time-steps at a reasonable length and preserving the numerical accuracy of the solution.

At present, we are following the collapse of small, 123 point-mass models to investigate the behaviour of our force-law, and to compare our results with analytical investigations of isothermal collapse.

Figure 1 through 4 indicate the free-fall collapse of an initially uniform density sphere. The distribution of point-masses in the X–Y plane is illustrated at fractions 0.0, 0.53, 0.76 and 0.91 of the theoretical free-fall time. 'Pooling' of points, a defect of point-mass models, can be seen in the later stages of the collapse. The addition of repulsive thermal forces will overcome this difficulty to a large extent.

The ability to represent shock-wave behaviour will be tested by allowing two stable 'clouds' to collide. Eventually, collapsing 500 point-mass models containing turbulent elements will be examined. The turbulence will be represented by superimposing random, but locally correlated, velocities on the inwardly directed collapse velocities.

Figure 5 indicates the generation of random but locally correlated velocity fields. The left-hand diagram shows random velocities generated on a two-dimensional grid according to a Maxwellian velocity distribution, truncated at 1.8 V_{rms}. The right-hand diagram shows how local correlation is introduced after three cycles of an iterative procedure which replaces the velocity at a grid point by a weighted mean of neighbouring velocities.

i.e. $$\mathbf{V}_i(\text{new}) = K \sum_{j=1}^{n_i} w_{ij}\mathbf{V}_j / \sum_{j=1}^{n_i} w_{ij},$$

where there are n_i grid points within some distance d of point i for which the weight $w_{ij}>0$. The factor K is used to preserve $\bar{V}^2$ for the distribution. The weighting function used was $w_{ij}=s_{ij}^2-2s_{ij}+1$, where $s_{ij}=r_{ij}/d$, d being 2.5 grid spacings.

Figure 6 shows the correlation function for velocity with distance before and after the introduction of local correlation. A measure of local correlation is obtained by calculating the velocity deviation in successive shells about each grid point with respect to the velocity at that grid point, and averaging each over all the grid points.

i.e. $$\Delta V(r) = \left\{\frac{1}{N} \sum_{i=1}^{N} \left[\sum_{j=1}^{n_i(r)} \frac{|\mathbf{V}_i - \mathbf{V}_j|^2}{n_i(r)\,\bar{V}^2}\right]\right\}^{1/2},$$

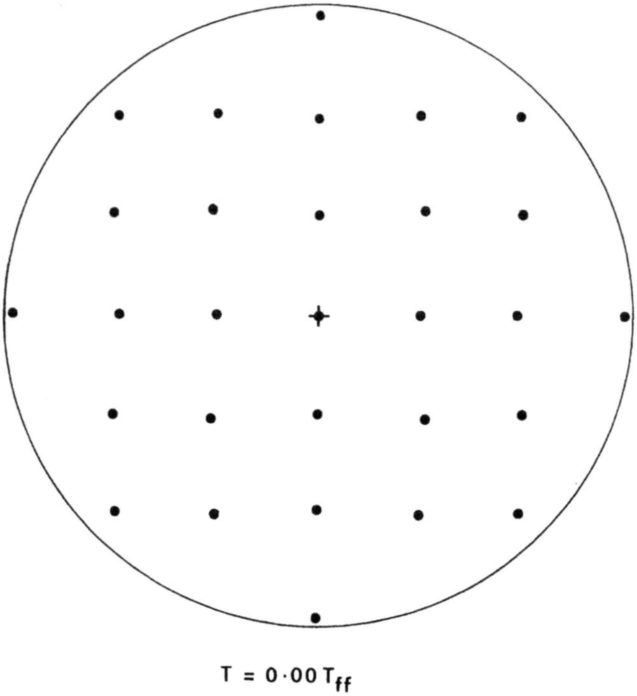

Fig. 1. Cloud Collapse, Sequence 1, Uniform Initial Density. $X-Y$ plane of 123 point-mass model. Time $= 0.00 T_{ff}$, Total mass $= 1.00$, Initial radius (circle) $= 1.00$.

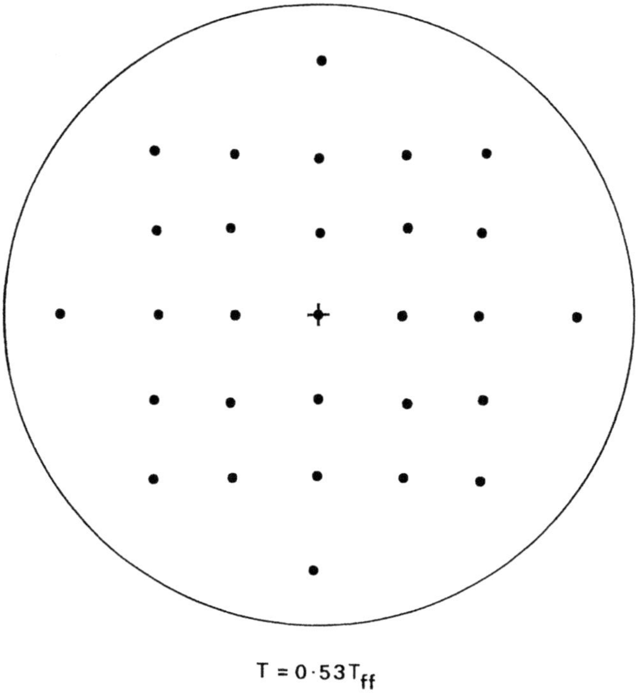

Fig. 2. Cloud Collapse, Sequence 1, Uniform Initial Density. $X-Y$ plane of 123 point-mass model. Time $= 0.53 T_{ff}$, Total mass $= 1.00$, Initial radius (circle) $= 1.00$.

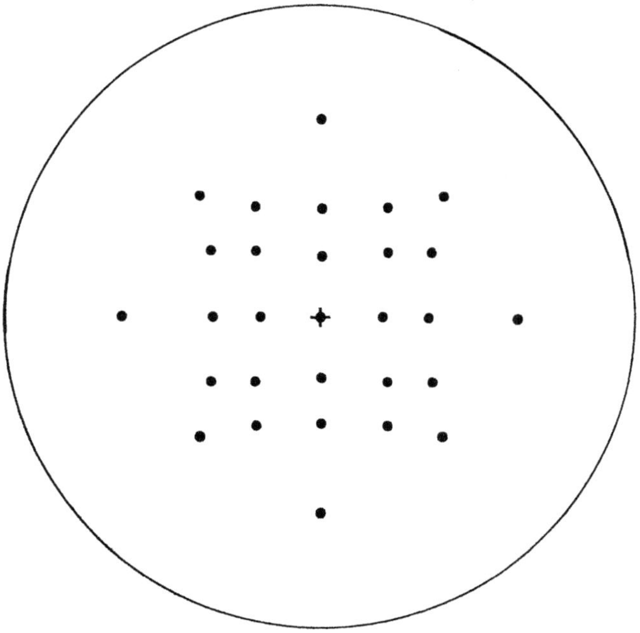

Fig. 3. Cloud Collapse, Sequence 1, Uniform Initial Density. $X-Y$ plane of 123 point mass model. Time $= 0.76 T_{ff}$, Total mass $= 1.00$, Initial radius (circle) $= 1.00$.

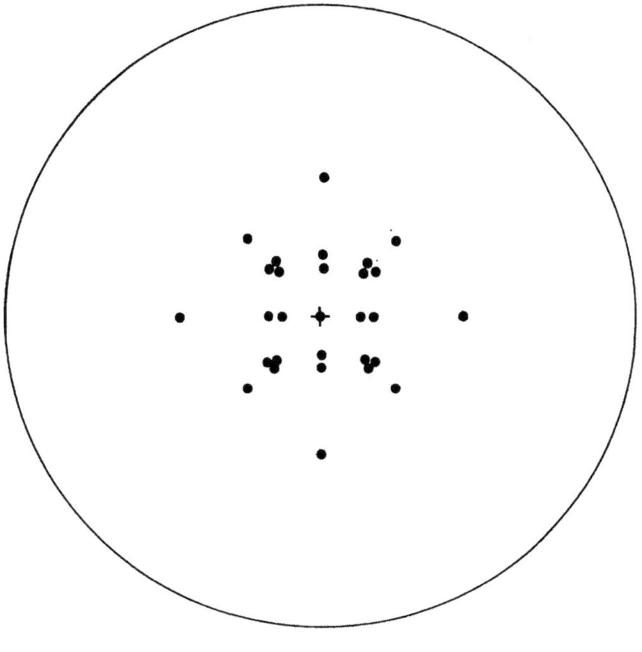

Fig. 4. Cloud Collapse, Sequence 1, Uniform Initial Density. $X-Y$ plane of 123 point-mass model. Time $= 0.91 T_{ff}$, Total mass $= 1.00$, Initial radius (circle) $= 1.00$.

Fig. 5a. Random velocities on a two-dimensional grid generated according to a Maxwellian velocity distribution.

Fig. 5b. Local correlation is introduced after three cycles of an iterative procedure which replaces the velocity at each grid point by a weighted mean of neighbouring velocities.

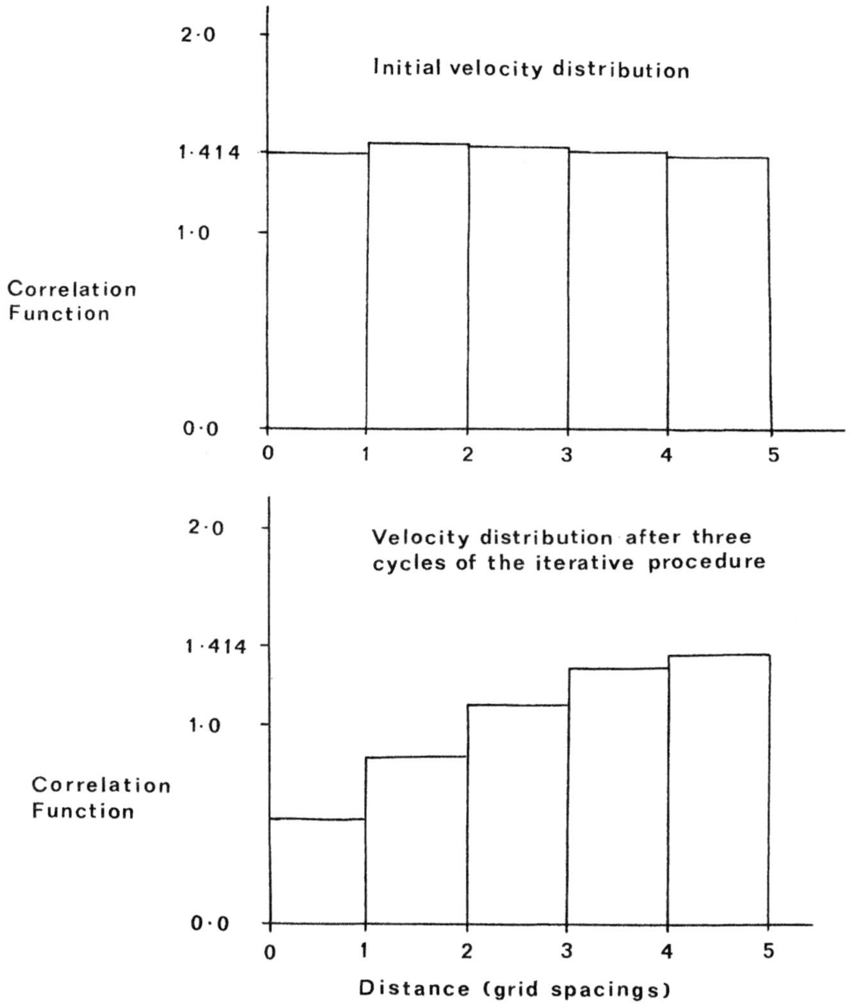

Fig. 6. Correlation function for velocity with distance from a grid point averaged over all the points.

where there are N grid points altogether, and $n_i(r)$ grid points in the shell r to $r+\Delta r$ about point i. For completely random velocities, $\Delta V(r)$ should be $\sqrt{2}$ in all shells.

As the collapse proceeds, the occurrence of high density regions, taken to be several point-masses coming together with large negative binding energies, will be looked for, as will self-generation of turbulence, fed by the overall collapse.

The proposed model will be too coarse for all the aspects of cluster formation mentioned at the beginning to be investigated. However, it may be possible to look at a particular region of the proto-cluster in more detail while using the coarse model to represent the behaviour of the proto-cluster outside the region of interest. This would get over the difficulty of using a greater number of point-masses initially, with the corresponding increase in computing time. We feel that the model will certainly point the way forward to a better understanding of the dynamical and physical processes involved in cluster formation.

References

Aarseth, S. J.: 1971, *Astrophys. Space Sci.* **14**, 118.
Dormand, J. R. and Woolfson, M. M.: 1971, *Monthly Notices Roy. Astron. Soc.* **151**, 307.
Grzędzielski, S.: 1966, *Monthly Notices Roy. Astron. Soc.* **134**, 109.
Woolfson, M. M.: 1964, *Proc. Roy. Soc.* **A282**, 485.

PART IV

THE THEORY OF GENERAL PERTURBATIONS

THEORY OF GENERAL PERTURBATIONS

G. HORI

Dept. of Astronomy, University of Tokyo, Tokyo, Japan

Abstract. A theory of general perturbations based on Lie series is presented together with its predecessor, the von Zeipel method. Some drawbacks of the von Zeipel method are shown together with the motivation for an improved theory by a pedagogical purpose. It is shown that perturbation theories based on canonical transformations and averaging principles yield the same results (through the second order) irrespective of the implicit or explicit type of the transformations. The motion of a near earth satellite in an orbit with small eccentricity and inclination, and that with small angular momentum are presented as two examples of the use of the theory.

1. Canonical Variables

The theory is applicable only for canonical systems, and any dynamical system we consider has to be presented in the canonical form, $\mathrm{d}x/\mathrm{d}t = \partial F(x, y, t)/\partial y$, $\mathrm{d}y/\mathrm{d}t = -\partial F(x, y, t)/\partial x$ with canonical variables $x = (x_1, x_2, \ldots)$, $y = (y_1, y_2, \ldots)$. Typical sets of canonical variables used in celestial mechanics are as follows.

Delaunay variables:

$$\begin{array}{lll} L = \sqrt{\mu a}\,, & G = \sqrt{\mu a(1-e^2)}\,, & H = \sqrt{\mu a(1-e^2)}\cos I \\ l = \text{mean anomaly}\,, & g = \omega\,, & h = \Omega \end{array} \tag{1}$$

where a, e, I, ω, Ω, are osculating Kepler elements and μ is a constant. In Kepler motion, l is linear in t and all the others are constant.

Poincaré variables (modified):

$$\begin{array}{lll} L\,, & \sqrt{2(L-G)}\,\sin(g+h)\,, & \sqrt{2(G-H)}\,\sin h \\ l+g+h\,, & \sqrt{2(L-G)}\cos(g+h)\,, & \sqrt{2(G-H)}\cos h \end{array}$$

Unnamed:

$$\begin{array}{lll} H\,, & \sqrt{2(L-G)}\cos l\,(=x_1)\,, & \sqrt{2(G-H)}\cos(l+g)\,(=x_2) \\ l+g+h\,(=\lambda)\,, & \sqrt{2(L-G)}\,\sin l\,(=y_1)\,, & \sqrt{2(G-H)}\,\sin(l+g)\,(=y_2)\,. \end{array} \tag{2}$$

Hill's variables:

$$\begin{array}{lll} \dot{r} & G & H \\ r & f+g\,(=u) & h\,, \end{array} \tag{3}$$

where f stands for true anomaly and r is the radial distance.

2. Von Zeipel Method (Brouwer, 1959)

(1) The theory based on Lie series comes out as an improvement of the von Zeipel method, so we first consider this original version. In terms of Delaunay variables, the

B. D. Tapley and V. Szebehely (eds.), Recent Advances in Dynamical Astronomy, 231–249.

equations of motion of perturbed Kepler motion are

$$\frac{\mathrm{d}}{\mathrm{d}t}(L, G, H) = \frac{\partial F}{\partial (l, g, h)}, \quad \frac{\mathrm{d}}{\mathrm{d}t}(l, g, h) = -\frac{\partial F}{\partial (L, G, H)} \tag{4}$$

with the Hamiltonian

$$F = \frac{\mu^2}{2L^2} + R(L, G, H, l, g, h),$$

and R (the disturbing function) being periodic in each of l, g, and h. When R depends on t explicitly, the equations of motion take the form

$$\frac{\mathrm{d}}{\mathrm{d}t}(L, G, H, K) = \frac{\partial F}{\partial (l, g, h, k)}, \quad \frac{\mathrm{d}}{\mathrm{d}t}(l, g, h, k) = -\frac{\partial F}{\partial (L, G, H, K)}$$

with

$$F = \frac{\mu^2}{2L^2} - K + R(L, G, H, l, g, h, k)$$

where $k(=t)$ is the fourth angular variable with K as its conjugate momentum

The disturbing function, denoted by F_1, is supposed to be a small quantity of the first order. In many cases, however, R is expanded in powers of a small parameter of the first order, and we may write

$$R = F_1 + F_2 + \cdots.$$

In what follows we discuss the case where R does not depend on t explicitly, and $\mu^2/2L^2$ is denoted by $F_0(L)$.

(2) By a suitable canonical transformation $L, G, H, l, g, h \rightarrow L', G', H', l', g', h'$, the equations of motion (4) may be reduced to a more accessible form, or even to an integrable form. Let the new equations of motion be

$$\frac{\mathrm{d}}{\mathrm{d}t}(L', G', H') = \frac{\partial F^*}{\partial (l', g', h')}, \quad \frac{\mathrm{d}}{\mathrm{d}t}(l', g', h') = -\frac{\partial F^*}{\partial (L', G', H')}$$

with the new Hamiltonian

$$F^*(L', G', H', l', g', h') = F(L, G, H, l, g, h). \tag{5}$$

We consider a canonical transformation developed by a generating function,

$$\begin{aligned} S &= L'l + G'g + H'h + S_1(L', G', H', l, g, h) + S_2 + \cdots : \\ (L, G, H) &= \frac{\partial S}{\partial (l, g, h)} = (L', G', H') + \frac{\partial S_1}{\partial (l, g, h)} + \frac{\partial S_2}{\partial (l, g, h)} + \cdots \\ (l', g', h') &= \frac{\partial S}{\partial (L', G', H')} = (l, g, h) + \frac{\partial S_1}{\partial (L', G', H')} + \frac{\partial S_2}{\partial (L', G', H')} + \cdots, \end{aligned} \tag{6}$$

where the subscripts in S denote the orders of magnitude. The substitution of Equations (6) into both sides of (5) results in an identity in the mixed variables L', G', H', l, g, and h. After that F and F^* are developed in powers of the small parameter and we have, after Taylor expansion

Order 0

$$F_0(L') = F^*(L') \tag{7}$$

Order 1

$$\frac{\mathrm{d}F_0}{\mathrm{d}L'}\frac{\partial S_1}{\partial l} + F_1(L', G', H', l, g, h) = F_1^*(L', G', H', l, g, h) \tag{8}$$

Order 2

$$\frac{\mathrm{d}F_0}{\mathrm{d}L'}\frac{\partial S_2}{\partial l} + \frac{1}{2}\frac{\mathrm{d}^2F_0}{\mathrm{d}L'^2}\left(\frac{\partial S_1}{\partial l}\right)^2 + \frac{\partial F_1}{\partial L'}\frac{\partial S_1}{\partial l} + \frac{\partial F_1}{\partial G'}\frac{\partial S_1}{\partial g} + \frac{\partial F_1}{\partial H'}\frac{\partial S_1}{\partial h} +$$
$$+ F_2(L', G', H', l, g, h) = \frac{\partial F_1^*}{\partial l}\frac{\partial S_1}{\partial L'} + \frac{\partial F_1^*}{\partial g}\frac{\partial S_1}{\partial G'} + \frac{\partial F_1^*}{\partial h}\frac{\partial S_1}{\partial H'} +$$
$$+ F_2^*(L', G', H', l, g, h).$$

(3) In each order the equation has two unknowns, S_j and F_j^*, and we might have ambiguity in the determination of S_j and F_j^*. But F_j^* should be chosen so that the partial derivatives of S_j with respect to the mixed variables remain as small as the jth order for the period of time we consider, and so that the Taylor expansion carried out above is valid. This is the least requirement for the determination of S_j and F_j^*, but also found sufficient in many cases.

Let us consider the equation of the first order (8). Since F_1 is assumed periodic in each of l, g, and h, the only choice for F_1^* is

$$F_1^* = F_1^*(L', G', H', -, g', h') = \langle F_1(L', G', H', l, g, h)\rangle_{g, h\to g', h'} \tag{9}$$

when our theory is required to be valid for an indefinite value of l, where $\langle\ \rangle$ denotes the average with respect to l. S_1 is then determined as

$$S_1 = -\left(\frac{\mathrm{d}F_0}{\mathrm{d}L'}\right)^{-1}\int (F_1 - \langle F_1\rangle)\,\mathrm{d}l.$$

Similarly we have in the second order

$$F_2^* = \left\langle \frac{1}{2}\frac{\mathrm{d}^2F_0}{\mathrm{d}L'^2}\left(\frac{\partial S_1}{\partial l}\right)^2 + \frac{\partial F_1}{\partial L'}\frac{\partial S_1}{\partial l} + \frac{\partial F_1}{\partial G'}\frac{\partial S_1}{\partial g} + \frac{\partial F_1}{\partial H'}\frac{\partial S_1}{\partial h} - \right.$$
$$\left. - \frac{\partial F_1^*}{\partial g}\frac{\partial S_1}{\partial G'} - \frac{\partial F_1^*}{\partial h}\frac{\partial S_1}{\partial H'} + F_2\right\rangle_{g, h\to g', h'}$$
$$S_2 = -\left(\frac{\mathrm{d}F_0}{\mathrm{d}L'}\right)^{-1}\int\left(\frac{1}{2}\frac{\mathrm{d}^2F_0}{\mathrm{d}L_2'^2}\left(\frac{\partial S_1}{\partial l}\right)^2 + \frac{\partial F_1}{\partial L'}\frac{\partial S_1}{\partial l} + \cdots + \right. \tag{10}$$
$$\left. + F_2 - \left\langle\frac{1}{2}\frac{\mathrm{d}^2F_0}{\mathrm{d}L'^2}\left(\frac{\partial S_1}{\partial l}\right)^2 + \frac{\partial F_1}{\partial L'}\frac{\partial S_1}{\partial l} + \cdots + F_2\right\rangle\right)\mathrm{d}l.$$

Therefore the principle for determination of F^* and S is usually stated as follows:

choose S in such a way that l' is eliminated from F^*. But note that the motion of an artificial satellite near the critical inclination, or in a hyperbolic orbit can be handled only by a more flexible method. (Hori, 1960, 1961).

In the usual cases where F^* is determined by Equations (7), (9), (10),..., the new equations of motion become

$$\frac{\mathrm{d}}{\mathrm{d}t}(G', H') = \frac{\partial F^*}{\partial (g', h')}, \quad \frac{\mathrm{d}}{\mathrm{d}t}(l', g', h') = -\frac{\partial F^*}{\partial (L', G', H')} \tag{11}$$

with the Hamiltonian

$$F^* = F_0^*(L') + F_1^*(L', G', H', -, g', h') + \\ + F_2^*(L', G', H', -, g', h') + \cdots$$

and with a new integral of motion

$$L' = \text{const}. \tag{12}$$

The system (11) has two degrees of freedom instead of the three degrees present in the original system (4), and this is the gain attained by the canonical transformation.

If F_1^* is independent of both g' and h', we apply a second canonical transformation $L', G', H', l', g', h' \to L', G'', H'', l'', g'', h''$:

$$(L', G', H') = \frac{\partial S^*}{\partial (l', g', h')}, \quad (l'', g'', h'') = \frac{\partial S^*}{\partial (L', G'', H'')} \tag{13}$$

with the generating function

$$S^* = L'l' + G''g' + H''h' + S_1^*(L', G'', H'', -, g', h') + \cdots,$$

and eliminate g'' and h'' from the new Hamiltonian F^{**}. The process is similar to the preceding one except that $F_1^*(L', G', H')$ plays the roll of $F_0(L)$, and that the pair S_1^* and F_2^{**} is determined by a second order equation of the form

$$\frac{\partial F_1^*}{\partial G''}\frac{\partial S_1^*}{\partial g'} + \frac{\partial F_1^*}{\partial H''}\frac{\partial S_1^*}{\partial h'} + \sum C_2^{ij}(L', G'', H'')\cos(ig' + jh') = F_2^{**}$$

where $F_2^*(L', G'', H'', -, g', h')$ is replaced by the Fourier series. Then, the only choice of F_2^{**} and S_1^* is

$$F_2^{**} = C_2^{00}(L', G'', H'')$$
$$S_1^* = -\sum{}' \frac{C_2^{ij}(L', G'', H'')}{i(\partial F_1^*/\partial G'') + j(\partial F_1^*/\partial H'')}\sin(ig' + jh').$$

(4) Since the new Hamiltonian depends only on L', G'', H'', the new equations of motion are immediately integrated and give the solution

$$L', G'', H'' = \text{const}, \quad (l'', g'', h'') = -\frac{\partial F^{**}}{\partial (L', G'', H'')}t - (l_0'', g_0'', h_0''). \tag{14}$$

Six constants L', G'', H'', l''_0, g''_0, h''_0 provide the integration constants of the original equations of motion (4). The coefficients of t in l'', g'', h'' are severally called the mean motions of l'', g'', h''.

When L', G'', H'', l'', g'', and h'' are given, we first obtain G', H', l', g', h' by solving Equations (13) by means of iterations or successive approximations, and then obtain L, G, H, l, g, h by solving Equations (6) by similar means. Finally we can calculate any quantity of interest if the quantity is expressed in terms of Delaunay variables. So, considerable effort is required after finding the solution (14). The elimination or diminution of this additional work was the motive for developing the new theory.

3. Improved Method (Hori, 1966)

(1) Let x, y be a set of canonical variables, and consider a canonical transformation $x, y \to \xi, \eta$:

$$x = \xi + \frac{\partial}{\partial y}\tilde{S}_1(\xi, y) + \frac{\partial}{\partial y}\tilde{S}_2(\xi, y) + \cdots$$
$$y = \eta - \frac{\partial}{\partial \xi}\tilde{S}_1(\xi, y) - \frac{\partial}{\partial \xi}\tilde{S}_2(\xi, y) - \cdots \qquad (15)$$

where $\tilde{S} = \sum \xi_i y_i + \tilde{S}_1(\xi, y) + \tilde{S}_2(\xi, y) + \cdots$

is the generating function of the canonical transformation.

If $\tilde{S}_j$ stands for $[\tilde{S}_j(\xi, y)]_{y\to\eta}$, and $\tilde{S}_{j\eta}$, $\tilde{S}_{j\xi}, \ldots$ for $\partial\tilde{S}_j/\partial\eta$, $\partial\tilde{S}_j/\partial\xi$, ..., the explicit equations for x, y in terms of ξ, η are given by

$$x = \xi + \tilde{S}_{1\eta} + \tilde{S}_{2\eta} - \sum \tilde{S}_{1\eta\eta_i}\tilde{S}_{1\xi_i} + \cdots$$
$$y = \eta - \tilde{S}_{1\xi} - \tilde{S}_{2\xi} + \sum \tilde{S}_{1\xi\eta_i}\tilde{S}_{1\xi_i} + \cdots$$

and further we have, for any function $f(x, y)$ of x and y, the equation

$$\begin{aligned} f(x, y) = f(\xi, \eta) &+ \sum_i f_{\xi_i}\left(\tilde{S}_{1\eta_i} + \tilde{S}_{2\eta_i} - \sum_j \tilde{S}_{1\eta_i\eta_j}\tilde{S}_{1\xi_j}\right) + \\ &+ \sum_i f_{\eta_i}\left(-\tilde{S}_{1\xi_i} - \tilde{S}_{2\xi_i} + \sum_j \tilde{S}_{1\xi_i\eta_j}\tilde{S}_{1\xi_j}\right) + \\ &+ \tfrac{1}{2}\sum_i\sum_j f_{\xi_i\xi_j}\tilde{S}_{1\eta_i}\tilde{S}_{1\eta_j} + \sum_i\sum_j f_{\xi_i\eta_j}\tilde{S}_{1\eta_i}(-\tilde{S}_{1\xi_j}) + \\ &+ \tfrac{1}{2}\sum_i\sum_j f_{\eta_i\eta_j}(-\tilde{S}_{1\xi_i})(-\tilde{S}_{1\xi_j}) + \cdots. \end{aligned}$$

But if we introduce Poisson brackets, $f(x, y)$ can be written in the form

$$\begin{aligned} f(x, y) = f(\xi, \eta) &+ \{f, \tilde{S}_1\} + \{f, \tilde{S}_2 - \tfrac{1}{2}\sum \tilde{S}_{1\xi_i}\tilde{S}_{1\eta_i}\} + \\ &+ \tfrac{1}{2}\{\{f, \tilde{S}_1\}, \tilde{S}_1\} + \cdots. \end{aligned}$$

Therefore, after obtaining $\tilde{S}_1(\xi, y) + \tilde{S}_2(\xi, y) + \cdots$ by the von Zeipel method, if we introduce $S_1(\xi, \eta)$, $S_2(\xi, \eta), \ldots$ by

$$S_1 = \tilde{S}_1, \quad S_2 = \tilde{S}_2 - \tfrac{1}{2}\sum \tilde{S}_{1\xi_i}\tilde{S}_{1\eta_i}, \ldots, \qquad (16)$$

any quantity we consider is directly calculated by the formula

$$f(x, y) = f(\xi, \eta) + \{f, S_1\} + \{f, S_2\} + \tfrac{1}{2}\{\{f, S_1\}, S_1\} + \cdots. \tag{17}$$

In particular, by $f(x, y) \equiv x$ or y, we have the transformation

$$\begin{aligned} x &= \xi + S_{1\eta} + S_{2\eta} + \tfrac{1}{2}\{S_{1\eta}, S_1\} + \cdots \\ y &= \eta - S_{1\xi} - S_{2\xi} - \tfrac{1}{2}\{S_{1\xi}, S_1\} - \cdots. \end{aligned} \tag{18}$$

(2) The introduction of $S_1, S_2, \ldots$ and the use of Equation (17) may reduce the amount of additional work mentioned at the end of Section 2 especially when $S_1, S_2, \ldots$ are calculated directly, rather than with the help of $\tilde{S}_1, \tilde{S}_2, \ldots$ Equation (17) has a further advantage which is attributable to the invariance of Poisson brackets: the right side of Equation (17) can be evaluated with the use of any set of canonical variables. Let x, y be Delaunay variables and ξ, η the primed Delaunay variables, and suppose $f(x, y)$ is r. Then Hill's set (3) may be the best choice for the evaluation of r. If $\dot{r}'$, r', u' have the same relations to the primed Delaunay variables as $\dot{r}$, r, u to Delaunay variables, we have

$$\begin{aligned} r &= r' + \{r', S_1\} + \{r', S_2\} + \tfrac{1}{2}\{\{r', S_1\}, S_1\} + \cdots = \\ &= r' - S_{1\dot{r}'} - S_{2\dot{r}'} - \tfrac{1}{2}\{S_{1\dot{r}'}, S_1\} + \cdots. \end{aligned}$$

Or, if we prefer the use of $\tilde{S}_1, \tilde{S}_2, \ldots$,

$$r = r' - \tilde{S}_{1\dot{r}'} - \tilde{S}_{2\dot{r}'} + \tilde{S}_{1\dot{r}'r'}\tilde{S}_{1\dot{r}'} + \tilde{S}_{1\dot{r}'u'}\tilde{S}_{1G'} + \tilde{S}_{1\dot{r}'h'}\tilde{S}_{1H'}$$

because of Equations (16). The first order part of this relation was noted by Izsak (1963).

The transformations (15) and (18) are equivalent if $\tilde{S}_1(\xi, y)$, $\tilde{S}_2(\xi, y), \ldots$ and $S_1(\xi, \eta)$, $S_2(\xi, \eta), \ldots$ are related by Equations (16). If S_2 is chosen instead, say as $\tilde{S}_2$, the transformation (18) is not equivalent to (15), but is still canonical, and in fact equivalent to the canonical transformation

$$\begin{aligned} x &= \xi + \frac{\partial}{\partial y}\tilde{S}_1(\xi, y) + \frac{\partial}{\partial y}\bar{S}_2(\xi, y) + \cdots \\ y &= \eta - \frac{\partial}{\partial \xi}\tilde{S}_1(\xi, y) - \frac{\partial}{\partial \xi}\bar{S}_2(\xi, y) - \cdots \end{aligned}$$

with

$$\bar{S}_2(\xi, y) = \tilde{S}_2(\xi, y) + \tfrac{1}{2}\sum \tilde{S}_{1\xi_i}\tilde{S}_{1y_i}.$$

This fact leads to a possibility of finding $S_1, S_2, \ldots$ directly. But we first generalize Equations (17) and (18) to higher orders.

If the operator D_S is introduced by

$$D_S f(\xi, \eta) = \{f(\xi, \eta), \quad S(\xi, \eta)\}, \quad D_S^n f(\xi, \eta) = D_S(D_S^{n-1} f(\xi, \eta)),$$

the transformation $x, y \to \xi, \eta$

$$x = \xi + \sum_{n=1} \frac{1}{n!} D_S^n \xi , \qquad y = \eta + \sum_{n=1} \frac{1}{n!} D_S^n \eta$$

is canonical for any function $S = S(\xi, \eta)$, and further we have

$$f(x, y) = f(\xi, \eta) + \sum_{n=1} \frac{1}{n!} D_S^n f(\xi, \eta) . \quad \text{(Lie series)}. \tag{19}$$

We assume that S is sufficiently small and that the series (17), (18) converge. The inverse transformation $\xi, \eta \to x, y$ is easily obtained: if $\bar{S}$ stands for $[-S(\xi, \eta)]_{\xi, \eta \to x, y}$, we have

$$f(\xi, \eta) = f(x, y) + \sum \frac{1}{n!} D_{\bar{S}}^n f(x, y) .$$

If S itself is partitioned according to orders of magnitude as

$$S = S_1 + S_2 + \cdots$$

the terms of order n in the right side of Equation (19) are

$$\sum_{m=1}^{n} \sum_{j_1 + j_2 \cdots + j_m = n} \frac{1}{m!} \prod_{\alpha=1}^{m} D_{S_{j_\alpha}} f(\xi, \eta) . \tag{20}$$

(3) Let the equations of motion be

$$\frac{\mathrm{d}x_j}{\mathrm{d}t} = \frac{\partial F}{\partial y_j} , \qquad \frac{\mathrm{d}y_j}{\mathrm{d}t} = - \frac{\partial F}{\partial x_j}$$

with the Hamiltonian

$$F = F_0(x, y) + F_1(x, y) + F_2(x, y) + \cdots .$$

By a canonical transformation generated by

$$S = S_1(\xi, \eta) + S_2(\xi, \eta) + \cdots$$

new equations of motion are obtained

$$\frac{\mathrm{d}\xi_j}{\mathrm{d}t} = \frac{\partial F^*}{\partial \eta_j} , \qquad \frac{\mathrm{d}\eta_j}{\mathrm{d}t} = - \frac{\partial F^*}{\partial \xi_j} \tag{21}$$

with the new Hamiltonian

$$F^* = F_0^*(\xi, \eta) + F_1^*(\xi, \eta) + F_2^*(\xi, \eta) + \cdots = F(x, y) .$$

Applying the expansion formula (19) with $f(x, y) \equiv F(x, y)$ to the right side of this equation, we find

$$\begin{array}{ll} \textit{Order } 0 & F_0 = F_0^* \\ \textit{Order } 1 & \{F_0, S_1\} + F_1 = F_1^* \\ \textit{Order } 2 & \{F_0, S_2\} + \tfrac{1}{2}\{\{F_0, S_1\}, S_1\} + \{F_1, S_1\} + F_2 = F_2^* \\ \ldots & \\ \ldots & \end{array} \tag{22}$$

$$\textit{Order } n \qquad \sum_{\nu=n}^{1} \sum_{m=1}^{\nu} \sum_{j_1+j_2+\cdots+j_m=\nu} \frac{1}{m!} \prod_{\alpha=1}^{m} D_{S_{j_\alpha}} F_{n-\nu} + F_n = F_n^* . \tag{23}$$

The equation of order 2 is replaced by

$$\{F_0, S_2\} + \tfrac{1}{2}\{F_1 + F_1^*, S_1\} + F_2 = F_2^*$$

with the help of the equation of order 1.

We next introduce a parameter τ by

$$\frac{\mathrm{d}\xi_j}{\mathrm{d}\tau} = \frac{\partial}{\partial \eta_j} F_0(\xi, \eta), \qquad \frac{\mathrm{d}\eta_j}{\mathrm{d}\tau} = -\frac{\partial}{\partial \xi_j} F_0(\xi, \eta) \quad \text{(Auxiliary system)}$$

which are the equations of unperturbed motion except for the change of notation, ξ, η, τ in place of x, y, t. Their general solution may be written

$$\xi_j = \xi_j(\tau + C_1, C_2, \ldots, C_{2n}), \qquad \eta_j = \eta_j(\tau + C_1, C_2, \ldots, C_{2n}), \qquad (j = 1, 2, \ldots, n) \tag{24}$$

$C_1, C_2, \ldots C_{2n}$ being $2n$ constants of integration with respect to τ. Inversely, $2n$ quantities $\tau + C_1, C_2, \ldots, C_{2n}$ may be expressed in terms of ξ, η by solving $2n$ Equations (24) (in principle at least):

$$\tau + C_1 = \zeta_1(\xi, \eta), \qquad C_j = \zeta_j(\xi, \eta), \qquad (j = 2, 3, \ldots, 2n). \tag{25}$$

However as we see below equations for ζ are not required but only derivatives of ζ in ξ and η, which are obtainable from Equations (24).

The equation of order 1 now takes the form

$$-\frac{\mathrm{d}S_1}{\mathrm{d}\tau} + F_1(\xi(\tau + C_1, \ldots, C_{2n}), \quad \eta(\tau + C_1, \ldots, C_{2n})) = F_1^* . \tag{26}$$

In usual cases where F_1 is multiply periodic in τ, S_1 is so chosen that F_1^* is free from τ:

$$F_1^* = \langle F_1 \rangle (C_2, C_3, \ldots C_{2n}) = F_1^*(\xi, \eta) \tag{27}$$

$$S_1 = \int (F_1 - \langle F_1 \rangle)\, \mathrm{d}\tau = S_1(\xi, \eta) \tag{28}$$

where $\langle\ \rangle$ denotes the average with respect to τ.

We note that, in this case, any other choice of F_1^* yields a term in S_1 which is linear in τ, and this may endanger the convergence of the Lie series.

Next, $\{F_1 + F_1^*, S_1\}$ is evaluated and here the derivatives of ζ in ξ, η are required.

(If the solution (25) is available it is of course better to make use of them.) The equation of order 2 then gives

$$F_2^* = \langle \tfrac{1}{2}\{F_1 + F_1^*, S_1\} + F_2 \rangle = F_2^*(\xi, \eta) \tag{29}$$

$$S_2 = \int \left(\tfrac{1}{2}\{F_1 + F_1^*, S_1\} + F_2 - \langle \tfrac{1}{2}\{F_1 + F_1^*, S_1\} + F_2 \rangle\right) \mathrm{d}\tau\,, \tag{30}$$

and so on for equations of higher orders.

The new equations of motion (21) admit an integral

$$F_0^*(\xi, \eta) = \text{const} \tag{31}$$

in addition to

$$F^*(\xi, \eta) = F(x, y) = \text{const}$$

because

$$0 = \frac{\mathrm{d}F^*}{\mathrm{d}\tau} = \{F^*, F_0\} = -\{F_0, F^*\} = -\frac{\mathrm{d}F_0}{\mathrm{d}t} = -\frac{\mathrm{d}F_0^*}{\mathrm{d}t}.$$

We note that the integral (31) reduces to (12) if $F_0^*(\xi, \eta) = \mu^2/2\xi_1^2$ with ξ, η as the primed Delaunay variables.

(4) The new equations of motion (21) might not admit immediate solution. We then consider a second canonical transformation $\xi, \eta \rightarrow p, q$ with a generating function

$$S^*(p, q) = S_1^* + S_2^* + \cdots,$$

and apply the expansion formula

$$f(\xi, \eta) = f(p, q) + \{f, S_1^*\} + \{f, S_2^*\} + \tfrac{1}{2}\{\{f, S_1^*\}, S_1^*\} + \cdots \tag{32}$$

to the right side of $F^*(\xi, \eta) = F^{**}(p, q)$ or

$$F_0^*(\xi, \eta) + F_1^*(\xi, \eta) + \cdots = F_0^{**}(p, q) + F_1^{**}(p, q) + \cdots$$

where $F^{**}(p, q)$ is the new Hamiltonian.

But here, because of the integral (31), we have

$$F_0^*(\xi(p, q), \eta(p, q)) = F_0^{**}(p, q)$$

and therefore the Formula (32) is applied to the left side of

$$F_1^*(\xi, \eta) + F_2^*(\xi, \eta) + \cdots = F_1^*(p, q) + F_2^*(p, q) + \cdots$$

We see that $F_1^*(\xi, \eta)$ plays the roll of $F_0(x, y)$ of the preceeding procedure. S_1^* is then determined together with F_2^{**} by the equation of order 2, similar to the situation encountered in the von Zeipel method.

The purpose of the second canonical transformation is to eliminate, from $F^{**}(p, q)$, a parameter τ^* introduced by

$$\frac{\mathrm{d}p_j}{\mathrm{d}\tau^*} = \frac{\partial F_1^*}{\partial q_j}, \quad \frac{\mathrm{d}q_j}{\mathrm{d}\tau^*} = -\frac{\partial F_1^*}{\partial p_j}, \quad F_1^* = [F_1^*(\xi, \eta)]_{\xi, \eta \rightarrow p, q}.$$

The procedure is exactly the same as the previous one. After the elimination of τ^*, we have a new integral

$$F_1^{**}(p, q) = \text{const} \tag{33}$$

in addition to

$$F_0^{**}(p, q) = F_0^*(\xi, \eta) = \text{const}$$

and

$$F^{**}(p, q) = F^*(\xi, \eta) = F(x, y) = \text{const}.$$

We note that $f(x, y)$ is directly given in terms of p, q by a single formula

$$\begin{aligned} f(x, y) = f(p, q) &+ \{f, S_1 + S_1^*\} + \{f, S_2 + S_2^*\} + \\ &+ \tfrac{1}{2}\{\{f, S_1 + S_1^*\}, S_1 + S_1^*\} + \tfrac{1}{2}\{f, \{S_1, S_1^*\}\} + \cdots \end{aligned} \tag{34}$$

where S_j stands for $[S_j(\xi, \eta)]_{\xi, \eta \to p, q}$.

The inverse formula is

$$\begin{aligned} f(p, q) = f(x, y) &+ \{f, -S_1 - S_1^*\} + \{f, -S_2 - S_2^*\} + \\ &+ \tfrac{1}{2}\{\{f, -S_1 - S_1^*\}, -S_1 - S_1^*\} + \\ &+ \tfrac{1}{2}\{f, \{-S_1, -S_1^*\}\} + \cdots \end{aligned} \tag{35}$$

if S_j, S_j^* stand for $[S_j(\xi, \eta)]_{\xi, \eta \to x, y}$, $[S_j^*(p, q)]_{p, q \to x, y}$ respectively.

Equation (35) may be used to calculate p, q when initial values for x, y are given. Equations (34) and (35) are not available when canonical transformations are given in implicit form as in (6) and (13). In addition, it is obvious to see that the determination of F^*, S in the first procedure and of F^{**}, S^* in the second one is independent of the choice of canonical variables owing to the invariance of the Poisson brackets: F^*, S, F^{**}, S^* obtained with the use of Delaunay variables have the same values as those obtained with the use of, say, Poincaré variables, so that we can easily switch from one to another set of canonical variables in any stage of procedure. This is an additional, but definite, advantage of the new method.

4. Comparison with a Theory Based on Implicit Canonical Transformations

Consider a theory based on implicit canonical transformations, like von Zeipel method. If the transformation is given by Equation (15), substitute these in x and η of equation $F(x, y) = F^*(\xi, \eta)$. After Taylor expansion, we have

$$\begin{aligned} &\textit{Order } 0 \quad F_0(\xi, y) = F_0^*(\xi, y) \\ &\textit{Order } 1 \quad \{F_0(\xi, y), \quad \tilde{S}_1(\xi, y)\} + F_1(\xi, y) = F_1^*(\xi, y) \\ &\textit{Order } 2 \quad \{F_0(\xi, y), \quad \tilde{S}_2(\xi, y)\} + \tfrac{1}{2}\sum F_{0\xi_i\xi_j}\tilde{S}_{1y_i}\tilde{S}_{1y_j} + \\ &\qquad + \sum F_{1\xi_i}\tilde{S}_{1y_i} + F_2(\xi, y) - \tfrac{1}{2}\sum F^*_{0\eta_i\eta_j}\tilde{S}_{1\xi_i}\tilde{S}_{1\xi_j} - \\ &\qquad - \sum F^*_{1\eta_i}\tilde{S}_{1\xi_i} = F_2^*(\xi, y). \\ &\cdots \end{aligned} \tag{36}$$

If a parameter τ' is introduced by

$$\frac{d\xi_j}{d\tau'} = \frac{\partial}{\partial y_j} F_0(\xi, y), \quad \frac{dy_j}{d\tau'} = -\frac{\partial}{\partial \xi_j} F_0(\xi, y),$$

the equation of order 1 is written as

$$-\frac{d\tilde{S}_1}{d\tau'} + F_1(\xi(\tau' + C_1, \dots), y(\tau' + C_1, \dots)) = F_1^*(\xi, y)$$

which leads to

$$F_1^* = \langle F_1 \rangle_{y \to \eta} \tag{37}$$

$$\tilde{S}_1 = \int (F_1 - \langle F_1 \rangle)\, d\tau' = \tilde{S}_1(\xi, y). \tag{38}$$

In the equation of order 2, we note the relation

$$\begin{aligned}\tfrac{1}{2}\sum F_{0\xi_i\xi_j}\tilde{S}_{1y_i}\tilde{S}_{1y_j} + \sum F_{1\xi_i}\tilde{S}_{1y_i} - \tfrac{1}{2}\sum F^*_{0\eta_i\eta_j}\tilde{S}_{1\xi_i}\tilde{S}_{1\xi_j} - \sum F^*_{1\eta_i}\tilde{S}_{1\xi_i} &= \\ = \tfrac{1}{2}\{F_1 + F_1^*, \tilde{S}_1\} - \{F_0, \tfrac{1}{2}\sum \tilde{S}_{1\xi_i}\tilde{S}_{1y_i}\} &= \\ = \tfrac{1}{2}\{F_1 + F_1^*, S_1\} + \frac{d}{d\tau'}\tfrac{1}{2}\sum \tilde{S}_{1\xi_i}\tilde{S}_{1y_i}&,\end{aligned}$$

then we have

$$F_2^* = \langle \tfrac{1}{2}\{F_1 + F_1^*, S_1\} + F_2(\xi, y)\rangle_{y \to \eta} \tag{39}$$

and

$$\begin{aligned}\tilde{S}_2 = \int (\tfrac{1}{2}\{F_1 + F_1^*, S_1\} + F_2 - \langle \tfrac{1}{2}\{F_1 + F_1^*, S_1\} + F_2\rangle)\, d\tau' + \\ + \tfrac{1}{2}\sum \tilde{S}_{1\xi_i}\tilde{S}_{1y_i}.\end{aligned} \tag{40}$$

The comparison of these results (36)–(40) with the corresponding ones (22), (27), (28), (29), and (30) shows that F_0^*, F_1^*, F_2^* are the same, and that

$$\begin{aligned}\tilde{S}_1(\xi, y) &= [S_1(\xi, \eta)]_{\eta \to y} \\ \tilde{S}_2(\xi, y) &= [S_2(\xi, \eta)]_{\eta \to y} + \frac{1}{2}\sum \frac{\partial \tilde{S}_1(\xi, y)}{\partial \xi_i}\frac{\partial \tilde{S}_1(\xi, y)}{\partial y_i}.\end{aligned} \tag{41}$$

These relations are equivalent to (16). Therefore, we see that the theories of general perturbations based on canonical transformations and on the averaging principle give the same result through the second order regardless of whether the transformation is implicitly given by (15) or explicitly by (17). (Hori, 1970; Yuasa, 1971; Mersman, 1971). But the important difference in the two theories is that S_j and F_j^* obtained by the latter are canonically invariant, while $\tilde{S}_j$ and F_j^* obtained by the former are generally not. In fact Equation (41) shows that $\tilde{S}_2(\xi, y)$ is not invariant because of the second terms in the right side of the equation. Equation (41) also shows the change in S_2 when we switch from one to another set of canonical variables: if we switch from x, y to

x', y' and correspondingly from ξ, η to ξ', η', we have

$$\begin{aligned}\tilde{S}_2(\xi, y) - \frac{1}{2}\sum \frac{\partial \tilde{S}_1(\xi, y)}{\partial \xi_i}\frac{\partial \tilde{S}_1(\xi, y)}{\partial y_i} &= [S_2(\xi, \eta)]_{\eta\to y} = \\ &= [S_2(\xi(\xi', \eta'), \eta(\xi', \eta'))]_{\eta'\to y'} = \\ &= \tilde{S}_2'(\xi', y') - \frac{1}{2}\sum \frac{\partial \tilde{S}_1'(\xi', y')}{\partial \xi_i'}\frac{\partial \tilde{S}_1'(\xi', y')}{\partial y_i'}. \end{aligned} \tag{42}$$

This transformational behavior in the canonical case is a special case of the 'covariant property' introduced in the following paper by Kirchgraber.

5. Motion of an Artificial Satellite with Small Eccentricity and Inclination

(1) Let the potential U of an oblate Earth be

$$U = \frac{\mu}{r}\left\{1 - \sum_{n=2} \frac{J_n R^n}{r^n} P_n(\sin\beta)\right\} \tag{43}$$

where R is the equatorial radius of the Earth and β, the latitude of the satellite, is referred to the Earth's equator, which is taken as the reference plane. In terms of canonical set (2), $U - \mu/r$ is written in the form

$$U - \frac{\mu}{r} = \sum_{n=2}^{\infty}\sum_{abcd} F_{abcd}^{(n)}(H)\, x_1^a y_1^b x_2^c y_2^d \qquad (a, b, c, d = 0, 1, 2, \ldots) \tag{44}$$

where

$$b + c = 0, 2, 4, \ldots \tag{45}$$

and in addition

$$c + d = 0, 2, 4, \ldots \quad (\text{if } n \text{ is even}).$$

Equation (45) is due to that U does not change by a simultaneous change of

$$g \to 180° - g \quad \text{and} \quad l \to -l.$$

The equations of motion have the Hamiltonian

$$F = F_0(x, y) + F_1(x, y) + F_2(x, y)$$

$$F_0(x, y) = \frac{\mu^2}{2[H + \frac{1}{2}(x_1^2 + y_1^2 + x_2^2 + y_2^2)]^2}$$

$$F_1(x, y) = -\frac{\mu J_2 R^2}{r_3} P_2(\sin\beta) = \sum_{abcd} F_{abcd}^{(2)}(H)\, x_1^a y_1^b x_2^c y_2^d \tag{46}$$

$$F_2(x, y) = -\sum_{n=3} \frac{\mu J_n R^n}{r^{n+1}} P_n(\sin\beta) = \sum_{n=3}\sum F_{abcd}^{(n)}(H)\, x_1^a y_1^b x_2^c y_2^d \tag{47}$$

and admit an integral $H = \text{const}$ because λ is absent in F.

(2) We consider a canonical transformation $H, x_1, x_2, \lambda, y_1, y_2 \to H, \xi_1, \xi_2, \lambda', \eta_1, \eta_2$ by a generating function $S(H, \xi_1, \xi_2, -, \eta_1, \eta_2) = S_1 + S_2 + \cdots$, and then a second transformation $H, \xi_1, \xi_2, \lambda', \eta_1, \eta_2 \to H, p_1, p_2, \lambda'', q_1, q_2$ by $S^*(H, p_1, p_2, -, q_1, q_2) = S_1^* + S_2^* + \cdots$. Let a set of canonical variables L', G', H, l', g', h' be introduced by

$$\begin{aligned} \xi_1 &= \sqrt{2(L' - G')} \cos l', & \xi_2 &= \sqrt{2(G' - H)} \cos (l' + g'), & & \\ \eta_1 &= \sqrt{2(L' - G')} \sin l', & \eta_2 &= \sqrt{2(G' - H)} \sin (l' + g'), & \lambda' &= l' + g' + h', \end{aligned} \tag{48}$$

and a second set $L', G'', H, l'', g'', h''$ by

$$\begin{aligned} p_1 &= \sqrt{2(L' - G'')} \cos l'', & p_2 &= \sqrt{2(G'' - H)} \cos (l'' + g''), & & \\ q_1 &= \sqrt{2(L' - G'')} \sin l'', & q_2 &= \sqrt{2(G'' - H)} \sin (l'' + g''), & \lambda'' &= l'' + g'' + h''. \end{aligned} \tag{49}$$

By the first canonical transformation generated by S, we eliminate τ or l' from the new Hamiltonian F^* where l' is related to ξ, η by (48). With F_1 of the form

$$F_1 = \sum_{abcd} F_{abcd}^{(2)}(H)\, \xi_1^a \eta_1^b \xi_2^c \eta_2^d, \qquad (b + c = 0, 2, 4, \ldots)$$

F_1^* and S_1 determined by Equations (27) and (28) have similar forms as F_1, but $b+c=0, 2, 4, \ldots$ (F type) in F_1^* and $1, 3, 5, \ldots$ (S type) in S_1. In fact F_1^* is a part of F_1, and $\mathrm{d}S_1/\mathrm{d}\tau$ is equal to an F-type quantity $F_1^* - F_1$. We note that the operator

$$\frac{\mathrm{d}}{\mathrm{d}\tau} = \frac{\mu^2}{[H + \frac{1}{2}(\xi_1^2 + \eta_1^2 + \xi_1^2 + \eta_2^2)]^3} \times$$
$$\times \left(\frac{\partial}{\partial \lambda'} - \eta_1 \frac{\partial}{\partial \xi_1} + \xi_1 \frac{\partial}{\partial \eta_1} - \eta_2 \frac{\partial}{\partial \xi_2} + \xi_2 \frac{\partial}{\partial \eta_2} \right)$$

changes one type to the other when operating on quantities independent of λ'. We also note that the Poisson brackets of the same type quantities are S-type and those of the different type quantities are F-type.

Then $\{F_1 + F_1^*, S_1\}$ is an F-type, and so is F_2^*. S_2 is then an S-type. Since the procedure of obtaining F_j^*, $S_j (j=3, 4, \ldots)$ is made up of the same operations – evaluation of multiple Poisson brackets $\{\cdots\{F\text{-type}, S\text{-type}\}, \ldots, S\text{-type}\}$, and averaging and integration in τ – F_j^* and S_j are respectively F-type and S-type quantities for any j.

By the second transformation generated by S^*, we eliminate τ^* or g'' from the new Hamiltonian F^{**} where g'' is related to p, q by (49). This is possible because $F_1^*(\xi, \eta)$ is found to be free from g' as well as l' and h'. By a similar discussion as before we see that F_j^{**} and S_j^* are respectively F-type and S-type quantities in p, q for any j.

Since F^{**} is free from all angular variables l'', g'', and h'', it depends on p, q only through $p_1^2 + q_1^2$, $p_2^2 + q_2^2$, and has the form

$$\begin{aligned} F^{**} &= \sum C_{ab}(H)\, (p_1^2 + q_1^2)^a (p_2^2 + q_2^2)^b = \\ &= \sum C_{ab}(H)\, 2^{a+b} (L' - G'')^a (G'' - H)^b, \qquad (a, b = 0, 1, 2, \ldots). \end{aligned}$$

Then L', G'', H are constant, and l'', g'', h'' are linear functions of t with the mean motions $-\partial F^{**}/\partial L'$, $-\partial F^{**}/\partial G''$, $-\partial F^{**}/\partial H$ respectively. Therefore, l'', g'', h'' are well defined quantities regardless of the values of integration constants L', G'', H. The same holds for p_1, q_1, p_2, q_2, and λ''. If $f(p, q)$ is any function of p, q not singular at $p_1 = q_1 = 0$ or $p_2 = q_2 = 0$ ($L' = G''$ or $G'' = H$), $f(\xi, \eta)$ is given by Equation (32) which is valid at $L' = G''$ or $G'' = H$ respectively. Similarly we can conclude that $f(x, y)$ given by Equation (34) is valid at $L' = G''$ or $G'' = H$. $\mathbf{r}(p, q)$ or $\mathbf{r}(L', G'', H, l'', g'', h'')$ is not singular at $L' = G''$ or $G'' = H$, so is $\mathbf{r}(x, y)$, the perturbed position vector of the satellite. Because of the invariance of Poisson brackets, however, their evaluation can be carried out even with the use of the set L', G'', H, l'', g'', h''. Therefore the expression of $\mathbf{r}(L, G, H, l, g, h)$ in terms of these variables is valid at $L' = G''$ or $G'' = H$.

(3) If $f(p, q)$ is a polynomial of S-type, Equation (34) shows that the expression of $f(x, y)$ consists of only S-type terms, and we then have

$$f(x, y) = 0 \quad \text{at} \quad L' = G'' = H.$$

Similarly we see that, if $f(p, q)$ is F-type, $f(x, y)$ consists of F-type terms, and then reduces to a non-vanishing term depending on H at $L' = G'' = H$.

Let $f(p, q)$ be p_1(F-type), q_1(S-type), p_2(S-type), and q_2(F-type) in turn. Then we have

$$x_1 = x_1^{(0)}(H), \quad y_1 = 0, \quad x_2 = 0, \quad y_2 = y_2^{(0)}(H) \quad \text{at} \quad L' = G'' = H,$$

which are equivalent to

$$\begin{aligned} l &= \begin{cases} 0 & (x_1^{(0)} > 0) \\ 180^\circ & (x_1^{(0)} < 0) \end{cases}, \quad \sqrt{2(L - G)} = |x_1^{(0)}(H)|. \\ l + g &= \begin{cases} 90^\circ & (y_2^{(0)} > 0) \\ 270^\circ & (y_2^{(0)} < 0) \end{cases}, \quad \sqrt{2(G - H)} = |y_2^{(0)}(H)|. \end{aligned} \tag{50}$$

Equations (50) show that the osculating $e(=\sqrt{1 - G^2/L^2})$ and $I(=\cos^{-1} H/G)$ are non-vanishing constants, and the apsidal line of the satellite orbit is perpendicular to the nodal line, and the satellite stays at one of the apses. Since the nodal line moves with the constant mean motion, $-\partial F^{**}/\partial H$, the satellite moves on a circle of radius $a(1 \pm e) \cos I$ parallel to the Earth's equator (Figure 1). This fact was pointed out by

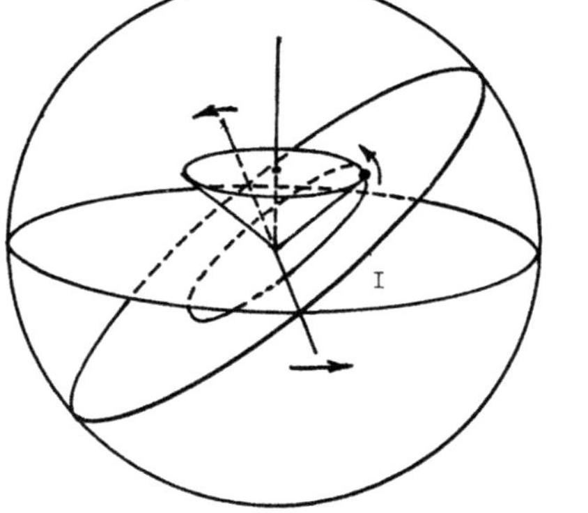

Fig. 1. Geometry for orbiting satellite.

Lyddane for the case of the equatorial motion (Lyddane, 1963), which is possible when the Earth has equatorial symmetry.

Remark: The circular motion specified by Equation (50) is obtained also as the stationary solution of the equations of motion

$$\frac{\mathrm{d}}{\mathrm{d}t}(L, G) = \frac{\partial F}{\partial (l, g)}, \quad \frac{\mathrm{d}}{\mathrm{d}t}(l, g, h) = -\frac{\partial F}{\partial (L, G, H)}$$

with the Hamiltonian

$$F = \frac{\mu^2}{2L^2} + U - \frac{\mu}{r} = \frac{\mu^2}{2L^2} - \sum_{n=2} \frac{\mu J_n R^n}{r^{n+1}} P_n(\sin\beta).$$

In fact the equations, $\partial F/\partial l = \partial F/\partial g = 0$, give the first and third of (50), and if $l=0$, $g=90°$ is the case, say, the equations, $\partial F/\partial L = \partial F/\partial G = 0$, yield the following relations among the osculating but constant a, e, and I:

$$e = -\sum_{n=2} \frac{(n+1) J_n R^n}{[a(1-e)]^n} P_n(\sin I), \quad \sin I = \sum_{n=2} \frac{(n+1) J_n R^n}{[a(1-e)]^n} P_{n+1}(\sin I), \tag{51}$$

which give e and I in terms of $R/[a(1-e)]$, and are essentially the same as the second and fourth of Equations (50). The uniform rotation of the nodal line is given by $\mathrm{d}h/\mathrm{d}t = -\partial F/\partial H$. The other three cases can be treated in a similar way. If $J_n (n \geqq 4)$ are omitted, Equations (51) give

$$e = \tfrac{3}{2} J_2 \left[\frac{R}{a(1-e)}\right]^2, \quad \text{and} \quad \sin I = \tfrac{3}{2} J_3 \left[\frac{R}{a(1-e)}\right]^3. \tag{52}$$

Since $l'' = \tan^{-1}(q_1/p_1)$ is singular at $p_1 = q_1 = 0$, Equation (34) with $f(p, q) \equiv l''$ is not valid at $L' = G''$. The value of l is evaluated however from $l = \tan^{-1}(y_1/x_1)$ when x_1 and y_1 are evaluated by Equation (34) with $f(p, q) \equiv p_1$ and q_1 respectively. We note that l librates around 90° or 180° if x_1 does not vanish, which is the case when $L' - G''$ and $G'' - H$ are very small. Similarly $l + g = \tan^{-1}(y_2/x_2)$ has a libration around 90° or 270° if y_2 has a constant sign since x_2 oscillates around 0.

(4) The libration of l occurs even with a finite value of the inclination constant $\cos^{-1}(H/G'')$, when the eccentricity constant $(1 - G''^2/L'^2)^{1/2}$ is very small. A suitable set of canonical variables for this case may be

$$\begin{pmatrix} H & G & x_1 = \sqrt{2(L-G)} \cos l \\ h & y = l + g & y_1 = \sqrt{2(L-G)} \sin l \end{pmatrix}, \tag{53}$$

which is transformed to the set $H, G'', p_1, h'', y'', q_1$ after the two canonical transformations generated by S and S^*.

Let us consider only J_2 and J_3 terms for simplicity. Borrowing Brouwer's S_1 and

S_1^*, expressed in terms of H, G'', p_1, h'', y'', q_1, we find

$$\begin{aligned}
x_1 &= p_1 + \{p_1, S_1 + S_1^*\} = \\
&= p_1 + \frac{\mu^2 J_2 R^2}{2G''^3}\left[\left(-\tfrac{1}{2} + \tfrac{3}{2}\theta^2\right)\left(\frac{3}{\sqrt{G''}} + \frac{9p_1}{2G''}\right) + \tfrac{3}{2}(1-\theta^2) \times \right. \\
&\quad \left. \times \left\{\left(\frac{5}{3\sqrt{G''}} + \frac{2p_1}{G''}\right)\cos 2y'' - \frac{7q_1}{G''}\sin 2y''\right\}\right] + \frac{\mu^2 J_2 R^2}{2G''^3} \times \\
&\quad \times \frac{(1-\theta^2)(1-15\theta^2)}{8(1-5\theta^2)}\left(\frac{p_1}{G''}\cos 2y'' + \frac{q_1}{G''}\sin 2y''\right) - \\
&\quad - \frac{\mu (J_3/J_2) R}{2G''\sqrt{G''}}\,(1- {}^2)^{1/2}\sin y'' \\
y_1 &= q_1 + \{q_1, S_1 + S_1^*\} = \\
&= -q_1 \frac{\mu^2 J_2 R^2}{2G''^3}\left[\left(-\tfrac{1}{2} + \tfrac{3}{2}\theta^2\right)\frac{9q_1}{2G''} + \tfrac{3}{2}(1-\theta^2)\left\{\left(\frac{2}{3\sqrt{G''}} - \frac{4p_1}{G''}\right)\sin 2y'' + \right.\right. \\
&\quad \left.\left. + \frac{2q_1}{G''}\cos 2y''\right\}\right] + \frac{\mu^2 J_2 R^2}{2G''^3}\cdot\frac{(1-\theta^2)(1-15\theta^2)}{8(1-5\theta^2)}\left(\frac{p_1}{G''}\sin 2y'' - \right. \\
&\quad \left. - \frac{q_1}{G''}\cos 2y''\right) + \frac{\mu (J_3/J_2) R}{2G''\sqrt{G''}}(1-\theta^2)^{1/2}\cos y''
\end{aligned}$$

where θ stands for H/G''.

Therefore, for $p_1 = q_1 = 0$ $(L' = G'')$, we have

$$\begin{aligned}
x_1 &= \frac{J_2}{2}\left(\frac{R}{a'}\right)^2 \sqrt{L'}\left[-\tfrac{3}{2} + \tfrac{9}{2}\theta^2 + \tfrac{5}{2}(1-\theta^2)\cos 2y''\right] - \\
&\qquad - \frac{1}{2}\frac{J_3}{J_2}\frac{R}{a'}\sqrt{L'}\,(1-\theta^2)^{1/2}\sin y'' \\
y_1 &= -\frac{J_2}{2}\left(\frac{R}{a'}\right)^2 \sqrt{L'}\,(1-\theta^2)\sin 2y'' + \\
&\qquad + \frac{1}{2}\frac{J_3}{J_2}\frac{R}{a'}\sqrt{L'}\,(1-\theta^2)^{1/2}\cos y''
\end{aligned} \tag{54}$$

where $a' = L'^2/\mu$.

Equations (54) show that x_1 is positive so that l librates if θ satisfies the inequality

$$-\tfrac{3}{2} + \tfrac{9}{2}\theta^2 > \tfrac{5}{2}(1-\theta^2) + \frac{|J_3|\,a'}{J_2^2 R}(1-\theta^2)^{1\}2}. \tag{55}$$

On the other hand l has a mean motion when x_1 can vanish. If x_1 has four zeros while y'' increases from 0° to 360°, l has the same mean motion as $2y''$. But $l + g (= y)$ has the mean motion of y'', so that g has the mean motion of $-y''$. Therefore in this case, the orbit of the satellite has the apsidal angle 90°, and is similar to an ellipse with the center of attraction at the center. If x_1 has two zeros, l has the mean motion of y'' and g has no mean motion. The orbit has the apsidal angle 180° and is similar to an ellipse with the center of attraction at one of its two foci. This latter is the case when inclination constant or $1-\theta^2$ is rather small but not vanishingly small.

6. Motion of an Artificial Satellite with Small Angular Momentum

(1) We again borrow Brouwer's S_1 (denoted by S_{1B} here)

$$S_{1B} = \frac{\mu^2 J_2 R^2}{2G''^3}\left\{\left(-\tfrac{1}{2}+\tfrac{3}{2}\theta^2\right)\left(f''-l''+e''\sin f''\right)+\tfrac{3}{4}\left(1-\theta^2\right)\times\right.$$
$$\left.\times\left[\tfrac{1}{2}\sin\left(2f''+2g''\right)+\frac{e''}{2}\sin\left(f''+2g''\right)+\frac{e''}{6}\sin\left(3f''+2g''\right)\right]\right\},$$

where $e''(=\sqrt{1-G''^2/L'^2})$ is the eccentricity constant and f'' is related to e'' and l'' as f (true anomaly) does to e and l.

When the angular momentum constant G'' is small, S_1 becomes large and so do the perturbations. This is because the satellite approaches close to the center of attraction at perigee, and, at apogee, its velocity is small. At the extreme case of $G''=0$, the perturbations become infinite at perigee (we suppose the Earth has zero volume), but not at the apogee even when its velocity vanishes. If we are concerned with the motion only near the apogee, then, because of the finite size of the Earth, Brouwer's S_1, which is obtained by the averaging principle assuming the revolutional motion for orbital motion about the Earth, is too much. In this case we need not and should not use the averaging procedure in obtaining S_1 and F_1^*. Instead we can choose the reference orbit as an Keplerian ellipse, and S_1 so as to vanish at the apogee. We then have

$$F_1^* = 0 \tag{56}$$

$$S_1 = \int F_1\,d\tau = \frac{\mu^2 J_2 R^2}{2G''^3}\left\{\left(-\tfrac{1}{2}+\tfrac{3}{2}\theta^2\right)\left(f''-\pi+e''\sin f''\right)+\right.$$
$$+\tfrac{3}{4}\left(1-\theta^2\right)\left[\tfrac{1}{2}\sin\left(2f''+2g''\right)+\frac{e''}{2}\sin\left(f''+2g''\right)+\right.$$
$$\left.\left.+\frac{e''}{6}\sin\left(3f''+2g''\right)+\left(\tfrac{2}{3}e''-\tfrac{1}{2}\right)\sin 2g''\right]\right\}. \tag{57}$$

(2) The first order perturbation in r, say, is then obtained by

$$r = r'' - \frac{\partial S_1}{\partial \dot{r}''} =$$
$$= r'' - \frac{\mu J_2 R^2}{2G''^2}\left\{\left(-\tfrac{1}{2}+\tfrac{3}{2}\theta^2\right)\left(\frac{\cos f''}{e''}+1\right)+\tfrac{1}{2}\left(1-\theta^2\right)\times\right.$$
$$\times\left[-\cos\left(2f''+2g''\right)+\left(\frac{3}{2e''}-1\right)\cos\left(-f''+2g''\right)+\right.$$
$$\left.\left.+\left(\frac{3}{2e}-3\right)\cos\left(f''+2g''\right)\right]\right\}, \tag{58}$$

where r'' is the radius vector of a Kepler motion with constant elements

$$a'\left(=\frac{L'^2}{\mu}\right),\quad e''\left(=\sqrt{1-\frac{G''^2}{L'^2}}\right),\quad \theta\left(=\frac{H}{G''}\right),\quad g'',\quad h''$$

and $l''=(\mu^2/L'^3)t+\text{const.}$

While S_{1B} gives

$$r = r - \frac{\partial S_{1B}}{\partial \dot{r}''} =$$

$$= r'' - \frac{\mu J_2 R^2}{2G''^2}\left\{(-\tfrac{1}{2}+\tfrac{3}{2}\theta^2)\left(\frac{1-\sqrt{1-e''^2}}{e''}\cos f'' + \frac{2r''}{a'\sqrt{1-e''^2}}\right) - \right.$$

$$\left. - \tfrac{1}{2}(1-\theta^2)\cos(2f''+2g'')\right\} \tag{59}$$

where r'' is the radius vector of a slowly precessing and rotating ellipse with $\dot{g}''\neq 0$, $\dot{h}''\neq 0$.

For the set of constants $a=0.6R$, $e=0.85$, we find

$$\frac{\partial S_1}{\partial \dot{r}''} = 3.00\, J_2 R\left\{\begin{pmatrix}-0.15\\-0.18\end{pmatrix}(-\tfrac{1}{2}+\tfrac{3}{2}\theta^2) + \tfrac{3}{2}(1-\theta^2)\times\right.$$

$$\left.\times\left[\begin{pmatrix}-0.15\\-0.18\end{pmatrix}\cos 2g'' + \begin{pmatrix}0.00\\0.00\end{pmatrix}\sin 2g''\right]\right\}$$

$$\frac{\partial S_{1B}}{\partial \dot{r}''} = 3.00\, J_2 R\left\{\begin{pmatrix}5.78\\5.77\end{pmatrix}(-\tfrac{1}{2}+\tfrac{3}{2}\theta^2) + \tfrac{3}{2}(1-\theta^2)\times\right.$$

$$\left.\times\left[\begin{pmatrix}-0.31\\-0.33\end{pmatrix}\cos 2g'' + \begin{pmatrix}-0.13\\0.00\end{pmatrix}\sin 2g''\right]\right\}$$

where the upper values are at $f=f_0=170°$ and the lower values at the apogee. The motion starts at $f=f_0$ and ends at $f=360°-f_0=190°$ (Figure 2). Because of the choice of S_1, we can expect that perturbations of the second order, which are roughly proportional to S_1^2, are also reduced considerably.

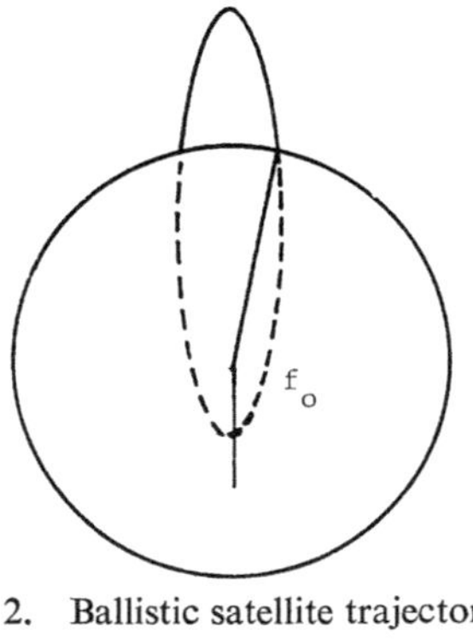

Fig. 2. Ballistic satellite trajectory.

An example of the application of the theory to problems in plasma physics is found in Kawakami (1970).

References

Brouwer, D.: 1959, *Astron. J.* **64**, 378.
Brouwer, D. and Clemence, G. M.: 1961, *Methods in Celestial Mechanics,* Chap. 17, Academic Press.
Hori, G.: 1960, *Astron. J.* **65**, 291; 1961, *Astron. J.* **66**, 258.
Hori, G.: 1966, *Publ. Astron. Soc. Japan* **18**, 287.
Hori, G.: 1970, *Publ. Astron. Soc. Japan* **22**, 191.
Izsak, I. G.: 1963, *Astron. J.* **68**, 559.
Kawakami, I.: 1970, *J. Phys. Soc. Japan* **28**, 505.
Lyddane, R. H.: 1963, *Astron. J.* **68**, 555.
Mersman, W. A.: 1971, *Celes. Mech.* **3**, 384.
Yuasa, M.: 1971, *Publ. Astron. Soc. Japan* **23**.

THE ALGORITHM OF THE INVERSE FOR LIE TRANSFORM

J. HENRARD

Facultés Universitaires de Namur, Namur, Belgium

Abstract. The algorithm for Lie transforms proposed here reduce the amount of computation to be carried out in particular but typical problems of Perturbation Theory. In conjunction with the formulas for inverting and composing Lie transforms it is shown to reduce by a factor of three the number of Poisson brackets to be evaluated in a part of the implementation of the Analytical Lunar Theory.

1. Introduction

The shortcomings of the so-called von Zeipel's method in the theory of general perturbations have long been recognized. Basically, they stem from two characteristics of the underlying transformation theory based upon a generating function in mixed variables.

(1) From a practical point of view, the technique does not lead easily to explicit algorithms by which to transform to any order functions or vector fields.

(2) More importantly, the technique is not canonically invariant and because of this, runs into difficulties when used in conjunction with angles-actions variables. Singularities for small values of the actions are introduced and ways of suppressing them at any order are yet to be designed.

In an attempt to overcome this last difficulty, one could define directely, from the symplectic character of their Jacobian, the canonical transformations to be used in perturbation theory (Henrard, 1965). This technique has been used with success in several instances (Deprit *et al.*, 1967; Deprit and Deprit, 1971) but suffers from a lack of simplicity.

The method of Lie transforms proposed by Hori (1966) and Deprit (1969) does much in overcoming these inconveniences. It is canonically invariant and the general algorithm proposed by Deprit is well adapted to automatic operations on a computer. However in many instances, the implementation of the technique of Lie transforms involves more computations than von Zeipel's method or than the direct construction of canonical transformations.

The advantages of the Lie transform technique far outweighs this slight inconvenience. Nevertheless, it is worthwile to investigate the possibility of modifying the algorithms in order to reduce the amount of computation in typical situations. We propose in Section 3 of these notes such a modification that we call the algorithm of the inverse for reasons that will be obvious to the reader. In Section 5, we analyze a typical situation, arising in the Lunar theory and show how the number of Poisson brackets to be computed can be decreased by a factor of three by using this algorithm.

2. The Group of Lie Transforms

Let us consider canonical transformations depending on a parameter ε and whose

B. D. Tapley and V. Szebehely (eds.), Recent Advances in Dynamical Astronomy, 250–259.

multipliers are unity:

$$y \to x = X(y, t, \varepsilon). \tag{1}$$

We assume that the transformations are analytic in a domain containing a cylinder of the form $\mathscr{D}X(-\varepsilon^*, +\varepsilon^*)$ where $\mathscr{D}$ is some domain of the $2n+1$ dimensional space yxt and $\varepsilon^*>0$. Such a transformation will be called close to the identity if for $\varepsilon=0$, the transformation reduces to the identity.

This set of transformations forms the group of canonical transformation relevant to perturbations theory. We will show that it can be generated by canonical Lie transforms.

Let us define a canonical Lie transformation as the transformation $y \to x = X(y, t, \varepsilon)$ where $X(y, t, \varepsilon)$ is the solution of the canonical system of differential equations,

$$\frac{\mathrm{d}x}{\mathrm{d}\varepsilon} = JW_x(x, t, \varepsilon) \tag{2}$$

with initial conditions $x(\varepsilon=0)=y$. We will note the transform of any function $f(x, t, \varepsilon)$ under this transformation by

$$g(y, t, \varepsilon) = \mathscr{L}(W) f(y, t, \varepsilon) = f(X(y, t, \varepsilon), t, \varepsilon). \tag{3}$$

Given a canonical transformation with multiplier unity, its inverse is also canonical with multiplier unity and is thus canonical not only with respect to t but also with respect to ε.

The remainder function $W(x, t, \varepsilon)$ (with respect to ε) of this inverse transformation is given by

$$\frac{\partial}{\partial\varepsilon} X(y, t, \varepsilon) = JW_x. \tag{4}$$

But this is nothing else than Equation (2) when y and t are independent of ε. Furthermore if the given transformation $X(y, t, \varepsilon)$ is close to the identity, the vector y represent the initial conditions of the system (2) of differential equations.

Conversely it is obvious that for any analytical function $W(x, t, \varepsilon)$, the Lie transform $\mathscr{L}(W)$ generated by the solution of (2) is close to the identity.

The basic operations of the group of Lie transforms, the inversion and the combination, can be represented by simple operations on their generating functions.

Let us consider the following diagram of transformations

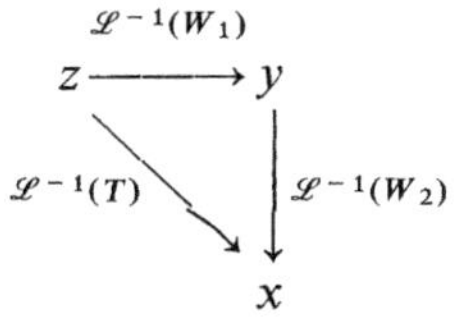

As we have seen, the remainder of the inverse of a Lie transform $\mathscr{L}^{-1}(T)$ is the generating function T. Hence applying the rules of composition of remainder functions we have.

$$T(z, t, \varepsilon) = W_2(Y(z, t, \varepsilon), t, \varepsilon) + W_1(z, t, \varepsilon) \tag{5}$$

where $Y(z, t, \varepsilon)$ represents the inverse of the transformation generated by W_1. Writing the identity in a more compact form, we obtain

$$\boxed{T = \mathscr{L}^{-1}(W_1)\, W_2 + W_1}\,. \tag{6}$$

If we assume now that $\mathscr{L}^{-1}(W_2) = \mathscr{L}(W_1)$, the resulting transformation $\mathscr{L}^{-1}(T)$ is the identity. Its remainder function is thus zero and the identity (6) can be rewritten as

$$\boxed{W_2 = -\mathscr{L}(W_1)\, W_1}\,. \tag{7}$$

This last formula yields the generator W_2 of the inverse of the Lie transform generated by W_1.

3. The Algorithm of the Inverse

The basic identity on which algorithms for Lie transforms are based is the Taylor series expansion

$$g(y, t, \varepsilon) = \sum_{n \geqslant 0} \frac{\varepsilon^n}{n!} [D^n f(x, t, \varepsilon)]_{\substack{\varepsilon=0 \\ x=y}} \tag{8}$$

where the operator D is the Lie derivative.

$$Df = \frac{\partial f}{\partial \varepsilon} + \{f; W\} \tag{9}$$

composed of the partial derivative of a function with respect to ε and of its Poisson bracket with the generating function W. The general algorithm proposed by Deprit is based upon the recursive computation of the auxiliary functions

$$D^n f = D[D^{n-1} f]. \tag{10}$$

We shall base the algorithm of the inverse upon the direct expansion of

$$[D^n f]_{\substack{\varepsilon=0 \\ x=y}} = [D^{n-1} Df]_{\substack{\varepsilon=0 \\ x=y}} \tag{11}$$

in terms of a recursive set of basic operators.

For reason of convenience, let us first develop what we could call the direct algorithm.

Direct algorithm: The coefficient of the Lie transform

$$g(y, t, \varepsilon) = \sum_{n \geqslant 0} \frac{\varepsilon^n}{n!} g_n(y, t) \tag{12}$$

of an analytic function

$$f(x, t, \varepsilon) = \sum_{n \geqslant 0} \frac{\varepsilon^n}{n!} f_n(x, t) \tag{13}$$

under the transformation generated by the function

$$W(x, t, \varepsilon) = \sum_{n \geqslant 0} \frac{\varepsilon^n}{n!} W_{n+1}(x, t) \tag{14}$$

can be computed recursively by the formula

$$\boxed{g_n = \sum_{j=0}^{n} \binom{j}{n} M_j f_{n-j}} \tag{15}$$

where the differential operators $M_j = M_j(W)$ are defined by

$$\boxed{\begin{aligned} &M_0 = 1 \\ &M_j = \sum_{i-1}^{j} \binom{i-1}{j-1} M_{j-i} L_i \quad \text{for} \quad j > 0 \end{aligned}}\,. \tag{16}$$

The differential operators $L_i = L_i(W)$ are defined by $L_i f = \{f;\, W_i\}$.

Let us prove the proposition by induction. It is evident for $n = 0$ and we assume that (15) holds for $n < N$. By the identity (8) we have

$$g_N = [D^N f]_{\varepsilon=0} = \left[D^{N-1} \frac{\partial f}{\partial \varepsilon}\right]_{\varepsilon=0} + [D^{N-1}\{f;\, W\}]_{\varepsilon=0}\,. \tag{17}$$

Using Equation (15) which holds by assumption for $n < N$ we have:

$$\left[D^{N-1} \frac{\partial f}{\partial \varepsilon}\right]_{\varepsilon=0} = \sum_{j=0}^{N-1} \binom{j}{N-1} M_j f_{N-j} \tag{18}$$

and also

$$[D^{N-1}\{f;\, W\}]_{\varepsilon=0} = \sum_{j=0}^{N-1} \binom{j}{N-1} M_j \left[\frac{\partial^{N-j-1}}{\partial \varepsilon^{N-j-1}} \{f;\, W\}\right]_{\varepsilon=0}. \tag{19}$$

Taking into account the rule of partial differentiation;

$$\frac{\partial^K}{\partial \varepsilon^K} \{f;\, W\} = \sum_{i=0}^{K} \binom{i}{K} \left\{\frac{\partial^{K-i} f}{\partial \varepsilon^{K-i}};\, \frac{\partial^i W}{\partial \varepsilon^i}\right\} \tag{20}$$

we obtain:

$$[D^{N-1}\{f;W\}]_{\varepsilon=0} = \sum_{j=0}^{N-1}\sum_{i=0}^{N-j-1}\binom{j}{N-1}\binom{i}{N-j-1} M_j L_{i+1} f_{m-i-j-1}\,. \tag{21}$$

Changing the summation from j and i to $r=j+i+1$ and $s=i+1$ and taking into account the induction rule of the operators M_j, we obtain

$$[D^{N-1}\{f;W\}]_{\varepsilon=0} = \sum_{r=1}^{N}\binom{r-1}{N-1} M_r f_{m-r}\,. \tag{22}$$

Adding the right hand members of (18) and (22) we check that the identity (15) holds for $n=N$. Hence by induction this identity holds for every n.

As a matter of fact, the operators M_j we just introduced are similar to the operator obtained by Faa de Bruno (1855) for the higher derivatives of a function depending upon another function. The operators of Faa de Bruno have been applied by Musen (1965) in order to extend to any order the formalisms of Krylov-Bogoliubov and of Poincaré–von Zeipel. The formalism proposed by Musen is very much similar to the formalism we just described.

As it stands, the algorithm does not lead to an easy implementation. Indeed while the operators M_j can be constructed by recurrence, the same is not true of the functions $M_j f_K$.

However when the formalism is formulated in terms of the inverse of a Lie transform, this operational difficulty disappears because the operators M and L commutes in Formula (16).

Algorithm of the inverse: The coefficient of the Lie transform

$$g(y,t,\varepsilon) = \sum_{n\geqslant 0}\frac{\varepsilon^n}{n!}\, g_n(y,t) \tag{23}$$

of an analytic function

$$f(x,t,\varepsilon) = \sum_{n\geqslant 0}\frac{\varepsilon^n}{n!}\, f_n(x,t) \tag{24}$$

under *the inverse of the transformation* generated by the function

$$V(x,t,\varepsilon) = \sum_{n\geqslant 0}\frac{\varepsilon^n}{n!}\, V_{n+1}(x,t) \tag{25}$$

can be computed recursively by the formula

$$\boxed{g_n = \sum_{j=0}^{n}\binom{j}{n} M_j f_{n-j}} \tag{26}$$

where the differential operators $M_j = M_j(W)$ are defined by

$$\boxed{\begin{aligned} &M_0 = 1 \\ &M_j = -\sum_{i=1}^{j} \binom{i-1}{j-1} L_i M_{j-i} \quad \text{for} \quad j > 0 \end{aligned}}\,. \tag{27}$$

The differential operators L_i are now related to the inverse generator by $L_i f = \{f;\ V_i\}$.

In order to show that the algorithm of the inverse is a consequence of the direct algorithm, let us apply the algorithm of the inverse to the inverse of the transformation defined by the direct algorithm. Hence in the algorithm of the inverse, the functions g and f correspond respectively to the functions f and g in the direct algorithm. The operators L_i have the same definition in both algorithms while the operators M_j in the algorithm of the inverse correspond now to the inverse transformation. To avoid confusion, we shall rename them

$$N_j(V) = M_j(W). \tag{28}$$

The product of the two transformation thus defined is the identity. Hence comparing (26) and (15) we obtain the two identities

$$\sum_{i=0}^{j} \binom{i}{j} M_{j-i} M_i = \sum_{i=0}^{j} \binom{i}{j} M_{j-i} N_i = 0. \tag{29}$$

It remains to show that the operators N_j are indeed the result of the recurrence (27). Let us prove the proposition by induction. It is evident for $j=0$ and we assume that (27) holds for $j<n$. From the second of the equalities (29) we obtain

$$M_n + N_n + \sum_{i=1}^{n-1} \left\{ \binom{i}{n-1} + \binom{i-1}{n-1} \right\} M_{n-i} N_i = 0 \tag{30}$$

which is equivalent to

$$M_n + N_n + \sum_{i=0}^{n-1} \binom{i-1}{n-1} \{M_i N_{n-i} + M_{n-i} N_i\} = 0. \tag{31}$$

Taking into account that the recurrence Formulae (16) and (27) hold for the operators M_i and N_i with $i<n$. we obtain from (31)

$$\begin{aligned} 0 = M_n + N_n + &\sum_{i=1}^{n-1} \binom{i-1}{n-1} \{L_i N_{n-i} - M_{n-i} L_i\} \\ + &\sum_{i=2}^{n-1} \sum_{k=1}^{i-1} \binom{i-1}{n-1} \binom{k-1}{i-1} \{M_{i-k} L_k M_{n-i} - M_{n-i} L_k N_{i-k}\}. \end{aligned} \tag{32}$$

In the set (i, k) of the indices of the second sum, we define the involution $(i, k) \to \to (i', k')$ such that

$$k = k' ; \quad i - k = n - i' ; \quad n - i = i' - k' . \tag{33}$$

We check that if we arrange the second sum of (32) by pair defined through the involution (33), the sum of each pair is identically zero; thus the sum itself vanishes and we obtain

$$M_n + N_n + \sum_{i=1}^{n-1} \binom{i-1}{n-1} \{L_i N_{n-i} - M_{n-i} L_i\} = 0 . \tag{34}$$

But Formula (16) is valid to any order; hence Equation (34) is identical to Formula (27) for $j = n$ and the recurrence step is concluded.

4. Advantages of the Algorithm of the Inverse

The algorithm of the inverse can be immediately written under the form of a recurrence involving functions rather than operators. This form is more suitable for computation.

Defining:

$$f_{i,j} = M_i f_j \tag{35}$$

Formulae (26) and (27) are now written

$$\begin{aligned} & f_{0,j} = f_j \\ & f_{i,j} = - \sum_{k=1}^{i} \binom{k-1}{i-1} L_k(V) \, f_{i-k,j} \quad \text{for} \quad i \neq 0 \end{aligned} \tag{36}$$

$$g_n = \sum_{j=0}^{n} \binom{j}{n} f_{j,n-j} \tag{37}$$

The progress of the recurrence is summarized in Figure 1.

$$\begin{array}{lllll} g_0 <= & f_0 & & & \\ & \downarrow & & & \\ g_1 <= & f_{1,0} & f_1 & & \\ & \downarrow & \downarrow & & \\ g_2 <= & f_{2,0} & f_{1,1} & f_2 & \\ & \downarrow & \downarrow & \downarrow & \\ g_3 <= & f_{3,0} & f_{2,1} & f_{1,2} & f_3 \end{array}$$

Fig. 1. Triangle for the inverse algorithm.

Each column of the triangle is independent of the others and is computed recursively by using Formula (36). Formula (37) then gives g_n as the weighted sum of the elements belonging to the nth row.

The first advantage of the algorithm of the inverse is that it can be used backward as well as forward. Indeed Formula (37) can be written as

$$f_n = g_n - \sum_{j=1}^{n} \binom{j}{n} f_{j,n-j} \tag{38}$$

giving the last element in a row as a weighted sum of all the elements preceding it. As the last element in a row is also the first one of its column, Formula (36) can then be used in order to complete this column. Hence the algorithm of the inverse can be used to compute indifferently the direct or the inverse Lie transform corresponding to a given generator $V(x, t, \varepsilon)$.

But the algorithm of the inverse has another advantage. In particular but typical situations, it can reduce drastically the number of operations to be performed. This stems from the fact that each column is independent. If the first element on any particular column vanishes, so does the entire column.

In typical situations, we have to transform either an Hamiltonian function which depends linearly on the small parameter ε or a function of the phase space, independent of this small parameter. As a result, only the first two columns or in the second case, only the first column of the triangle have to be computed. The number of Poisson brackets to be computed is respectively n^2 and $n(n+1)/2$ if the transformation is performed up to order n.

This compare with the $(n(n+1)(n+2))/6$ Poisson brackets necessary to implement other algorithms for Lie transforms.

5. Example of the Use of the Algorithm of the Inverse

Let us consider the following example which arise in the development of the Analytical Lunar Ephemeris (Deprit *et al.*, to appear). Given the Hamiltonian function $H(x, \varepsilon)$, linear in ε, we wish to average it first over the monthly terms, them over the annual terms by two succesive canonical transformations.

We thus have to define two generating functions $W_M(x, \varepsilon)$ and $W_A(y, \varepsilon)$ in such a way that in the following diagram:

$$H(x, \varepsilon) \xrightarrow{\mathscr{L}(W_M)} H'(y, \varepsilon) \xrightarrow{\mathscr{L}(W_A)} H''(z, \varepsilon) \tag{39}$$

H' (resp. H'') does not contain terms of period equal to or less than a month (respectively than a year).

The transformations are to be defined up to order 20 and three functions of the original coordinates (namely the longitude, the latitude and the sine parallax) have to be expressed in terms of the averaged variables z.

Using the straightforward algorithm originally proposed by Deprit, or the one proposed by Mersman, the number of non vanishing Poisson brackets to be computed is as follows:

define W_M	:	1.350	Poisson brackets
define W_A	:	1.540	Poisson brackets
apply $\mathscr{L}(W_M)$ to 3 functions	:	4.620	Poisson brackets
apply $\mathscr{L}(W_A)$ to 3 functions	:	4.620	Poisson brackets
Total	:	12.130	Poisson brackets

The simplified algorithms we have presented and the fact that it is possible to compute the generator of the product of two transformations can be combined here to decrease the number of Poisson brackets to be computed by a factor of three.

Using the algorithm of the inverse, we first define V_M the generator of the inverse of the transformation eliminating the monthly terms. The resulting Hamiltonian, H' is no longer linear in ε, so that we use any algorithm to define W_A, the generator of the transformation eliminating the annual terms.

As we wish to use the algorithm of the inverse to transform the three functions independent of ε, we then compute T the generator of the inverse of the transformation $\mathscr{L}(W_M) \circ \mathscr{L}(W_A)$. Formulas (6) and (7) give

$$T = V_A + \mathscr{L}^{-1}(V_A)\, V_M = \mathscr{L}(W_A)\,(V_M - W_A). \tag{40}$$

The results of the procedure is the following:

define V_M	:	400	Poisson brackets
define W_A	:	1.540	Poisson brackets
define T	:	1.540	Poisson brackets
apply $\mathscr{L}^{-1}(T)$ to three functions	:	630	Poisson brackets
		4.110	Poisson brackets

6. Conclusions

The possibility of performing algebraic manipulations on a computer has affected considerably the criteria by which one judges the quality of a general perturbation algorithm. A good algorithm should certainly be valid to any order and build recursively from a few simple formulae.

The Lie transform method leads easily to such algorithms.

The main algorithm proposed originally by Deprit and then by Mersman in connection with Hori's version of the Lie transform performs generally well but can be improved in particular but typical instances. Such an improvement is proposed here.

References

Deprit. A.: 1969, *Celes. Mech.* **1**, 12–30.

Deprit, A. and Deprit, A.: 1967, *Astron. J.* **72**, 173–179.
Deprit, A., Henrard, J., and Rom, A.: 1967, *Icarus* **6**, 381–406.
Henrard, J.: 1965, Thesis, University of Louvain.
Henrard, J.: 1970, *Celes. Mech.* **3**, 107–120.
Hori, G.: 1966, *Publ. Astron. Soc. Japan* **18**, 287–296.
Mersman, W. A.: 1970, *Celes. Mech.* **3**, 81–89.
Musen, P.: 1965, *J. Astron. Sci.* **12**, 129–134.

THE PROPERTY OF COVARIANCE IN HORI'S NONCANONICAL PERTURBATION THEORY

U. KIRCHGRABER
Seminar für Angewandte Mathematik der E.T.H., Zürich, Switzerland

Abstract. The present paper is strongly based on the work of Hori (Hori, 1966, 1971). In his 1966 paper Hori introduced the idea of Lie series in the field of canonical perturbation theory, in his 1971 article he applied the same idea to noncanonical autonomous systems of differential equations.

While the canonical invariance of Hori's 1966 theory was already mentioned by Hori in the 1966 publication, it seems that the analogous property in the noncanonical case, the covariance with respect to arbitrary transformations, has not yet been recognized.

Let us briefly characterize the property of covariance and its consequences.

Consider some perturbed problem, given in terms of some set of variables. Consider the attached perturbation equations and assume them to be solved. Consider next the same problem, now in terms of some different set of variables and the corresponding perturbation equations. The property of covariance allows us to establish the relation between the solutions of the two sets of perturbation equations.

It is possible to base an integration procedure of the perturbation equations on the property of covariance. In order to emphasize the difference between the usual integration procedure of the perturbation equations and the one presented in this paper, let us recall the usual approach:

A perturbation problem is given in terms of some variables, say cartesian coordinates. Making use of the fact that the unperturbed problem is assumed to be solvable, a new set of variables, the elements, are introduced and the perturbed problem is formulated in terms of these variables. By solving the attached perturbation equations a near identical transformation is performed. The very final step is the solution of the remaining, the so called longperiodic, system:

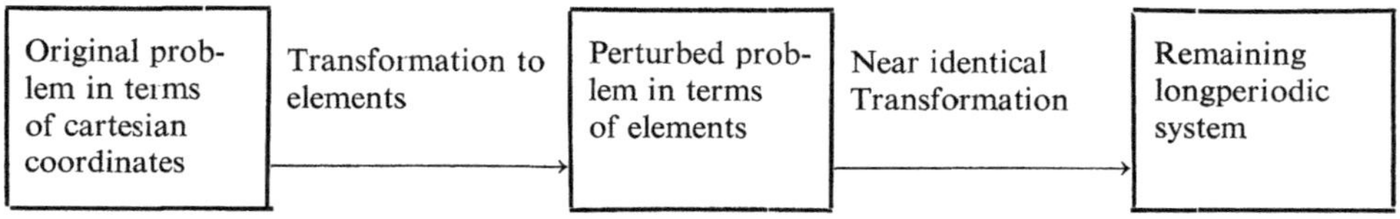

Thus: first we perform the transformation to elements and afterwards the near identical transformation is established.

It is natural to ask if it is not possible to perform *first* a near identical transformation.

If we write down the perturbation equations in terms of the original variables they are, in general, fairly complicated and not easy to handle.

By virtue of the property of covariance, however, we are not obliged to solve these

B. D. Tapley and V. Szebehely (eds.), Recent Advances in Dynamical Astronomy, 260–261.

perturbation equations in terms of the original variables. We can choose a more convenient set of variables, for instance switch to elements. The perturbation equations expressed in terms of the elements are treated in the usual manner (elimination of short periodic terms).

This procedure generates a near identical transformation and a remaining long-periodic system of differential equations, expressed in terms of 'slightly-changed-elements'.

The property of covariance permits one to formulate an equivalent near identical transformation and an equivalent remaining system of differential equations in terms of 'slightly-changed-original-variables'.

This very system of differential equations must still be solved and for this purpose it may be useful to transform it to elements. The resulting system of differential equations is obviously identical with the remaining longperiodic system of the usual approach.

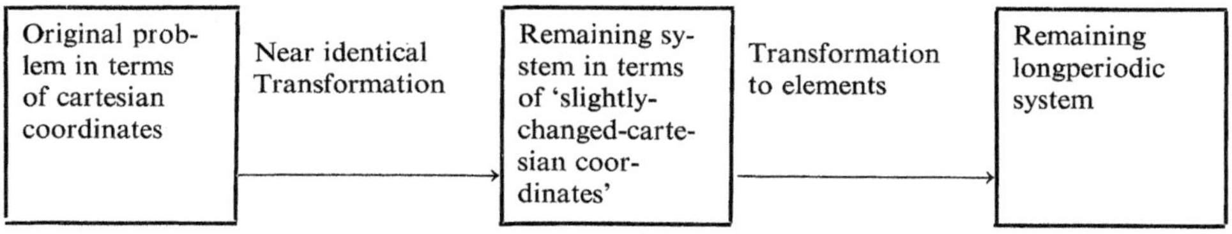

Thus, first we perform the near identical transformation, afterwards the transformation to elements is applied.

Therefore, in the present approach the two basic transformations are interchanged.

What is the use of this interchange?

It is known from the canonical case, that the interchange of the two basic steps allows one to remove certain difficulties which are produced by the fact that the set of elements used are singular. The author expects that, by means of the present approach, similar difficulties in the noncanonical case become removable.

More details are presented in the author's paper: The Transformational Behaviour of Perturbation Theories.

References

Hori, G.: 1966, *Publ. Astron. Soc. Japan* **18**, 287.
Hori, G.: 1971, *Publ. Astron. Soc. Japan* **23**, 567.
Kirchgraber, U.: 1973, *Celest. Mech.* **7**, 474. 'The Transformational Behaviour of Perturbation Theories', to appear in *Celest. Mech.*

SUBDYNAMICS IN CLASSICAL MECHANICS

J. RAE

Faculté des Sciences, Université Libre de Bruxelles, Belgium

Abstract. An introductory account is given of the idea of subdynamics as developed by Prigogine and his coworkers. In the first few sections we outline the usual form of the theory as it applies to classical Hamiltonian systems with a finite number of degrees of freedom. The later parts of the lectures provide an alternative approach to the same ideas, an approach which shows more clearly the connection with established methods of perturbation theory such as the method of Lie series.

1. Introduction

Since about 1960 a theory of non-equilibrium statistical mechanics, for both classical and quantum systems, has been developed by Prof. I. Prigogine and others at Bruxelles (Prigogine, *et al.*, 1962, 1968, 1969, 1970a, b). The theory is based on a mechanical, rather than stochastical, point of view and was originally intended for discussion of the long time behaviour of systems with many degrees of freedom. However, it has lately become apparent that many of the ideas and results apply also to systems with few degrees of freedom and, in particular, give an approach to classical mechanics along the lines pioneered by Koopman (1931).

In the following I will restrict myself to classical Hamiltonian systems, although various generalizations are possible, and discuss mainly the particular aspect known as subdynamics. First I will adumbrate the theory as it was developed originally with statistical mechanics in mind (most of the existing literature takes this standpoint (Prigogine, *et al.*, 1969, 1970a, b)) and then cover some of the ground again by an equivalent, but superficially quite different, approach (Rae, 1972a, b): this latter approach should be more familiar to the participants at his meeting and will, I hope, make connection with some of the other lectures delivered here.

2. The Liouville Equation and Perturbation Expansion

Our starting point is the Liouville equation for the time evolution of a phase function f

$$i \frac{\partial f(t)}{\partial t} = Lf(t) \tag{1}$$

where (i) f is a function over the phase space of the system, expressed for example, in action-angle coordinates $J_1 \ldots J_N$, $\alpha_1 \ldots \alpha_N$; for definiteness we may think of f as an element of the Hilbert space $\mathscr{L}$ of functions square integrable over phase space, but since much of what follows is on a formal level this is of no great importance.

(ii) the *linear* operator L is i times the Poisson bracket with the Hamiltonian (the i

B. D. Tapley and V. Szebehely (eds.), Recent Advances in Dynamical Astronomy, 262–269.

is for self-adjointness of L)

$$L = i\{H, \ \} = i\sum_{j=1}^{N}\left(\frac{\partial H}{\partial \alpha_j}\frac{\partial}{\partial J_j} - \frac{\partial H}{\partial J_j}\frac{\partial}{\partial \alpha_j}\right) \tag{2}$$

with periodic boundary conditions in the angles.

The solution of (1) can be written in terms of the initial function $f(0)$ and the resolvent operator of L

$$f(t) = U(t)\, f(0) = \frac{1}{2\pi i}\int_C e^{-izt}\,\frac{1}{z - L}\, f(0)\,\mathrm{d}z \tag{3}$$

where the contour C lies above the real z axis (which contains the spectrum of L).

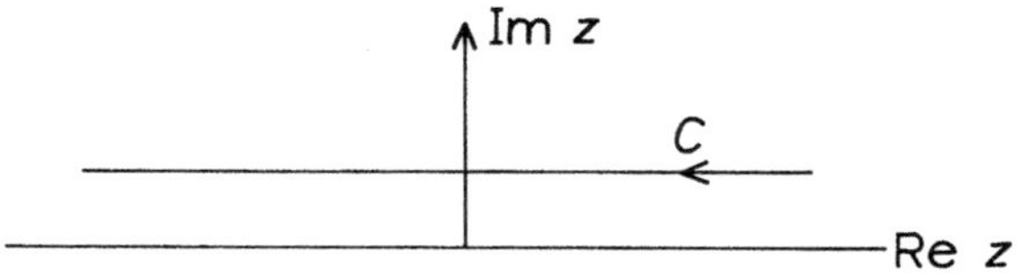

Fig. 1.

The right hand side of (3) can, of course, be evaluated only for very special hamiltonians (e.g. independent simple harmonic oscillators). In practice what one usually encounters is a Hamiltonian of the form $H = H_0 + \varepsilon H_1$ where H_0 is in some sense solvable and εH_1 in some sense small so that perturbation theory may be applied. If we write the corresponding decomposition of L as $L_0 + \varepsilon L_1$, the most naive perturbation expansion which may be applied

$$\frac{1}{z - L} = \frac{1}{z - L_0} + \frac{1}{z - L_0}\,\varepsilon L_1\,\frac{1}{z - L} = \sum_{M=0}^{\infty}\frac{1}{z - L_0}\left(\varepsilon L_1\,\frac{1}{z - L_0}\right)^M \tag{4}$$

is of little interest for the following reason. L_0 has a (degenerate) zero eigenvalue so that terms like z^{-m} occur in the right hand side of (4). On making the Laplace inversion these give rise to terms like t^m so that high order terms in (4) give large powers of ε multiplied by large powers of t. If one is interested in small ε and long times such secular terms give trouble. This difficulty can be avoided in the following way. We introduce the nullspace projector P of L_0, that is the projector onto those functions f such that $L_0 f = 0$. (For example, if $H_0 = H_0(J_1 \cdots J_N) = \sum \omega_j J_i$ with incommensurable frequencies $\omega_1 \cdots \omega_N$ then

$$L_0 = -\, i \sum_j \omega_j \frac{\partial}{\partial \alpha_j}$$

and

$$P = \frac{1}{(2\pi)^N} \int_0^{2\pi} d\alpha_1 \int_0^{2\pi} d\alpha_2 \ldots \int_0^{2\pi} d\alpha_N \Big) .$$

With $Q = 1 - P$ we now see that the operator $Q(1/(Z - QLQ))Q$ *can* be expanded without difficulty

$$Q \frac{1}{z - L_0 - \varepsilon Q L_1 Q} Q = \sum_{M:0}^{\infty} Q \frac{1}{z - L_0} Q \left(\varepsilon L_1 Q \frac{1}{z - L_0} Q \right)^M \tag{5}$$

and the L_0 in the denominator cannot now give zeros. Thus, if we first express $(z - L)^{-1}$ in terms of operators in which L_0 occurs only in the form above, these operators may be expanded without risk. A little algebraic manipulation gives

$$\frac{1}{z - L} = \mathscr{P}(z) + [P + \mathscr{C}(z)] \frac{1}{z - PLP - \psi(z)} [P + \mathscr{D}(z)] \tag{6}$$

where

$$\psi(z) = PLQ \frac{1}{z - QLQ} QLP \quad \text{'collision operator'} \tag{7}$$

$$\mathscr{C}(z) = Q \frac{1}{z - QLQ} QLP \quad \text{'creation operator'} \tag{8}$$

$$\mathscr{D}(z) = PLQ \frac{1}{z - QLQ} Q \quad \text{'destruction operator'} \tag{9}$$

$$\mathscr{P}(z) = Q \frac{1}{z - QLQ} Q \quad \text{'propagator of correlations'}. \tag{10}$$

The operators (7)–(10) may be expanded in terms of $(z - L_0)^{-1}$ and the presence of the Q excludes z^{-m} terms. In the original approach to the theory (Prigogine, 1962, 1968) in terms of resummation of diagrams this exclusion was called the irreducibility condition: the diagram expansion is also the source of the names given to these operators. Note also that although for reasons of perturbation theory we take P as the nullspace projector, the form (6) holds for any choice of projector P.

3. Subdynamics

We now want to look at (6) in more detail. The most slowly evolving behaviour corresponds to $z \to 0$ so an important role is played by the nature of the operators in a neighborhood of the origin (in the z-plane). Working within the perturbation framework it is enough to assume that L_0 has no point spectrum (eigenvalues) in a neighborhood of $z = 0$ (other than the point $z = 0$ itself). In this case all the operators ψ, $\mathscr{C}$, $\mathscr{D}$, $\mathscr{P}$ are regular near $z = 0$ when interpreted as expansions of the type (5). (These operators are always defined and regular above the real axis. If L_0 has a smooth continuous spectrum near 0 they can be defined in the lower $\frac{1}{2}$-plane by analytic continuation (Prigogine, 1962, 1968). If L_0 has no spectrum near 0 except $Z=0$ this continuation is not necessary). We introduce the notion of subdynamics in the following way. From (3) and (6)

$$U(t)\, f(0) = \frac{1}{2\pi i} \int\limits_C e^{-izt} \left\{ \mathscr{P}(z) + [P + \mathscr{C}(z)] \frac{1}{z - PLP - \psi(z)} \times \right.$$
$$\left. \times [P + \mathscr{D}(z)] \right\} f(0)\, dz. \qquad (11)$$

The 'most slowly evolving' part of this expression is separated by the following procedure: $1/(z-PLP-\psi(z))$ is written $\sum_{n=0}^{\infty}(PLP+\psi(z))^n/z^{n+1}$ and only the contribution coming from the singularity at $z=0$ in each term is kept. This gives rise to a certain evolution operator $\Sigma(t)$ acting on $f(0)$. All other contributions to the evolution are collected into an operator $\hat{\Sigma}(t)$ so that

$$U(t)\, f = \hat{\Sigma}(t)\, f + \hat{\Sigma}(t)\, f. \qquad (12)$$

The explicit forms of these operators have been calculated and may be found in Prigogine *et al.* (1969, 1970a), George (1967). $\Sigma(t)$ satisfies the (almost obvious) semigroup property

$$\Sigma(t + t') = \Sigma(t)\, \Sigma(t') \qquad t, t' \geqslant 0. \qquad (13)$$

For this reason the evolution defined by $\Sigma(t)$ is referred to as a subdynamics. Following standard ideas for semigroups one obtains an associated projection operator by taking the limit $t \to 0+$

$$\Pi = \lim_{t \to 0} \Sigma(t). \qquad (14)$$

This projector can be shown to commute with L and, most importantly, has the property

$$\Sigma(t) = e^{-itL}\Pi = \Pi e^{-itL}. \qquad (15)$$

The original proof of this can be found in Balescu *et al.* (1969): for our simpler context we will give a different argument later. Once (15) has been established it follows immediately from (12) that $\hat{\Sigma}$ is also a semigroup (in fact e^{-itL} $(1-\Pi)$) and that the subdynamics in the Π and $1-\Pi$ projected subspaces of $\mathscr{L}$ are independent. In this way we see that the asymptotic motion governed by $\Sigma(t)$ (in statistical mechanics this is the long time motion because it relaxes most slowly: in our case it is better regarded as the slowly evolving part) can be looked on as the projection under Π of the exact $U(t)$ motion.

What is the content of the subspace $\Pi\mathscr{L}$? First, we can show that it contains all invariants of the motion. Let g be invariant: $Lg=0$

$$\therefore \frac{1}{z} g = \frac{1}{z - L} g.$$

From (11) we see that the P part of this becomes

$$\frac{1}{z} Pg = \frac{1}{z - PLP - \psi(z)} (Pg + \mathscr{D}(z)\, Qg) \tag{16}$$

which on rearranging and taking $z \to 0$ is

$$[PLP + \psi(0)]\, Pg = 0. \tag{17}$$

The Q part gives

$$\frac{1}{z} Qg = \mathscr{P}(z)\, Qg + \mathscr{C}(z) \frac{1}{z - PLP - \psi(z)} (Pg + \mathscr{D}(z)\, Qg) =$$

$$\overset{(16)}{=} \mathscr{P}(z)\, Qg + \mathscr{C}(z) \frac{1}{z} Pg.$$

Rearrangement and $z \to 0$ gives

$$Qg = \mathscr{C}(0)\, Pg. \tag{18}$$

Now

$$\Pi g = \frac{1}{2\pi i} \int_0 [P + \mathscr{C}(z)] \sum_{n:0}^{\infty} \frac{(PLP + \psi(z))^n}{z^{n+1}} [P + \mathscr{D}(z)]\, g\, \mathrm{d}z =$$

$$\overset{(16)}{=} \frac{1}{2\pi i} \int_0 [P + \mathscr{C}(z)] \frac{1}{z} Pg = Pg + \mathscr{C}(0)\, Pg \overset{(18)}{=} g.$$

This demonstrates that all invariants lie in $\Pi \mathscr{L}$.

In the simplest systems this may be the entire content of $\Pi\mathscr{L}$ but in more complicated systems there may be other, time evolving, functions in $\Pi\mathscr{L}$. Here we merely quote a couple of simple examples representing more or less the extreme cases.

(A) Consider a simple integrable system with Hamiltonian $H(I_1, \ldots, I_N, \alpha_1, \ldots, \alpha_N)$ for which there exists a canonical transformation $(I, \alpha) \leftrightarrow (J, \beta)$ such that the new Hamiltonian is cyclic in β. i.e.

$$H(I, \alpha) \leftrightarrow \tilde{H}(J).$$

In these circumstances the subdynamics is trivial. The space $\Pi\mathscr{L}$ consists entirely of invariants, that is it is the space of functions of $J_1 \ldots J_N$ alone and there is no evolution (Prigogine *et al.*, 1971).

(B) Consider now a Hamiltonian of the form $H_0 = \sum_{j=1}^{N} \omega_j I_j$ with constant frequencies ω_j which are degenerate in the sense that for every i, j ω_i/ω_j is a rational number. Such an H_0 has, of course, the N constants of motion I_j but, due to the degeneracy, has an extra $N-1$ 'accidental' constants which are certain combinations of the angles α_j. If we now add to H_0 a small perturbation $\varepsilon H_1(I, \alpha)$ these former constants in general evolve in time. Such systems have a canonical transformation (Kruskal, 1962)

$$(I, \alpha) \leftrightarrow (\phi, z_1, z_2, \ldots, z_{2N-1})$$

such that the equations of motion take the form

$$\frac{d\phi}{dt} = 1 + \varepsilon f(\mathbf{z}, \varepsilon)$$

$$\frac{d\mathbf{z}}{dt} = \varepsilon \mathbf{g}(\mathbf{z}, \varepsilon)$$

so that one angle varies on a 'fast' time scale and all other variables vary slowly, i.e. proportionally to ε. In this case it turns out that the Π subspace consists of all functions of the variables z (Rae and Davidson, 1973).

At this point, having introduced the basic ideas of subdynamics it would be reasonable in a longer lecture course to develop the subject further and demonstrate, for example, how these methods provide a useful method for calculating, in a systematic way, the adiabatic invariant known to exist in system (B) above (Rae and Davidson, 1973; MacNamara and Whiteman, 1967). This, however, would necessitate a considerable number of definitions and new notation, much of it motivated from statistical mechanics. As many of the participants at this meeting may be unfamiliar with this background I think it is better to take another approach, namely to develop the above ideas once again from a different standpoint. This new point of view will, I hope, indicate more clearly to the present audience the close connection of the above notions to the cluster of ideas associated with the names Hamilton-Jacobi method, Lie series method, the averaging method etc.

4. Alternative Approach

The method which follows exists in the literature in several slightly different forms: we follow here Kummer (1971a, b) which also give the connection with the averaging method etc.

One starts from a Hamiltonian $H = H_0 + \varepsilon H_1$ and introduces, as before, the Liouville operators L and L_0, the nullspace projector P and $Q = 1 - P$. The problem can now be stated in the form: can one find a function $K(\varepsilon)$ in the nullspace of L_0 and a function $\omega(\varepsilon)$ such that

$$e^{i\varepsilon\{\omega,\cdot\}}(H_0 + \varepsilon H_1) = H_0 + \varepsilon K(\varepsilon)? \tag{19}$$

(We abbreviate the exponential Poisson bracket operation $e^{i\varepsilon\{\omega,\cdot\}}$ by V.) Supposing this can be done, what will be gained? Well, the Liouville Equation (1) will now be solved by

$$f(t) = V^{-1} e^{-it\{H_0 + \varepsilon K, \cdot\}} V f(0) = V^{-1} e^{-it(L_0 + \varepsilon \tilde{L}_1)} V f(0) \tag{20}$$

where L_0 and $\tilde{L}_1$ *now commute*: this makes the exponential operator very easy to handle. With the above motivation we now must show that such V and K can be found. Before outlining the construction we need a preliminary result and an assumption.

Preliminary result: If $x_0 \in P\mathscr{L}$ and $y_0 \in P\mathscr{L}$ then $\{x_0, y_0\} \in P\mathscr{L}$.

Proof. By Jacobi's identity

$$\{H_0, \{x_0, y_0\}\} = - \{y_0, \{H_0, x_0\}\} - \{x_0, \{y_0, H_0\}\} = 0 .$$

Assumption: If $X_0 \in P\mathscr{L}$ and $Y \in Q\mathscr{L}$ then $\{X_0, Y\} \in Q\mathscr{L}$.

This seems to be satisfied in all cases of interest. The required ω and K are found by an inductive argument using a formal expansion in ε. Equation (19) is equivalent to

$$K(\varepsilon) + iL_0\omega = H_1 + \varepsilon X \tag{21}$$

with

$$X = \frac{1}{\varepsilon}(\exp i\varepsilon\{\omega, \cdot\} - 1)\, H_1 + \frac{1}{\varepsilon^2}(\exp i\varepsilon\{\omega, \cdot\} - i\varepsilon\{\omega, \cdot\} - 1)\, H_0 .$$

We expand the various terms in this equation in powers of ε, e.g.

$$K(\varepsilon) = \sum_{n=0}^{\infty} \varepsilon^n K^{(n)}$$

and separate the successive orders. The lowest order is ($\varepsilon=0$)

$$K^{(0)} + iL_0\omega^{(0)} = H_1 .$$

This can be solved for $K^{(0)}$ and $\omega^{(0)}$ by projection with P and Q to give

$$K^{(0)} = PH_1 ; \qquad \omega^{(0)} = \frac{-i}{L_0} QH_1 .$$

Suppose now that one has determined $K^{(0)},\ldots, K^{(n-1)}$ and $\omega^{(0)},\ldots, \omega^{(n-1)}$. Then $X^{(n-1)}$ is known from (22) and the equation for the nth order is

$$K^{(n)} + iL_0\omega^{(n)} = X^{(n-1)}$$

which is solved by

$$K^{(n)} = PX^{(n-1)} ; \qquad \omega^{(n)} = \frac{-i}{L_0} QX^{(n-1)} .$$

Induction gives ω, K to any desired order. The ω, K obtained in this way are not unique but all possibilities are related in a simple way (Kummer, 1971a). We may make a standard choice by putting $P\omega=0$. (Although we did not go into this above, the same freedom exists for Prigogine's theory and has played a role in its development: (Prigogine *et al.*, 1970b; Rae, 1972a.)) These formal series for V and K are closely related to the operators introduced in our previous sections: we notice that

$$PVf(t) = P\, e^{-it(L_0 + \varepsilon L_1)} Vf(0) = e^{-it\varepsilon P\tilde{L}_1 P} PVf(0)$$

so that the component PVf evolves independently of QVf. This can be rewritten

$$V^{-1}PVf(t) = e^{-it\varepsilon V^{-1}P\tilde{L}_1 PV} \cdot V^{-1}PVf(0) \equiv \Sigma(t)\, f(0)$$

Further, the projector $\Pi = \lim_{t\to 0+} \sum(t) = V^{-1}PV$ obviously has the property $\Pi L = L\Pi$. The operators $\sum(t)$ and Π introduced in this way are identical to those of Section 3 as may be checked dircetly by comparison of the relevant series expansions in ε. (For more details on this see Rae (1972a, b).

References

Balescu, R., *et al.*; 1969, *Bull. Acad. Roy. Belg. Cl. Sci.* **55**, 1055.
George, Cl.: 1967, *Bull. Acad. Roy. Belg. Cl. Sci.* **53**, 623.
Koopman, B. O.: 1931, *Proc. Nat. Acad. Sci.* **17**, 315.
Kruskal, M.: 1962, *J. Math. Phys.* **3**, 806.
Kummer, M.: 1971a, *Nuovo Cimento* **1B**, 123.
Kummer, M.: 1971b, *J. Math. Phys.* **12**, 4.
McNamara, B. and Whiteman, K. J.: 1967, *J. Math. Phys.* **8**, 2029.
Prigogine, I.: 1962, *Non-Equilibrium Statistical Mechanics*, Interscience, New York.
Prigogine, I.: 1968, in N. J. Zabusky (ed.), *Topics in Nonlinear Physics*, Springer-Verlag, New York.
Prigogine, I., *et al.*: 1969, *Physica* **45**, 418.
Prigogine, I., *et al.*: 1970a, *Proc. Nat. Acad. Sci.* **65**, 789.
Prigogine, I., *et al.*: 1970b, *Proc. Nat. Acad. Sci.* **66**, 709.
Prigogine, I., *et al.*: 1971, *Physica* **56**, 25.
Rae, J.: 1972a, *Lettere al Nuovo Cimento* **3**, 520.
Rae, J.: 1972b, *Physica* **64**, 36.
Rae, J. and Davidson, R.: 1973, to appear in, *J. Math. Phys.*

PART V

THE SOLAR SYSTEM AND ORBITAL RESONANCES

ASTRONOMICAL SYSTEMS OF UNITS, CO-ORDINATES AND TIME-SCALES

G. A. WILKINS

Royal Greenwich Observatory, England

Abstract. The systems of units, co-ordinates and time-scales that are currently in use in dynamical astronomy are reviewed, and attention is drawn to the needs for simplification and clarification, and for explicit statements of the relationships between astronomical and SI units. It is concluded that ephemeris time, as it is currently defined, does not provide a satisfactory time scale for use in dynamical astronomy. It is recommended that the astronomical unit of time should be one day of 86400 SI seconds.

1. Introduction

1.1. Objectives

My first objective is to review the systems of units, co-ordinates and time-scales that are currently in use in dynamical astronomy, with particular emphasis on the difficulties they present. Some of these difficulties are inherent in the subject, but other arise from confusing terminology or from our failure to find a satisfactory synthesis of the systems of classical astronomy and of modern physics.

My second objective is to stimulate discussion of the possible forms of such a synthesis by presenting some conclusions that I have drawn from my analysis of the current situation. My views have been very much influenced by the ideas and comments of the members of the IAU Commission 4 Working Group on Units and Time scales, but I must emphasise that the Group has not yet agreed on any recommendations about future systems.

1.2. Restrictions

At this time I am limiting myself to a consideration of the systems that are appropriate for use in the observation and computations of the motions of bodies in the solar system. The scales of stellar systems are so different that different units might be justified; but these units should be directly linked to the astronomical and physical units discussed here. I also recognise that for the purposes of mathematical analysis and development other specialised systems are useful, but it is, I believe, desirable that the final results should normally be transformed to an appropriate conventional system. The purely theoretical treatment of the character of the motions of bodies in a hypothetical system may appear to be an exception, but even here the use of familiar units may help many readers.

Apart from a few asides, I am using the Newtonian concepts of space and time throughout. I hope that relativistic effects can be taken into account later, if necessary, by a refinement of the treatment and that they will not necessitate any radical change in approach.

B. D. Tapley and V. Szebehely (eds.), Recent Advances in Dynamical Astronomy, 273–286.

1.3. Background

At the present time there is a bewildering variety of astronomical reference systems (i.e. units, co-ordinates, and time-scales) in use. I believe that the widespread adoption of a well-founded set of internationally agreed recommendations to simplify and clarify the terminology and to reduce the number of systems in general use would be of great benefit to professional astronomers, as well as to non-specialists who wish to use or understand astronomical publications and concepts. In particular, I believe that it is important that astronomers should use SI units except in those cases where more appropriate units are a consequence of the physical nature of the situation. Even in these cases the relationships between the astronomical units and the corresponding SI units should be as explicit as possible.

In dynamical astronomy we are primarily concerned with the units of mass, length and time. For many purposes, it is customary to use the Gaussian system of astronomical units in which the unit of mass is the mass of the Sun, the unit of length is approximately the mean distance of the Earth from the Sun, and the unit of time is the day. There is no internationally-agreed statement of the definition of the astronomical unit of length, and there is considerable controversy as to whether the astronomical unit of time should be based on the ephemeris second or the atomic second. These definitions are being examined by Working Groups of IAU Commission 4 as part of a much wider examination of the desirability of making changes in the IAU system of astronomical constants (e.g. in the constant of precession) and in the forms of presentation of astronomical ephemerides.

Measures of both angle and time are expressed in a wide variety of different ways, leading to a proliferation of conversion constants and tables, but there appears to be little prospect of obtaining general agreement to recommendations for preferred usage, and these problems are not discussed further here.

The examination of the arguments for the use of the Gaussian system of astronomical units cannot be separated from the consideration of the factors that influence the adoption of a system of astronomical constants for use in the reduction of astronomical data from one reference system to another. In fact it seems to me that the converse is also true, and that any revised formulation of the IAU system must be preceded by the adoption of the definitions of any non-SI units that is uses. I also believe that the choice of both units and constants should be made after a careful study of the essential needs of current astronomy. Such a study is likely to lead to the inclusion of some new constants and the elimination of others that are no longer of value.

Most of the astronomical observations with which we are concerned have been made from the Earth's surface and are referred to axes that rotate and move with the Earth and to a time-scale that is dependent on the rotation of the Earth. On the other hand, the equations of motion take their simplest form when referred to an inertial set of axes and a uniform time-scale. Well-defined methods and coefficients for the transformation (or reduction) of co-ordinates and times from one system to another are required so that observations made at different places and times can be satisfactorily used together, and so that predictions (ephemerides) based on theories can be prepared. Even though it is now possible to make observations of radar travel times, etc., with repsect to an atomic time scale, similar reduction procedures are involved. Hence, although the main emphasis is on the choice of units for future use, it seems best to start by considering the co-ordinate systems and time-scales that are currently in use. I hope that it will then be possible to see which features are essential to a proper description and analysis of solar system dynamics, and so should be formally recommended for general use.

2. Co-ordinate Systems

2.1. At a given time and place on the Earth's surface two directions can be uniquely established: firstly, the direction of the zenith, as defined by the direction of gravity; and, secondly, the direction of the north (or south) pole as determined by the direction of the axis of rotation of the Earth. The angle between these directions is the co-latitude of the place. The plane containing these two directions is used to define the origins of azimuthal co-ordinates in the planes at right angles to these directions. Azimuth in the horizontal plane is normally measured from the north point, but hour angle in the equatorial plane is measured from the local meridian, i.e. from the south.

In the short term, the declinations of stars are constant, but owing to the rotation of the Earth the hour angles increase at a uniform rate. By measuring the azimuthal co-ordinate in the equatorial plane from an arbitrary fixed direction (whose hour angle increases at the same uniform rate) we obtain sidereal hour angle which, in the short term, is constant. This arbitrary direction is conventionally chosen to be the direction of the (vernal) equinox, defined by the line of intersection of the equator and the ecliptic (i.e. the mean plane of the Earth's orbit). The complement of sidereal hour angle is right ascension, and the hour angle of the equinox is local sidereal time. (This direction of measurement is such that the right ascensions of the Sun, Moon and planets are normally increasing functions of time.) Co-ordinates based on the ecliptic as the fundamental reference plane are also used for some purposes; the inclination of the ecliptic to the equator is known as the obliquity of the ecliptic.

In order to complete the specification of the principal relationships between these co-ordinate systems it is necessary to specify the rotational orientation of the Earth with respect to the (fixed) direction of the equinox at any instant of universal time. (In practice, the observation of this orientation is used to determine universal time.) This is done by adopting a polynomial expression for the Greenwich hour angle of the equinox (i.e. Greenwich sidereal time) as a function of universal time. This expression is such that the mean sun transits the Greenwich meridian at noon, universal time. Local sidereal time differs from Greenwich sidereal time by the longitude of the place with respect to the Greenwich meridian.

The transformations between these various angular co-ordinate systems can be easily derived from the standard formulae of spherical trigonometry. Alternatively, each pair of angles can be replaced, with some redundancy that is useful for checking purposes, by three direction cosines, and the transformation can then be made by matrix multiplication methods.

2.2. The preceding brief description of the principal angular co-ordinate systems used in dynamical astronomy ignores a number of factors that affect the directions of the reference axes and together increase very greatly the complexity of the situation. The largest effect is that of luni-solar precession of the equator which causes a motion of the equinox in space so that the equatorial co-ordinates of a fixed direction depend upon the time for which the position of the equinox is specified. In addition, the following effects must also be taken into account in work of the highest precision:

(a) The difference between the direction of gravity and the direction to the centre of the Earth.

(b) The variations in (geographical) latitude and longitude caused by polar motion, i.e. by the free nutation of the axis of figure of the Earth with respect to the axis of rotation.

(c) The forced luni-solar nutation which gives quasi-periodic variations in the obliquity and in the position of the true equinox with respect to the mean equinox, which moves steadily on account of precession.

(d) Planetary precession of the ecliptic which again causes steady changes in the obliquity and the direction of the equinox.

(e) Secular, quasi-periodic and irregular variations in the rate of rotation of the Earth which mean that hour angles do not increase at a uniform rate, so that neither sidereal time nor universal time are uniform time systems.

2.3. Apart from atmospheric refraction, there are two other fundamental factors that affect the direction in which an astronomical object is seen that are relevant to this discussion of astronomical co-ordinate systems. They are parallax and (stellar) aberration, and they depend, respectively, on the position and motion of the observer in space. In order that observations made at different places and times may more easily be used together it is therefore customary to apply corrections to the angular co-ordinates so as to change the origin of co-ordinates from one observer to the centre of the Earth and then, if appropriate, to the centre of mass of the solar system. If the distance of the object is not large in comparison with the change in the position of the origin, it is usual to allow for the parallax effects by working in rectangular co-ordinates.

The motion of the observer may be considered to consist of two components – motion with respect to the geocentre, which gives diurnal aberration, and the motion of the geocentre with respect to the centre of mass of the solar system, which gives annual aberration. The correction for the motion of a solar system object during the light-time is often combined with the correction for annual aberration, and the sum is often referred to as planetary aberration, but the physical reason for the light-time correction is quite different from that for annual aberration even though the mathematical form is similar. A further complication arises from the fact that part of the correction for annual aberration is (almost) constant for each star and so is, at present, assumed to be included in the catalogue mean place. I hope that this assumption will not be carried forward into the next fundamental catalogue since, in practice, it causes confusion and involves extra computation.

It must be emphasised that the direction of a star (or other object) at a given time may be referred to a co-ordinate system that is appropriate to a different time (or epoch). This is particularly true of co-ordinates given in observational catalogues, but in fundamental catalogues corrections for the proper motions of the stars are normally included.

2.4. Even those who are unfamiliar with the details of the usage of these various co-ordinate systems will, I hope, be convinced that extreme care is necessary in high-precision work to ensure that the basis of any tabulation of astronomical co-ordinates is clearly stated. Unfortunately, some words, such as 'apparent' and 'true' are not used consistently and incomplete phrases are often used. Furthermore, it is important that the numerical values of the quantities used in the transformations should be self-consistent and the same values should normally be used by all astronomers – any departures from the use of the conventionally adopted values should be clearly stated. Unfortunately, the IAU system of astronomical constants that was adopted in 1964 does not provide a complete specification of all the quantities concerned. In particular, the precession transformations are not determined uniquely by a single constant. Some of the expressions in common use are not rigorous and so discrepancies between different methods of calculation occur.

The introduction of new methods of observing and the increasing availability of powerful computing facilities suggest that traditional practices and forms of presentation need to be examined carefully to see where changes are desirable. For example, in the apparent ephemerides of the planets it might be desirable to tabulate round-trip travel times in light-seconds, rather than true distances in astronomical units of length. The use of day numbers for the transformation of angular co-ordinates has been superseded in some establishments by the use of rigorous matrix methods even though this entails conversions between the angular co-ordinates and direction cosines, and the calculation of rotation matrices from basic formulae; this technique eliminates the need to form mean places for the beginning of the year but requires a knowledge of the position and velocity of the Earth with respect to a standard reference frame. Proposals for changes from traditional systems and practices must be subjected to detailed scrutiny before being adopted for general use.

2.5. So far, we have been primarily concerned with angular co-ordinates but it is relevant to introduce here some aspects of the question of the choice of units of measurement of distance. From geometrical considerations Kepler was able to deduce the relative sizes of the planetary orbits and hence to show that the cubes of the sizes are proportional to the squares of the periods of the orbital motions. This relationship was later justified, and refined, by the use of Newton's theory of gravitation. Since the periods could be measured accurately (in terms of the day or year) it was possible to determine relative distances to a correspondingly high precision, but the determination of the distances in terrestrial units was limited by the accuracy with which the small parallaxes of the planets could be measured. It may be noted that the measurement of the constant of aberration coupled with a measurement of the speed of light in terrestrial units, can also be used to determine the distance scale, but the smallness of the angular aberrational displacements severely limits the precision of the result. By radar techniques it is now possible to determine distances directly in light-seconds, and so once again we find that distance determinations depend upon measurements of time intervals and that the value of the speed of light is an important constant for astronomical purposes. This will become an absolute constant in SI units if the SI units of length and time are re-defined by adopting arbitrary values for the wavelength and frequency of the same monochromatic radiation.

3. Time Scales

3.1. In dynamical astronomy we require to use three quite distinct time scales: a rotational time scale, such as universal time and sidereal time, is required to specify the rotational orientation of the Earth and to identify the instants of observation when no other scale is available; the use of a dynamical time scale, such as ephemeris time, is implied in every theoretical or numerical development and in current attempts to establish a uniform time scale for events before 1956; an atomic time scale, such as international atomic time (IAT) provides a precise, uniform, time scale for the identification of events since 1956 and for comparison with dynamical time scales. Our main concern here is whether the recommended dynamical time scale should continue to be ephemeris time as now defined, or whether a different scale that is more closely compatible with IAT should be introduced. The arguments concerning this question cannot be understood without an appreciation of the general nature of time scales and of the particular systems that are currently in use.

For astronomical purposes we are primarily interested in determining the times at which events occur rather than in measuring, as in laboratory physics, the intervals of time required for the completion of particular processes. (The 'time' of an event is to be understood to include the identification of the day as well as the time of day of its occurrence. The term 'epoch' is sometimes used in this sense but it is also used with other meanings and so it seems preferable to continue to use the word time, but to qualify it whenever the context does not make its meaning clear.) It is therefore not sufficient merely to define a unit of time; it is necessary also to specify the procedure that should be adopted in order that a time may be assigned to an observed event. These procedures depend on both the fundamental principle of the time scale and on the choice of 'clock' that is used to establish, or 'realize', the time scale.

In the absence of any absolute standards we can judge the likely quality of a time scale against the following criteria:

(a) Simplicity in principle – the fundamental principle of the time scale should be clear and free from ambiguity.

(b) Simplicity in definition – the parameters that specify the initial setting and rate of the clock should be defined clearly and directly.

(c) Simplicity in operation – the procedures for realizing the scale, i.e. for maintaining and for 'reading' the clock, should be as simple as possible.

(d) Continuity – the clock should run continuously over long periods without any significant risk of loss of scale.

(e) Accessibility – the reading of the clock should be readily available at any time.

(f) Comparability (or precision) – independent readings of the clock should agree well with each other. (This property is also referred to as stability.)

(g) Uniformity – this is the most important criterion, and is assessed by the absence of evidence of non-uniformity, which may arise either from an inappropriate choice of fundamental principle or from imperfections in the clock, and which is only fully apparent when a better time scale is available for comparison. Non-uniformity of a scale may otherwise be suspected if independent processes timed against the scale show greater variability than would be expected.

3.2. For general purposes we require a time scale that corresponds to the alternation of day and night; such a time scale therefore depends on both the rotation of the Earth and the revolution of the Earth around the Sun. Since the rate of rotation of the Earth is unpredictably variable, and since the theory of the motion of the Earth around the Sun can only be satisfactorily developed in terms of a uniform dynamical system of time, it is necessary to introduce the concept of a fictitious mean sun whose motion in right ascension can be treated as a known function of the time scale that is to be determined. It is thus possible to define universal time (UT) so that it is numerically related to Greenwich mean sidereal time (GMST) by a conventional formula. GMST (i.e. the Greenwich hour angle of the mean equinox) is itself calculated from the local observed sidereal time, which is traditionally deduced from observations of the transits of stars.

In order that UT should be free from irregularities due to polar motion small corrections have to be applied during the reduction of the observational data; at present these corrections are applied in UT, thus making necessary a distinction between UT 0 and the more uniform UT 1, but it would be more logical and less confusing to the non-specialist to apply the correction in the reduction from local to Greenwich mean sidereal time. The actual correction depends on the adopted definition of the mean pole and, unfortunately, full international agreement on the definition to be used by the national time services has not yet been obtained.

Corrections for the predicted seasonal variations in the rate of rotation of the Earth can be applied to UT 1 in order to give a more uniform time scale known as UT 2. The unit of even this scale varies both irregularly and secularly and so it is no longer used as the basis of time signal emissions. It would therefore be possible, and in my view desirable, to discontinue the recognition of the UT 0 and UT 2 scales, so that the term universal time and the notation UT refer unambiguously to the scale now denoted by UT 1.

3.3. The SI second has been defined in such a way that it is equal to the ephemeris second to within the uncertainty of the determination of the latter over a few years. It is shorter than the current value of the universal second by about 1 part in 4×10^7, so that an atomic time scale, based on the counting of SI seconds, will gain on the universal time scale by about 1 s per year. In order that time signal emissions may be based on the nominally constant value of the SI second and yet may provide universal time to an accuracy that is appropriate for navigation and surveying, etc., there was international agreement to the introduction of a compromise time scale, known as coordinated universal time (UTC), as from 1972 January 1.0. This scale differs from international atomic time (IAT) by an exact number of seconds but step adjustments of 1 s each (known as leap seconds) are made from time to time to keep UTC within 0.7 s of UT 1. This compromise system appears to be as satisfactory as could be expected, and I know of no good reason to change it. There is, however, a confusion in terminology as the name Greenwich mean time (GMT) is used for both UTC and UT 1. (GMT is occasionally still used in the sense of Greenwich mean solar time reckoned from noon, rather than midnight, at Greenwich, as was the practice before 1925. It is also occasionally used for the scale defined by the hour angle of actual mean sun as distinct from the hour angle of fictitious mean sun that is used in the definition of UT.) There seems, however, no doubt that the term GMT should not be used in astronomy for the precisely defined UT 1 system and that it will continue to be widely used for the basic scale of the time signals (i.e. for UTC) in ordinary civil life.

It is clear that UT does not provide a suitable scale for theoretical use in dynamical astronomy, although, of course, UT is required in connection with the recording and reduction of observational data. Its non-uniformity is only significant over very long periods or in connection with measurements of high precision.

3.4. The international atomic scale (IAT) is based on the weighted mean of the atomic time scales of a number of independent time services. The fundamental epoch of the scale has been defined so that IAT and UT 2 were coincident on 1958 January 1 at 0 h, but differences between IAT and UT are now on record as far back as 1956. The existence of systematic differences between IAT and each of the independent scales from which it is derived shows that IAT may drift from the ideal atomic time scale implied by the counting of SI seconds. Furthermore, it is realized that the atomic transition frequencies depend on the strength of the gravitational field, and so the scale generated by a single clock depends on its location and will vary as the Earth rotates and moves around the Sun. However, it must be emphasised that the accumulated effect of these practical and relativistic departures from non-uniformity is extremely small in comparison with those of the rotational and dynamical time scales of astronomy. Further improvements in precision are likely and, barring a world catastrophe, the future continuity of the scale seems assured. Some tightening up of the scale may be introduced, but such changes should make the scale more uniform and precise. Nothing, however, can make an atomic time scale available for events before 1956. Although it is possible to adjust a dynamical time scale so that the unit is nominally equal to the SI second, it would be misleading to refer to such a scale as an atomic time scale. Thus, even if we were to adopt IAT as the future reference scale for dynamical astronomy, it will still be necessary to define a procedure for realizing a dynamical time scale that can be used for events before 1956. It is, however, appropriate to examine the present definition of ephemeris time and to decide whether the scale so defined is suitable or whether some new definition of a dynamical time scale is to be preferred.

3.5. Ephemeris time (ET) has been described in the following terms – "Ephemeris time is a uniform measure of time depending for its determination on the laws of dynamics. It is the independent variable in the gravitational theories of the Sun, Moon, and planets, and the argument for the fundamental ephemerides in the Ephemeris." (*Explanatory Supplement to the Astronomical Ephemeris*, London, 1961, p. 69). The formal definitions are, however, much more prosaic:

"Ephemeris time is reckoned from the instant, near the beginning of the calendar year AD 1900, when the geometric mean longitude of the Sun was 279° 41′ 48.″04, at which instant the measure of ephemeris time was 1900 January 0d 12h precisely."
"The (ephemeris) second is the fraction 1/31 556 925.9747 of the tropical year for 1900 January 0 at 12 h ephemeris time."

These definitions are based on the expression for the mean longitude of the Sun given by Newcomb in his *Tables of the... Sun*; they replaced an earlier definition for the quantity ΔT, the difference between ET and UT, in terms of the departure (B) of the value of the mean longitude of the Moon (as deduced from observations) from that calculated from Brown's *Tables of the... Moon*. Subsequently, Brown's theory was amended with the intention of reducing it to a gravitational theory in which the measure of time is ephemeris time; ephemerides based on this amended theory (with some subsequent improvements) have been published in *The Astronomical Ephemeris* since 1960, and have been used in the determination of ephemeris time.

Thus, we see that the practical realization of ephemeris time is several stages removed from the basic definition of the ideal scale – there are three principal assumptions involved:

(a) that ephemerides based on Newcomb's theory of the Sun are consistent with the fundamental definitions;

(b) that the determination from observations of the Sun, Moon and planets of the corrections necessary to make the time scale of Brown's theory of the Moon identical with ephemeris time has been made accurately;

(c) that the periodic perturbations in Brown's theory have been correctly determined.

None of these assumptions is fully satisfied, and in particular there is considerable doubt about the second since it now appears that Spencer Jones' determination of the effects of tidal friction on the Moon's orbital motion may be significantly in error. If this is true, then ephemeris time, as determined from observations of the Moon, is not uniform, and the currently tabulated values of ΔT give a false impression of the variations in the rate of rotation of the Earth.

The definition of the unit of ephemeris time was expressed in terms of the tropical year, rather than the sidereal year, in order that the observational determination of the unit should not be affected by the uncertainties in our knowledge of the constant of precession. On the other hand, the reduction from observed values of the longitude of the Sun to geometric mean longitudes does involve the value of the constant of aberration so that, in principle, the change in the adopted value of this constant introduces a discontinuity into the determination of ephemeris time. It is clear that the present definitions of the unit and epoch of ephemeris time are not simple in principle, but involve complex concepts, such as 'geometric mean longitude' and 'tropical year', that can only be fully defined in terms of particular procedures for the analysis of observations.

A further criticism that has been made of the present realization of ephemeris time is that the zero points of the star catalogues used in deriving the observed co-ordinates of the Sun and Moon do not coincide with the equinox defined by the equator and the ecliptic, and that, moreover, the differences are not constant. It is, however, not clear to me whether these effects can be taken into account by simple changes in the reduction procedures or whether more fundamental amendments are required.

3.6. The conclusion that I draw from the arguments of the preceding section is that ephemeris time as it is currently defined and determined does not meet the criteria listed above, except that of continuity, and so does not provide a satisfactory time scale for dynamical astronomy. I believe, however, that it is necessary to look for an alternative set of definitions for a standard dynamical time scale. The dynamical time scale would only be used for long-term investigations, such as for setting up a uniform time scale for the period before 1956, since it is evident that IAT is much more suitable for short-term use. (IAT is already used for orbital computations for spacecraft and in the tabulation of observational data.) By matching the dynamical scale to the current IAT scale, it would be possible to use IAT directly for extrapolation, thus overcoming the objection that a dynamical scale is relatively inaccessible. (This is, in fact, already done for ET.) The definition of the unit and fundamental epoch must be simple in principle and should not be based on a particular theory of one astronomical body. It will have to be recognized, however, that the implementation of the definition will, of necessity, involve a series of successive approximations as new observations are made and better theories are evolved. I can see no reason why artificial bodies, as well as natural bodies, should not be used, nor why the concept of dynamical time should be limited to the Newtonian theory of gravitation. The possibility of allowing for the variability of the constant of gravitation should be considered if this appears to be the only way by which a common time scale can be used for both dynamics and atomic physics.

It may seem that I am being inconsistent in attempting to maintain a distinction between dynamical time and atomic time while at the same time suggesting that the current definition of ephemeris time should be dropped in favour of a new definition that reduces the practical differences between the scales. I believe, however, that the formal distinction is necessary since the actual time scale of a new dynamical theory of the motion of a body can only be established in retrospect. This fact would be concealed if ephemerides were to claim to be on the scale of IAT. But it is problematical whether it is better to use a new name for a modified version of an old concept or to continue to use the old name and risk the confusion that this may cause.

4. The System of Astronomical Units

4.1. The astronomical units of length, mass and time are based on the mean radius of the Earth's orbit, the mass of the Sun and the period of rotation of the Earth on its axis. (The term unit of 'length' will be used here to match the usage in the SI system, although the term unit of 'distance' is normally used in astronomy.) They are such that for most applications in dynamical astronomy the quantities concerned can be conveniently expressed in these units without the use of high or low multiples of the basic units. More significantly, this system took advantage of the fact that the relative distances and masses of the bodies of the solar system could be determined with very much higher precision than could their values in terms of arbitrary terrestrial units, such as the metre and the kilogram. At the time that the system first came into use the terrestrial unit of time was also determined astronomically from the apparent diurnal motions of the Sun and stars. It is, however, now possible to measure some astronomical distances to a high precision in terms of the distance travelled by electromagnetic radiation in unit time (for example, in terms of the light-second) and intervals of time can be measured with very high precision using atomic standards.

The advantages of using a single coherent system of units for science and technology are being increasingly recognised, and SI units are now replacing many sets of units previously used in specialised fields. It is therefore an appropriate time to review the system of astronomical units and to decide whether it is desirable to retain it as it stands, to modify it to make it appropriate to modern conditions, or to withdraw IAU recognition from it. (Any individual astronomer would, of course, be at liberty to continue to use the set of units that he considers most appropriate to his work.) It is my opinion that the second course of action should be followed. In particular, I suggest that the astronomical unit of time should be based on the SI second, and that the statement of the IAU system of constants should show clearly the relationships between the astronomical and SI units.

4.2. The Gaussian System of Astronomical Units

The currently adopted form of the system of astronomical units is referred to here as the 'Gaussian' system, since it depends on the adoption of a fixed value, originally determined by Gauss in 1809, for the constant of gravitation appropriate to the units of the system. It is this special feature that justifies the term 'system' and the continued use of these units in dynamical astronomy.

The basis of the Gaussian system is most easily seen by considering an isolated system of two particles of mass M and m revolving around their centre of mass. The equations of motion for the two particles about the centre of mass can be combined to give the following equation for the motion of m relative to M

$$\frac{d^2\mathbf{r}}{dt^2} = -\frac{k^2(M+m)}{r^3}\mathbf{r},$$

where $\mathbf{r}$ is the vector from M to m, and k^2 corresponds to the constant normally known as the constant of gravitation and often denoted by G. The relative orbit of the particles is an ellipse and the period of revolution (P) is given by

$$\left(\frac{2\pi}{P}\right)^2 a^3 = k^2(M+m),$$

where a is the semi-major axis of the relative ellipse. In astronomical terminology a is known as the mean distance and $2\pi/P$ is known as the mean motion (in radians per unit time) and is usually denoted by n. Now Gauss attempted to determine the actual value of k from the motion of the Earth (or more precisely the centre of mass of the Earth and Moon) moving around the Sun when the unit of mass is the mass of the Sun (i.e. $M=1$), the unit of length is the same as the mean distance of the Earth's orbit (i.e. $a=1$), and the unit of time is the mean solar day. This required an estimate of the mass of the Earth and the Moon as well as of the length of the sidereal year. The particular numerical value ($k=0.017\,202\,098\,950$) that he obtained is very widely used and so even when it was realized that Gauss' estimates were significantly in error it was decided (*Trans. IAU* **6**, 20, 336, 357, 1939) that the value of k should be treated as an absolute constant, even though this implies that the mean distance of the Earth from the Sun is no longer exactly equal to the unit of length. In effect, the unit of length became the major semi-axis of the unperturbed orbit of a particle of negligible mass moving around the Sun with a mean motion of k radians per unit time, but, in practice, this does not provide a useful definition of the unit.

The adoption of a fixed value for k means that the numerical development of theories and of ephemerides can be carried out once and for all to a high precision in terms of these astronomical units. For some purposes these units suffice, but there is an increasing need to express astronomical distances and time-intervals in SI-based units, although even now the relative masses of the Sun and planets can be determined much more precisely than can the individual masses in, say, kilograms since the value of G in SI units is only known to a low precision. If a Gaussian system is used for the basic theories and ephemerides it is a straightforward matter to convert from astronomical units to more appropriate units for particular applications, choosing the conversion factors to best meet the users' requirements. On the other hand, if a Gaussian system is not used, it may still be necessary to apply correction factors, even though an ephemeris may nominally be expressed in SI units, in order to obtain the highest precision, and there will be the danger that this fact may be overlooked or the factors chosen incorrectly as the true basis of the ephemeris may not be obvious. This disadvantage applies also to any ephemeris prepared before the adoption of Gauss' value for k, and would also apply if the astronomical constant of gravitation were again to be treated as a quantity to be determined from observation.

Even if these arguments for the retention of a Gaussian system of units are accepted, it is still necessary to adopt a new unit of time in place of the mean solar day, which is now known to be unpredictably variable. There are two obvious possibilities – either the day of the ephemeris time (ET) scale can be used (as has been assumed in the 1964 IAU system of astronomical constants) or a day of 86400 SI seconds can be used. The latter has several advantages over the former: for example:

(a) it eliminates the confusion that would arise from the use of two distinct but nearly equal units, especially in such quantities as the speed of light;

(b) the conversion between astronomical units and SI units will only require the determination and application of two, and not three, conversion factors;

(c) it avoids the practical uncertainties that would be associated with the use of the ephemeris second, in terms of which time intervals can only be measured indirectly in arrears to a much lower precision then is now required for some applications.

If the astronomical unit of time is defined to be a day of 86400 SI seconds, it will, of course, necessarily imply the use of a scale of dynamical time that differs from ephemeris time as now defined in terms of the motion of the Sun. However, in the same way that when ephemeris time was introduced it was possible to assume that the unit of time used in theories previously developed was the ephemeris day, so it will be possible to use existing theories and ephemerides as if the unit of time is a day of 86400 SI seconds. The mean motion of the Sun will become once again a quantity to be determined from observations, but unless it eventually proves to be impossible to reconcile the atomic and dynamical time scales no other change will be necessary.

4.3. IAU SYSTEM

Since the values in the IAU system of astronomical constants depend on the choice of units, it is essential that any non-SI units used should be clearly defined in the statement of the system. The statement of these definitions and of the relevant constants could be as follows:

(1) The astronomical unit of mass (aum) = mass of the Sun.

(2) The astronomical unit of time (day) = 86400 SI seconds.

In what follows the terms day and second will be used without qualification.

(3) Gaussian gravitational constant, $k=0.017202098950$, k^2 has the dimensions of the (newtonian) constant of gravitation G, i.e. $L^3\ M^{-1}\ T^{-2}$.

(4) The astronomical unit of length (aul) is that length for which k takes the value specified in (3) when the units of measurement are the astronomical units of length, mass and time. (It is almost the mean distance of the Earth from the Sun.)

(5) Measure of 1 aul in metres $A = 149.600 \times 10^9$

(6) Speed of light in metres per second $c = 299.792500 \times 10^6$

(7) Light-time for unit distance (1 aul) $\tau_A = 499^{s}.012$
$= 0^{d}.0057756$
(8) Speed of light in aul per day $c = 173.142$
(9) Constant of gravitation in SI units $G = 6.670 \times 10^{-11}$
(10) Measure of 1 aum in kilograms $S = 1.990 \times 10^{30}$.

Many details of the presentation of a revised system require further examination. The symbols 'aul' and 'aum' are only intended for temporary use; it is hoped that better proposals will be forthcoming, but is must be admitted that it will be difficult to find new symbols that will be apposite and yet will not clash with other symbols of the SI system. The values given are merely those of the 1964 system; it seems likely that a new value of the speed of light will be adopted soon for international use and so consideration will have to be given to the question of whether the values of this and other constants should also be changed. It is arguable too that τ_A, rather than A, should be treated as exact. The constants 9 and 10 were not given in the 1964 system, but it is desirable to add them in order to complete the statement of the relationships between astronomical and SI units. The value of constant 10, for example, is required (to low precision) for some astronomical purposes such as in estimating mean densities of planets. Some other constants may be omitted but, for the reasons given in Section 2, I believe that other constants should be added so that the relationships between the various astronomical co-ordinate systems are clearly defined.

5. Conclusions

In preparing this material I have frequently found myself questioning astronomical terminology and procedures that, although they are normally correctly used, do not fit into any logical pattern of development and that must therefore be very confusing to those who are not familiar with them. I have not attempted to provide a full set of references for further reading. The *Explanatory Supplement to the Astronomical Ephemeris* (London, Her Majesty's Stationery Office, 1961) provides much detail and references, but unfortunately it often illustrates only too well the difficulties of present practices. I hope that the changes that are now being considered will lessen, rather than increase, the number of inconsistencies between the various aspects of these complex topics.

MASSES OF THE PLANETS AND SATELLITES

R. L. DUNCOMBE

U.S. Naval Observatory, Washington, D.C., U.S.A.

Abstract. This paper reviews work by P. K. Seidelmann, W. J. Klepczynski, and the author on the general problem of the determination of the masses of the principal planets by systematic comparison of observations with theory. It also presents a complete compilation of the published values of planetary and satellite mass determinations (see Tables I and II). A provisional discussion of the values of the masses of the principal planets was presented at IAU Colloquium No. 9, Heidelberg, August 1970 (Klepczynski *et al.*, 1971).

TABLE I

Derived values of the masses of the principal planets

	Newcomb (1898) (currently adopted)	Clemence (1964)	Kulikov (1965)	Duncombe *et al.* (1973)
Mercury	6000000	6110000 ± 60000	6127000 ± 60000	5972000 ± 45000
Venus	408000	408539 ± 18	408120 ± 190	408520 ± 9
Earth-Moon	329390	328906 ± 9	328546 ± 37	328900.12 ± 0.20
Mars	3093500	3050000	3087000 ± 4200	3098709 ± 9
Jupiter	1047.355[a]	1047.41 ± 0.03	1047.394 ± 0.027	1047.357 ± 0.005
Saturn	3501.6	3499.6 ± 0.6	3498.85 ± 0.3	3498.1 ± 0.4
Uranus	22869	22930 ± 9	22929 ± 9	22759 ± 87
Neptune	19314	19070 ± 31		19332 ± 27
Pluto	360000[a]	400000 ± 60000		3000000 ± 500000

[a] From Eckert *et al.* (1951)

The present paper comprises a critical discussion of all previous determinations of the planetary and satellite masses and their use in the formation of a final set which may be adopted as the basis for new planetary and satellite ephemerides (Duncombe *et al.*, 1973). The following tables summarize the derived values of the planetary and satellite masses.

TABLE II

Satellite mass determinations

Planet	Satellite	Mass	Mean error
Earth			
	Moon	E/M = 81.3033 Satellite/Planet	± 0.0001
Mars			
	Phobos	2.7×10^{-8}	
	Deimos	4.8×10^{-9}	

B. D. Tapley and V. Szebehely (eds.), Recent Advances in Dynamical Astronomy, 287–288.

Table II (Continued)

Planet	Satellit	Mass	Mean error
Jupiter			
	V	$18. \times 10^{-10}$	
	I Io	4.153×10^{-5}	$\pm 0.343 \times 10^{-5}$
	II Europa	2.508×10^{-5}	$\pm 0.049 \times 10^{-5}$
	III Ganymede	8.079×10^{-5}	$\pm 0.011 \times 10^{-5}$
	IV Callisto	4.797×10^{-5}	$\pm 0.417 \times 10^{-5}$
	VI	8.5×10^{-10}	
	VII	0.35×10^{-10}	
	X	0.010×10^{-10}	
	XII	0.007×10^{-10}	
	XI	0.020×10^{-10}	
	VIII	0.077×10^{-10}	
	IX	0.015×10^{-10}	
Saturn	I Mimas	6.59×10^{-8}	$\pm 0.15 \times 10^{-8}$
	II Enceladus	1.48×10^{-7}	$\pm 0.61 \times 10^{-7}$
	III Tethys	1.095×10^{-6}	$\pm 0.22 \times 10^{-6}$
	IV Dione	2.39×10^{-6}	$\pm 0.053 \times 10^{-6}$
	V Rhea	3.2×10^{-6}	$\pm 5.6 \times 10^{-6}$
	VI Titan	2.4619×10^{-4}	$\pm 0.0029 \times 10^{-4}$
	VII Hyperion	2×10^{-7}	
	VIII Iapetus	3.94×10^{-6}	$\pm 1.93 \times 10^{-6}$
	IX Phoebe		
	RINGS	1/23269	
Uranus			
	V Miranda	1×10^{-6}	
	I Ariel	15×10^{-6}	
	II Umbriel	6×10^{-6}	
	III Titania	50×10^{-6}	
	IV Oberon	29×10^{-6}	
Neptune			
	I Triton	3.3×10^{-3}	$\pm 1.96 \times 10^{-3}$
	II Nereid		

References

Duncombe, R. L., Klepczynski, W. J., and Seidelmann, P. K.: 1973, *Fundamentals Cosmic Physics*, in press.

Eckert, W. J., Brouwer, D., and Clemence, G. M.: 1951, *Astron. Pap. Am. Ephemeris* **12**.

Klepczynski, W. J., Seidelmann, P. K., and Duncombe, R. L.: 1971, *Celes. Mech.* **4**, 253.

PLANETARY EPHEMERIDES

R. L. DUNCOMBE
U.S. Naval Observatory, Washington, D.C., U.S.A.

Abstract. The historical development of planetary ephemerides is reviewed and newly determined elements for the principal planets presented. An hypothesis regarding an exterior unknown planet is examined.

In discussing planetary ephemerides, and their underlying theories of planetary motion, we must consider them with respect to their role in fundamental astronomy. The object of fundamental astronomy is to determine the position and motion of the fundamental reference system and the apparent position of objects with respect to this reference system. To do this, observations are made principally with specialized instruments known as transit instruments and vertical circles or, in combination, as meridian transit circles. More recently these instruments have been augmented by the photographic zenith tube and the astrolabe. From the observations of the stars we derive the position of the equatorial plane and the instantaneous axis of rotation of the Earth. From the observations of the Sun, Moon, and planets, we derive the position of the ecliptic plane and the vernal equinox. The stellar observations provide the positions of the brighter system of stars at a particular epoch.

It requires five to ten years to observe a catalog of stars and several years more to complete the reduction and discussion. The planetary observations continue with each catalog. Final reductions of the planetary observations furnish corrections to the tabular positions of the planets and to the stars adopted for comparison. The comparison of these independent catalogs of stars taken at widely different epochs furnish the proper motions of the stars and the combination of these catalogs forms a fundamental catalog, such as the present FK4. The first use of a planetary ephemeris, therefore, is to enable the observer to predict the position of the object and later to serve as a standard of comparison for the observations.

With the exception of the last 15 yr, man's knowledge of the motions of the bodies of the solar system has been deduced from observations in two dimensions only. With the advent of radar and laser techniques, it has been possible for him to directly observe the third dimension. Astronomers in the past, therefore, have been constrained to infer the planetary orbits from observations of the projected motion of these objects on the celestial sphere. Since the motion of a principal planet is confined to within a few degrees of the plane of the ecliptic, the description of its position at any instant can be conveniently indicated by its longitude measured along the ecliptic from some fiducial point; by its latitude, that is, its distance above or below the ecliptic plane; and by its distance from the center of mass or radius vector. Of these position coordinates, however, only two – longitude and latitude – could actually be measured by the astronomer as he observed the position of the body projected against the celestial sphere. The radius vector was derived from the planets observed period by

B. D. Tapley and V. Szebehely (eds.), Recent Advances in Dynamical Astronomy, 289–308.

Kepler's Third Law. In developing planetary tables, therefore, primary emphasis was placed on representation of the longitude, since this was the most rapidly changing angle. Second in importance came the representation of the latitude. Of least importance was the expression representing the radius vector.

The early attempts to represent planetary motions were entirely empirical culminating in Kepler's brilliant deduction of his three empirical laws of planetary motion from the extensive series of observations made by Tycho Brahe. It was not until the introduction of Newton's law of gravitation and the development of celestial mechanics that it became possible to derive general theories of planetary motion based on dynamical principles.

Although tables of Sun, Moon, and planet positions had appeared sporadically the *Connaissance de Temps* which began publication in 1679 was, I think, the first systematic production of an ephemeris on a regular sustained basis, followed by the *British Nautical Almanac* in 1767. Quoting from the first edition written by Neville Maskelyne, the *Astronomer Royal*, "The Commissioners of longitude, in pursuance of the powers vested in them by a late act of Parliament, present the public with the Nautical Almanac and Astronomical Ephemeris for the year 1767 to be continued annually; a work which must greatly contribute to the improvement of astronomy, geography, and navigation."

Leverier's theories of the motions of Mercury, Venus, Earth and Mars epoch 1850 formed the first systematic application of dynamical principles to the motions of these bodies. These theories suffered from two defects however. They were not based on a consistent set of planetary masses and Leverier using only Newtonian mechanics was unable to account for the secular motions of the perihelia. Therefore, he augmented his dynamical theories by nondynamical terms in order to agree with observations. Leverier's theories remained as the basis for the ephemerides of Mercury, Venus, Earth and Mars in the *Connaissance de Temps* until 1960. However, they were superseded in other national ephemerides in 1900 by the planetary theories of Simon Newcomb. Newcomb's theories of Mercury, Venus, Earth and Mars were based on dynamical principles and incorporate a uniform system of planetary masses in all four theories. Using Newtonian mechanics, he was unable to represent the observed motions of the perihelia, and so after a lengthy discussion was forced to assume that gravitation toward the Sun is not exactly as the inverse square of the distance. That is, in the expression for the gravitation between two bodies of masses m and m' and at distance r, $f=mm'/r^n$, the exponent n of r is not exactly 2 but $2+\delta$, δ being a very small fraction. On this hypothesis, the perihelion of each planet will have a direct motion found by multiplying its mean motion by $\frac{1}{2}\delta$.

In the theories of the four inner planets, Newcomb used the value of $n=2.0000001612$ and augmented the secular motion of the perihelia of all four planets by the corresponding contributions.

Thus far we have not distinguished between planetary theories and planetary tables. Before the introduction of modern computing machinery it was an almost impossible task to evaluate a position directly from the theory of the four inner planets. The theory of the Earth, for instance, is comprised of mean elements, their secular variations, and approximately 310 periodic terms representing the planetary and lunar perturbations and the nutation. The theory of Venus is comprised of mean elements and approximately 200 periodic terms. To facilitate deriving a position of one of the four inner planets, tables based on the theories were formed. These tables in the case of the Sun are only 38 in number and in the case of Venus only 27. The first few tables give arguments as a function of the time which are then used to enter the remaining tables to derive a position of the planet. It is these tables, based on the theories of Simon Newcomb, that presently form the basis for the published ephemerides of the four inner planets.

Although these tables are formed with great care, particularly the ones representing the longitudes, they do represent a slight dilution of the accuracy inherent in the planetary theories. As an example, a harmonic synthesis of Newcomb's theory of the Sun differs, at maximum, from an evaluation of his tables by amounts of 0.″13 in longitude and 0.″05 in latitude and -11×10^{-7} AU in radius vector. For Venus $\Delta\lambda=0.''09$, $\Delta\beta=0.''03$, $\Delta rv=-1\times10^{-7}$ AU. In Δrv the deviation is always in the sense that the tables give a smaller value than the theory.

Planetary theories of the type developed by Newcomb are termed general perturbation theories. With the introduction of modern computing equipment it has become possible to generate special perturbation theories, that is, to numerically integrate the equations of motion. This method lends itself readily to machine evaluation because of the repetitive nature of the computations. With the introduction of larger and faster electronic calculators, it has become the most widely used method of generating planetary ephemerides. A drawback of the special perturbation method is that it yields little information on the internal structure of the particular problem, since it produces only the perturbed position and velocity vectors of the planet. With the general perturbation theory, however, the effects of the individual perturbing bodies are readily evident and the method enjoys the additional advantage that small adjustments can be made to the mean elements of the theory without having to restructure the entire theory.

I think you can gather from my remarks that from the dynamicist's point of view, although we use the method of special perturbations, we prefer to form ultimately general perturbation theories for the motions of the planets.

The test of any planetary theory is its ability to represent observations. Soon after the introduction of Newcomb's tables of Mars into the national ephemerides, it was noticed that they failed to represent observations of the planet by amounts great enough to indicate an error in the tables. It was concluded by F. E. Ross in 1912 that the eccentricity of the orbit adopted by Newcomb was in error by about two thirds of a second of arc; the consequent error in the geocentric place being nearly 4″ at a close opposition. Ross compared observations of Mars with Newcomb's tables and obtained new values of the elements and also observational values of their secular variations. He calculated corrections to Newcomb's tables and these corrections have since been incorporated in the ephemeris of the planet. In this manner the ephemeris has been brought into closer agreement with current observations but the introduction of the observed secular variations has robbed the ephemeris of any claim it might have to consistency with the laws of gravitation.

In 1950 Eckert *et al.* produced a simultaneous numerical integration of the orbits of the five outer planets to replace Newcomb's and Hill's tables, which, particularly in the case of Saturn, were becoming degraded in accuracy. In 1960 these special perturbation theories were introduced as the basis for the ephemerides of the outer planets printed in the American Ephemeris.

Since we make our observations from a moving platform in space, any error in our own position is reflected as an error in the position of the observed object. It is necessary, therefore, to analyze observations over a long period of time in order to separate deficiencies in the theory of the Earth, from deficiencies in the theory of the observed planet. Although Newcomb's tables of Mercury, Venus, and Earth have proven more adequate than this tables of Mars, certain deficiencies are known to exist. The secular variations incorporated into these theories were derived within the framework of Newtonian mechanics. The observed excess motion of the perihelion of Mercury, which could not be accounted for by any admissible change in the adopted planetary masses, made it necessary for Newcomb, after an extensive investigation, to augment the theoretical motions of the perihelia of the four inner planets by non-gravitational terms.

Since the introduction in 1900 of Newcomb's planetary tables as the basis for the national ephemerides, a large number of precise observations of these planets have been accumulated. The fundamental star systems to which these observations are referred are systematically more accurate than those used earlier. The advent of the theory of relativity with an explanation of the observed excess motions of the perihelia and our increased knowledge of the fluctuations in rate of rotation of the Earth have removed two of the obstacles Newcomb encountered in his investigations. Utilizing this new material, several investigations for the improvement of Newcomb's tables have been made. G. M. Clemence compared Newcomb's tables of Mercury with observations extending from 1765 to 1937 and derived improved values of the constants of the theory. (Clemence, 1943). Newcomb's tables of the Earth have been compared with observations over an extended period by Morgan and Scott and improved values of the elements derived (Morgan, 1933). Newcomb's tables of Venus have been compared with observations from 1750 to 1949 by Duncombe and improved values of the elements determined (Duncombe, 1958). Also, a numerical integration of the equations of the motion of the Earth by Paul Herget extending from 1920 to 2000 was compared with Newcomb's tables by Clemence and Duncombe (1957).

The small adjustments to the elliptic elements reflect the introduction of the precise modern observations, while the changes to the secular variations of the elements stem principally from improvements to the system of planetary masses used by Newcomb and to the use of a revised value of the precession. The largest discrepancy with observations in the theory of the Earth seems to be in the secular change of the obliquity. This difference was at first thought to arise from a deficiency in Newcomb's theory but more recently has been attributed to a systematic declination error in the 19th century star catalogs to which the Sun and planet observations were referenced. The comparison of the numerical integration with Newcomb's tables of the Earth indicates a further discrepancy in the latitude residuals. These residuals are due in part to Newcomb's omission of a periodic perturbation by Saturn having a coefficient of about 0".03. It is interesting to note, however, that Leverrier's expression includes this perturbation and that therefore, his expression for the latitude of the Earth is more accurate than Newcomb's.

Since we know of these corrections to the fundamental planetary theories, why haven't the ephemerides been changed? Following the adoption of the system of astronomical constants and theories of the Sun and planets early in 1900 for use in the national ephemerides, the Nautical Almanac Offices were hesitant to introduce changes. It was felt that no changes should be introduced so long as an annual astronomical ephemeris fulfilled the requirements (a) that the difference between the observed and computed position of the planet was so small that its square could be neglected and (b) that the ephemeris was founded on theory that is consistent with the accepted laws of gravitation.

Observations at different epochs can be compared with each other only if the ephemeris is homogeneous. I can state from personal experience that the comparisons of the observations of Mercury and Venus with theory were greatly facilitated by the long time span over which Newcomb's tables have formed a basis for the printed ephemerides.

Newcomb's tables of the four inner planets have been in use now for over 70 yr. For Mercury and Venus they do fulfill requirement (a) although for ultimate precision they do not fulfill requirement (b). Since Newcomb's theory of the Earth is defective in respect to the development of the general perturbations, any improvement in the geocentric ephemeris of Mercury and Venus must await the development of a new general theory of the motion of the Earth. This task was commenced by G. M. Clemence and will be continued by the Nautical Almanac Office.

On the other hand, Newcomb's tables of Mars as amended by Ross are so unsatisfactory that it is intended to replace them as soon as possible by the new theory by Clemence (1949). The definitive constants of this theory have been derived by R. E. Laubscher and the theory is now being evaluated to the year 2000.

With the beginning of the space age, there was a gradual accumulation of precise observational evidence indicating refinements to a number of the basic constants forming Newcomb's system. The advent of artificial satellites allowed improved determination of the form of the Earth's gravitational potential as well as refinement of geodetic parameters. The introduction of radar and laser observing techniques with their ability to directly probe the distance to other bodies in the solar system is adding significant refinement to our knowledge of the astronomical unit. Flights of instrumented spacecraft, such as the Mariner series, have improved our knowledge of the positions and masses of the inner planets. The results of these new observing techniques, plus the evidence already accumulated from transit circle observations catapulted the 1963 Paris Symposium on the System of Astronomical Constants. The new system resulting from this conference, known as the IAU System of Astronomical Constants, was formally adopted at the General Assembly of the IAU in 1964 and the effects of the new system were introduced into the national ephemerides commencing in 1968. The IAU System of Astronomical Constants and the manner in which their effect has been introduced into the AE are completely delineated in the Supplement to the AE 1968 and I will not discuss them here. There were however several constants carried unchanged into the new system with the recommendation they be considered later; among them are precession, and the masses of the principal planets.

These constants were the subject of discussion at Colloquium No. 9 of the International Astronomical Union held in Heidelberg 11 to 15 August 1970. As a consequence of the colloquium, and later action by Commission 4 of the IAU, Working Groups were set up to study the subjects of Precessional Constants, Units and Time Scales, and Planetary Ephemerides. It was proposed that the Working Group on Planetary Ephemerides be set up to specify the basis for the planetary ephemerides to be published in the almanacs for 1980 onwards, and that ephemerides on this basis be made available in machine-readable form at the earliest opportunity. The colloquium participants further recommended that there be no changes in the basis of the ephemerides published in the national almanacs before 1980 on the grounds that this delay before the possible introduction of new constants is required to allow adequate time (1) for the preparation of new fundamental catalogs and ephemerides, (2) for the subsequent preparation of derived data and explanatory material for publication and (3) for their printing, proofreading, and distribution well in advance of the year to which the data refer.

It is the mandate of Commission 4 that the Working Group on Planetary Ephemerides will consider the system of planetary masses to be used in a new set of fundamental ephemerides as well as other factors that it considers to be relevant to the adoption of a new set of ephemerides, including choice of orbital elements or starting values, coordinate systems, form of equations and precision of computation. I have been asked to serve as convenor of this working party and the other members are: J. Kovalevsky, V. Abalakin, D. O'Handley, C. Oesterwinter, B. Morando, A. Sinclair, J. Schubart, W. Klepczynski, P. Janiczek (Secretary).

But now let us turn to the accuracy of the presently published planetary ephemerides and the possible improvements that can be realized prior to the introduction of new theories in 1980. For Mercury, Venus and Earth the present accuracy is one tenth of a second of arc in longitude and approximately the same in latitude. Application of the corrections to the mean elements mentioned earlier would at least halve these longitude and latitude errors. For Mars the presently published ephemeris yields an accuracy of two seconds of arc in longitude and about one tenth of a second of arc in latitude. The use of Clemence's new theory reduces these to two or three hundredths of a second of arc. New general theories of the motion of Mercury, Venus and Earth based on the new system of masses will further improve the accuracy of the ephemerides. To facilitate this work, the Nautical Almanac Office, U.S. Naval Observatory is collecting and systematically reducing the observations necessary to determine the definitive values of the constants of these theories.

The possibility of a grand tour mission to the outer planets in the late 1970's aroused interest in the accuracy of the ephemerides of these planets. The outer planet ephemerides in the AE are based on a simultaneous numerical integration of their orbits by Eckert *et al.* (1951). In the past several years, a considerable amount of work has been done in the Nautical Almanac Office, U.S. Naval Observatory, by Seidelmann *et al.* on the masses and orbits of the outer planets, utilizing all of the observational data to the present time (Klepczynski *et al.*, 1970, 1971; Seidelmann *et al.*, 1969, 1971). Presented here are specific values for the mean errors of the ephemerides in the American Ephemeris and the errors associated with the more accurate research ephemerides which have recently been determined from observations.

The observational data used are: for Jupiter and Saturn, the Six-Inch Transit Circle observations from 1913 to 1968; for Uranus, the normal points of Wylie (1947) augmented by the Six-Inch Transit Circle observations to give a period of normal points from 1830 to 1968; for Neptune, the transit observations from 1846 to 1968 as discussed by Jackson (1973); and for Pluto, normal points formed from the observations as discussed by Cohen *et al.* (1967).

Table I gives the osculating elliptic elements which are the basis of the Eckert *et al.* (1951) numerical integration. The orbital elements and their errors are referred to the equator and equinox of 1950.0 and they are given for the epoch J.D. 2430000.5. These osculating elliptic elements define the orbit at that epoch and could be converted to rectangular coordinates and velocities to begin a numerical integration. The specific elements given are the mean anomaly, l, the argument of perihelion, ω, the longitude of the node, Ω, the inclination, i, the eccentricity, e, and the semi-major axis, a.

TABLE I

American Ephemeris – osculating elliptic elements for 243 0000.5

	Jupiter	Saturn	Uranus	Neptune	Pluto
l	29° 5′ 0″.096	318°29′31″.276	256°41′33″.515	133°46′57″.978	289°16′47″.700
ω	274°35′34″.956	336°27′ 3″.205	89°14′38″.424	270° 0′43″.841	113°20′31″.719
Ω	99°57′ 2″.186	113°14′42″.242	73°48′17″.521	131°16′44″.491	109°36′24″.672
i	1°18′27″.115	2°29′13″.115	0°46′23″.706	1°46′33″.407	17° 7′21″.357
e	0.0490137305	0.0562631702	0.0447360290	0.0118545587	0.2459387823
a	5.2043041446	9.5836692762	19.3160573025	29.9867909509	39.5181761979

The mean errors determined when this integration is compared to the observations are given in Table II. In addition to the errors in the elliptic elements, the error in the mean longitude, L, and the error in the mean latitude, B, at the epoch are given. For the semi-major axis, the mean longitude and the latitude at epoch, values are given in kilometers as well as in astronomical units and seconds of arc. The error in the

TABLE II
Mean errors of elliptic elements in American Ephemeris

	Jupiter	Saturn	Uranus	Neptune	Pluto
Δl	±0″.240	±0″.231	±0″.409	±2″.289	±59″.941
$\Delta\omega$	±0.996	±0.625	±2.903	±2.422	±59.478
$\Delta\Omega$	±0.967	±0.582	±2.875	±0.795	± 0.906
Δi	±0.023	±0.025	±0.036	±0.021	± 0.516
Δe	±0.0000000538	±0.0000000664	±0.0000000979	±0.0000001216	±0.0000199257
Δa	±0.0000000327 AU (±5 km)	±0.0000001812 (±27 km)	±0.0000006719 (±100 km)	±0.0000007826 (±565 km)	±0.0040007226 (±598.108 km)
ΔL	±0″.016 (±60 km)	±0″.018 (±125 km)	±0″.033 (±462 km)	±0″.059 (±1282 km)	±7″.486 (±214.400 km)
ΔB	±0.032 (±120 km)	±0.032 (±222 km)	±0.047 (±659 km)	±0.003 (±65 km)	±1″.107 (±31.700 km)

Where L is mean longitude and B is latitude at epoch.

semi-major axis is derived by Kepler's Third Law from the mean motion, determined from optical observations of angular position. However, due to the value used for the mass of Pluto, the ephemeris of Neptune deviates systematically from the observations by almost four seconds of arc at 1968.

The planetary masses determined by the authors from the observations specified above are listed in Table III under the heading 'Research Ephemerides'. The first column of Table III gives the reciprocal mass values which are used in the American Ephemeris and which are also the mass values adopted by the International Astronomical Union. The improvement in the representation of the observations is shown in Figures 1–6.

The osculating elliptic elements of the new 'research ephemerides' of the outer planets are given in Table IV. The mean errors of these elliptic elements are given in Table V. An indication of the improvement in the fit to the observations provided by the new 'research ephemerides' is given in Table VI. The first column gives the sums of the squares of the differences of the observations as compared to the ephemerides appearing in the American Ephemeris. The second column gives the sums of the squares of the residuals when the same observations are compared to the 'research ephem-

TABLE III
Reciprocal mass values

	American Ephemeris	Research Ephemerides
Jupiter	1047.355	1047.355
Saturn	3501.6	3498.7 ± 0.2
Uranus	22869	22692 ± 33
Neptune	19314	19349 ± 28
Pluto	360000	3000000 ± 500000

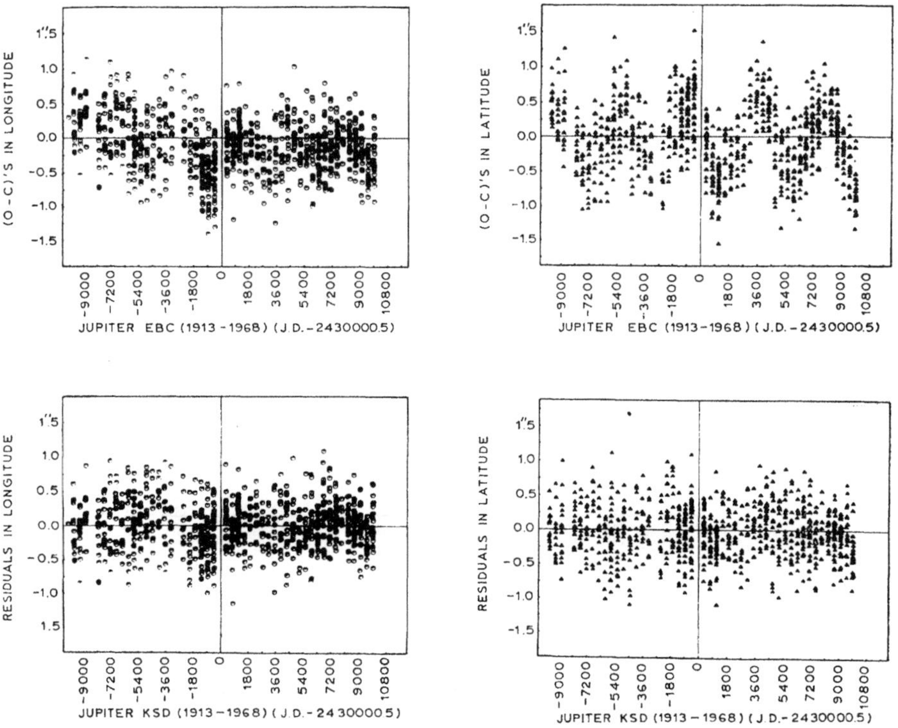
(O-C)'S IN LONGITUDE
JUPITER EBC (1913-1968) (J.D.-2430000.5)
(O-C)'S IN LATITUDE
JUPITER EBC (1913-1968) (J.D.-2430000.5)
RESIDUALS IN LONGITUDE
JUPITER KSD (1913-1968) (J.D.-2430000.5)
RESIDUALS IN LATITUDE
JUPITER KSD (1913-1968) (J.D.-2430000.5)

Fig. 1.

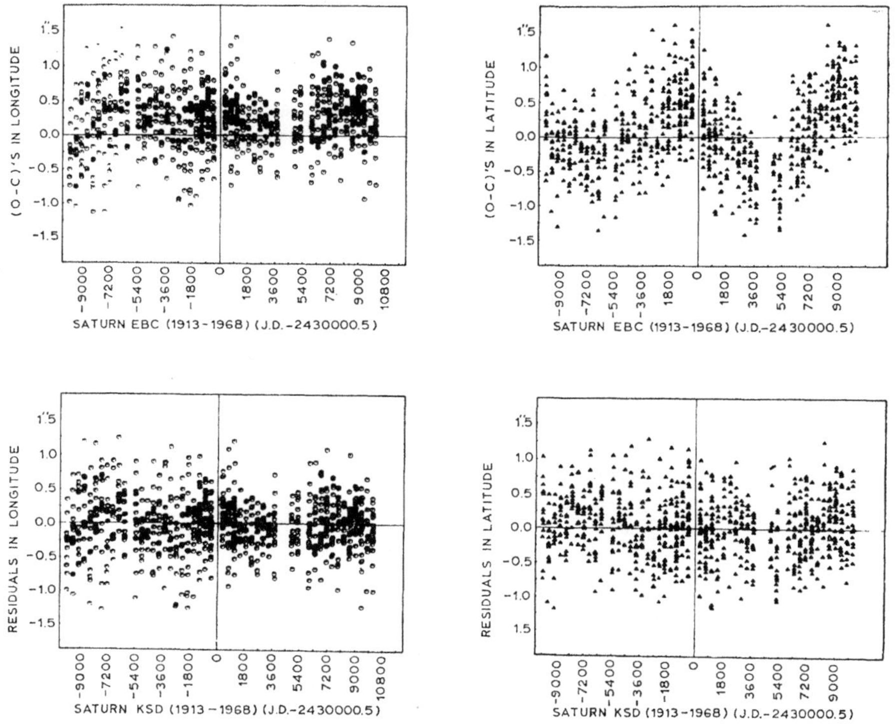
(O-C)'S IN LONGITUDE
SATURN EBC (1913-1968) (J.D.-2430000.5)
(O-C)'S IN LATITUDE
SATURN EBC (1913-1968) (J.D.-2430000.5)
RESIDUALS IN LONGITUDE
SATURN KSD (1913-1968) (J.D.-2430000.5)
RESIDUALS IN LATITUDE
SATURN KSD (1913-1968) (J.D.-2430000.5)

Fig. 2.

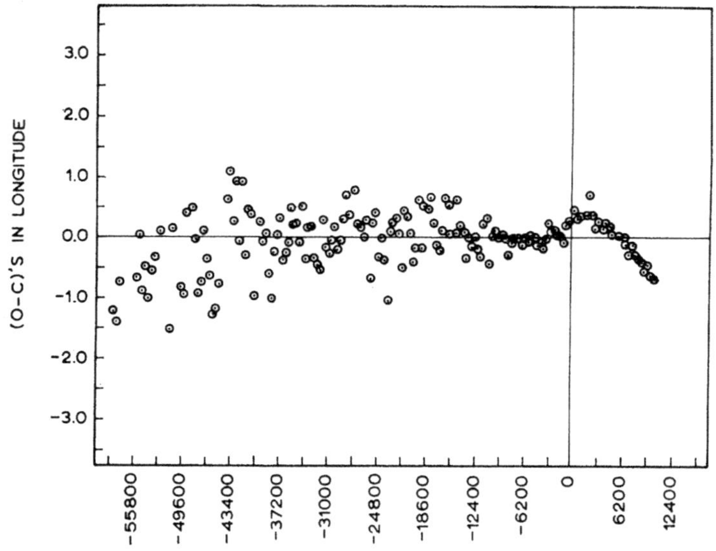

Fig. 3. Uranus, five outer planets (JD-2430000.5).

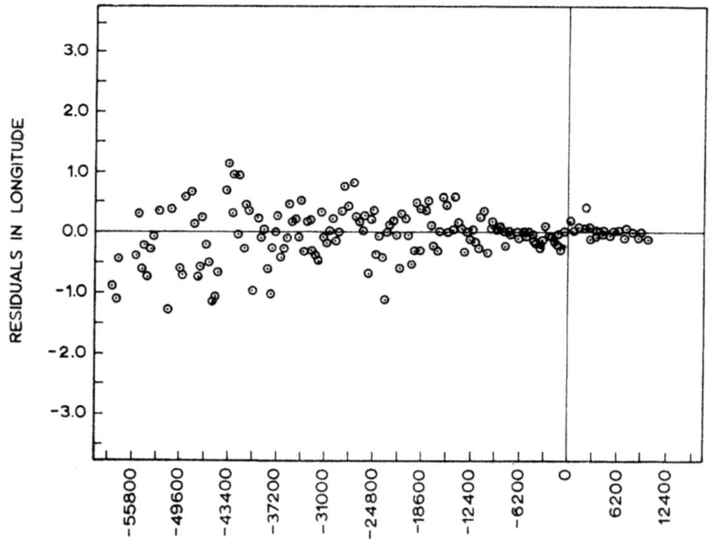

Fig. 4. Uranus 1781–1968 (JD-2430000.5).

erides'. These research ephemerides represent the accuracy that may be achieved at the present time utilizing optical observations of the outer planets. It is apparent that the mean errors of the elements of the research ephemerides have not been significantly reduced by the introduction of the newly determined masses. Although the use of the new masses has removed the systematic differences between theory and observa-

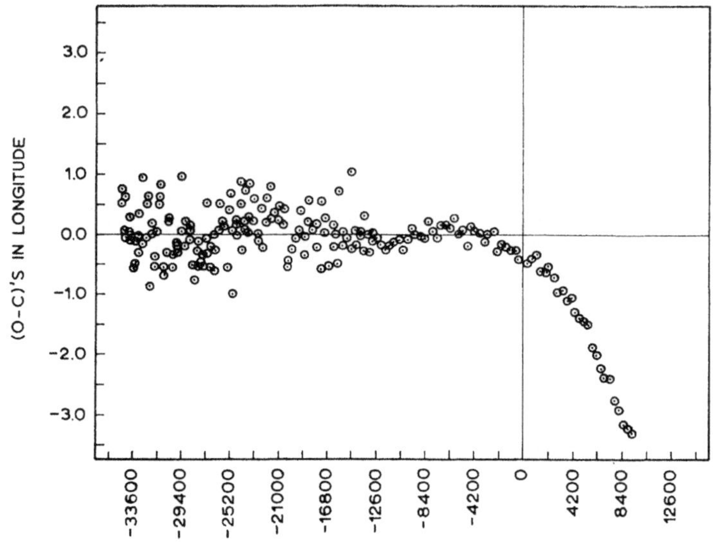

Fig. 5. Neptune, five outer planets (JD-2430000.5).

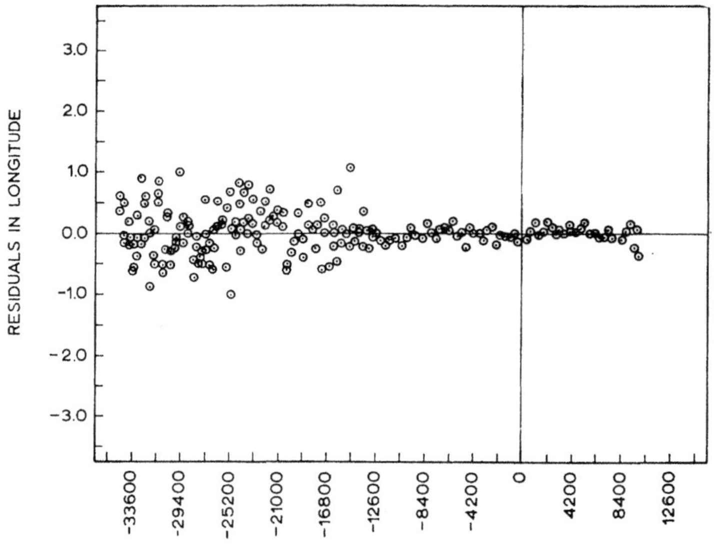

Fig. 6. Neptune 1846–1968 (JD–2430000.5).

tion, it does not diminish the random errors in the older observations which are primarily responsible for the size of the mean errors of the elements.

The prospect of radar observations over a significant portion of the outer planets' orbits offers a promise of increased accuracy in the determination of their motions, primarily for Jupiter. However, the long orbital periods involved in the other outer

TABLE IV

Elliptic osculating elements of Research Ephemerides for 2430000.5

	Jupiter	Saturn	Uranus	Neptune	Pluto
l	29° 4′57″.592	318°29′32″.032	256°41′41″.245	133°44′55″.031	289°21′ 5″.510
ω	274°35′26″.412	336°26′58″.466	89°14′24″.497	270° 3′39″.032	113°16′38″.202
Ω	99°57′13″.050	113°14′46″.475	73°48′23″.532	131°16′38″.320	109°36′23″.321
i	1°18′26″.726	2°29′12″.716	0°46′23″.578	1°46′33″.796	17° 7′22″.464
e	0.0490143422	0.0562636509	0.0447344038	0.0118574023	0.2460004277
a	5.2043055494	9.5836743386	19.3161021827	29.9871269920	39.5326598084

TABLE V

Mean errors of elliptic elements of the Research Ephemerides

	Jupiter	Saturn	Uranus	Neptune	Pluto
Δl	±0.189	±0″.192	±0″.357	±2″.082	±15″.563
$\Delta\omega$	±0.782	±0.516	±2.535	±2.204	±15.442
$\Delta\Omega$	±0.760	±0.480	±2.510	±0.723	± 0.235
Δi	±0.018	±0.021	±0.031	±0.019	± 0.134
Δe	±0.0000000423	±0.0000000548	±0.0000000855	±0.0000001109	±0.0000051949
Δa	±0.0000000257 AU (±4 km)	±0.0000001496 (±22 km)	±0.0000005866 (±88 km)	±0.0000034442 (±515 km)	±0.0010398701 (±157.538 km)
ΔL	±0″.013 (±49 km)	±0″.015 (±104 km)	±0″.028 (±392 km)	±0″.053 (±1.151 km)	± 1″.944 (±55.766 km)
ΔB	±0.025 (±94 km)	±0.027 (±187 km)	±0.041 (±574 km)	±0.003 (±65 km)	±0.200 (± 5.728 km)

Where L is mean longitude and B is latitude at epoch.

TABLE VI

Sums of the squares of the residuals

	American Ephemeris	Research Ephemerides
Jupiter		
long.	152.8631	107.6036
lat.	225.8087	125.7293
Saturn		
long.	201.1810	144.5148
lat.	257.9453	168.1212
Uranus		
long.	14.9376	11.3916
lat.	17.1340	14.3146
Neptune		
long.	9081.7789	6600.2477
lat.	4933.2571	5018.0830
Pluto		
long.	1204.1282	46.4719
lat.	108.2901	42.3776

planets requires the use of observations over an extended length of time. To date, there are observations of Neptune covering only 75% of the orbit, and for Pluto only a little more than 20% of the orbit has been observed. Further improvement in the accuracy of the ephemerides of the outer planets will require the passage of time.

In discussing the orbits of the major planets, it is intriguing to consider some of the evidence for the existence of additional planets. Ever since the discovery of the planet Pluto there has been persistent conjecture concerning the presence of a trans-Plutonian planet. The presence of such an additional object in the solar system was thought to be shown by the latitude residuals in the motion of Uranus and Neptune. Some residuals still persist in the latest adjustment of the planetary theories to observations and are unexplained. Again, the new theory of the planet Neptune fails to satisfy the prediscovery observations of 1795, producing a residual in orbital longitude of about 8″. Although these prediscovery observations are of questionable value, this large residual in longitude has been seized upon by several mathematicians to hypothesize the presence of a tenth planet. The inability to accurately predict the times of perihelion passage of Halley's Comet and other long-period comets has induced several investigators to hypothesize the presence of a trans-Plutonian planet. The most recent of these was an attempt (Brady, 1972) to reduce the residuals in the times of perihelion passage of Halley's Comet and other long-period comets by the introduction of a hypothetical trans-Plutonian planet of mass only slightly less than that of Jupiter and moving in a nearly circular, but highly inclined ($i=120°$) orbit at 60 AU from the Sun. The author acknowledges that the motions of many comets, of both long and short period, are affected by 'nongravitational forces' and that the perturbations by this distant planet obviously will not explain the anomalies in the motions of typical short period comets with their aphelion distances of some 4–6 AU. The problem with cometary motions is essentially a secular acceleration; the observed delay of about 4 days at each return of *P*/Halley with a 75 yr period is equivalent to a delay of 0.04 day for a comet of period 7.5 yr, which is in accord with the observed delays for short period comets.

It seems more reasonable to ascribe the anomalies of both short and long period comets to the same general causes, rather than to hypothesize a massive planet as the explanation of only long period comet residuals. P. K. Seidelmann, B. G. Marsden and H. L. Giclas have recently completed a study of the effect of the presence of planet X on the motions of the five planets Jupiter to Pluto and have attempted to fit this augmented planetary model to the observations of the known planets. They have shown that the planet does not appear on photographic plates, and that the observed motions of the outer planets do not permit planet X as hypothesized. The following is a report of their study.

The motions of the five planets Jupiter-Pluto, as well as of the hypothetical planet X, were integrated for an interval of 300 yr, and the integration was fitted to the observational data.

Based on the improved elements, a second integration was performed and again fitted to the observations. It was assumed that the effect of planet X on the four inner planets would be negligible. Further, no attempt was made to improve the masses of the outer planets to satisfy the effects of the hypothetical planet, for the residuals, particularly in latitude, were judged to be greatly in excess of what could be eliminated by any reasonable mass corrections. Table VII gives the sums of the squares of the residuals in longitude and latitude for each of the known outer planets, first for an integration of the known planets only and second with planet X included as described above.

TABLE VII

Sums of the squares of the residuals

	Known solar system	With planet X
Jupiter (1913–1970)		
long.	136.0536	172.5313
lat.	145.5703	374.6467
Saturn (1913–1968)		
long.	144.5148	876.2358
lat.	168.1212	1642.6147
Uranus (1830–1968)		
long.	11.3916	13644.3151
lat.	14.3146	3913.0325
Neptune (1846–1968)		
long.	5.4228	2464.3439
lat.	12.0129	3788.3330
Pluto (1914–1968)		
long.	46.8153	43.7333
lat.	42.5415	1298.9459

Since the hypothetical new planet has an orbit that is considerably inclined to the ecliptic, its effect on the latitude of the known planets would be very large. Figures 7–11 show this effect for Jupiter, Saturn, Uranus, Neptune, and Pluto, respectively. Even for Jupiter the residuals are quite unacceptable, while the changing ordinate-scale (in seconds of arc) for the other planets speaks for itself.

The residuals in longitude (also in seconds of arc) are shown in Figures 12–15 (Jupiter, Saturn, Uranus, and Neptune, respectively). Since the observations of Pluto cover only about one-fifth of a revolution, there is no detectable systematic effect on the longitude of this planet. It should be further noted that the presence of planet X causes the residuals in longitude of the 1795 observations of Neptune to increase to 49″.

Brady (1972) indicates that if the hypothetical planet has the albedo of Pluto and any reasonable density it will be of the 13th or 14th magnitude. Accepting this estimate of the magnitude, it seems rather surprising that a planet with an annual average motion of 46′ has been overlooked. It ought to have been discovered in the course of the Lowell Proper Motion Survey, for example, which covers the whole northern sky down to magnitude 16–17 and is reasonably complete for stars of annual proper

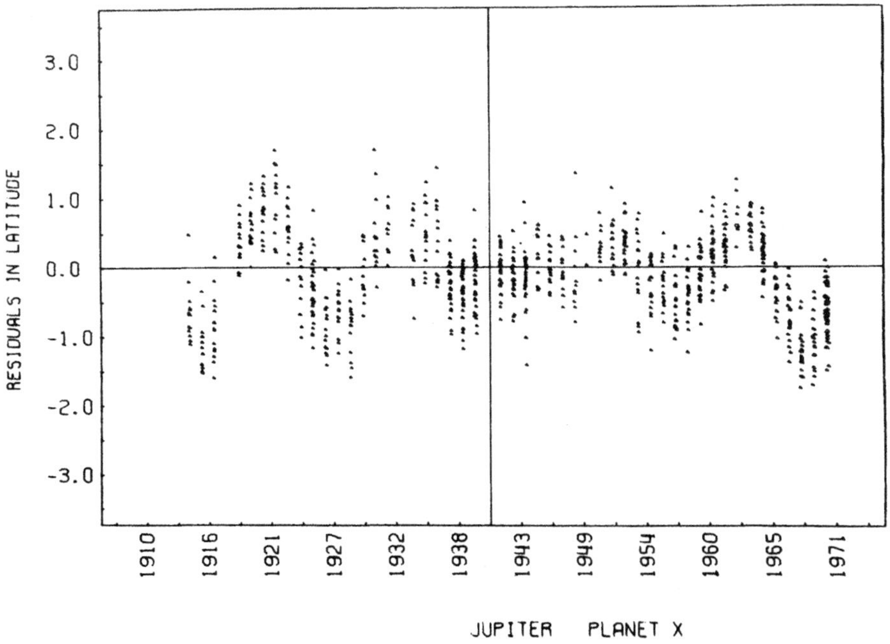

Fig. 7. Residuals (in seconds of arc) in the latitude of Jupiter caused by the presence of planet X.

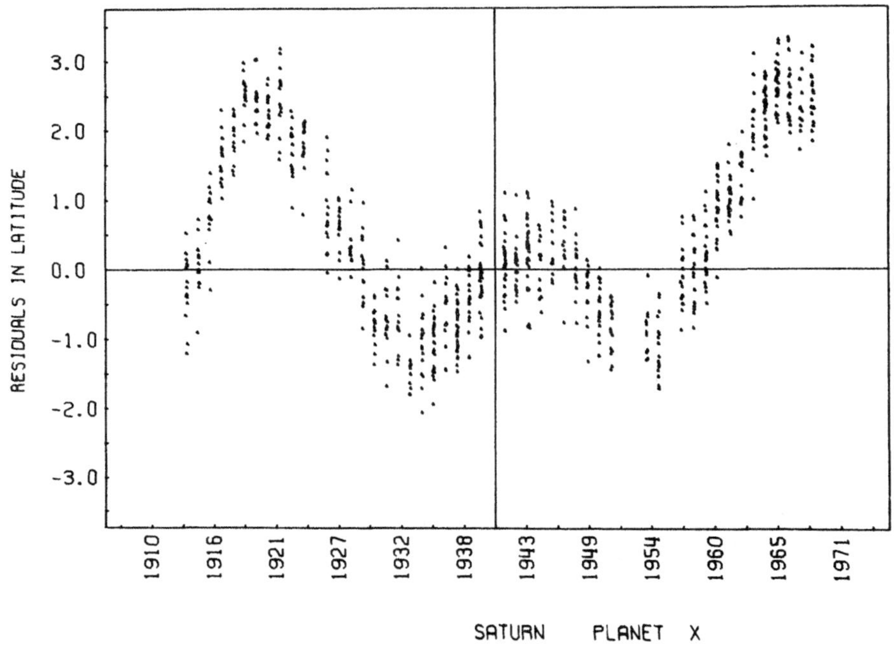

Fig. 8. Residuals in the latitude of Saturn.

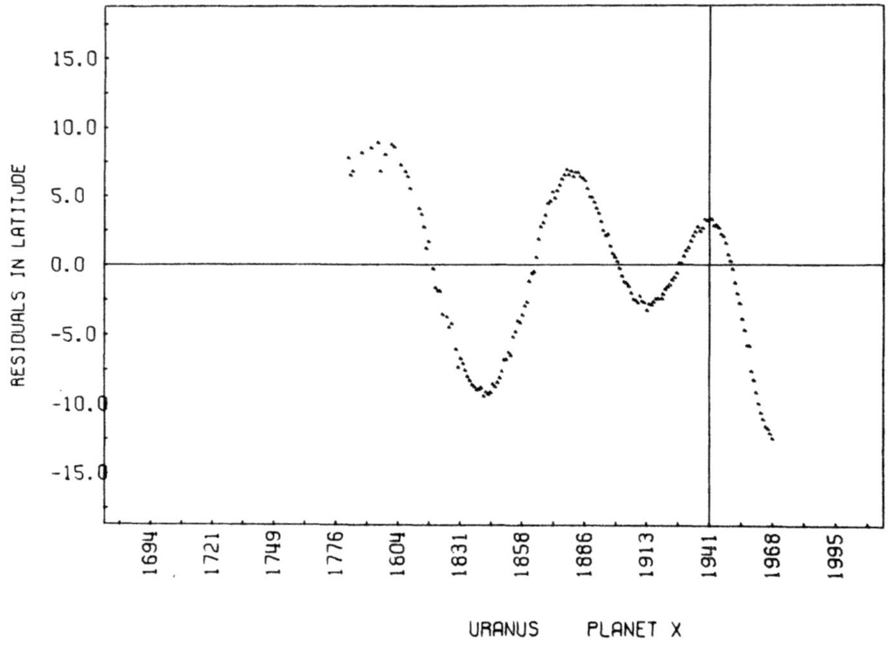

Fig. 9. Residuals in the latitude of Uranus.

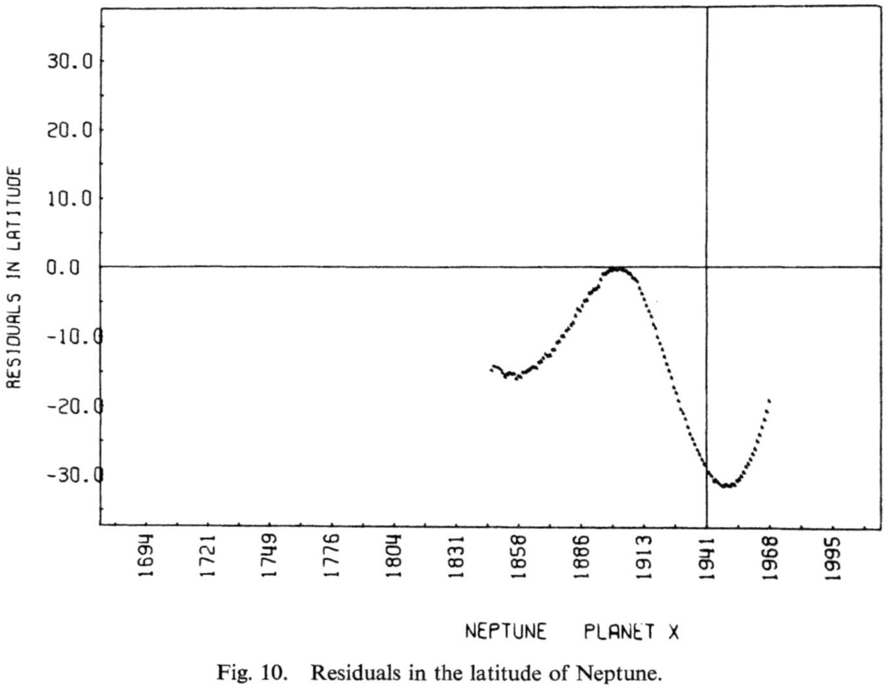

Fig. 10. Residuals in the latitude of Neptune.

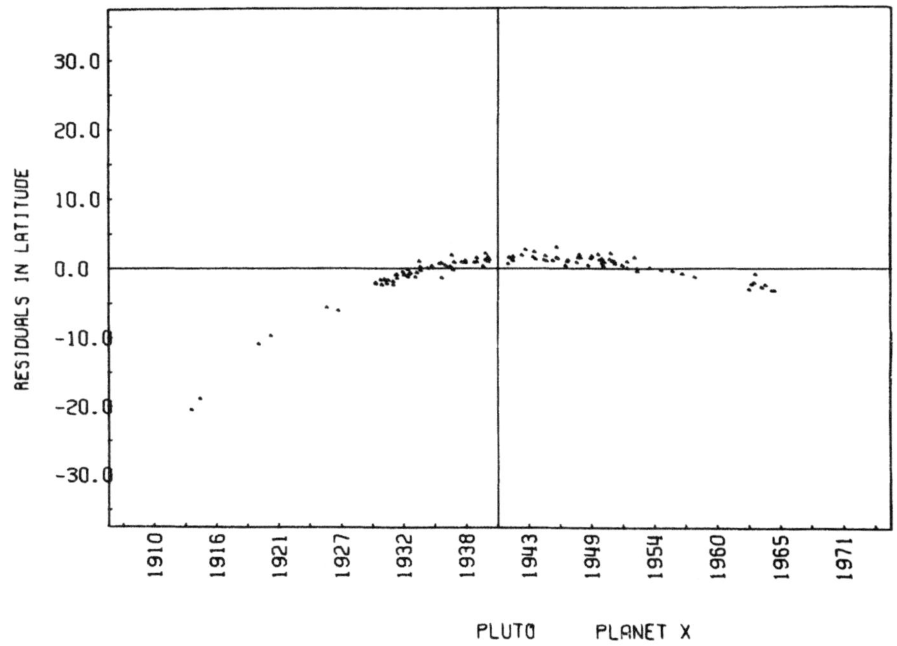

Fig. 11. Residuals in the latitude of Pluto.

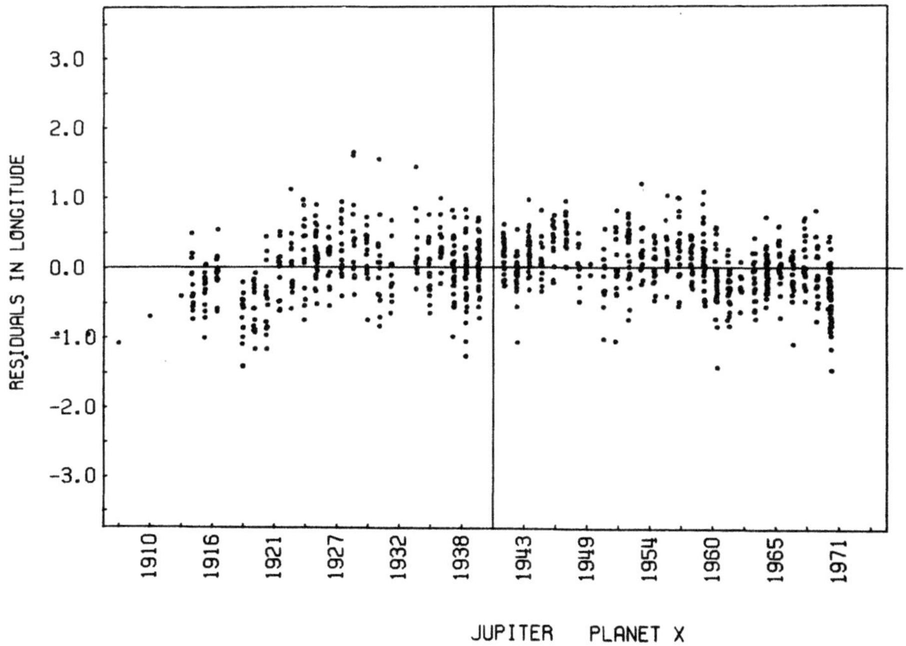

Fig. 12. Residuals (in seconds of arc) in the longitude of Jupiter caused by the presence of planet X.

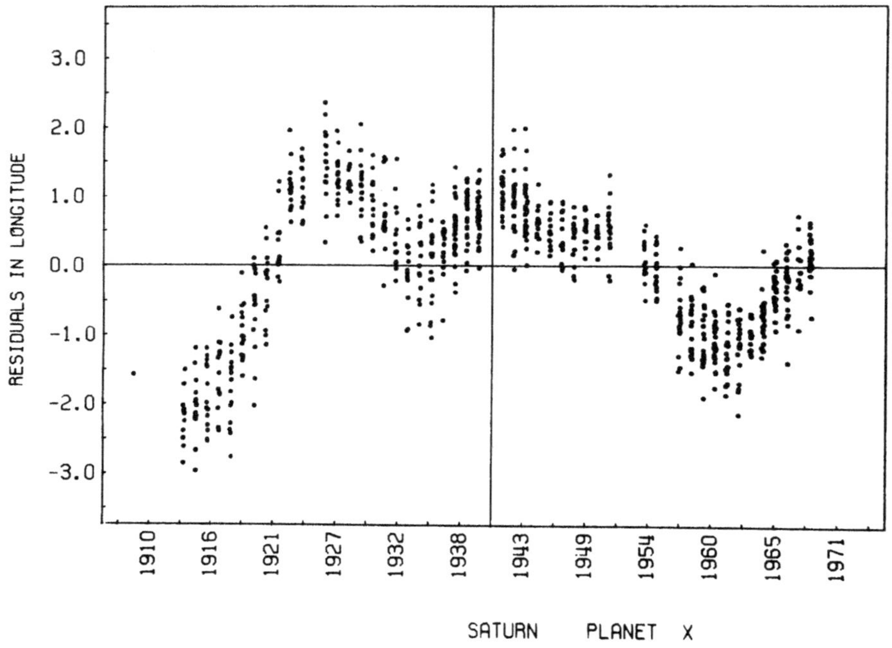

Fig. 13. Residuals in the longitude of Saturn.

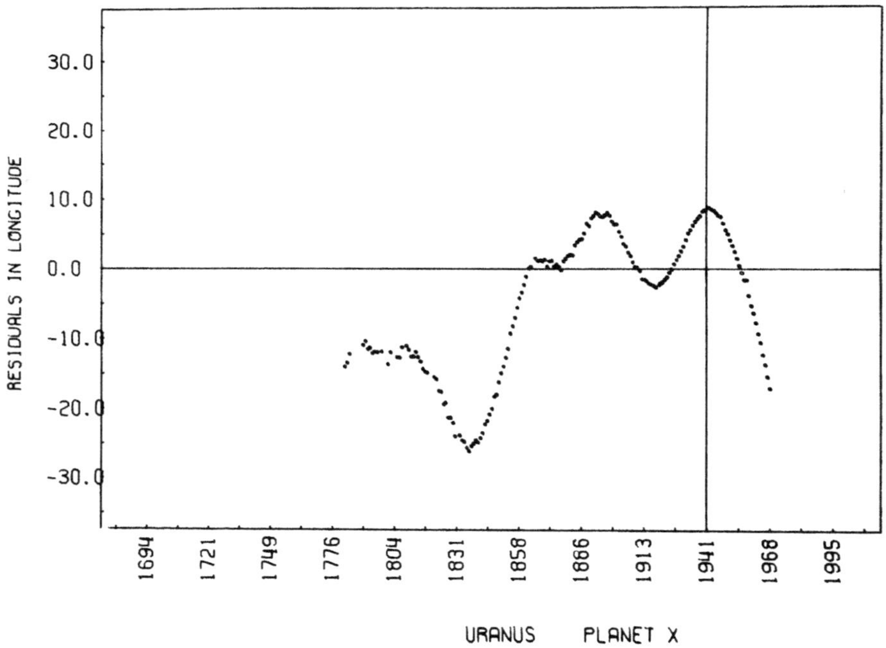

Fig. 14. Residuals in the longitude of Uranus.

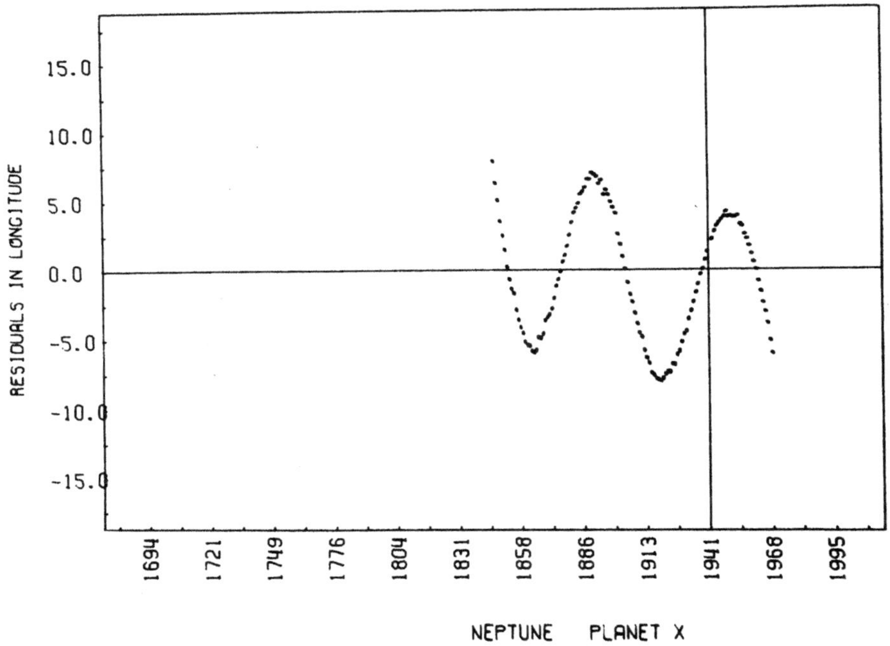

Fig. 15. Residuals in the longitude of Neptune.

motion 0″.27 and larger. Luyten has also made a rather exhaustive proper-motion survey of the northern sky with the 122-cm (48-in.) Palomar Schmidt.

The predicted position of the plant is contained on plates taken by Tombaugh at Lowell in 1940–41. These plates, centered at $\alpha = 5^h 50^m$, $\delta = +64°20'$ and covering an area of 12° by 15°, have now been carefully blink-examined to limiting magnitude 16.5, but no planet has been found. They have also been blinked against Lowell plates obtained of the same area in 1969. Re-examination of the notes made at the time these plates were blinked, as well as of those concerning the region centered at $\alpha = 2^h 00^m$, $\delta = +75°53'$ (where the planet should have been in 1968.8), has revealed no suspicious objects. All these modern plates have again been carefully blink-examined in their entirety to a limiting magnitude of 16.5 or fainter. It is extremely improbable that an object as bright as magnitude 13–14 could have escaped detection during all these surveys.

It must be concluded that while small or very distant unknown planets may exist, there is no evidence to support the presence of a planet having the mass, magnitude, mean distance, and orbital inclination hypothesized in Brady (1972). The anomalous motion of comet Halley must have some other cause.

References

Brady, J. L.: 1972, *Publ. Astron. Soc. Pac.* **84**, 314.
Clemence, G. M.: 1943, *Astron. Pap. Am. Ephemeris* **11**, Part 2.

Clemence, G. M.: 1949, *Astron. Pap. Am. Ephemeris* **11**, Part 2; 1961, **16**, Part 2.
Clemence, G. M. and Duncombe, R. L.: 1957, *Astron. J.* **62**.
Cohen, C. J., Hubbard, E. C., and Oesterwinter, C.: 1967, *Astron. J.* **72**.

Duncombe, R. L.: 1958, *Astron. Pap. Am. Ephemeris* **16**, Part 1.
Eckert, W. J., Brouwer, D., and Clemence, G. M.: 1951, *Astron. Pap. Am. Ephemeris* **12**.
Jackson, E. S.: 1973, *Astron. Pap. Am. Ephemeris* (in press).
Klepczynski, W. J., Seidelmann, P. K., and Duncombe, R. L.: 1970 *Astron. J.* **75**.
Klepczynski, W. J., Seidelmann, P. K., and Duncombe, R. L.: 1971, *Proceedings of the 17th Annual Meeting of the American Astronautical Society*.
Morgan, H. R.: 1933, *Astron. J.* **42**.
Seidelmann, P. K., Duncombe, R. L., and Klepczynski, W. J.: 1969, *Astron. J.* **74**.
Seidelmann, P. K., Klepczynski, W. J., Duncombe, R. L., and Jackson, E. S.: 1971, *Astron J.* **76**.
Wylie, L. R.: 1947, *Publ. U.S. Naval Obs.* Ser. 2, **15**, Part 3.

ORBITS OF THE NATURAL SATELLITES LECTURES ON SOME TYPES OF ORBITAL MOTION AMONGST THEM

P. J. MESSAGE

Dept. of Applied Mathematics, Univ. of Liverpool, Liverpool, England

Abstracts. The lectures outlined here describe various types of perturbed Keplerian motion which arise in the satellite systems of Jupiter and Saturn, dealing with Japetus, and near-commensurability cases in Saturn's system, as well as with the great satellites of Jupiter.

The aim of these two lectures is to describe the dominant features of a few of the different types of orbital motion to be found in the satellite systems of the major planets. At the outset of a precise study of the motion of a particular satellite, the choice and careful specification of the reference plane is important. In most cases, however, the motion is nearly planar, and the difficulty and interest of the problem is mostly concerned with the motion of the projection of the satellite on a suitably chosen reference plane, and the theory of the latitude is a relatively simple extension of the main theory. An important exception to this is the case of Saturn's satellite Japetus, the perturbations to whose motion by the other satellites, especially Titan, is of comparable importance to those by the Sun. Titan's orbit is inclined to the ecliptic at about 27°, and so the motion of Japetus cannot be even approximately planar over periods of a few centuries, and the theory of the latitude motions is of paramount importance. Apart from small short period effects, the pole of Japetus' osculating orbit describes an elliptical cone whose axis is perpendicular to a plane, known as the 'Laplacian plane', which may for practical purposes be regarded as fixed. The mathematical theory is given by Tisserand (1896, Chapter VI), and is briefly described in the first lecture.

Most of the other features of interest in the dynamics of Saturn's satellite system arise from the occurrence of close small integer near-commensurabilities between the orbital periods of pairs of satellites. The first case of this to arouse interest was that involving the satellite Hyperion, which, discovered in 1848, was found by Hall in 1884 to have a retrograde motion of the apse, of about 20.3° yr^{-1}, whereas the prediction from the usual procedure, of using the non-periodic terms in the expansion of the disturbing function, led to a forward motion of the apse. Newcomb showed in 1884 that the anomaly arose from the fact that the orbital period of Hyperion is nearly four-thirds that of Titan, in fact within 0.1%, that $\dot{\varpi}$ is approximately equal to $4n-3n'$ (accents denoting elements of Titan's orbit), and that the quantity $4\lambda-3\lambda'-\varpi=\theta$, called the 'critical argument', instead of increasing or decreasing approximately linearly in time, in fact oscillates, or 'librates', about the equilibrium value 180°, and so that in turn those cosine terms in the equation for $\mathrm{d}\varpi/\mathrm{d}t$ which have θ and its multiples as argument are not periodic, but make a non-zero contribution to the mean

B. D. Tapley and V. Szebehely (eds.), Recent Advances in Dynamical Astronomy, 309–311.

value of $d\varpi/dt$. In terms of the movement of Hyperion in space, Tisserand showed that the periodic fluctuation in the radial distance from Saturn due to Titan's perturbation exceeds in amplitude the free oscillation usually identified with the eccentricity, so that the observed apses are the maxima and minima of the radius vector under the perturbations of Titan. If the amplitude of the free oscillation had been zero, and Titan's orbit circular, then Hyperion's motion would have provided an example of a periodic solution of Poincaré's second sort in the restricted problem of three bodies, in which, at each conjunction of Hyperion with Titan, Hyperion is at apocenter. The free oscillation is reflected in the changes of the elliptic elements most significantly in the libration of θ, of amplitude about 36°, and in the corresponding periodic term in the longitude, of amplitude about 9°. The eccentricity of Titan's orbit gives rise to periodic terms in Hyperion's eccentricity and apse longitude, the amplitude in the latter being about 13.8°. The fullest treatment published so far is by Woltjer (1928).

The pair of satellite Enceladus and Dione in Saturn's system provide a further example of near-commensurability of orbital period, that of Dione being nearly half that of Enceladus. The mass of Enceladus is about $\frac{1}{12}$ that of Dione, so its effect on the latter is not negligible. The motion is close to a periodic solution of Poincaré's second sort in the general problem of three bodies, the forced eccentricity exceeding the free one in the case of Enceladus (as for Hyperion), but the free exceeding the forced for Dione. The outline of the theory given in the lectures is based on that given by Jeffreys (1953).

A further case of near 2:1 commensurability in Saturn's system is provided by Mimas and Tethys. In this case the critical argument is $4\lambda_M - 2\lambda_T - \Omega_M - \Omega_T$, and thus involves the node longitudes and not the apses. The motion represents the result of the superposition of a free oscillation on a periodic solution of Poincaré's third sort in the general three body problem. The theory is treated as in Tisserand (1896, Chapter VII). We owe our knowledge of the masses of most of Saturn's satellites to the existence of these near-commensurabilities of period, as a result of which many of the perturbations are augmented so as to be detectable in analysis of observations of satellite positions. There is a serious lack of recent observations of many satellites, however.

A different type of perturbation is presented by the satellite Rhea. Though there is no small integer commensurability important here, there is a libration of the difference of the apse longitudes of Rhea and Titan, due to the dominance in the secular variations of the eccentricity and apse of the forced motion due to Titan over the free motion. The treatment given is due to Woltjer (1922).

To find the most intricate commensurability relationship in the solar system we pass from the satellite system of Saturn to that of Jupiter, to consider the satellites Io, Europa, and Ganymede. There are two 2:1 near-commensurability relations, leading to forced oscillations ('great inequalities') of the type met with in Enceladus and Dione, linking Io to Europa, and the latter to Ganymede. These two relations are themselves linked, since $n_I - 2n_E = n_E - 2n_G$, and the motion of all three satellites approximates to one with a single period in a suitably rotating frame. Laplace showed that the quantity

$\theta = \lambda_I - 3\lambda_E + 2\lambda_G$, which is an argument in the second-order terms, has a stable equilibrium value of 180°. De Sitter developed a new mathematical theory of the motion of these satellites and Callisto, in which a periodic solution of this type, with uniformly moving apses, was the basis of the intermediary solution, the actual motion being derived from it by the superposition of various periodic fluctuations, including the free (or 'proper') eccentricities and latitude motions, and also free oscillation of Laplace's critical angle θ. The masses of the four great satellites of Jupiter are not known to a very great percentage precision, especially those of Io and Callisto, since, although the three great inequalities in longitude are well determined from observations, the free oscillations are too small and poorly determined to prevent the equations for the four masses and the oblateness co-efficient J_2 from forming a poorly conditioned set.

References

Jeffreys, H.: 1953, *Monthly Notices Roy. Astron. Soc.* **113**, 81–96.
Newcomb, S.: 1891, in *Astronomical Papers of the American Ephemeris*, Vol. III, pp. 347–371.
Sitter, W. de: 1918, *Ann. Sterrewacht Leiden* **12**, Part 1.
Tisserand, F.: 1896, *Traité de Mécanique Céleste*, Vol. IV, Gauthier-Villars, Paris.
Woltjer, J.: 1922, *Bull. Astron. Inst. Neth.* **1**, 175–6.
Woltjer, J.: 1928, *Ann. Sterrewacht Leiden* **16**, Part 3.

ON THE ORIGIN OF THE TITAN-HYPERION SYSTEM

G. COLOMBO

University of Padova and Smithsonian Astrophysical Observatory

and

F. A. FRANKLIN

Smithsonian Astrophysical Observatory

Abstract. The mean motions of the two satellites Titan and Hyperion are very nearly in the ratio 4/3. In order to examine some possibilities for the formation of this system, we present some calculations providing the range of initial orbital velocities that lead to stable librating solutions at the 4/3 resonance.

1. Introduction

The Saturnian satellites Titan and Hyperion form a system of two bodies whose mean motions are nearly in the ratio 4/3. Successive conjunctions of the two oscillate, or librate, about the aposaturnium of Hyperion with an amplitude of about 36° and with a period of about 640 days. Sample calculations we have carried out indicate that this libration is stable over many apsidal periods and there seems no reason to believe that it cannot persist over times comparable to the age of the solar system. The question then arises: How was this system formed, or how has the present stage been reached? The same question has also been posed in the case of the Mimas-Tethys and Enceladus-Dione satellite systems, both of which are examples of a 2/1 commensurability. Goldreich (1965) has provided a very likely answer for both of these cases by pointing out the importance of tidal interaction between Saturn and these satellites. Because the effect, on the semi-major axis of a satellite, of tides raised on the primary by the satellite depends on the satellite's mass and its distance from the primary, it provides a powerful mechanism for changing the ratio of the semi major axes of two inner satellites. Potentially then, it can convert an arbitrary non-commensurable ratio into a low order commensurable one. Arguments that certain commensurable cases once reached are stable in the presence of continued tidal interaction have also been given by Goldreich (1965). Some numerical calculations we have carried through indicate that this is indeed true even for unrealistically large tidal forces provided only that the initial eccentricities or eccentricity of the bodies is $\gtrsim 0.05$. Greenberg (1972) has worked on the problem analytically, discussing stable capture in great detail. We can conclude that tidal interaction provides a very likely mechanism that accounts for the formation of the Mimas-Tethys and Enceladus-Dione systems and that it is consistent with their continued existence.

For the Titan-Hyperion case, however, the greater distance of Titan from Saturn, despite the relatively large mass of that satellite, means that its semi-major axis can have been changed by tidal forces only very slightly during the age of the solar system. Adopting the numbers provided by Goldreich and Soter (1966) as characteristic values,

B. D. Tapley and V. Szebehely (eds.), Recent Advances in Dynamical Astronomy, 312–318.

we find, for example, that Titan's semi major axis has been increased by only 0.3% in 4.5×10^9 yr. Any change in the semi-major axis of Hyperion's orbit would have been much smaller. Thus, while a tidal interaction in presumably now slowly operating and may possible be partially responsible for Hyperion's relatively large eccentricity, it is unlikely to have had a major effect in generating the commensurability. We are therefore driven to the viewpoint that the presently observed 4/3 commensurability is strongly related to the initial state of the system. If this be so, it is natural to ask: How large is the volume in phase space occupied by stable librating solutions associated with the 4/3 resonance? If this volume is negligibly small, the origin of the Titan-Hyperion system in resonance becomes highly unlikely and one should look for some new mechanism to produce it. If librating solutions are a relatively probable state, we may conclude that it is likely that the system may have been formed in resonance. We shall try to present some answer to this question using the restricted three body problem, which in view of the small though unknown mass of Hyperion, is an appropriate model.

2. Calculations

For this study, we have used an advanced program for the integration of a large number of both massive and massless bodies developed at SAO by M. Lecar and R. Loeser and described elsewhere (Lecar and Loeser, 1972). Figure 1 presents a few

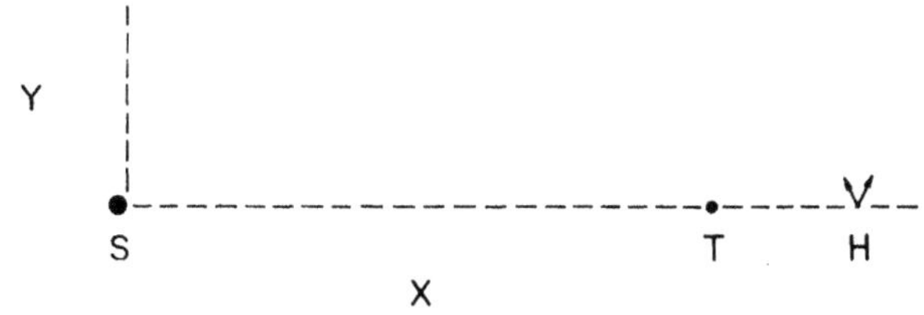

Fig. 1. Initial configuration of Saturn (S), Titan (T) and a sample massless body (H), a set of which is introduced with a spread in the velocities $\dot{X}$ and $\dot{Y}$ at the point X_0.

details of the calculations. Units are such that the (circular) velocity of the less massive primary is $2\pi(1+\mu)^{1/2}$ where μ is the mass ratio of the two massive bodies which are separated by unit distance. At $t=0$, a set of massless bodies is introduced at the position $Y=0$, $X=X_0$ with a spread in the velocities $\dot{X}$ and $\dot{Y}$. Values of X_0 determine the eccentricities of the orbits of the massless bodies. In order to avoid the case in which conjunctions of the secondary (Titan) and the massless body (Hyperion) occur at the latter's pericenter, X_0 must be greater than the semi-major axis of Hyperion. Thus, for the 4/3 resonance, $X_0 > 1.211$.

Integrations have been continued for many libration periods of the massless bodies and, in critical cases, for more than an apsidal period. For most calculations we have used a mass ratio, $\mu = 10^{-3}$, rather than the actual Titan-Saturn value of 1/4150. This effectively halves the machine time required since libration periods are proportional to $\mu^{-1/2}$. Librating objects can in general be readily recognized by the typical oscillations (librations) of successive conjunctions with the secondary about the line of apsides. They also show characteristic variations in a and e.

3. Results

Figures 2–5 are plots that show, for representative values of X_0, those values of $\dot{X}$ and $\dot{Y}$ that lead to librating solutions at the 4/3 and 3/2 resonances. To interpret our results, we first need to place some limits upon the velocity space in which orbits of interest may exist. We have done this somewhat arbitrarily by restricting consideration to orbits whose eccentricity is less than a certain value. Thus we refine the question posed

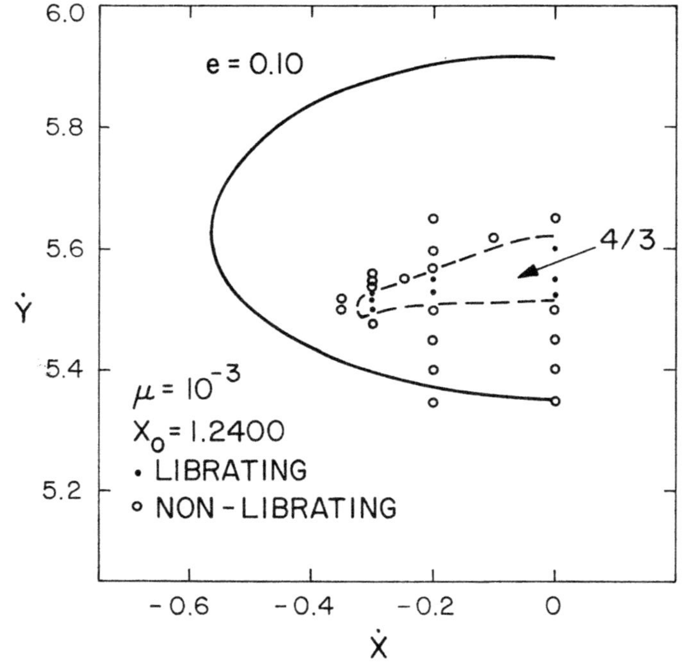

Fig. 2. A plot giving, for $X_0 = 1.24$ and an adopted Titan-Saturn mass ratio $\mu = 10^{-3}$, the values of $\dot{X}$ and $\dot{Y}$ for which stable librating orbits near the 4/3 resonance occur. The solid line marks the boundary inside of which orbits will have initial eccentricities less than 0.10.

earlier to ask what fraction of all orbits that pass through a given point on the X axis with eccentricities less than 0.1 at $t=0$ will librate at the 4/3 resonance. We have also assumed that the areas shown in Figures 2–5 can be used to define probabilities and that these probabilities do not depend critically upon the sampling region in position space – i.e., that no basic change would result had we mapped the librating region around $X=0$, $Y=Y_0$ instead of $X=X_0$, $Y=0$.

Figure 3 shows that, for a mass ratio of 10^{-3}, as X_0 falls below 1.33, the librating region for the 4/3 resonance begins to occupy a reasonable fraction of the area

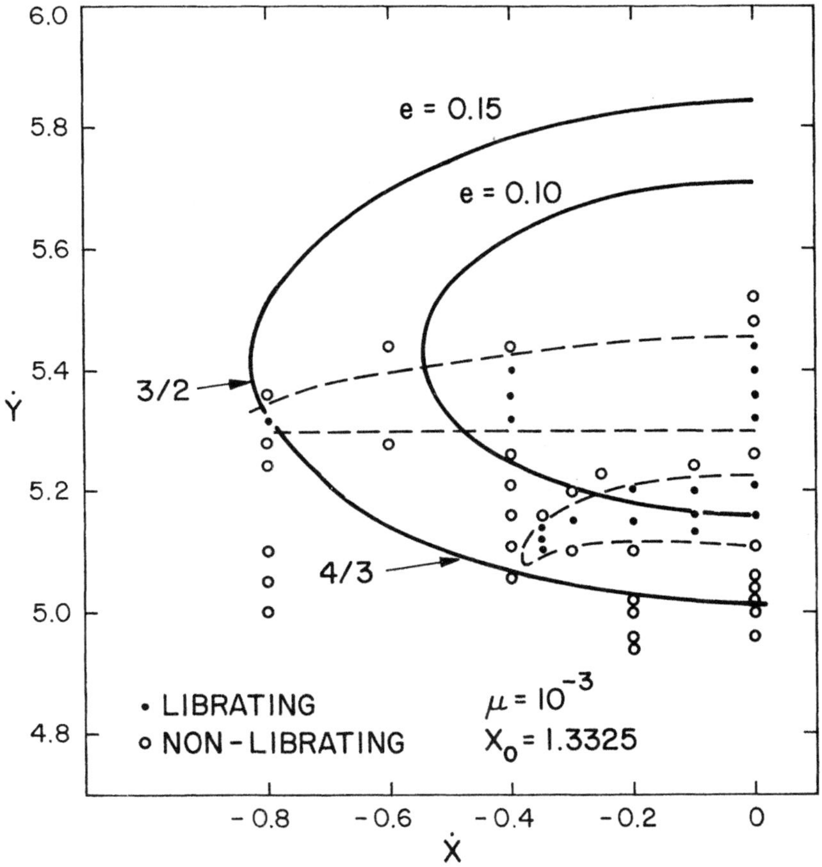

Fig. 3. The librating region near the 4/3 and 3/2 resonances for $X_0 = 1.3325$. Note here and in Figure 2 the absence of a stable librating region at the 5/4 resonance for this mass ratio.

defined by $e \leqslant .10$. As X_0 drops to 1.30, the librating region is essentially completely included and occupies about 14% of the total area. This situation prevails, as Figure 2 shows, at $X_0 = 1.240$, for which case the area occupied is still about 9%. It will continue at about the same value for slightly smaller X_0. Once X_0 decreases below a (4/3), or 1.211, then conjunction can occur at or near pericenter and the librating region will either vanish or be very small. Other studies (Lecar *et al.*, 1972) indicate that, for a mass ratio of 10^{-3}, a body may be permanent if its minimum distance from the less massive primary remains always >0.2 in our units.

We can now conclude, again relying on the case $\mu = 10^{-3}$, that in the region $1.21 \leqslant X_0 \leqslant 1.30$, about 10% of all orbits with $e \leqslant 0.10$ will show a stable libration at the 4/3 resonance. Use of the observed Titan-Saturn mass ratio of 1/4150 (Figure 5) indicates that in the actual situation the librating region increases in area leading to a somewhat greater probability. Figure 4 shows that the 4/3 resonance still persists at $X_0 = 1.44$ but that the librating bodies now have eccentricities too high (0.18) to be

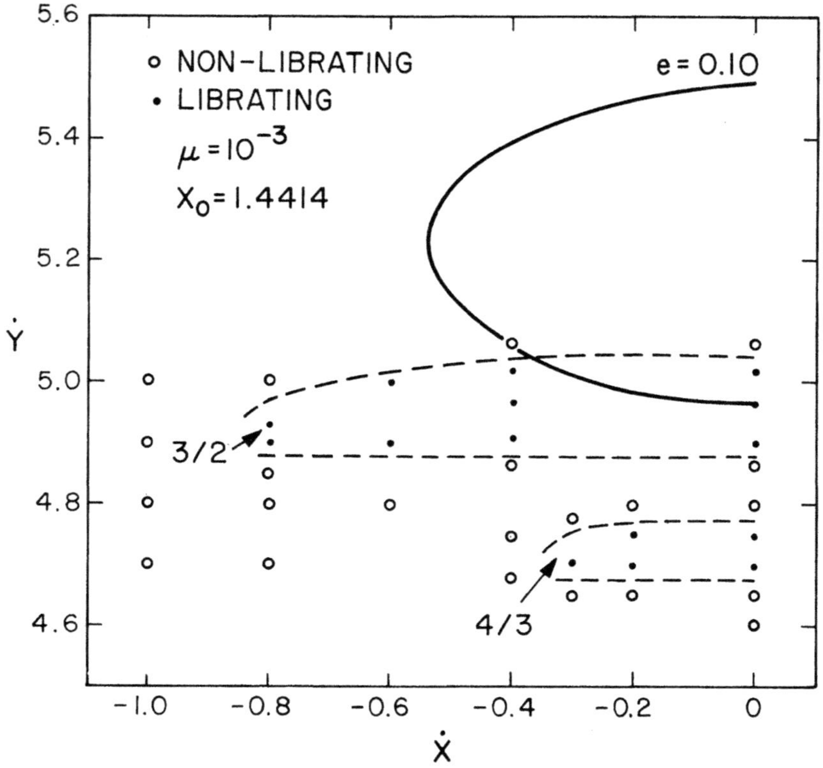

Fig. 4. As before for the case $X_0 = 1.4414$.

included in the $e \leqslant .10$ region. However, even for this value of X_0, and for larger ones, orbits of low eccentricity librating about the 3/2 resonance are quite probable. On the other hand, as is shown by the open circles representing non-librating orbits in Figure 2 at $\dot{Y} \cong 5.4$ and in Figure 3 at $\dot{Y} \cong 4.9$, the possible librating region associated with the 5/4 resonance appears to have vanished or at least become exceedingly small.

4. Concluding Remarks

We should like to stress that our results are limited and will need to be extended. We were anxious to establish whether, given the present Titan-Saturn mass ratio, the existence of a body librating at the 4/3 resonance could be considered a reasonable probable occurrence. The answer, as provided by Figure 2–5, appears to be yes. If, at an earlier time, Titan's mass relative to Saturn was much smaller, then the area of the librating region at 4/3 would apparently once have been greater (cf. Figures 3 and 5). This allows the possibility that the 4/3 (and 3/2) resonance with Titan was at one time populated with librating, Hyperion-like, objects that have become unstable as Titan's mass grew and the areas of the librating regions consequently diminished.

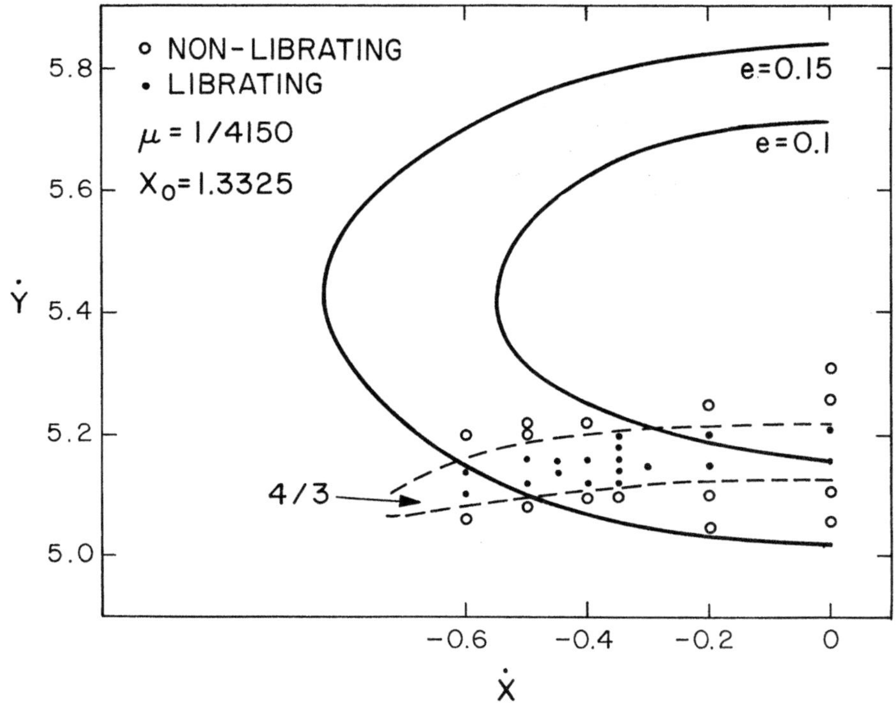

Fig. 5. Same as Figure 3, but with the observed Titan-Saturn mass ratio of 1/4150.

A similar behavior is shown by the Trojan points which would become unstable if the Jupiter-Sun mass ratio were increased to 0.038. In a later paper we shall examine this interesting question more precisely by allowing Titan's mass to vary with time. The possibility also remains that the mass ratio has not drastically changed.

In any event, the area of the 4/3 librating region calculated for the observed mass ratio of 1/4150 is still sufficiently large that the presence of 'one known Hyperion' is not improbable. It is however curious that Hyperion should be librating at the 4/3 resonance when the region associated with the 3/2 resonance (and presumably also with the 2/1 resonance) is substantially larger. Thus the 3/2 libration shown by the Pluto-Neptune system, whose location in the solar system resembles that of Titan and Hyperion in the Saturnian system, is in some sense the more probable. Recall also the somewhat similar situation in which the sole asteroid Thule librates at 3/4 Jupiter's period while some 40 bodies form the librating Hilda group at 2/3 that period. In view of the clear differences between this case and that of Titan-Hyperion, this comparison is only suggestive. These remarks seem to imply, although somewhat weakly inasmuch as we have available only a single satellite pair for discussion, that some process in the formation or evolution of the Titan-Hyperion system is still missing.

It is indeed a pleasure to thank Dr M. Lecar and Dr I. Shapiro for valuable discussions and Mr R. Loeser for great aid with the calculations.

References

Goldreich, P.: 1965, *Monthly Notices Roy. Astron. Soc.* **130**, 159–181.
Goldreich, P. and Soter, S.: 1966, *Icarus* **5**, 375–389.
Greenberg, R. J., 1972, 'Evolution of Orbit-Orbit Resonances in the Solar System', Unpublished Doctoral Thesis, Mass. Inst. of Technology, 110 pp.
Lecar, M. and Loeser, K. R.: 1972, in preparation.
Lecar, M., Colombo G., and Franklin, F.: 1972, Submitted to *Icarus*.

PLANETARY DISTANCES AND THE MISSING PLANET

M. W. OVENDEN

The Dept. of Geophysics and Astronomy, and The Institute of Astronomy and Space Science, The University of British Columbia, Vancouver, Canada

Abstract. Work (to be published elsewhere) leading to the formulation of The Principle of Least Interaction Action is summarized. The analysis of satellite orbits is here extended to systems with more than three satellites, and it is shown that the principle correctly predicts the major semi-axes of the five Uranian satellites and the five inner satellites of Jupiter to the precision of the known masses, allowing for the orbital eccentricities, and also the distribution of the five planets of Barnard's Star recently recognised.

Applied to the solar system, the principle is capable of representing all of the major semi-axes of the planets Mercury through to Neptune with an accuracy of the order of 1% provided that it is assumed that a mass of 90 $M_{\oplus}$ (where $M_{\oplus}$ is the mass of the Earth) existed in the asteroid belt from the beginning of the solar system until 1.6×10^7 yr ago, and then suddenly dissipated. A brief examination is made of the evidence from meteorites of such an event.

1. Introduction

Just two hundred years ago, Titius (1772) published a numerical rule which represented the relative major semi-axes of the then-known planets. His rule may be represented by the formula

$$a_i \propto 2^{(i-2)} \times 3 + 2^2$$

where a_i is the major semi-axis of the orbit of the planet with ordinal i, i being measured from Mercury outwards from the Sun. The formula gives the correct major semi-axes, to a few per cent, provided

(a) that for Mercury $i = -\infty$ instead of unity

(b) that the ordinal $i = 5$ was left vacant.

When Uranus was discovered (by William Herschel in 1781) it was found to fit closely the next orbital beyond Saturn, $i = 8$. Bode drew attention to the 'law', which became known as 'Bode's law'. The search for a missing planet for $i = 5$ led to the discovery of the first asteroid, Ceres, in 1801. Both Adams and Leverrier assumed in their calculations that 'Neptune' occupied the orbital $i = 9$; fortunately their predictions were insensitive to the assumed distance of the then unknown planet, because Neptune does not fit the law. The existence of a similar form of law for the satellite system of Saturn led to the discovery of Hyperion in 1848.

I have recently discussed the possible physical reasons for a quasi-harmonic distribution of planetary (or satellite) orbits (Ovenden, 1972). Suffice here to say that the similarities in the distribution laws for the planetary and satellite systems (Dermott, 1968) seems to rule out the 'chance' hypothesis (whatever that vague hypothesis may mean). The choice lies between

(i) physical processes operating at the time of formation of the solar system,

(ii) the effects of subsequent mutual perturbations, these perturbations being either

B. D. Tapley and V. Szebehely (eds.), Recent Advances in Dynamical Astronomy, 319–332.

point-mass perturbations or else perturbations involving dissipative forces (such as tidal action).

In my discussion already cited, I argued that mutual point-mass perturbations without dissipative forces were adequate to rearrange the planetary (and satellite) orbits on a sufficiently-short time scale. The results of this paper prove this conjecture to have been correct.

2. The Principle of Least Interaction Action

The intuitively-obvious conclusion that a system of planets (or satellites) will change its configuration slowly when the planets are far apart and quickly when they are close together and interacting violently, is confirmed by numerical integrations of simulated planetary systems (Hills, 1970; Ovenden, 1973; Ovenden *et al.*, 1973). This conclusion has been generalized by Ovenden (1973) to The Principle of Least Interaction Action, viz:

> A planetary or satellite system of N point masses moving solely under their mutual gravitational attractions spends most of its time close to a configuration for which the time-mean of the action associated with the mutual interactions of the planets or satellites is a minimum.

It has been shown (Ovenden, 1973; Ovenden *et al.*, 1973) that a minimum interaction action configuration is one for which the time-mean of the reduced disturbing function

$$\bar{R} = \sum_i \sum_{j>i} \overline{\frac{m_i m_j}{\varrho_{ij}}}$$

is a minimum (where ϱ_{ij} is the instantaneous separation of the ith and jth bodies). The theorem of Poincaré (1957) (see also Hori, 1960; Giacaglia and Nacozy, 1970), viz:

> A resonant gravitational system has the property that the disturbing function, averaged with respect to the critical or resonant argument, has a local minimum

means that, when the Principle of Least Interaction Action is applied to a two-satellite system, it leads to the resonant structures of the form

$$A_1 n_1 + A_2 n_2 + A_3 \dot{\omega} = 0$$

where the n's are the mean motions, and $\dot{\omega}$ is the angular velocity of either the apse line or the line of common nodes, according to the nature of the resonance. Since $\dot{\omega} \ll n_1, n_2$ and since (for the formula to be invariant to rotation of the local inertial frame) $\sum_i A_i = 0$, the Principle of Least Interaction Action leads directly to the near-commensurabilities of mean motion found to occur preferentially in the solar system (Roy and Ovenden, 1954) and explained by them (Roy and Ovenden, 1955) in a manner wholly consistent with the present arguments. These earlier results may thus be viewed as anticipatory confirmation of the Principle of Least Interaction Action.

To place the principle upon a more solid foundation, it is necessary

(a) to extend the analysis to more than two bodies

(b) to prove that the time-scales of evolution towards a minimum interaction action distribution are sufficiently short compared with the age of the solar system ($4.5 \times \times 10^9$ yr, see later).

The extension to three-satellite systems was made by Ovenden *et al.* (1973). They showed that the Laplacean triplets* of Uranus and Jupiter were not only close to a local minimum of the interaction action (as evidenced by the Laplace relationships) but that they were also close to *overall* minima. The calculation of $\bar{R}$ was made, of course, on the assumption that each triplet was an isolated gravitational system (together with the primary). Ovenden *et al.* (1973) also showed that, for both systems, the time-scale of evolution from an arbitrary configuration to the neighbourhood of the resonance was $\sim 10^7$–10^8 yr, and furthermore that the time required to approach resonance within the observed precisions of the two systems (3×10^{-4} for the Uranian triplet and 2×10^{-7} for the Jovian) was $\sim 5 \times 10^9$ yr for *both* systems.

It is the purpose of this paper to extend the techniques of Ovenden *et al.* (1973) to more than three satellites, and to apply the modified techniques to the planetary system.

3. Extension to More than Three Satellites

The essential problem is to arrive at an approximate sequence of configurations through which a given system must evolve. From such a sequence, the configuration of least interaction action *available to the system* may be determined, and from an estimate of the amount of energy available at any time for secular change, an estimate made of the time required for the system to evolve from one configuration of the sequence to the next.

To achieve this we make a number of approximations. In the first place, we assume the systems to be strictly coplanar. Secondly, since it is an observed fact that orbital eccentricities in the solar system are small, we may approximate the evolution (at least for limited periods from the present configurations) by circular orbits. For a three-satellite system, the sequence of quasi-circular, coplanar orbits may be completely identified using the integrals of energy and angular momentum.

Following Ovenden (1973) we set α_1, α_2, α_3 to be the radii of the circular orbits, and $x = \sqrt{\alpha_2/\alpha_1}$, $y = \sqrt{\alpha_3/\alpha_2}$. Assuming successive values of y, x is found from the solution to a quartic equation, and the scale-factor b found from the requirement of energy conservation. The quartic equation is derived from the supposition of constancy of both energy and angular momentum. A given sequence is *generated* by using the observed masses and major semi-axes of the satellite system as one member of the sequence. (Body 1 is the innermost body, body 3 the outermost.)

To estimate the time-scale of evolution through such a sequence, the following

* i.e those triplets showing the Laplace relationship between the mean motions, $n_1 - 3n_2 + 2n_3 \simeq 0$.

formula was derived (see Ovenden *et al.*, 1973):

$$t_{AB} = (P_{1g}/2\pi)\, m_1^{-2}\, e^{-1/2} \sum_A^B \left| \frac{F\, \delta\alpha_1}{\sqrt{\Delta R}} \right| \tag{1}$$

where P_{1g} is the observed period of body No. 1 (i.e. at the generating configuration);
m_1 is the mass of body No. 1 relative to the primary body;
$e^{1/2}$ is the mean of the square-roots of the eccentricities of the satellites

$$F = \sum_1^3 \mu_i (\delta\alpha_i/\delta\alpha_1)^2 \qquad (\mu_i = m_i/m_1)$$

$$\Delta\bar{R} = \left\{ \frac{1}{2\bar{R}} \sum_i^3 \sum_{j>i}^3 \frac{\mu_i\mu_j^3 + \mu_j\mu_i^3}{\varrho_{ij}} \right\} \{\bar{R} - \bar{R}_0\}.$$

Here $\bar{R}_0$ is the value of $\bar{R}$ at the minimum interaction configuration and $\delta\alpha_1$ is the change in α_1 from one member of the calculated sequence of configurations to the next, in units of α_1 *at the generating configuration.*

The extension to more than three satellites is not trivial. The two known integrals (energy and angular momentum) do not suffice to determine more than two quantities (in the above case, x for any assumed y, and the corresponding scale-factor b). The clue to the extension lies in the fact that, as far as locating the configuration of minimum interaction action, each triplet could be considered as an isolated system, notwithstanding the fact that simple calculation shows that the differential effects of the other satellites in the systems are far from negligible, and would produce energy changes in the triplets. The *relative configuration* of least interaction action for a given energy and angular momentum is determined primarily by the values of x and y, and only to a small extent by changes in the scale-factor b. The triplet results indicate that, at least in the neighbourhood of the minimum configuration, each triplet of a larger system evolves through a sequence of configurations (x, y) essentially similar to that through which it would have evolved if isolated, although the *scale factor* for each configuration must be adjusted so that the total energy of the whole system is conserved. Therefore I adopted the following procedure for generating an evolutionary sequence for a system of more than three satellites:

Consider a sequence (x, y) computed for the outermost triplet of the system. For each value of y, x is computed as in Ovenden *et al.* (1973) but the scale factor b is not evaluated. The value of x then becomes the y-value for the next triplet proceeding inwards (e.g. for a 6-satellite system, the first triplet would be 654, the next 543, etc.) This process is repeated $(N-2)$ times until the innermost triplet 321 has been incorporated. Then the scale factor is determined from the requirement of constant energy.

In principle, the procedure does not necessarily ensure conservation of angular momentum through the sequence. However, in practice, except when two satellites come very close together, the fractional error in the angular momentum of the system is less than 10^{-2}. (In all the figures in this paper, parts of any evolutionary sequence for which the fractional error in the angular momentum exceeds 5×10^{-3} are shown with broken lines.) Hence errors in angular momentum can be easily accommodated by orbits with small eccentricities, thus not violating the approximations of the quasi-circular analysis.

For each configuration of the sequence generated as above, $\bar{R}$ is calculated, and the overall minimum interaction action configuration found. (The computations given later are all restricted to evolutionary sequences for which the *order* of the planets or satellites in the system remains unchanged.) The generalization of Equation (1) for time-scales is trivial, merely requiring the redefinition of F as below and the corresponding change in $\Delta\bar{R}$.

$$F = \sum_{i=1}^{N} \mu_i(\delta\alpha_i/\delta\alpha_1)^2 .$$

The calculated time of evolution of a system from an 'arbitrary' configuration to (say) the present configuration obviously depends upon the choice of 'arbitrary' configuration. However, the evolution is (relatively) so rapid far from minimum interaction that usually only small changes in the estimated times are made for quite large variations in 'arbitrary' configuration.

I should emphasise that the analysis is clearly not rigorous, and makes no pretence to be so. We cannot afford the luxury of rigour if we are to consider evolution over times of the order of 10^9 revolutions or more. The validity of the approximations made should be judged only at the end of the paper, after the full picture has been described. I claim that the completely self-consistent, quantitative and accurate description given of the present planetary positions, together with totally independent confirmation from outside the domain of pure dynamics, gives compelling reasons for the acceptance of the essential correctness of the arguments here presented.

4. Application to the Uranian and Jovian Satellite Systems, and to Barnard's Star

The system of five Uranian satellites provides a good test of the procedure described above, in that the satellites have orbits with small eccentricities that are sensibly coplanar, and there are no other (known) major satellites to cause perturbations. The only limitation is that the mass of the innermost satellite, UV, is poorly known, being tabulated to only one significant figure*. These results are displayed in Figure 1, with distances on a logarithmic scale. Such a scale exaggerates discrepancies for the innermost bodies relative to those of the outermost bodies, but is physically more meaningful than a linear scale (in terms of energies).

In interpreting the results of Table I, the following points should be borne in mind:

(i) The masses of the satellites are more or less uncertain. The shaded area in

* Planetary and Satellite data are taken from *The Handbook of the British Astronomical Association for 1972.*

Figure 1 shows the changes in the minimum interaction distribution consequent upon varying the relative mass of UV from 1×10^{-6} (the tabulated value) to 2×10^{-6}.

(ii) The minimum interaction configuration is calculated on the circular approximation. The actual eccentricities will cause the system to occupy a local minimum in the neighbourhood of the calculated overall minimum so that exact agreement is not to be expected. The establishment of the Laplace relationship is the tending of the system towards a minimum of $\bar{R}$ that depends upon the eccentricities.

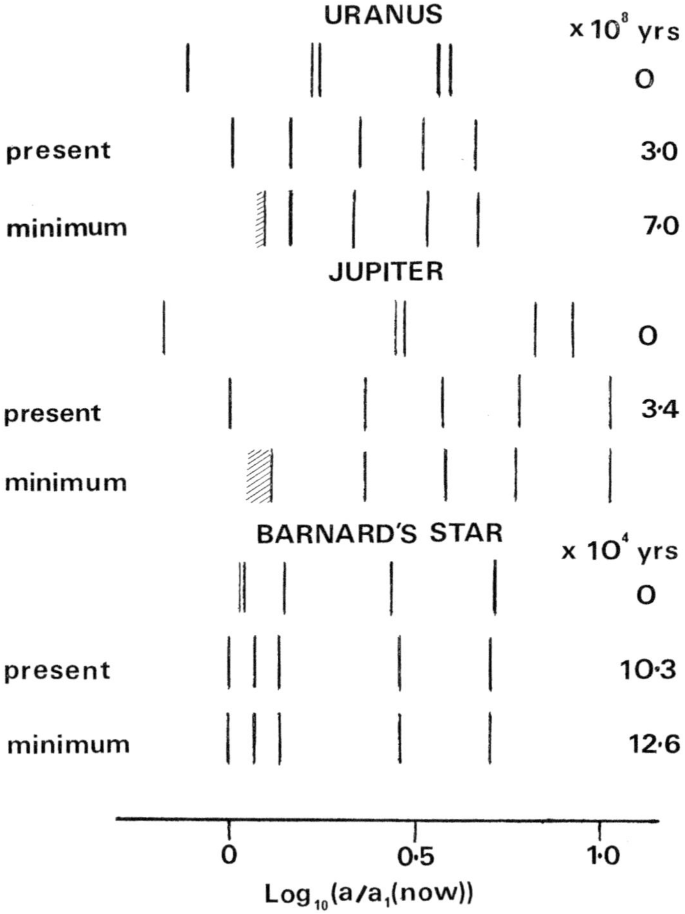

Fig. 1.

(iii) If the mass of one of the bodies (m_i, say) is reduced to zero, the evolutionary sequence calculated will be that of the variation of m_i only. In the case $m_i \to 0$, the minimum configuration necessarily duplicates exactly the major semi-axes of the remaining $(N-1)$ satellites, because m_i is unable to perturb them. However, such a case is immediately recognised in the calculations, for the range of configurations which may be generated by the technique used will involve negligible changes in the satellites other than i. The time-scale of such a degenerate sequence will be small, and on a longer time-scale the remaining $(N-1)$ satellites will proceed with their own undisturbed evolution. In Figure 1 we see that, for the Uranian system, UV is massive enough to cause significant evolution of the remaining satellites. The minimum interaction action configuration found is thus a true minimum for the system.

(iv) Since the time-mean reciprocal separation is evaluated (using a 20th-order polynomial) only to a relative precision of 10^{-4}, if one satellite is much less massive than the others, its location in the minimum interaction distribution will be poorly determined, since its contribution to the calculated $\bar{R}$ may become of the same order as the errors in calculating the larger contributions of the other satellites.

TABLE I

Uranian system

Satellite	e	m^{-1}
UV	0.00	1000000
UI	0.0028	67000
UII	0.0035	170000
UIII	0.0024	20000
UIV	0.0007	34000

Sample configurations

	a_{V}	a_{I}	a_{II}	a_{III}	a_{IV}	$\bar{R}$	T
	6.64	14.42	14.94	31.09	33.52	11.92	0.00
	6.74	13.84	15.66	30.49	35.74	10.61	0.73
	7.31	13.28	16.71	29.88	37.75	9.93	1.72
Present	8.72	12.82	17.86	29.30	39.19	9.59	2.99
	9.99	12.64	18.36	29.04	39.63	9.51	4.18
Minimum	10.83	12.57	18.55	28.92	39.76	9.50	(7.0)
	With $m^{-1}{}_{V} = 500000$						
	7.51	14.44	14.94	31.10	33.51	30.73	0.00
	7.55	13.83	15.71	30.47	35.87	27.29	0.51
	7.89	13.30	16.68	29.91	37.72	25.75	1.11
Present	8.72	12.82	17.86	29.30	39.19	24.90	1.84
	9.30	12.64	18.37	29.04	39.64	24.68	2.56
Minimum	10.29	12.44	18.93	28.73	40.01	24.60	(6.4)

e = present orbital eccentricity; m^{-1} = reciprocal relative mass;
a = major semi-axis, in units of 10^{-4} AU.
$\bar{R}$ = time-mean interaction action; arbitrary units, different in two parts of the table.
T = evolution time from first tabulated configuration, in units of 10^8 yr

Because of these limitations, it is concluded that the present configuration of the 5 Uranian satellites shown in Table I and Figure 1 is a good approximation to the minimum interaction action distribution available to the system, as is to be expected from a calculated evolutionary time-scale of 3×10^8 yr.

Table II shows similar results for the satellites V, I, II, III, IV of Jupiter. (The remaining satellites are far out and of very small (actually untabulated) mass.) In this case, no mass is quoted at all for JV. I have assumed a mass of 10^{-6}. The shaded area in Figure 1 shows the change in the minimum interaction distribution if m_v is changed to 2×10^{-6}. Again, the agreement of the observed system with the calculated minimum interaction distribution is good.

TABLE II

Jovian system

Satellite	e	m^{-1}				
JV	0.003	(1000000)				
JI	0.000	26200				
JII	0.001	40300				
JIII	0.0014	12200				
JIV	0.0074	19600				

Sample configurations

	a_{V}	a_{I}	a_{II}	a_{III}	a_{IV}	$\bar{R}$	T
	6.18	33.87	35.67	78.28	100.2	24.44	0
	6.82	31.17	39.01	75.93	112.4	21.61	0.96
	8.43	29.45	42.02	73.72	120.4	20.67	1.99
	10.52	28.56	43.97	72.24	124.3	20.37	2.85
Present	12.10	28.19	44.86	71.55	125.8	20.29	3.41
	13.70	27.94	45.49	71.04	126.8	20.26	4.06
Minimum	15.85	27.71	46.07	70.56	127.6	20.24	(10.0)
	With $m^{-1}{}_{\mathrm{V}} = (500000)$						
	8.09	33.90	35.70	78.34	100.3	6.207	0
	8.61	31.19	39.04	75.99	112.5	5.502	0.54
	8.99	30.48	40.17	75.16	115.8	5.392	0.75
	11.23	28.56	43.97	72.25	124.4	5.198	1.66
Present	11.65	28.37	44.42	71.89	125.1	5.187	1.82
	12.85	27.94	45.49	71.04	126.8	5.172	2.37
Minimum	13.62	27.73	46.02	70.60	127.5	5.168	(4.9)

e = present orbital eccentricity; m^{-1} = reciprocal relative mass;
a = major semi-axis, in units of 10^{-4} AU.
$\bar{R}$ = time-mean interaction action; arbitrary units, different in two parts of table
T = evolution time from first tabulated configuration, in units of 10^8 yr.

Finally, as an additional example of the application of the Principle of Least Interaction Action to a system of 5 satellites, we take the planetary system of Barnard's Star, as determined recently by Jensen and Ulrych (1972) using a maximum-entropy method of analysing the observations. The results are given in Table III and displayed in Figure 1.

I consider that these results establish the validity of the Principle of Least Interaction Action. The principle is now applied to the solar planetary system, with unexpected results.

TABLE III

System of Barnard's star

Planet	e (assumed)	m^{-1} [a]
BI	(0.01)	105.2
BII	(0.01)	100.0
BIII	(0.01)	90.9
BIV	(0.01)	192.3
BV	(0.01)	101.0

Sample configurations

	a_{I}	a_{II}	a_{III}	a_{IV}	a_{V}	$\bar{R}$	T
	1.005	1.024	1.328	2.543	4.846	5.480	0
	0.983	1.047	1.323	2.597	4.814	5.099	1.12
	0.974	1.059	1.319	2.622	4.793	5.019	1.98
	0.963	1.076	1.321	2.657	4.757	4.951	3.90
[a]Present	0.950	1.100	1.300	2.700	4.700	4.912	10.30
Minimum	0.946	1.109	1.295	2.715	4.676	4.909	(12.6)

e = *assumed* orbital eccentrcity; m^{-1} = reciprocal relative mass;
a = major semi-axis, in AU.
$\bar{R}$ = time-mean interaction action, arbitrary units.
T = evolution time from first tabulated configuration, in units of 10^{-4} yr.
[a] = data from Jensen and Ulrych (1972).

5. The Solar Planetary System

I first consider the system of Jupiter, Saturn, Uranus and Neptune, as the motions of these massive planets is hardly affected by the terrestrial planets, or Pluto. In fact, Pluto is not considered at all, because Nacozy and Diehl (1972) have shown that, considering the orbits of Jupiter, Saturn, Uranus and Neptune to be unchanging, Pluto is in a configuration corresponding to a local minimum of the interaction action.

In the lower part of Figure 2 I show the calculated evolution of the JSUN system. It is at once obvious that the system is far from the configuration of minimum interaction action, although the time-scale for evolution to the vicinity of minimum is $\sim 10^9$ yr. This contradiction may be removed very precisely by supposing that, for most of the life-time of the solar system, there was significant mass residing in the asteroid belt, which suddenly dissipated, in the relatively recent past. Without prejudice to later discussion, this material will be referred to as A.

As a first try, I suppose that A was exactly in the asteroid belt. In the upper part of Figure 2 I show the *minimum configurations* of the system AJSUN (with A at 2.8 AU at the generating configuration) as a function of the mass of A in units of the mass of the Earth. For my hypothesis to be tenable, the following stringent conditions must hold.

(i) There must be a minimum interaction configuration, corresponding to some value of the mass of A, M_A, which yields the same *absolute values* for the major semi-axes of JSUN as does *some* past configuration in the evolutionary sequence of JSUN, since,

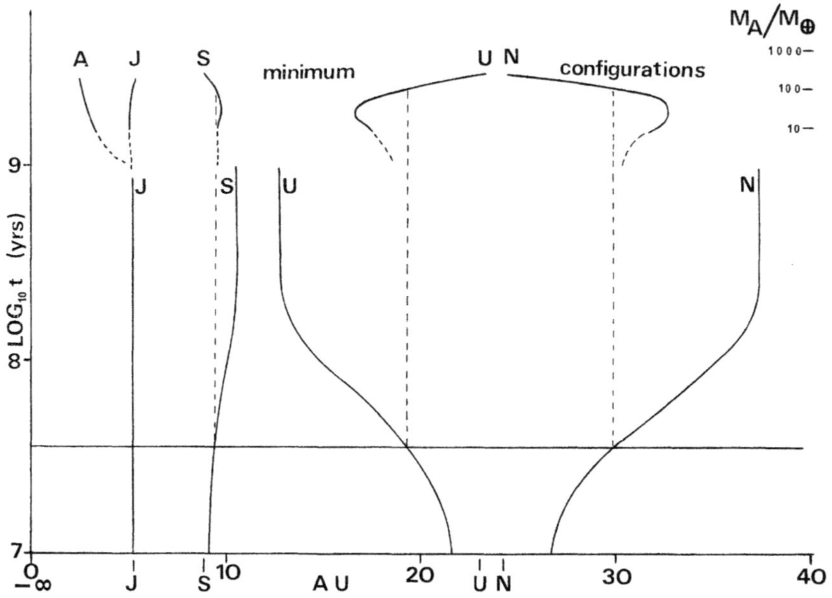

Fig. 2.

after the dispersal of A, JSUN began to evolve towards its new minimum interaction action configuration.

(ii) The time-scale for the evolution of AJSUN must be short enough for the minimum interaction configuration to have been established prior to the dispersal of A.

(iii) The time-scale for the evolution of the residual material between Jupiter and Mars must be sufficiently long for no appreciable change to have occurred in its major semi-axis (on the average) since the dispersal of A. (Had this not been true, it would have been necessary to try an iteration procedure whereby the assumed position of A was changed in such a way that the evolution of the residue after the dispersal would have given the present major semi-axis of 2.8 AU.)

(iv) The present configuration of the inner planets with Jupiter, MeVEMsJ, must correspond to the evolution of this system *from* the minimum interaction configuration of MeVEMsAJ *towards* the minimum configuration of the MeVEMsJ system. However, now both M_A and the epoch of dispersal of A are known, so that there is *no freedom of adjustment* at all for the MeVEMsJ system.

All these conditions are very accurately fulfilled if the mass of A is taken to have been 90 $M_\oplus$, with dissipation having occurred at 1.6×10^7 yr ago.

Since the dispersal of A took place relatively recently, the scale of Figure 2 is too coarse to show the process of matching the minimum configurations with the evolutionary sequence of JSUN. In Figure 3 we plot a small part of the JSUN sequence, now on a linear negative time-scale with the present being $t=0$, so that the indicated times are times since the dispersal of A. Superimposed upon this is the sequence of

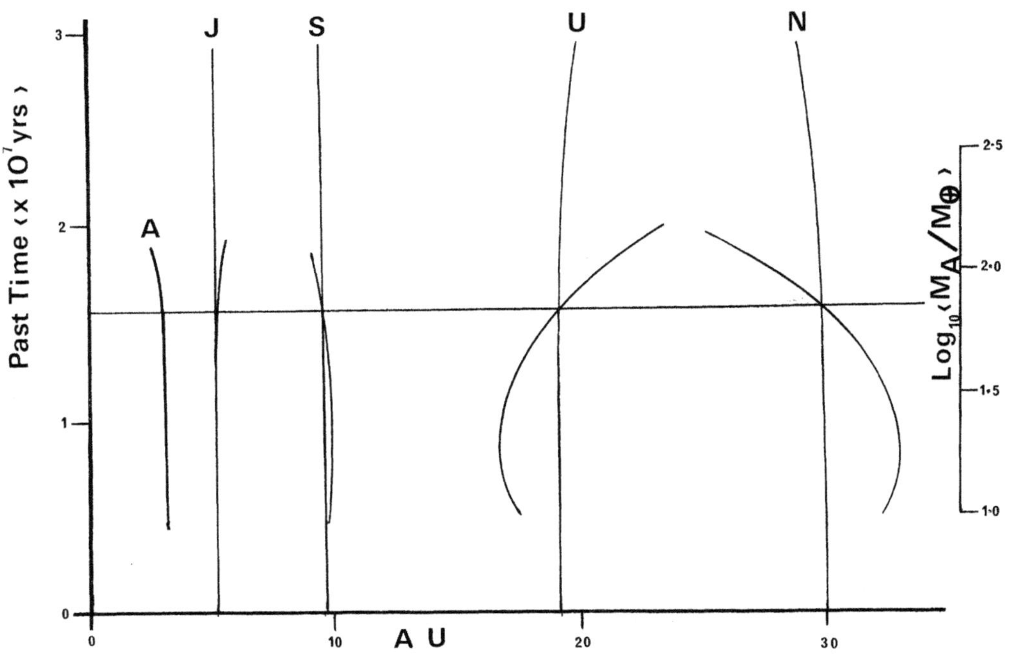

Fig. 3.

minimum configurations of AJSUN, as a function of the mass of A, correctly placed to give the best fit. The closeness of fit is indicated by the following comparison of the calculated major semi-axes (found from the condition that the intersections of the two graphs shall correspond to a definite past time):

$M_A = 90\ M_\oplus$	A	J	S	U	N	
Major semi-axes (AU)	2.794	5.211	9.509	19.46	29.71	Min. config.
	–	5.206	9.506	19.40	29.74	-1.6×10^7 yr
	–	5.203	9.539	19.18	30.06	Present
	–	5.15	10.5	12.7	37.3	Min. config. ($+2.8 \times 10^8$ yr)

In assessing the significance of this agreement, it must be recalled that both the sequence of minimum configurations as a function of M_A and the evolutionary sequence of JSUN are *one-parameter* sequences, yet the fit is to five absolute values of major semi-axis, since the *minimum* configuration must have A close to 2.8 AU (see next section). Also the epoch found for dispersal of A is not merely possible (i.e. in the past) but is also a *physically-significant* epoch (see Section 6).

Calculations have been made for the EMsAJ system with masses $\sim 0.1\ M_\oplus$ in the position of A, to represent the residue of the dispersed material. The evolution is such that, in 1.6×10^7 yr, changes ~ 0.05 AU only are to be expected in the major semi-axis of the residuum, thus justifying the placing of A at the present distance of the asteroid belt. In fact, it is not possible to get a fit with JSUN nearly as precise as that found above with A (at the generating configuration) in any position other than 2.8 ± 0.1 AU.

It might be thought, from the upper part of Figure 2, that another solution would be possible with very much smaller M_A. The solid lines here terminate when the contribution of A to the interaction action has become of the same order as the errors in evaluating $\bar{R}$. However, the discussion of Section 4, (iii), shows that in the limit as $M_A \to 0$, the minimum configuration will tend to the generating configuration, since all evolution of the other bodies would have been suppressed. Hence the 'solution' for very small M_A would simply be JSUN stationary in their present positions. Clearly, then, the insertion of material $\sim 0.1\ M_\oplus$ in the present asteroid belt could not make the present positions of JSUN correspond to a true minimum interaction distribution. We must therefore adopt 90 $M_\oplus$ as the mass of A.

Knowing now the mass of A and the epoch of its dispersal, we ask whether the present configuration of MeVEMsJ is such as to have evolved from the minimum configuration of MeVEMsAJ in a time of 1.6×10^7 yr. In fact the evolution of the MeVEMsJ system is so slow that no appreciable change should have taken place in the relative configuration of the system since adispersal. The agreement is good.

$M_A = 90\ M_\oplus$	Me	V	E	Ms	A	J
Minimum configuration	0.394	0.719	1.00	1.49	2.79	5.20
Present configuration	0.387	0.723	1.00	1.52	–	5.20
Minimum configuration	0.461	0.696	1.02	1.24		5.21

It is seen that the major discrepancies are for the planets with the largest eccentricities, Mercury and Mars. Their eccentricities are 0.205 and 0.093 respectively, so that the errors are smaller than the eccentricities, and are thus of the order to be expected if the actual contribution to $\bar{R}$ of these planets is approximated by calculations that assume circular orbits.

In addition, the time-scale for evolution of the system MeVEMsAJ from an arbitrary configuration to minimum $\sim 5 \times 10^8$ yr (short enough for the minimum to have been established) while the time-scale for the system MeVEMsJ to reach its new minimum configuration $\sim 7.6 \times 10^{11}$ yr.

The hypothesis thus serves to explain the absolute major semi-axes of the planets from Mercury through to Neptune with an accuracy $\sim 0.5\%$, save for the eccentric planets Mercury and Mars where the accuracy $\sim 2\%$, of the order to be expected from the circular approximation made.

6. Evidence from the Meteorites

I now turn to totally independent evidence, given by the physical structure of meteorites, of a 'sudden event' occurring at about the time found above for the dispersal of A. This question will be discussed more fully elsewhere. Here I wish simply to indicate the extent to which the dynamical conclusions are confirmed or contradicted by the meteorite data. (For a general reference on meteorite physics, see Wood (1968).)

(1) Rubidium-Strontium and Uranium-Lead isotropic distributions yield an average age of 4.5×10^9 yr, this being the time that has elapsed since the parent meteoritic material had existed in a well-mixed state. This is what is usually called 'the age of the solar system'.

(2) The crystalline structure of meteorites shows that they were once part of a larger body (or bodies). The crystalline structure yields cooling rates ~ 1–$10\,°C$ per 10^6 yr, which proves that the parent material must have lain at least 100 km below the surface of a larger body.

(3) The degree of heating evidenced in the physical state of the meteorite material seems to demand pre-meteorite bodies of planetary size, although this conclusion is controversial.

(4) The different chemical compositions of meteorites is usually taken to imply that the parent meteorite body (or bodies) was large enough for fractionation to have taken place in its gravitational field. The critical mass seems to lie between that of the Moon (unfractionated) and that of the Earth (fractionated).

(4) Cosmic-ray exposure ages: These are found from the isotopic residue of spallation reactions. Since cosmic rays can penetrate only about a metre or so of rocky material, the cosmic-ray exposure ages give the time since the meteorites became bodies of roughly their present sizes. The cosmic-ray exposure ages of chondrites and achondrites lie between 2.2×10^7 yr and 5.0×10^6 yr, with a sharper boundary at the long-time end and a tail to the distribution of ages at the short-time end. This is wholly consistent with the view that a large body broke into pieces, some already smaller than the critical cosmic ray exposure size, at 2.2×10^7 yr ago, and that the larger fragments were subsequently gradually broken-up by collisions. The agreement between the earliest chondrite exposure age and the time found for the dissipation of A is good. (However, it should be mentioned that some iron meteorites show much longer cosmic-ray exposure ages, $\sim 6 \times 10^8$ yr. This is difficult to understand on any hypothesis that the iron meteorites came from the same parent bodies as the chondrites, since the iron meteorites are stronger and would thus resist being broken up into pieces. Their cosmic-ray exposure ages would be expected to be smaller than those of the chondrites, not larger.)

7. Postscript

As far as the dynamical argument of this paper is concerned, it would seem that the mass of A could have been in the form of a single planet, several smaller planets, or indeed distributed in a ring. Actually, it is just within the limits of the arguments given in this paper that all of the 90 $M_\oplus$ might still be in the asteroid belt. The errors in the planetary distances then become somewhat larger, since the second row of the table on page 326 would then have to be taken as an approximation to the third row.

The main objection to this hypothesis is that observation has yielded an estimate $\sim 0.1\ M_{\oplus}$ for the mass of the asteroidal material. Also, I would have expected that this amount of matter, currently present, could have been detected from the very precise knowledge that we have of the current motions of the planets.

If the mass of A did indeed disperse suddenly, the hypothesis of a single original planet is more attractive than that of a ring; while we may know of no mechanism whereby a planet might 'explode' with sufficient violence to disperse 99.9% of its mass, it seems even harder to understand why a ring of matter should have existed undisturbed for 4.5×10^9 yr, and then suddenly dissipate 1.6×10^7 yr ago.

The following problems remain:

(i) What physical process caused the dissipation of A?

(ii) Only $\sim 0.1\ M_{\oplus}$ seems to remain in the asteroid belt. What has happened to the other 89.9 $M_{\oplus}$?

(iii) If the object A had been a planet, what was its name?

It is customary for the discoverer of a new astronomical object to have the privilege of naming it. I know of no precedent for the discoverer of a non-existent object, but I prehend to myself the privilege of naming A 'Planet AZTEX'. The name is derived from 'AeroSpace department of the university of Texas at Austin' whose hospitality I was enjoying when the hint of Aztex first appeared. The name has been 'corrupted' in recognition also of the hospitality I received at El Instituto de Astronomía de la Universidad Nacional Autónoma de México during the time when much of this investigation was being pursued.

Acknowledgements

My thanks are also due to the University of British Columbia for the granting to me of a year's leave of absence.

This work was supported in part by a grant from the National Research Council of Canada.

References

Dermott, S. F.: 1968, *Monthly Notices Roy. Astron. Soc.* **141**, 363.
Giacaglia, G. E. O. and Nacozy, P. E.: 1970. in Giacaglia (ed.), *Periodic Orbits, Stability and Resonances*, D. Reidel Publ. Co., Dordrecht, Holland, p. 96.
Hori, G.: 1960, *Astrophys. J.* **74**, 1254.
Jensen, O. G. and Ulrych, T.: 1972, Private communication.
Nacozy, P. E. and Diehl, R.: 1972, Private communication.
Ovenden, M. W.: 1973, *Vistas Astron.* **16**, in press.
Ovenden, M. W., Feagin, T., and Graf, O.: 1973, *Celes. Mech.* (in press).
Poincaré, H.: 1957, *Les Méthodes Nouvelles de la Mécanique Céleste*, Dover.
Roy, A. E. and Ovenden, M. W.: 1954, *Monthly Notices Roy. Astron. Soc.* **114**, 232.
Roy, A. E. and Ovenden, M. W.: 1955, *Monthly Notices Roy. Astron. Soc.* **115**, 296.
Wood, J. W.: 1968, *Meteorites and the Origin of Planets*, McGraw-Hill.

CHANGE OF INCLINATION IN PASSING THROUGH RESONANCE

R. R. ALLAN
Royal Aircraft Establishment, Farnborough, Hants., England

Abstract. The Earth's gravity anomalies can give rise to significant changes in inclination for a close satellite when its motion is commensurable with the Earth's rotation. For a decaying satellite passing through resonance, the change in inclination depends on the value of a resonant variable at exact commensurability (an essentially random quantity). Many different gravity coefficients may contribute significantly, with relative amplitudes which are highly dependent on inclination.

An approximate solution can be found in terms of Fresnel integrals, which is valid in the usual situation when the drag greatly exceeds the resonance forces. This shows that the inclination is almost equally likely to increase or decrease and that the total change is proportional to $(\text{drag})^{-1/2}$, i.e. to the time taken to pass through resonance.

The effect offers a way of deriving gravity coefficients of medium order (e.g. $m = 15$) from the observed magnitude and shape of the variation in inclination. The magnitudes of even higher order gravity coefficients obtained from some resonance with $\alpha = 2$ (e.g. 29/2) or even with $\alpha = 3$ (e.g. 44/3) might yield information on the depth of the sources of the high order gravity field.

1. Introduction

It is now well-established (Gaposchkin and Lambeck, 1971) that the Earth's gravity depends slightly but significantly on longitude, corresponding to maximum variations of between ± 50 m and ± 100 m in the height of the geoid. Since the Earth is rotating, the motion of an artificial satellite is not conservative, and energy and angular momentum are being continuously transferred between the Earth and the orbital motion of the satellite. Usually the longitude-dependent terms lead only to short period effects which on balance quickly average out to zero. This is not so, however, if the motion of the satellite is nearly commensurable with the rotation of the Earth, or, to be precise, if the path of the satellite repeats relative to the Earth.

Here we concentrate on one aspect of the exchange of angular momentum that can readily be observed, namely the change in inclination for close satellites passing through resonance (Allan, 1967a) as their orbits contract due to air drag. So far the effect has been observed and analysed (Gooding, 1971a; King-Hele and Hiller, 1972 and King-Hele, 1972) only for the 15/1 resonance, which occurs when the mean height reaches some value between 480 km and 660 km depending on the inclination. This height is such that a goodly number of satellites pass through the resonance, yet the drag is still sufficiently low that passage through resonance is slow enough to allow a significant change in inclination to build up. The effects will be similar, however, for other resonances such as 14/1 (or possibly 16/1 for a low-drag satellite), or for weaker resonances of the form 29/2, 31/2, etc., or for even weaker resonances of the form 43/3, 44/3, etc.

For the particular example of the 15/1 resonance, each successive ascending node is 24°W of the preceding one at exact commensurability, and the pattern repeats after

B. D. Tapley and V. Szebehely (eds.), Recent Advances in Dynamical Astronomy, 333–348.

one sidereal day (approximately, since there is a correction for regression of the orbital plane). Consider every fifteenth ascending node on the Earth: As the orbit decays towards resonance, the node is drifting slowly westwards but is decelerating, reaches a maximum westwards longitude at exact resonance, and then accelerates eastwards again. Thus at resonance the satellite threads its path amongst the gravity anomalies on the Earth in exactly the same way on successive days, so that whatever the net deflection is on one day, it is the same on the next day, and so builds up for as long as the ground track is effectively stationary.

This physical explanation reveals the important factors which influence the change in inclination:

(i) The appropriate gravity coefficients; or, what is equivalent, the distribution of gravity anomalies in magnitude and position (including depth).

(ii) How fast the satellite passes through resonance.

(iii) Where the ground track reverses, which is effectively a random variable.

(iv) The inclination; the satellite will get a very different view of the gravity anomalies if the inclination is, say, 10° different.

Clearly the effect offers a way of deriving at least statistically significant values of gravity coefficients of certain medium and high orders. In turn, these will help to determine the depth of the sources of the high order gravity field (Allan, 1972a)

2. Mathematical Development

The longitude-dependent part of the geopotential, from which the resonance effects arise, can be developed in terms of conventional osculating elliptic elements, a, e, I, Ω, ω and M, referred to the equator, in the standard form (Allan, 1967b; Kaula, 1966)

$$U' = (\mu/R) \sum_{l,m,p,q} \bar{J}_{lm} (R/a)^{l+1} \mathscr{R} i^{l-m} \bar{F}_{lmp}(I)\, G_{lpq}(e) \exp(i\psi_{lmpq}), \tag{1}$$

where the limits on the summations are

$$2 \leqslant l \leqslant \infty, \quad 1 \leqslant m \leqslant l, \quad 0 \leqslant p \leqslant l, \quad -\infty \leqslant q \leqslant \infty, \tag{2}$$

and

$$\psi_{lmpq} \equiv (l-2p)\,\omega + (l-2p+q)\,M + m\,(\Omega - n_0 t - \phi_{lm}), \tag{3}$$

where $n_0 t$ is the sidereal hour angle. In (1), $\mu = G \times$ the mass of the Earth, and R is the mean equatorial radius. The gravity parameters are given as the normalized amplitude $\bar{J}_{lm}$ and the phase ϕ_{lm}, where the normalization is defined (Kaula, 1966) by taking the normalized spherical harmonics as

$$N_{lm} P_l^m (\cos\theta) \begin{matrix}\cos\\ \sin\end{matrix}\, m\phi, \quad 0 \leqslant m \leqslant l,$$

with

$$N_{lm}^2 = (2 - \delta_{m0})(2l+1) \frac{(l-m)!}{(l+m)!}, \tag{4}$$

chosen so that the integral of the square of a normalized function over the unit sphere is 4π. The normalized $\bar{C}$- and $\bar{S}$- gravity coefficients are then given by

$$\bar{C}_{lm} = \bar{J}_{lm} \cos m\phi_{lm}, \qquad \bar{S}_{lm} = \bar{J}_{lm} \sin m\phi_{lm}.$$

In (1), the normalized inclination function $\bar{F}_{lmp}(I)$ is as defined previously (Allan, 1967b) except that it includes the normalization factor N_{lm}, and the complex factor has been removed and written explicitly in (1); thus

$$\bar{F}_{lmp}(I) = N_{lm} \frac{(l+m)!}{2^l p!(l-p)!} \sum_k (-)^k \binom{2l-2p}{k} \times \\ \times \binom{2p}{l-m-k} c^{3l-m-2p-2k} s^{m-l+2p+2k}, \tag{5}$$

where $c = \cos\frac{1}{2}I$, $s = \sin\frac{1}{2}I$, and the summation is over all permissible values of k, i.e. from $k = \max(0, l - m^{-2}p)$ to $k = \min(l-m, 2l-2p)$. The eccentricity function $G_{lpq}(e)$, which is exactly as defined by Kaula (1966), is simply a more convenient notation for the Hansen coefficient (Plummer, 1918) $X_{l-2p+q}^{-l-1,\, l-2p}(e)$, and is of order $e^{|q|}$.

The axially-symmetric part of the potential does not play an essential rôle in the resonance, although it affects quite markedly the heights (and periods) at which the resonances occur for different inclinations. A discussion of this point can be found elsewhere (Allan, 1972b) and will not be considered here.

Equations for the variation of the elements can now be derived in general form from the disturbing function (1) using the standard procedure of Lagrange's planetary equations, and in particular

$$\mathrm{d}I/\mathrm{d}t = n(1-e^2)^{-1/2} \sum \bar{J}_{lm} (R/a)^l \mathscr{R} i^{l-m+1} \times \\ \times \{(l-2p)\cot I - m \operatorname{cosec} I\} \bar{F}_{lmp}(I)\, G_{lpq}(e) \exp(i\psi_{lmpq}), \tag{6}$$

where the summation is over l, m, p, q with the limits given by (2).

In dealing with close resonances, the orbits are necessarily nearly circular so that a first approximation will be given by taking (6) only to lowest order in e. Since $G_{lpq}(e)$ is of order $e^{|q|}$, it follows from (6) that the lowest order part is of zero order and is given simply by taking $q=0$ in (6); since also $G_{lp0}(e) = 1 + 0(e^2)$, (6) reduces to

$$\mathrm{d}I/\mathrm{d}t = n \sum_{l,m,p} \bar{J}_{lm} (R/a)^l \mathscr{R} i^{l-m+1} \{(l-2p)\cot I - m \operatorname{cosec} I\} \times \\ \times \bar{F}_{lmp}(I) \exp(i\psi_{lmp0}) + 0(e), \tag{7}$$

where the so-far neglected part of order e arises from terms with $q = \pm 1$. The arguments of the zero order terms are of the form

$$\psi_{lmp0} \equiv (l-2p)(\omega + M) + m(\Omega - n_0 t - \phi_{lm}) \tag{8}$$

and resonance effects will arise if an argument of this form is slowly-varying in other words if

$$\alpha(\dot{\omega} + \dot{M}) \simeq \beta(n_0 - \dot{\Omega}), \tag{9}$$

where α and β are some pair of mutually prime integers. To interpret (9), the exact condition for commensurability is that the satellite performs β nodal periods while the Earth rotates α times relative to the *precessing* satellite orbit plane. After this interval the path of the satellite relative to the Earth repeats exactly, which is the physical reason for the resonance effects. Alternatively the approximate condition is that the mean motion of the satellite is β/α times the angular velocity of the Earth.

On comparing (8) and (9), the argument of the general term in (7) will be slowly varying near the β/α commensurability if l, m, p satisfy the conditions

$$l - 2p = \alpha\gamma, \qquad m = \beta\gamma, \qquad \gamma = 1, 2, 3, \ldots. \tag{10}$$

Given α and β there is an infinite sequence of terms satisfying the resonance conditions (10) and also the inequalities

$$2 \leqslant l \leqslant \infty, \qquad 1 \leqslant m \leqslant l, \qquad 0 \leqslant p \leqslant l. \tag{11}$$

Also the appropriate resonant variable near the β/α commensurability, suggested clearly by (9), is simply

$$\Phi_{\alpha B} = \alpha(\omega + M) + \beta(\Omega - n_0 t). \tag{12}$$

For close resonances $n \gg n_0$, and the longest possible period for a non-resonant term is therefore approximately $2\pi/n_0$, i.e. about one sidereal day. Thus non-resonant terms lead only to short-period effects, and to investigate the behaviour near resonance it is only necessary to retain the terms satisfying the resonance conditions (10). From (8) and (10), the argument ψ_{lmp0} for a general resonant term reduces to $\gamma\Phi_{\alpha\beta} - m\phi_{lm}$. Thus (7) can be written as

$$\begin{aligned} \mathrm{d}I/\mathrm{d}t = &- n\left(\frac{\beta - \alpha\cos I}{\sin I}\right) \mathscr{R} \sum{}^{*} \bar{J}_{lm} \exp(-im\phi_{lm}) (R/a)^l \bar{F}_{lmp}(I) \times \\ &\times i^{l-m+1}\gamma \exp(i\gamma\Phi_{\alpha B}) + 0(e) + \text{short-period terms}, \end{aligned} \tag{13}$$

where the asterisk on the summation sign denotes that only resonant terms are included. Furthermore all the terms with any given value of γ can be combined so that (13) can be written as

$$\frac{\mathrm{d}I}{\mathrm{d}t} = n\left(\frac{\beta - \alpha\cos I}{\sin I}\right) \sum_{\gamma = 1, 2, 3, \ldots} A_{\alpha\beta;\gamma}\gamma \sin(\gamma\Phi_{\alpha\beta} - \delta_{\alpha\beta;\gamma}), \tag{14}$$

where the 'lumped' amplitudes $A_{\alpha\beta;\gamma}$ and phases $\delta_{\alpha\beta\gamma;}$ are given by

$$A_{\alpha\beta;\gamma} \exp(i\delta_{\alpha\beta;\gamma}) = \sum_{\gamma\,\text{fixed}}^{*} (\bar{C}_{lm} + i\bar{S}_{lm}) (R/a)^l (-i)^{l-m} \bar{F}_{lmp}(I), \tag{15}$$

where we have reverted to using $\bar{C}$- and $\bar{S}$- gravity coefficients, and the summation is now over all terms satisfying (10) and (11) with a particular value of γ. In practice only the 'lumped' amplitude with $\gamma = 1$ will be significant; in the analysis of the Ariel 3 observations Gooding (1971a) found no detectable effects attributable to the amplitudes with $\gamma = 2$, 3, etc.

Since $\alpha<\beta$, the form of the summation in (15) differs slightly depending on whether $(\beta-\alpha)\gamma$ is even or odd. The reason is that, from (10), $(l-m)$ must have the same parity as $(\beta-\alpha)\gamma$ since $2p$ is necessarily even; the condition $l\geqslant m$ then imposes an extra constraint. The possible resonant terms, starting from the lowest value of l, are given by the following schemes:

case (i), $(\beta-\alpha)\gamma$ *even*

$$\begin{array}{lll} l=\beta\gamma & m=\beta\gamma & p=\tfrac{1}{2}(\beta-\alpha)\gamma \\ l=\beta\gamma+2 & m=\beta\gamma & p=\tfrac{1}{2}(\beta-\alpha)\gamma+1 \\ \cdots\cdots & \cdots\cdots & \cdots\cdots\cdots\cdots \end{array}$$

case (ii), $(\beta-\alpha)\gamma$ *odd*

$$\begin{array}{lll} l=\beta\gamma+1 & m=\beta\gamma & p=\tfrac{1}{2}(\beta-\alpha)\gamma+\tfrac{1}{2} \\ l=\beta\gamma+3 & m=\beta\gamma & p=\tfrac{1}{2}(\beta-\alpha)\gamma+\tfrac{3}{2} \end{array}$$

We shall illustrate this by writing out the first few terms in (14) and (15) for the 15/1 and 16/1 resonances. For simplicity the suffices α and β are dropped from $A_{\alpha\beta;\gamma}$ and $\delta_{\alpha\beta;\gamma}$.

For the 15/1 *resonance:*

$$\frac{\mathrm{d}I}{\mathrm{d}t}=n\left(\frac{15-\cos I}{\sin I}\right)\{A_1\sin(\Phi_{1,15}-\delta_1)+2A_2\sin(2\Phi_{1,15}-\delta_2)+\cdots\},\tag{16}$$

where

$$A_1\exp(i\delta_1)=(\bar{C}_{15,15}+i\bar{S}_{15,15})(R/a)^{15}\bar{F}_{15,15,7}-\\ -(\bar{C}_{17,15}+i\bar{S}_{17,15})(R/a)^{17}\bar{F}_{17,15,8}+\cdots,\tag{17}$$

and

$$A_2\exp(i\delta_2)=(\bar{C}_{30,30}+i\bar{S}_{30,30})(R/a)^{30}\bar{F}_{30,30,14}-\\ -(\bar{C}_{32,30}+i\bar{S}_{32,30})(R/a)^{32}\bar{F}_{32,30,15}+\cdots,\tag{18}$$

etc.

For the 16/1 *resonance:*

$$\frac{\mathrm{d}I}{\mathrm{d}t}=n\left(\frac{16-\cos I}{\sin I}\right)\{A_1\sin(\Phi_{1,16}-\delta_1)+2A_2\sin(2\Phi_{1,16}-\delta_2)+\cdots\},\tag{19}$$

where

$$iA_1\exp(i\delta_1)=(\bar{C}_{17,16}+i\bar{S}_{17,16})(R/a)^{17}\bar{F}_{17,16,8}-\\ -(\bar{C}_{19,16}+i\bar{S}_{19,16})(R/a)^{19}\bar{F}_{19,16,9}+\cdots,\tag{20}$$

and

$$A_2\exp(i\delta_2)=(\bar{C}_{32,32}+i\bar{S}_{32,32})(R/a)^{32}\bar{F}_{32,32,15}-\\ -(\bar{C}_{34,32}+i\bar{S}_{34,32})(R/a)^{34}\bar{F}_{34,32,16}+\cdots,\tag{21}$$

etc.

Expressions like the foregoing hold for all resonances whether close or distant, but for distant resonances the factor $(R/a)^l$ ensures that the first term is almost always much more important than the succeeding terms. For close resonances, however, R/a is only slightly less than 1, and a special situation arises. For on average the normalised gravity coefficients are also decreasing only slowly with increasing l. The simplest general estimate of the magnitudes is Kaula's 'rule-of-thumb' (Kaula, 1966) according to which the rms values of $\bar{C}_{lm}$ and $\bar{S}_{lm}$ vary as $10^{-5}/l^2$. The geoid solution of Gaposchkin and Lambeck (1971) bears out the validity of this simple estimate (Gaposchkin and Lambeck, 1971 and Allan, 1972b). If, as seems likely, the higher order gravity field arises from many independent randomly-distributed density variations, then the individual coefficients $\bar{C}_{lm}$ and $\bar{S}_{lm}$ will *a priori* be Gaussianly distributed with zero mean.

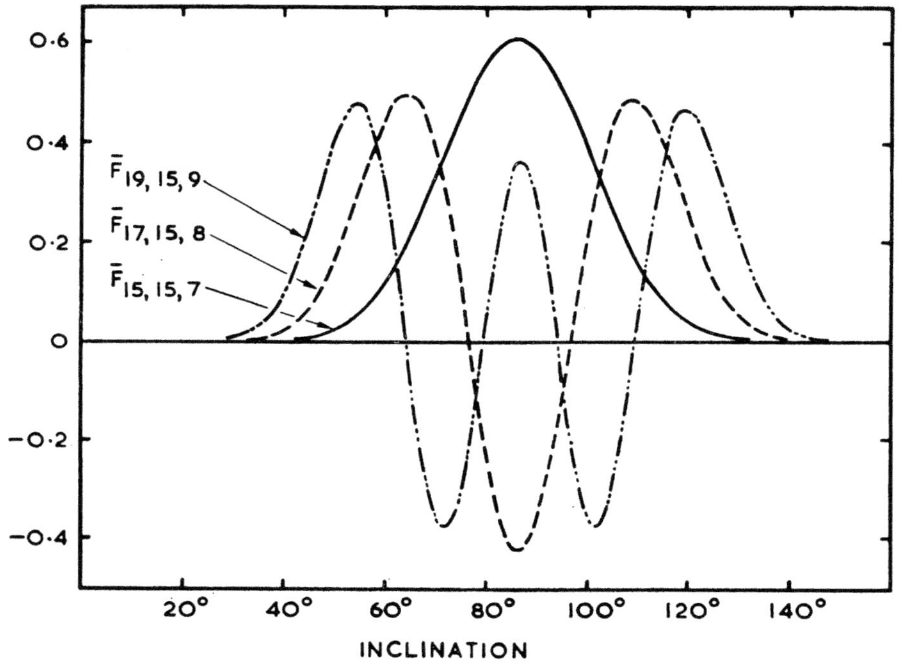

Fig. 1. Normalised inclination functions for 15/1 resonance. In this instance, with $(\beta-\alpha)$ even, all the functions are approximately symmetric.

The remaining factors in the 'lumped' amplitudes are the normalised inclination functions, and the first three functions in (17) and (20) for the 15/1 and 16/1 resonances are shown in Figures 1 and 2 respectively. There is a clear systematic difference between the two cases which stems from the distinction between $(\beta-\alpha)\gamma$ even and odd; the functions in Figure 1 are all approximately symmetric, and those in Figure 2 approximately anti-symmetric, in all cases about an inclination somewhat below 90°. To emphasise the distinction, explicit forms of the inclination functions for the lowest value of l (given by $\gamma=1$) for the two cases $(\beta-\alpha)$ even and odd are given below. The summation in (5) reduces to either one or two terms only for the lowest l-value.

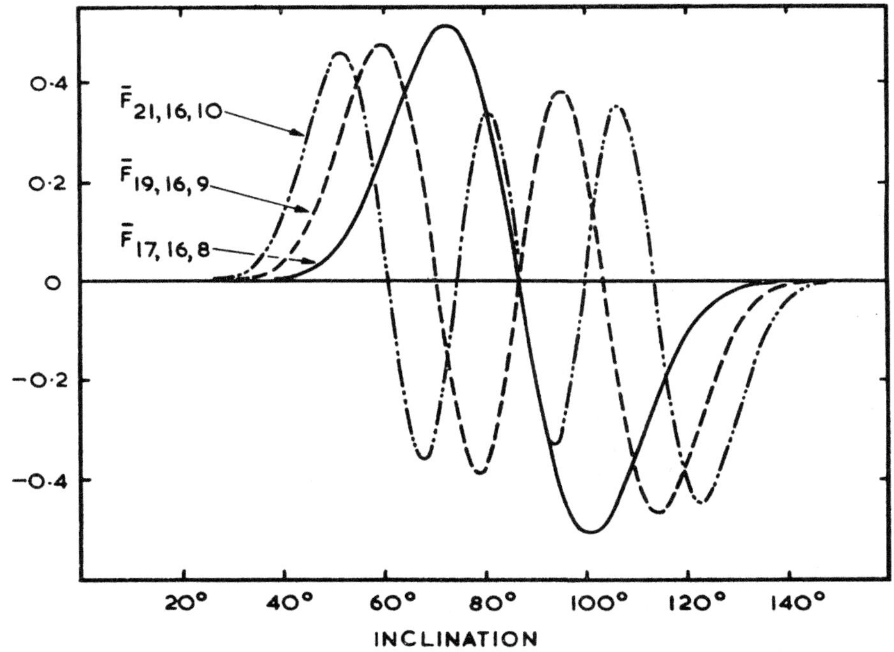

Fig. 2. Normalised inclination functions for 16/1 resonance. Here, with $(\beta-\alpha)$ odd, all the functions are approximately anti-symmetric.

(i) For $(\beta-\alpha)$ even, the term of lowest l-value is given by $l=\beta$, $m=\beta$, $p=\frac{1}{2}(\beta-\alpha)$, and the corresponding inclination function is

$$\bar{F}_{\beta,\beta,(\beta-\alpha)/2} = N_{\beta\beta}\frac{(2\beta)!}{2^{\beta}[\frac{1}{2}(\beta-\alpha)]!\,[\frac{1}{2}(\beta+\alpha)]!}\,c^{\beta+\alpha}s^{\beta-\alpha}. \tag{22}$$

This function has a single maximum at $I=\cos^{-1}(\alpha/\beta)$, i.e. slightly below 90° for a close resonance, as is shown by the curve $\bar{F}_{15,15,7}$ in Figure 1

(ii) For $(\beta-\alpha)$ odd, the term of lowest l-value is given by $l=\beta+1$, $m=\beta$, $p=\frac{1}{2}(\beta+1-\alpha)$, and the corresponding inclination function is

$$\bar{F}_{\beta+1,\beta,\frac{1}{2}(\beta+1-\alpha)} = N_{\beta+1,\beta}\frac{(2\beta+1)!}{2^{\beta+1}[\frac{1}{2}(\beta+1-\alpha)]!\,[\frac{1}{2}(\beta+1+\alpha)]!}\times$$
$$\times\, c^{\beta+\alpha}s^{\beta-\alpha}\{(\beta+1-\alpha)\,c^2-(\beta+1+\alpha)\,s^2\}. \tag{23}$$

Here the part within curly brackets can be written as $(\beta+1)\cos I-\alpha$, and the function therefore *vanishes* for $I=\cos^{-1}[\alpha/(\beta+1)]$, i.e. slightly below 90° for close resonance. There is a maximum at lower inclination and a minimum at higher inclination, the maximum being always slightly larger in absolute value. An example is given by the curve $\bar{F}_{17,16,8}$ for 16/1 resonance in Figure 2. As might be suspected, the two types of function are related; in fact the second type is the derivative of the first.

Similar explicit expressions for the succeeding functions become fairly complicated, but they are readily evaluated on a computer, for which purpose (5) is eminently suited. Alternatively Gooding's recurrence relation (Gooding, 1971b), which applies specifically for functions related in this way, can be used. In Figure 1 the first function $\bar{F}_{15,15,7}$ has a single maximum at $I=\cos^{-1}(1/15)\simeq 86°11'$, and successive functions have three, five, etc., extrema. Similarly in Figure 2, the first function $\bar{F}_{17,16,8}$ has a zero at $I=\cos^{-1}(1/17)\simeq 86°38'$, and both a maximum and a minimum, while succeeding functions have four, six, etc., extrema. The central extrema in Figure 1 and the central zeros in Figure 2 do not exactly coincide, and a formal proof of this is easily constructed using Gooding's recurrence relation (Gooding, 1971b). The following conclusions hold for both types of resonance:

(i) Successive functions oscillate increasingly wildly with inclination. In general $|\bar{F}_{lmp}|$ has $(l-m+1)$ maxima, as indeed has $|P_l^m(\cos\theta)|$. This might have been predicted by considering the form of the tesseral harmonic $P_l^m(\cos\theta)e^{im\phi}$ and how a satellite in an inclined circular orbit extracts an average therefrom.

(ii) All the maxima are roughly comparable, falling off only slowly with increasing l.

(iii) Successive functions are significantly non-zero over a wider and wider range of inclination.

We may therefore end this section by drawing the following conclusions on the 'lumped' amplitudes and phases:

(i) A fair number of terms will contribute significantly to the amplitudes A_1, A_2, etc., and one should not expect that the largest contribution comes from the first term in the sum (or even the second, or the third...).

(ii) The amplitudes A and phases δ will be highly dependent on inclination. Indeed there would be some point in considering them simply as functions of inclination without attempting to resolve them into gravity coefficients of different degrees.

(iii) On average, A_1 will be considerably larger than A_2, which in turn will be considerably larger than A_3, etc.

(iv) On average the amplitude A_1 for, say, the 31/2 resonance will be comparable with A_2 for the 15/1 and 16/1 resonances; in other words resonances with $\alpha=1$ will be, on average, considerably stronger than those with $\alpha=2$, which in turn will be considerably stronger than those with $\alpha=3$, etc.

3. Approximate Solution

It follows from the definition (12) that the resonant variable is almost entirely controlled by the changes in the mean motion. For a circular orbit a drag acceleration D, taken as positive if opposed to the motion, will produce a change in mean motion given by $\mathrm{d}n/\mathrm{d}t=3D/a$. The resonance forces also contribute via the transfer of energy, and it can be shown (Allan, 1967a, 1972b), to lowest order in e, and considering only the resonant terms with $\gamma=1$, that

$$\mathrm{d}^2\psi/\mathrm{d}t^2 = -k^2\sin\psi + F\,, \tag{24}$$

where

$$\psi = \Phi_{\alpha\beta} - \delta_{\alpha\beta;\,1} + \pi\,, \tag{25}$$

$$k^2 = 3n^2\alpha^2 A_{\alpha\beta;\,1}\,, \tag{26}$$

and

$$F = 3\alpha D/a\,, \tag{27}$$

where $A_{\alpha\beta;1}$ and $\delta_{\alpha\beta;1}$ are as defined in (15), and ψ is simply the resonant variable referred to a more convenient origin. To the same approximation and in the same form, the change of inclination induced by the resonance forces is

$$\Delta I = -\,n\,\frac{(\beta - \alpha\cos I)}{\sin I}\,A_{\alpha\beta;\,1}\int \sin\psi\,\mathrm{d}t\,, \tag{28}$$

and clearly depends on the behaviour of the resonant variable which is determined by (24).

The simplest situation arises when $F \gg k^2$, which will usually be the case except for very low drag satellites, or for the more distant strong resonances such as 13/1 or 14/1. When the resonant contribution in (24) is negligible compared with the drag, the mean motion increases linearly with time, and ψ varies quadratically, falling to some minimum value ψ_0 (at the moment of exact commensurability $\dot{\psi}=0$) and then rising again. Clearly the most critical contribution to the integral $\int \sin\psi\,\mathrm{d}t$ in (28) arises from the part near ψ_0 where ψ is changing slowly. Also the integrand is symmetric about ψ_0 so that the parts before and after exact commensurability make equal contributions to the integral.

To derive an approximate result for the change in inclination, set

$$\psi \simeq \psi_0 + \tfrac{1}{2}Ft^2\,, \tag{29}$$

taking $t=0$ at exact commensurability. For the integral in (28) we therefore have

$$\int_0^t \sin\psi(t')\,\mathrm{d}t' = \sin\psi_0\int_0^t \cos\left(\tfrac{1}{2}Ft'^2\right)\mathrm{d}t' + \cos\psi_0\int_0^t \sin\left(\tfrac{1}{2}Ft'^2\right)\mathrm{d}t'\,.$$

Taking the Fresnel integrals as defined by Jahnke *et al.* (1960), namely

$$C(z) = \sqrt{\left(\frac{2}{\pi}\right)}\int_0^{\sqrt{z}} \cos t^2\,\mathrm{d}t\,,\qquad S(z) = \sqrt{\left(\frac{2}{\pi}\right)}\int_0^{\sqrt{z}} \sin t^2\,\mathrm{d}t\,,$$

it follows that

$$\int_0^t \sin\psi(t')\,\mathrm{d}t' = \sqrt{\left(\frac{\pi}{F}\right)}\left\{\sin\psi_0 C\left(\tfrac{1}{2}Ft^2\right) + \cos\psi_0 S\left(\tfrac{1}{2}Ft^2\right)\right\}. \tag{30}$$

Using the asymptotic forms (Jahnke *et al.*, 1960)

$$C(z) \sim \frac{1}{2} + \frac{\sin z}{\sqrt{(2\pi z)}} + \cdots, \qquad S(z) \sim \frac{1}{2} - \frac{\cos z}{\sqrt{(2\pi z)}} + \cdots,$$

we find that, for t large,

$$\int_0^t \sin\psi(t')\, dt' \sim \sqrt{\left(\frac{\pi}{2F}\right)} \sin\left(\psi_0 + \frac{\pi}{4}\right) - \frac{1}{Ft}\cos\left(\psi_0 + \tfrac{1}{2}Ft^2\right) + \cdots. \tag{31}$$

Finally, from (28), the total change in inclination in passing through the resonance (from $t = -\infty$ to $t = +\infty$) is given by

$$\Delta I \sim -\frac{(\beta - \alpha\cos I)}{\sin I} nA_{\alpha\beta;1} \sqrt{\left(\frac{2\pi}{F}\right)} \sin\left(\psi_0 + \frac{\pi}{4}\right), \qquad F \gg k^2, \tag{32}$$

and therefore depends strongly on ψ_0. Since $\beta > \alpha$ for close resonances, the inclination *increases* for $-5\pi/4 < \psi_0 < -\pi/4$, with the greatest increase for $\psi_0 = -3\pi/4$. Similarly the inclination *decreases* for $-\pi/4 < \psi_0 < 3\pi/4$, with the greatest decrease for $\psi_0 = \pi/4$. It must be emphasised that ψ_0 is a *random* variable. For $F/k^2 \gg 1$, all values of ψ_0 will be equally probable, so that the inclination is equally likely to increase or decrease in passing through resonance. The approximate result (32) also shows that the total change in inclination is proportional to $F^{-1/2}$, i.e. to the time taken to pass through resonance.

The approximate result (32) also serves to settle one troublesome question. So far we have proceeded as if there were only a finite number of discrete resonances, but in reality the ratio n/n_0, or more strictly the ratio $(\dot{\omega} + \dot{M})/(n_0 - \dot{\Omega})$, can be approximated as closely as desired by the ratio of two integers β/α. In other words a decaying satellite is *always* passing through some resonance. This argument is more usefuly turned the other way round as follows: In decaying from n_1 to n_2 $(> n_1)$, no matter how close these values are, the satellite has passed through infinitely many resonances, albeit of very high order and therefore very weak. If the change in inclination tended to have one sign rather than the other, one might suspect that the total change could be non-zero, e.g. roughly proportional to $(n_2 - n_1)$. Since the change is equally likely to have either sign, however, the probable total change in inclination is zero. This is of particular relevance in attempting to deduce an average rotational speed for the upper atmosphere from the change in inclination of decaying satellite orbits (King-Hele *et al.*, 1970).

The most interesting situation arises if a low drag satellite encounters a strong resonance so that F and k^2 are comparable, and the resonance term $-k^2 \sin\psi$ in (24) significantly affects the mean motion and the behaviour of the resonant variable. The conclusion, which held for $F \gg k^2$, that the inclination is equally likely to increase or decrease, is no longer valid; it is still true that ψ_0 is a random variable, but it is not uniformly distributed. This question has been discussed in more detail elsewhere

(Allan, 1972b), with the conclusion that an increase in inclination is more likely than a decrease, but any decrease, if it does occur, is likely to be larger and could be very much larger. How these two opposing effects balance out could be found by numerical integration for various values of the parameters F/k^2 and ψ_0.

4. Derivation of Gravity Coefficients

Clearly the change in inclination of a satellite passing through resonance offers a way of deducing at least 'lumped' values of gravity coefficients of a given order m. As already emphasised, the total change in inclination depends on the random variable ψ_0, i.e. the value of the resonant variable at exact commensurability, and can have either sign. Thus the total change in inclination can give, at most, a lower limit to a 'lumped' value. Instead one needs at least some measure of the shape of the curve of inclination against time.

This situation is illustrated in Figure 3. Suppose that F is known and that $F \gg k^2$ so that the resonant variable behaves according to (29). Suppose also that the total change in inclination is known, and is positive (to afford an easier comparison with the Fresnel

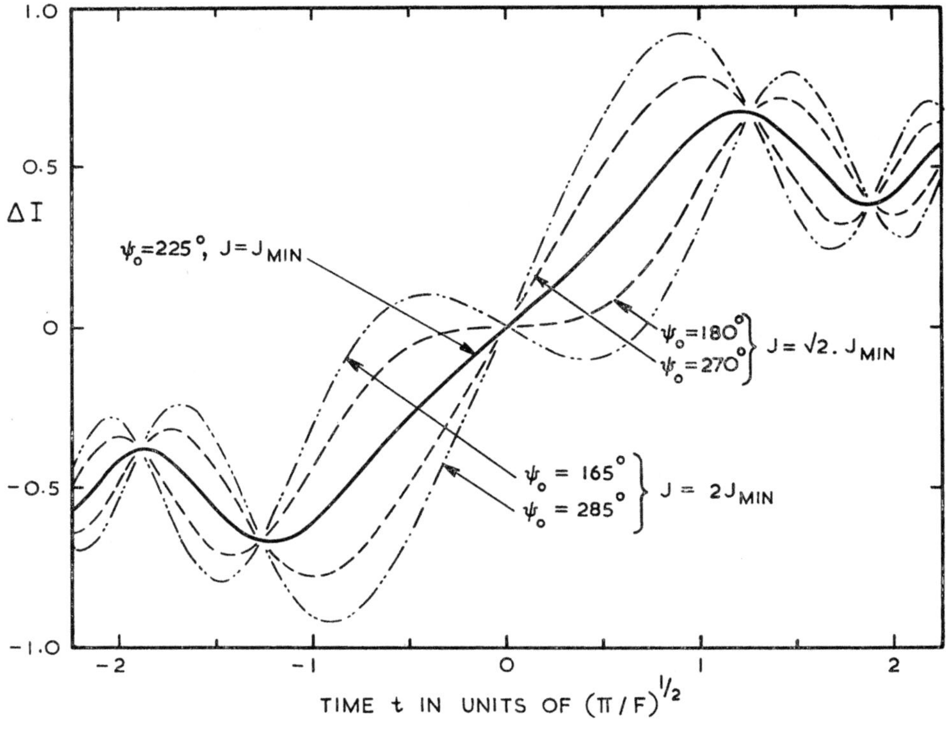

Fig. 3. How the inclination varies for different values of the resonant variable at exact commensurability. All the curves give the same total change (a unit increase) in inclination, but correspond to different 'lumped' amplitudes.

integrals in standard form), and is taken for simplicity as unity in some arbitrary units. From (32) an increase in inclination implies that ψ_0 lies between 135° and 315°. If the 'lumped' amplitude $A_{\alpha\beta;1}$ is fixed, the greatest increase occurs when $\psi_0 = 225°$: the complete variation of I, found by combining (28) and (30), is shown by the continuous curve in Figure 3. The same total change in inclination also occurs for other values of ψ_0 if the 'lumped' amplitude is suitably increased. Thus the broken curves in Figure 3 give the variation of inclination for $\psi_0 = 180°$ and 270°, corresponding to an increase of $\sqrt{2}$ in the 'lumped' amplitude, and the chain-dotted curves show the change for $\psi_0 = 165°$ and 285° corresponding to an increase of 2 in the 'lumped' amplitude. It is clear that all the curves in Figure 3 intersect at a sequence of points, which are in fact the roots of $C(\frac{1}{2}Ft^2) = S(\frac{1}{2}Ft^2)$; from (28) and (30), when the Fresnel integrals are equal, the change in inclination depends only on the product $A_{\alpha\beta;1}(\sin\psi_0 + \cos\psi_0)$, which is determined by the total change in inclination according to (32).

For the particular example of Ariel 3, the points in Figure 4 show the observed values of the inclination (after subtracting luni-solar corrections) and the resonant variable at three-day intervals (after Gooding, 1971a). The first point is that $\Phi_{1,15}$ appears to be changing quadratically corresponding to a mean acceleration $dn/dt \simeq \simeq 0.093°\ \text{day}^{-2}$; the lack of any obvious departure from quadratic variation implies (i) that any variations in drag average out, and (ii) that $F/k^2 \gg 1$. The inclination falls by about 0°.02 from a maximum near MJD 39,840 to a minimum near MJD 39,930, while the resonant variable falls from about 200° to a minimum at ~107° and rises again to about 200°; this implies that $\Phi_{1,15} \simeq 200°$ corresponds to $\psi = 180°$. A least squares fit (Gooding, 1971a) of *all* the observed values of inclination (after luni-solar corrections), including also a linear decrease in inclination with time due to the 'super-rotation' of the upper atmosphere (King-Hele *et al.*, 1970), yields the broken curve in Figure 4. The fit is very good in the central region, but is not so close beyond one or two cycles of oscillation (see Figure 3 of Gooding, 1971a); the fitted curve, however, illustrates very well the decaying oscillations typical of the Fresnel integrals. If the resonance effect is attributed entirely to the (15, 15) harmonic, Gooding's analysis gives the following 'lumped' values:

$$\bar{S}_{15,15} = (-1.99 \pm 0.12) \times 10^{-8}, \qquad \bar{C}_{15,15} = (-0.77 \pm 0.08) \times 10^{-8}$$

with a correlation coefficient −0.83. After disentangling the phase according to (17) and (25), this solution corresponds to $\psi = \Phi_{1,15} - 22°$, i.e. $\psi = 180°$ does indeed correspond to $\Phi_{1,15} \simeq 200°$. The scale of ψ is shown on the right of Figure 4, and the minimum value of ψ is ~85°; in other words the random variable ψ_0 is fairly near the value 45° which gives a maximum decrease in inclination in the approximate solution.

Strictly speaking, the above 'lumped' values apply to linear combinations of coefficients of order 15 and degrees 15, 17, 19 etc., as in (17). It would be possible to obtain values for other linear combinations of the same coefficients from other satellites at other inclinations as they pass through the 15/1 commensurability. One might then attempt to determine individual values by solving a set of simultaneous linear equations. There are a number of difficulties, however, which suggest that the values so obtained are likely to be of largely statistical significance. Thus the Ariel 3 results (Gooding, 1971a) reveal the presence of significant systematic errors. For about the first eight cycles of oscillation after passing through resonance, the observed values of inclination are beautifully in phase with the fitted curve but the amplitude is larger than the fitted curve by approaching a factor of two. If only these observed points had been available, one would have derived correspondingly larger values for the 'lumped' coefficients. After about eight cycles of oscillation, the observed values show a sharply increased scatter coincident with a marked increase in the standard deviation of the orbit determinations.

There seem to be two possible sources for such systematic errors. The first, and simpler possibility is that there is a systematic bias depending on ψ in the value of inclination determined from the observations. Since the values of ψ determine the position of the ground track on the earth, the orbit determination is based on very similar observations whenever the resonant variable returns to the same value.

Anothei possible source of systematic error lies in the neglected resonant terms. Here the higher order terms with $\gamma=2$ which contain the arguments $2\Phi_{1,15}$ and the coefficients $\bar{J}_{30,30}$, $\bar{J}_{32,30}$ etc., can probably be disregarded; Gooding (1971a) has included terms of the form $\begin{smallmatrix}\cos\\ \sin\end{smallmatrix} 2\Phi_{1,15}$ in his data analysis and concluded that they have no significant effect on the goodness of fit and give negligible values for the coefficients. There remain the terms of higher order in e neglected in (7) and (13). The terms with $q\neq 0$ in (6) are of order $e^{|q|}$. Near the β/α commensurability the resonance conditions for these terms are, from (8),

$$l - 2p + q = \alpha\gamma\,, \qquad m = \beta\gamma\,, \qquad \gamma = 1, 2, 3, \ldots, \tag{33}$$

and the argument of the general resonant term reduces to $\gamma\Phi_{\alpha\beta} - m\Phi_{lm} - q\omega$.

In other words there are two sequences of companion resonances of increasing order in e, one sequence on each side of the main commensurability. The most important subsidiary resonances are those of order e with the arguments $\Phi_{\alpha\beta} \pm \omega$.

To be specific, for the 15/1 commensurability, taking $\alpha=1$, $\beta=15$ (and $\gamma=1$), the resonant terms of order e are given by the following scheme:

$q=+1$			$q=-1$		
$l=16$,	$m=15$,	$p=8$	$l=16$,	$m=15$,	$p=7$
$l=18$,	$m=15$,	$p=9$	$l=18$,	$m=15$,	$p=8$
$l=20$,	$m=15$,	$p=10$	$l=20$,	$m=15$,	$p=9$
........etc.........			........etc.........		

The same set of gravity coefficients is involved for both $q=+1$ and $q=-1$; it differs from the set for $q=0$, but on balance the two sets of coefficients will be comparable in magnitude. Similarly the two new sets of inclination functions for $q=\pm 1$ differ from the set for $q=0$; in fact for 15/1 resonance the functions for $q=0$ (shown in Figure 1) are of the 'symmetric' type, while those for $q=\pm 1$ are of the 'anti-symmetric' type and are virtually indistinguishable from the functions shown in Figure 2. The inclination functions are therefore comparable in magnitude and behaviour, and the only significant difference between the sets of terms lies in the Hansen coefficients. Thus it follows from Appendix B of Allan (1967b) that

$$G_{lp,+1} = \tfrac{1}{2}(3l - 4p + 1)\,e + 0(e^3), \quad G_{lp,-1} = \tfrac{1}{2}(4p - l + 1)\,e + 0(e^3), \tag{34}$$

whereas, of course, $G_{lp0} = 1 + 0(e^2)$. In particular,

$$G_{16,8,1} = \tfrac{17}{2}e + 0(e^3), \quad G_{18,9,1} = \tfrac{19}{2}e + 0(e^3), \dots \text{ etc.,}$$
$$G_{16,7,-1} = \tfrac{13}{2}e + 0(e^3), \quad G_{18,8,-1} = \tfrac{15}{2}e + 0(e^3), \dots \text{ etc.}$$

As an average over the l-values involved for 15/1 resonance, the Hansen coefficients $G_{lp,\pm 1}(e)$ can therefore be taken as approximately $10\,e$. On balance the expected contribution to dI/dt from all resonant terms of order e will be $10\sqrt{2}e$ times the expected contribution of the zeroth order terms. Since the two sets of terms of order e involve the same gravity coefficients, they will interfere, either cancelling or reinforcing one another depending on the values of ω and I, it would be more prudent to replace the factor $10\sqrt{2}e$ by $20e$ simply. In that case the terms of order e will on average contribute 10% of the total change in inclination if the eccentricity is 0.005. At first sight it might appear strange that the terms of order e are proportionately more effective than the zeroth order terms, but the physical reason is that the eccentricity alters the timing of the orbit and so distorts the ground track over the earth.

The terms of order e will be of proportionately greater importance if the zeroth order effect is anomalously low as appears to be the case for Ariel 3; if the magnitudes of the high degree coefficients are in accord with the geoid solution of Gaposchkin and Lambeck (1971) or Kaula's 'rule-of-thumb' (Kaula, 1966), the 'lumped' amplitude for Ariel 3 is lower than expected by at least a factor of three (Allan, 1972b). Since the eccentricity is 0.007, the terms of order e would be expected to contribute $\pm 40\%$ of the total change in inclination, and could easily account for the factor of two discrepancy in the amplitude away from resonance. Indeed the contribution of the terms of order e could equally well be anomalously high. Moreover any discussion in terms of the magnitudes of the total changes in inclination is over-simplified since the subsidiary resonances do not coincide with the main resonance. For Ariel 3 the behaviour of the arguments $\Phi_{1,15} \pm \omega$ is shown by the broken curves at the bottom of Figure 4, so that the subsidiary resonances are centred near the first turning points of the inclination curve, and could therefore cause a damaging distortion in the variation of inclination.

Gravity coefficients of higher degree obtained from higher order resonances of the form 29/2 (or 31/2) or even possibly 44/3 etc., will give an estimate of the depth of the sources of the high order gravity field (Allan, 1972a). Since only a knowledge of the distribution of the magnitudes is required, approximate 'lumped' values are perfectly acceptable. According to Figure 3 of Allan (1967a), the expected strengths of the 31/2 and 29/2 resonances should be smaller than the 15/1 resonance by factors of 20 and 50

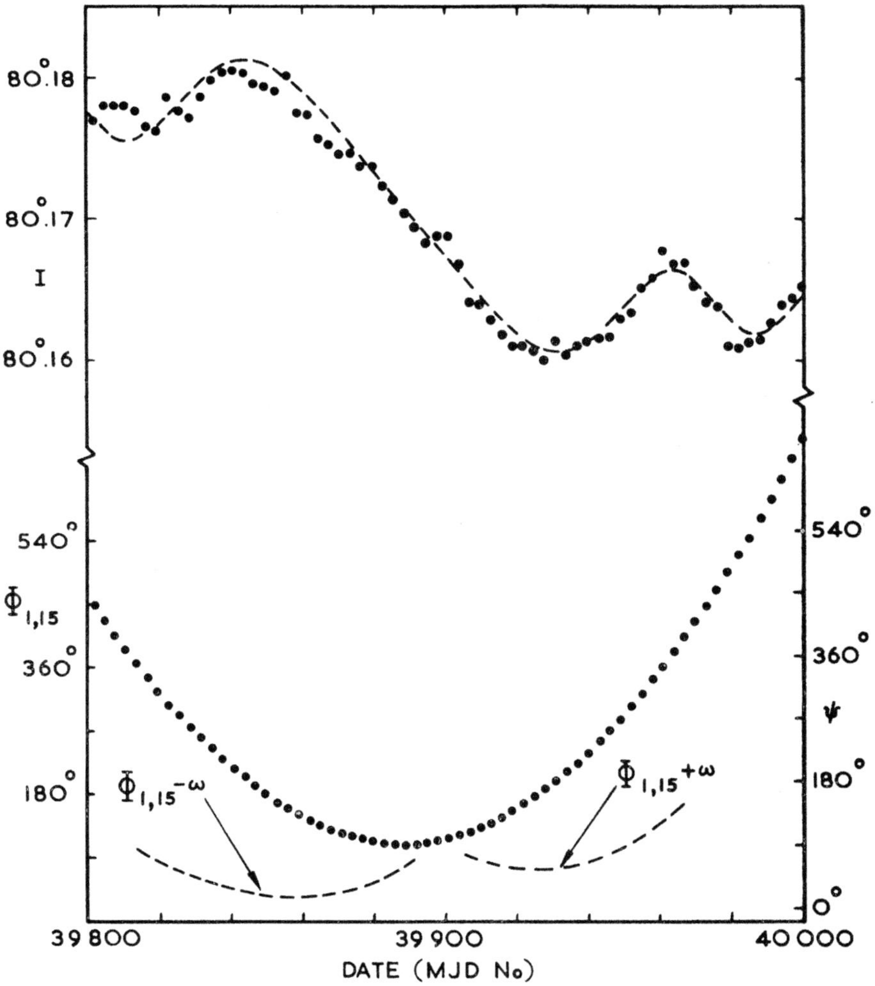

Fig. 4. Orbital inclination and resonant variable for Ariel 3 (1967-42A). The upper points give the observed values of inclination (after luni-solar corrections) together with the fitted curve. The lower points show the resonant variable $\Phi_{1,15}$ (left-hand scale), with the corresponding values of ψ (derived using the 'lumped' phase of the fitted curve) on the right-hand scale. The broken curves at the bottom show the resonant variables $\Phi_{1,15} \pm \omega$ for the subsidiary resonances of order e.

respectively. Since the strength of the 15/1 resonance for Ariel 3 is lower than average by a factor of at least three, changes in inclination approaching 0°.005 and 0°.002 might be expected for the 31/2 and 29/2 resonances for a satellite with the same dn/dt as Ariel 3 (at the 15/1 resonance). In fact Ariel 3 also passed through the 31/2 resonance (at a height of ~380 km corresponding to its 80° inclination), but so much faster that there is unlikely to be any detectable effect. Thus the practical case might well be the 29/2 resonance at a height ranging from about 650 km to about 750 km (between 50° and 120° inclination); with the lower air density at this greater height, the slower passage through resonance should give an easily detectable change in inclination.

Acknowledgement

Crown Copyright reserved. Reproduced with the permission of the Controller, Her Majesty's Stationery Office.

References

Allan, R. R.: 1967a, *Planetary Space Sci.* **15**, 53.
Allan, R. R.: 1967b, *Planetary Space Sci.* **15**, 1829.
Allan, R. R.: 1972a, *Nature Phys. Sci.* **236**, 22.
Allan, R. R.: 1972b, *Planetary Space Sci.* **21**, 205.
Gaposchkin, E. M. and Lambeck, K.: 1971, *J. Geophys. Res.* **76**, 4855.
Gooding, R. H.: 1971a, *Nature Phys. Sci.* **231**, 168.
Gooding, R. H.: 1971b, *Celes. Mech.* **4**, 91.
Jahnke, E., Emde, F., and Lösch, F.: 1960, *Tables of Higher Functions*, Sixth (revised) edition, McGraw-Hill Book Co., Inc., New York.
Kaula, W. M.: 1966, *Theory of Satellite Geodesy*, Blaisdell Publ. Co., Waltham, Mass., U.S.A.
King-Hele, D. G.: 1972, *Nature Phys. Sci.* **238**, 13.
King-Hele, D. G. and Hiller, H.: 1972, *Nature Phys. Sci.* **235**, 130.
King-Hele, D. G., Scott, D. W., and Walker, D. M. C.: 1970, *Planetary Space Sci.* **18**, 1433.
Plummer, H. C.: 1918, *An Introductory Treatise on Dynamical Astronomy*, pp. 44–46, C.U.P.: republished by Dover, (1960), New York.

PART VI

TRAJECTORY DETERMINATION AND THE MOTION OF RIGID BODIES

REVIEW OF NON-NUMERICAL USES OF COMPUTERS*

M. S. DAVIS

University of North Carolina, Chapel Hill, N.C., U.S.A.

The field of non-numerical uses of computers has become so vast that today it is only feasible to give a partial list of the most useful and interesting applications, e.g.

(1) Commercial, PERT (Program Evaluation and Review Technique) is one of the earliest applications in construction, network theory, etc.) Airline scheduling is another modern, indispensable application.

(2) Traffic Control.

(3) Information Retrieval.

(4) Mechanical Translation and Analysis of Natural Languages, Linguistics.

(5) Printing and Communication. Editing and photocomposition.

(6) Design, Numerical Control, Pattern Analysis.

(7) Theorem Proving. Problem Solving.

(8) Literal Mathematics on Digital Computers.

(9) Learning and Teaching, Computer Aided Instruction.

(10) Game-playing, Game-learning, Executives Games.

(11) Construction of Psychological Models: Value Systems, Belief Systems.

(12) Cryptanalysis, Security of Information.

(13) Music Research, Composition, Simulation and Creation of new instruments (sounds).

(14) Construction of indexes, bibliographies, catalogues.

(15) Artificial Intelligence. Simulation of Human Cognitive Processes.

(16) Operations Research.

In particular, we shall be concerned with No. 8, Literal Mathematics on Digital Computers. It is well-known that analog computers in the nature of 'differential analyzers' reached their zenith during World War II in solving a variety of differential equations primarily in ballistics. Interestingly, James Thomson (Scott, 1960), the brother of Lord Kelvin, in 1875 invented an integrator which was used as the prototype by Vannevar Bush at MIT in 1931. However, the shortcomings of analog computers for general-purpose computations quickly relegated them to a limited number of highly specialized applications as soon as high-speed digital electronic computers appeared on the horizon.

The earliest application to digital computers probably took place in 1953 when two students, Kahrimanian and Nolan, the first at Temple University and the second at MIT, independently wrote masters' theses on carrying out differentiation analytically on a digital computer.

At an early conference in Celestial Mechanics (Davis, 1958), discussions took place

* Most figures and tables are from the original sources.

B. D. Tapley and V. Szebehely (eds.), Recent Advances in Dynamical Astronomy, 351–391.

concerning the feasibility of constructing general-purpose compilers which would be able to deal with the principal theories in celestial mechanics in a literal fashion. In those days the main limitations on building such programs were the small memories and slow execution times of the extant computers. Compiler technology was then in its infancy. Since then computers have improved several orders of magnitude in both their capacities and speeds. In modern applications the capacity of a computer severely bounds the problem and requires very prudent management of the system's resources. Nanosecond speeds are sufficiently great nowadays so as not to impede the solution of most consequential problems.

As to compilers and programs to carry out literal developments on digital computers, the 1960's saw the development and proliferation of a large number. Table I gives a partial list of symbol manipulative languages (Sammet, 1966). These languages break down into four main classes, with overlapping functions, in some applications.

TABLE I

Languages for non-numerical applications

ALGAN	A set of FORTRAN routines to solve sets of linear equations and find polynomial factors common to two given polynomials.
ALGEM	A set of SLIP coded routines to manipulate polynomials with variable exponents.
ALGY	A system to manipulate expressions.
ALMS	A system to do differentiation.
ALPAK	A system of subroutines to manipulate polynomials and rational functions.
ALTRAN	An extension of FORTRAN which incorporates some of the ALPAK capability.
AMBIT	A language to express a certain type of string manipulation.
AUTOMAST (= AUTOMS)	A system to solve systems of ordinary differential equations.
AXLE	A language to express a certain type of string manipulation.
COGENT	A syntax-directed compiler which has been used for writing programs within the scope of this bibliography.
COMIT	A string-processing language, originally designed for use by people doing mechanical translation, but which is useful for other applications as well.
CONVERT (= CONVRT)	A language (based on LISP) and routines to handle transformation rules of the type as in COMIT and SNOBOL.
CORAL	A list processing system for handling graphical data structures.
DYSTAL	A system of subroutines added to FORTRAN to do a limited type of list processing.
FLIP	A notation and language (and routines) for expressing string transformations, such as those in COMIT or SNOBOL, in LISP.
FLPL	FORTRAN-List-Processing Language. A set of subroutines added to FORTRAN in connection with work on geometry theorem proving.
FORMAC	An extension of FORTRAN which does many types of manipulation of mathematical expressions.
FRMALG (= Formula ALGOL)	An extension of ALGOL to allow formal data types and their manipulations, and string and list processing capabilities.
GRAD (= GRAD Assistant)	A LISP program to do algebraic manipulation.
IPL	Early versions of a list processing language.
IPL-V	A list processing system whose commands are similar in spirit to those of an assembly language, but for manipulating lists.

Table I (Continued)

L6	A system to manipulate list structures.
LIPL	A linearized version of IPL-V.
LISP	A list processing system with emphasis on recursion and formalism.
MAGPAP (= Magic Paper)	A system to permit man-machine interaction in the manipulation of mathematical expressions.
MANIP	A set of FORTRAN programs to manipulate expressions.
MTHLAB (= MATHLAB)	A system to permit man-machine interaction in the manipulation of mathematical expressions.
METEOR	A set of LISP functions to perform some of the manipulations done by COMIT.
NUSPK (= NU-SPEAK)	A set of list processing routines added to FORTRAN or FAP
PANON	A language for symbol manipulation based on a particular extension of Markov Algorithms.
PM	A system of subroutines to manipulate polynomials with variable size integral coefficients.
REFCO (= REFCO III)	A set of subroutines to do list processing.
SAINT	A program to do formal integration.
SLIP	A system of subroutines added to FORTRAN to do list processing.
SNOBOL	A string manipulating system.
SYMBLG (= SYMBOLANG)	A set of subroutines useful for handling polynomials and rational functions.
TIPL	A program to check IPL-V programs.
TRAC	A man-machine string manipulating language.
TREET	A list processing system.
VPRPAK	A set of subroutines to do variable precision and rational arithmetic.

(1) List Processing Ex. LISP 1.5, IPL-V, SLIP.

(2) String Manipulation, Ex. COMIT, SNOBOL.

(3) Symbol Manipulation This, in effect, is the same as No. 2 and sometimes describes all four categories.

(4) Formula or Mathematical Manipulation Ex. MATHLAB, FORMAC, ESP, TRIGMAN.

All classes of language have been used in dynamical applications and it is important to have some conception of the nature of these types.

In list processing languages, related groups of information called lists can be manipulated in various ways. Because of the lack of *a priori* information on the length of the lists which may 'pulsate' in size, an associative memory structure is utilized, i.e., elements of the list are not stored consecutively but rather each element points towards its successor (which in some list structures may point back to its predecessor). The pointers are necessarily part of the data but the programmer need not be concerned with these details, only with how to manipulate the lists (the system takes care of most pointing information).

Let us look at an associative type memory in greater detail. In so-called general formula manipulators, expressions are stored in some level of Łukasiewicz or Polish notation rather than 'infix' notation which is the usual representation. This is a parenthesis free notation most expeditious for internal, machine representation's of algebraic expressions. For example, in LISP a 'Cambridge Polish' notation is used. In ordinary left-hand Polish notation $+++ABCD$ means $((A+B)+C)+D$. In Cambridge Polish notation this is written as $+ABCD$. Thus, a binary operator is 'used up'.

Let the representation for $X**2$ in schematic form be

```
**
|
X-2
```

To represent this in memory the following scheme is used:

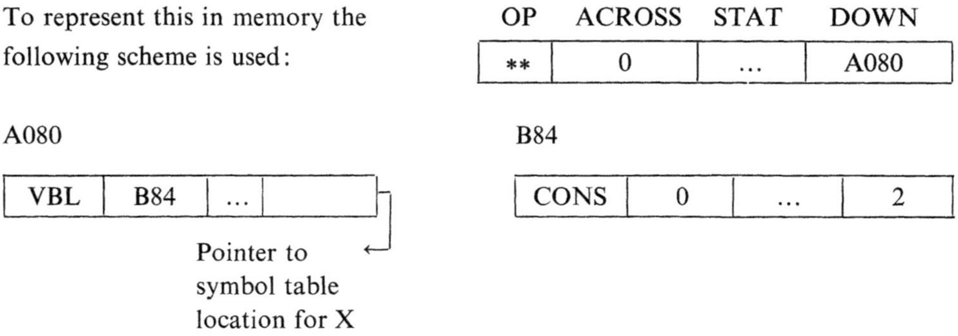

Fig. 1. Representation of X**2 in memory.

Figure 1 indicates that a word (in PL/I FORMAC it is actually a double word) is divided into designators and pointers. The first part may designate an operator, variable or constant, for instance. The second part of the word points horizontally and the last part points vertically. The meaning of zero is obvious. Other examples of built-in and defined functions as well as other expressions in schematic notation are (the reader can easily picture the equivalent memory distribution of words and pointers):

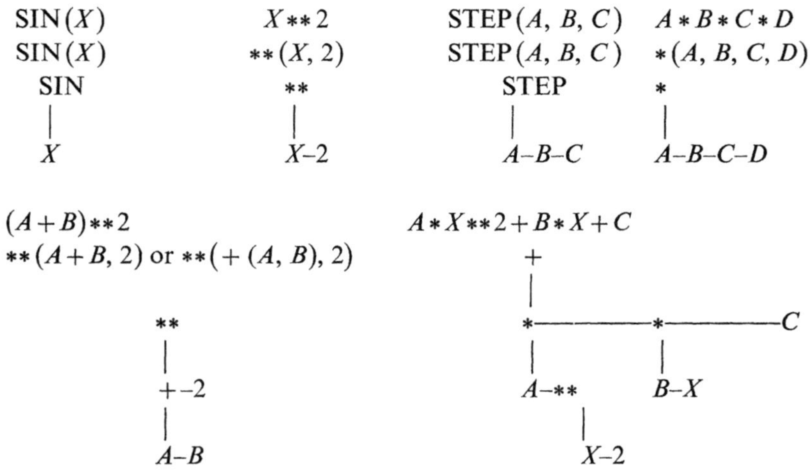

Fig. 2. Scheme for other expressions.

The power of the associative memory is suggested by the following. Suppose in Figure 2 $X=f(t)$. The two words containing X point towards the symbol table where the current value of X is stored. To replace X in the quadratic expression all that needs to be done is to change the pointers so that they point instead to $f(t)$. Addition or deletion of terms are again accomplished by changes in pointers. The question of simplification of expressions is a very difficult matter which has different solutions in different systems.

In the course of a short time of running a list-processing program, particular elements of lists may be erased, or entire lists themselves. This necessarily fragments memory to a point where the system has to determine anew where all the available cells are. This process, called 'garbage collection' is a system function which takes place automatically when it is needed. It may, however, be time consuming. Further, it is obvious that the price for such flexibility of storage is loss of storage itself to make the associative relations possible.

String manipulative languages allow the programmer to operate on strings (concatenations of allowable characters in the alphabet of the language). The principal manipulations are: matching of patterns, insertion of patterns, deletions of patterns (where patterns may also be individual characters). Modern languages like SNOBOL are fairly easy to learn and have a great deal of flexibility. As we shall see, SNOBOL was used by Jefferys to make TRIGMAN a more readily legible language.

Our main concern will be with mathematical manipulators which are popularly called formula manipulation languages, or algebraic manipulators, or mechanized operators. It will be very instructive to review what may have been the first paper deal with Poisson series (Herget and Musen, 1959). The philosophy adopted by the authors – and indeed one of the most productive positions that one would even take nowadays – was to develop a special-purpose program to produce and operate on Poisson series whose general term was written

$$N\cdot 10^n x_1^a x_2^b x_3^c x_4^d x_5^e x_6^f x_7^g \,{\cos \atop \sin}\,(hA_1 + iA_2 + jA_3 + kA_4 + lA_5), \tag{1}$$

In Deprit's notation the program represents a (7,5)-system, i.e., there are seven polynomial variables and five trigonometric arguments. However, in practice only one argument l (the mean anomaly) and one polynomial variable were used in the paper.

Of course, power series are subsumed when $h=i=j=k=l=0$. This work was done on an IBM 650 with a 2000 10-decimal digit word memory. Each term of the series is punched on a card in the form

$$\text{OO IDEN } aa\ bb\ cc\ dd\ ee\ ff\ gg\cdot \overset{\pm}{N}NNNNNNNnn\ hh\ ii\ jj\ kk.\overset{\pm}{l}l.$$

where OO is the order of the term, IDEN, is a 4-digit identification, aa, bb,..., gg are exponents of the polynomial in (1) in excess 50 notation. The sign over N is the sign of the mantissa and the sign over l means a cosine term if + and a sine term if −.

In multiplication of series the multiplicand series is stored in memory and the multiplier series passes through one term (card) at a time. Truncation on order takes place. The cards are sorted on all the arguments, $a, b, \ldots, g, h, i, \ldots, l$ and combination takes place in another run with the simplified product series as output.

Herget and Musen computed the Bessel functions as in the form shown in Figure 3 for computation. This is nothing more than the method of computing a polynomial by nesting.

$$J_k(ie) = \left(\frac{ie}{2}\right)^k \left[\frac{1}{0!}\frac{1}{k!} - \left(\frac{ie}{2}\right)^2 \left[\frac{1}{1!}\frac{1}{(k+1)!} - \left(\frac{ie}{2}\right)^2 \left[\frac{1}{2!}\frac{1}{(k+2)!} - \left(\frac{ie}{2}\right)^2\right.\right.\right.$$

$$\left.\left.\left.\times\left[\cdots\right]\right]\right]\right] = \left[\left[\left[\left[\frac{1}{n!} \times \frac{1}{(k+n)!}\right] \times \left(\frac{-i^2e^2}{4}\right) + \frac{1}{(n-1)!}\right.\right.\right.$$

$$\left.\times \frac{1}{(k+n-1)!}\right] \times \left(\frac{-i^2e^2}{4}\right) + \cdots\right]$$

$$\left.\times \left(\frac{-i^2e^2}{4}\right) + \frac{1}{0!} \times \frac{1}{k!}\right] \times \left(\frac{ie}{2}\right)^k.$$

Fig. 3. Computation of Bessel function by Nesting.

By well known relations the following series are then computed:

$$a/r,\ (a/r)^2,\ (a/r)^3,\ (r/a)^2,\ (a/r)^2\ (r/a)^2 = 1\ (ck),$$
$$\cos f,\ \sin f,\ \cos 2f,\ \sin 2f,\ r/a,\ \cos E,\ \sin E.$$

Figure 4 shows partial output of these calculations. The first exponent is for the coefficient, the second is for e, the eccentricity and the third index – under COS or SIN is the multiple of the mean anomaly l.

Herget and Musen's program exemplifies several important points. First, it is a specially designed program to solve a specific set of problems. Second, it takes peculiar advantage of the particular machine to optimize the operations. Third, it requires special knowledge to run and the output requires special knowledge to interpret (though trivial in this instance) to use again as input. Such special-purpose programs have the highest level of efficiency as to best management of memory, and operating speeds are minimized. In more advanced compilers, even though the programs may be special purpose, the designer may have to be intimately acquainted with the hardware, operating systems and special languages (such as LISP), or, at least, the techniques of list-processing and symbol manipulative languages, in order to construct a special-purpose compiler. The designer of a general-purpose compiler has a problem of a greater order of magnitude. He has to find the greatest balance between the greatest generality, most prudent use and management of system resources, and ability for his system to solve not only textbook problems but problems of consequence. Present technology, which is still hampered by the total amount of core space available, has been unable to provide the most general purpose compilers that can be used to solve consequential problems.

a/r SER 1		E	COS	$(a/r)^2$ SER 2		E	COS	$(a/r)^3$ SER 3		E	COS	$(a/r)^2(r/a)^2=1$ SER 4		E	COS
−.16954210	51	57	55	+.42916666	51	54	54	+.96249999	51	54	54	−.10000000	44	54	54
+.75688440	50	59	55	−.16125001	51	56	54	+.16124999	51	56	54	−.10000000	44	56	54
−.19710531	50	61	55	+.11618056	51	58	54	+.56187500	51	58	54	−.30000000	44	58	54
+.34219670	49	63	55	+.37987790	50	60	54	+.53213797	51	60	54	+.00000000	50	60	54
−.42774588	48	65	55	+.45986010	50	62	54	+.58828257	51	62	54	+.30000000	44	62	54
+.40506242	47	67	55	+.41592520	50	64	54	+.62882220	51	64	54	+.00000000	50	64	54
−.30138572	46	69	55	+.39153550	50	66	54	+.66805520	51	66	54	+.10000000	44	66	54
+.20250000	51	56	56	+.37008040	50	68	54	+.70505990	51	68	54	−.90000000	44	68	54
−.26035714	51	58	56	+.57135417	51	55	55	+.13851562	52	55	55	+.10000000	44	55	55
+.14645091	51	60	56	−.36069878	51	57	55	−.16233724	51	57	55	+.20000000	44	57	55
−.48816964	50	62	56	+.21062479	51	59	55	+.72948352	51	59	55	−.10000000	44	59	55
+.10983816	50	64	56	+.90259300	49	61	55	+.52169362	51	61	55	−.30000000	44	61	55
−.17973518	49	66	56	+.48952280	50	63	55	+.61513130	51	63	55	+.20000000	44	63	55
+.22466898	48	68	56	+.39588170	50	65	55	+.64830410	51	65	55	−.70000000	44	65	55
+.25531468	51	57	57	+.38038310	50	67	55	+.68702990	51	67	55	+.40000000	44	67	55
−.39095058	51	59	57	+.36017000	50	69	55	+.72299150	51	69	55	+.90000000	44	69	55
+.26606360	51	61	57	+.76437500	51	56	56	+.19793750	52	56	56	−.10000000	44	56	56
+.10864264	51	63	57	−.67758927	51	58	56	−.78107138	1	58	56	−.10000000	45	58	56
+.30247096	50	65	57	+.40555249	51	60	56	+.10889901	52	60	56	−.50000000	44	60	56
−.61754488	49	67	57	−.63002220	50	62	56	+.44018420	51	62	56	+.20000000	44	62	56
+.96986218	48	69	57	+.63675410	50	64	56	+.65763280	51	64	56	+.20000000	44	64	56
+.32507936	51	58	58	+.35610990	50	66	56	+.66448720	51	66	56	+.00000000	50	66	56
−.57791886	51	60	58	+.37303540	50	68	56	+.70583320	51	68	56	−.60000000	44	68	56
+.46233512	51	62	58	+.10258898	52	57	57	+.28130925	52	57	57	+.00000000	50	57	57
−.22416248	51	64	58	−.11706620	52	59	57	−.18761248	52	59	57	+.00000000	50	59	57
+.74720826	50	66	58	+.77613399	51	61	57	+.18267087	52	61	57	+.10000000	45	61	57
−.18392819	50	68	58	−.22767056	51	63	57	+.18668390	51	63	57	−.50000000	44	63	57
+.41704180	51	59	59	+.10954314	51	65	57	+.74993310	51	65	57	+.50000000	44	65	57
−.84450964	51	61	59	+.24579980	50	67	57	+.66953210	51	67	57	−.90000000	44	67	57
+.77733270	51	63	59	+.37882100	50	69	57	+.72583890	51	69	57	+.13000000	45	69	57
−.43724968	51	65	59	+.13799678	52	58	58	+.39806027	52	58	58	+.00000000	50	58	58
+.17027512	51	67	59	−.19254719	52	60	58	−.37240579	52	60	58	+.10000000	45	60	58
−.49258160	50	69	59	+.14441040	52	62	58	+.32680674	52	62	58	−.80000000	44	62	58
+.53822888	51	60	60	−.57772094	51	64	58	−.44792850	51	64	58	+.11000000	45	64	58
−.12232475	52	62	60	+.22996709	51	66	58	+.97757170	51	66	58	−.10000000	44	66	58
+.12742162	52	64	60	−.73217600	49	68	58	+.63918130	51	68	58	−.10000000	44	68	58
−.81680524	51	66	60	+.18593040	52	59	59	+.56126802	52	59	59	−.10000000	45	59	59
+.36464522	51	68	60	−.30661132	52	61	59	−.67430390	52	61	59	+.00000000	50	61	59
+.69801356	51	61	61	+.26034198	52	63	59	+.59657066	52	63	59	−.18000000	45	63	59
−.17595759	52	63	61	−.12802456	52	65	59	−.18732842	52	65	59	−.29000000	45	65	59
+.20471988	52	65	61	+.51572729	51	67	59	+.15371195	52	67	59	+.10000000	45	67	59
−.14744707	52	67	61	−.94145700	50	69	59	+.50707150	51	69	59	+.80000000	44	69	59
+.74337892	51	69	61	+.25082627	52	60	60	+.78905516	52	60	60	+.00000000	50	60	60
+.90888310	51	62	62	−.47718759	52	62	60	−.11560294	53	62	60	+.00000000	50	62	60
−.25169070	52	64	62	+.45590443	52	64	60	+.10837776	53	64	60	+.40000000	45	64	60
+.32360236	52	66	62	−.26277027	52	66	60	−.48588103	52	66	60	+.26000000	45	66	60
−.25888188	52	68	62	+.11472865	52	68	60	+.28520885	52	68	60	+.40000000	44	68	60
+.11874718	52	63	63	+.33870215	52	61	61	+.11065171	53	61	61	+.00000000	50	61	61
−.35836204	52	65	63	−.73008756	52	63	61	−.19110191	53	63	61	−.10000000	45	63	61
+.50469322	52	67	63	+.77852092	52	65	61	+.19383001	53	65	61	−.11000000	46	65	61
−.44423516	52	69	63	−.51202999	52	67	61	−.10803300	53	67	61	−.60000000	45	67	61
+.15559449	52	64	64	+.24696023	52	69	61	+.57905026	52	69	61	+.70000000	44	69	61
−.50827532	52	66	64	+.45772023	52	62	62	+.15483800	53	62	62	+.00000000	50	62	62
+.77829662	52	68	64	−.11023521	53	64	62	−.30776420	53	64	62	+.70000000	45	64	62
+.20438514	52	65	65	+.13012491	53	66	62	+.34012445	53	66	62	+.70000000	45	66	62
−.71854154	52	67	65	−.95975672	52	68	62	−.22183860	53	68	62	+.10000000	45	68	62
+.11887636	53	69	65	+.61895163	52	63	63	+.21626650	53	63	63	+.00000000	50	63	63
+.26906066	52	66	66	−.16469844	53	65	63	−.48597499	53	65	63	+.00000000	50	65	63
−.10129342	53	68	66	+.21357317	53	67	63	+.58557662	53	67	63	−.50000000	45	67	63
+.35488138	52	67	67	−.17446471	53	69	63	−.43293489	53	69	63	+.20000000	45	69	63
−.14244544	53	69	67	+.83741410	52	64	64	+.30157369	53	64	64	+.00000000	50	64	64
+.46887186	52	68	68	−.24396315	53	66	64	−.75562997	53	66	64	+.20000000	46	66	64

Fig. 4. Partial output of Herget and Musen's program.

Also in the class of special-purpose programs is an early program of Jirauch and Westerwick (1964) to obtain the equations of motion in complicated dynamical systems. The left hand side of Lagrange's equations

$$\frac{\mathrm{d}}{\mathrm{d}t}\left(\frac{\partial T}{\partial \dot{q}_i}\right) - \frac{\partial T}{\partial q_i} = - \frac{\partial U}{\partial q_i}$$

are calculated in literal form. For example in Figure 5, a massive satellite undergoes yaw, pitch and role. Table II gives the program in the very special language created

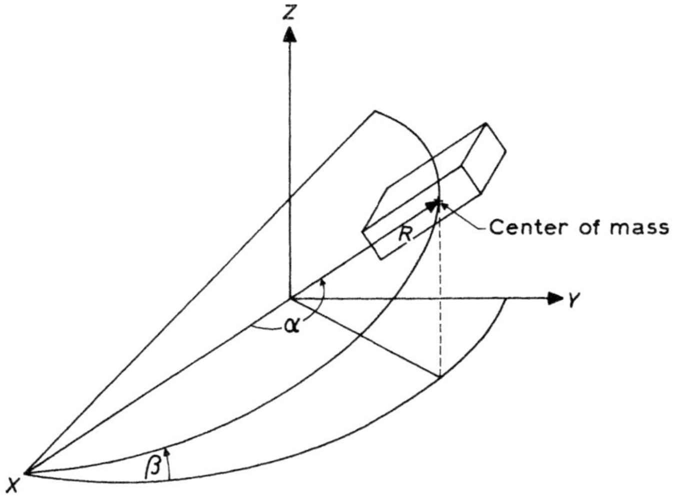

Fig. 5. A body in orbit undergoing yaw, pitch and roll.

TABLE II

Program for massive body in orbit

```
GENERAL EQUATION OF MOTION     JOB NO. 3271
        * THIS IS FOR THE EQUATIONS OF A SATELLITE WITH MASS.
        * BETA IS THE TILT ANGLE OF THE ORBIT
        * ALPHA IS THE POLAR ANGLE OF THE ORBIT.
        * R IS THE DISTANCE FROM THE CENTER OF THE EARTH TO THE
        *     CENTER OF MASS OF THE SATELLITE.
        * PSI IS THE YAW ANGLE OF THE SATELLITE.
        * THETA IS THE PITCH ANGLE OF THE SATELLITE.
        * PHI IS THE ROLL ANGLE OF THE SATELLITE.
        * XB IS THE X COORDINATE OF A POINT IN THE BODY (FOR
        *     INTEGRATION)
        * YB IS THE Y COORDINATE OF A POINT IN THE BODY (FOR
        *     INTEGRATION)
        * ZB IS THE Z COORDINATE OF A POINT IN THE BODY (FOR
        *     INTEGRATION)
        J OOO   Z OVO   X OOO   Z OOO   Y AAA   P
        BETA    ALPHA   PHI        THETA   PSI
                                           XB
                R                          YB
                                           ZB
```

Table II (Continued)

```
              THIS IS ELEMENT NUMBER 1 OF THE P MATRIX
- SIN  ALPHA COS BETA R
+ COS  PSI COS THETA COS ALPHA COS BETA XB
- COS  PSI COS THETA COS ALPHA COS BETA XB
- COS  PSI SIN  THETA COS PHI SIN ALPHA COS BETA XB
+ COS  PSI SIN  THETA SIN PHI SIN BETA XB
- SIN  PSI SIN  PHI SIN ALPHA COS BETA XB
- SIN  PSI COS PHI SIN BETA XB
- SIN  THETA COS ALPHA COS BETA YB
- COS  THETA COS PHI SIN ALPHA COS BETA YB
+ COS  THETA SIN PHI SIN BETA YB
+ SIN  PSI COS THETA   COS ALPHA COS BETA ZB
- SIN  PSI SIN  THETA COS PHI ALPHA COS BETA ZB
+ SIN  PSI SIN THETA  SIN  PHI SIN BETA ZB
+ COS  PSI COS PHI SIN BETA ZB

             THIS IS ELEMENT NUMBER 2 OF THE P MATRIX
+ COS  ALPHA R
+ COS  PSI COS THETA SIN ALPHA XB
+ COS  PSI SIN  THETA COS PHI COS ALPHA XB
+ SIN  PSI SIN  PHI COS ALPHA XB
- SIN  THETA SIN ALPHA YB
+ COS  THETA COS PHI COS ALPHA YB
+ SIN  PSI COS THETA SIN ALPHA ZB
+ SIN  PSI COS THETA SIN ALPHA ZB
+ SIN  PSI SIN THETA COS PHI COS ALPHA ZB
- COS  PSI SIN PHI COS ALPHA ZB

             THIS IS ELEMENT NUMBER 3 OF THE P MATRIX
+ SIN  ALPHA SIN BETA R
- COS  PSI COS THETA COS ALPHA SIN BETA XB
+ COS  PSI SIN THETA COS PHI SIN ALPHA SIN BETA XB
+ COS  PSI SIN  THETA SIN PHI COS BETA XB
+ SIN  PSI  IN  PHI SIN  ALPHA SIN BETA XB
- SIN  PSI COS PHI COS BETA XB
+ SIN  THETA COS ALPHA SIN BETA YB
+ COS  THETA COS PHI SIN ALPHA SIN BETA YB
+ COS  THETA SIN PHI COS BETA YB
- SIN  PSI COS THETA COS ALPHA SIN BETA ZB
+ SIN  PSI SIN  THETA COS PHI SIN ALPHA SIN BETA ZB
+ SIN  PSI SIN  THETA SIN  PHI COS BETA ZB
- COS  PSI SIN  PHI SIN  ALPHA SIN BETA ZB
+ COS  PSI COS PHI COS BETA ZB
```

by the authors. All statements marked with * are comments. The program is actually 5 statements (cards) long. It consists of the set of rotations R_j and translations T_j from the center of mass system of the satellite to the inertial system X, Y, Z, thus

$$P = R_1[T_1 + R_2[T_2 + \quad + R_8[T_8] \quad]].$$

The program computes the rotations and translations by an internal set of routines and will print out the transformed matrix of the coordinates (the P-matrix, in this case), if the command P is given (it is given as the last character of the first statement). The program, in effect, says: J is a constant rotation β about the Y-axis (inertial), followed by 000, no translation, followed by Z, a rotation α about the Z-axis (inertial), followed by OVO a translation along the Y axis, and so forth. XB, YB, ZB are coordinates of the center of the body which is at a distance R from the center of the Earth.

After the P-matrix is developed internal subroutines perform the differentiations with respect to the coordinates and the velocities and combine terms. A partial listing of output is given in Table III. The equation with respect to r is

$$+ 2\ddot{r} - 2r\dot{\alpha}^2$$

while the first terms with respect to α are

$$+ 4r\dot{r} + 2r^2\ddot{\alpha} - 4\cos^2\psi\sin\theta\cos\theta\cdot x_B^2\dot{\alpha}\dot{\theta}.$$

Thus the output is in quite readable form.

TABLE III

Partial output of satellite problem

```
THE LAGRANGE EQUATION WITH RESPECT TO R
+ 2 R..
- 2 R ALPHA. ALPHA.

THE LAGRANGE EQUATION WITH RESPECT TO ALPHA
+ 4 R R. ALPHA.
+ 2 R R ALPHA..
- 4 COSCOS PSI SINCOS THETA XB XB ALPHA. THETA.
- 4 SINCOS PSI COSCOS THETA XB XB ALPHA. PSI.
+ 2 COSCOS PSI COSCOS THETA XB XB ALPHA..
- 4 COSCOS PSI SINSIN THETA SINCOS PHI XB XB ALPHA. PHI.
+ 4 COSCOS PSI SINCOS THETA COSCOS PHI XB XB ALPHA. THETA.
- 4 SINCOS PSI SINSIN THETA COSCOS PHI XB XB ALPHA. PSI.
+ 2 COSCOS PSI SINSIN THETA COSCOS PHI XB XB ALPHA..
- 4 SINCOS PSI SIN THETA SINSIN PHI XB XB ALPHA. PHI.
+ 4 SINCOS PSI SIN THETA COSCOS PHI XB XB ALPHA. PHI.
+ 4 SINCOS PSI COS THETA SINCOS PHI XB XB ALPHA. THETA.
- 4 SINSIN PSI SIN THETA SINCOS PHI XB XB ALPHA. PSI.
+ 4 COSCOS PSI SIN THETA SINCOS PHI XB XB ALPHA. PSI.
+ 4 SINCOS PSI SIN THETA SINCOS PHI XB XB ALPHA..
+ 4 SINSIN PSI SINCOS PHI XB XB ALPHA. PHI.
+ 4 SINCOS PSI SINSIN PHI XB XB ALPHA. PSI.
+ 2 SINSIN PSI SINSIN PHI XB XB ALPHA..
+ 4 COS PSI SINSIN THETA XB YB ALPHA. THETA.
- 4 COS PSI COSCOS THETA XB YB ALPHA. THETA.
+ 4 SIN PSI SINCOS THETA XB YB ALPHA. PSI.
- 4 COS PSI SINCOS THETA XB YB ALPHA..
- 8 COS PSI SINCOS THETA SINCOS PHI XB YB ALPHA. PHI.
- 4 COS PSI SINSIN THETA COSCOS PHI XB YB ALPHA. THETA.
+ 4 COS PSI COSCOS THETA COSCOS PHI XB YB ALPHA. THETA.
- 4 SIN PSI SINCOS THETA COSCOS PHI XB YB ALPHA. PSI.
+ 4 COS PSI SINCOS THETA COSCOS PHI XB YB ALPHA..
- 4 SIN PSI COS THETA SINSIN PHI XB YB ALPHA. PHI.
+ 4 SIN PSI COS THETA COSCOS PHI XB YB ALPHA. PHI.
- 4 SIN PSI SIN THETA SINCOS PHI XB YB ALPHA. THETA.
```

Table III (Continued)

```
+ 4 COS PSI COS THETA SINCOS PHI XB YB ALPHA. PSI.
+ 4 SIN PSI COS THETA SINCOS PHI XB YB ALPHA..
- 8 SINCOS PSI SINCOS THETA XB ZB ALPHA. THETA.
- 4 SINSIN PSI COSCOS THETA XB ZB ALPHA. PSI.
+ 4 COSCOS PSI COSCOS THETA XB ZB ALPHA. PSI.
+ 4 SINCOS PSI COSCOS THETA XB ZB ALPHA..
- 8 SINCOS PSI SINSIN THETA SINCOS PHI XB ZB ALPHA. PHI.
+ 8 SINCOS PSI SINCOS THETA COSCOS PHI XB ZB ALPHA. THETA.
- 4 SINSIN PSI SINSIN THETA COSCOS PHI XB ZB ALPHA. PSI.
+ 4 COSCOS PSI SINSIN THETA COSCOS PHI XB ZB ALPHA. PSI.
+ 4 SINCOS PSI SINSIN THETA COSCOS PHI XB ZB ALPHA..
+ 4 COSCOS PSI SIN THETA SINSIN PHI XB ZB ALPHA. PHI.
- 4 COSCOS PSI SIN THETA COSCOS PHI XB ZB ALPHA. PHI.
- 4 COSCOS PSI COS THETA SINCOS PHI XB ZB ALPHA. THETA.
+ 16 SINCOS PSI SIN THETA SINCOS PHI XB ZB ALPHA. PSI.
- 4 COSCOS PSI SIN THETA SINCOS PHI XB ZB ALPHA..
- 4 SINSIN PSI SIN THETA SINSIN PHI XB ZB ALPHA. PHI.
+ 4 SINSIN PSI SIN THETA COSCOS PHI XB ZB ALPHA. PHI.
+ 4 SINSIN PSI COS THETA SINCOS PHI XB ZB ALPHA. THETA.
+ 4 SINSIN PSI SIN THETA SINCOS PHI XB ZB ALPHA..
- 8 SINCOS PSI SINCOS PHI XB ZB ALPHA. PHI.
+ 4 SINSIN PSI SINSIN PHI XB ZB ALPHA. PSI.
- 4 COSCOS PSI SINSIN PHI XB ZB ALPHA. PSI.
- 4 SINCOS PSI SINSIN PHI XB ZB ALPHA..
```

The seven additional pages of output are deleted.

So far we have dealt with special-purpose programs to solve specific problems. A somewhat wider class of compilers are those which are of a more general nature but still confine themselves to special techniques. The best known in this class are polynomial manipulators. Among the earliest successful systems was the ALPAK system (Brown *et al.*, 1963) built at Bell Telephone Laboratories. In dealing with literal series developments, a hard choice must usually be made in the design stage as to whether the coefficients are to be represented in REAL form or as rational coefficients, so that no approximations are present in the coefficient representation. Herget and Musen chose the REAL form, as have most builders of Poisson series manipulators. The advantages of rational numbers are obvious, but there is a price to pay in terms of space and time. In ALPAK the choice was made for rational number representation. Irrational numbers can be used, as in any mathematical processor, in symbolic form.

One of the principal problems in the field of symbolic manipulation is the representation of the mathematical expressions. This is of supreme importance for purposes of simplification, reduction, proof of equivalence to zero (or equivalence of expressions) and organization of the manipulator. Furthermore, management of series becomes a much simpler problem and memory space may be reduced to an enormous extent. Many special-purpose manipulators lend themselves to such canonical representation, particularly if they deal with Poisson series. ALPAK belongs to this class. Figure 6 shows the canonical representation of a polynomial as a $k\times(k+1)$ matrix for a polynomial in k variables.

$3x^2 + 2xyz - 5yz^2$			
	x	y	z
3	2	0	0
2	1	1	1
-5	0	1	2

Fig. 6. Matrix representation of variables.

The internal representation in Figure 7 shows that the name of the polynomial is made to point to a header which in turn has pointers for the exponent's precision, coefficients and the exponents themselves.

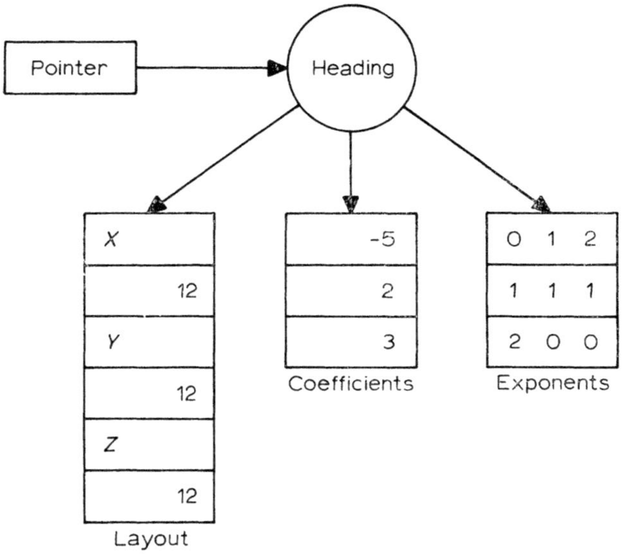

Fig. 7. Internal representation of a polynomial.

TABLE IV

Language of ALPAK

(*i*) *Basic Operations*				
POLADD	R,P,Q	$R = P + Q$	add	(a)
POLSUB	R,P,Q	$R = P - Q$	subtract	(b)
POLMPY	R,P,Q	$R = P*Q$	multiply	(c)
POLDIV	R,P,Q,NODIV	$R = P/Q$	divide (if divisible)	(d)
POLSST	G,F(LISTP) (LISTV)	$G = F$(LISTV = LISTP)	substitute	(e)
POLDIF	Q,P,X	$Q = \partial P/\partial X$	differentiate	(f)
POLZET	P	skip *iff* $P = 0$	zero test	(g)
POLNZT	P	skip *iff* $P \neq 0$	nonzero test	(h)
POLEQT	P,Q	skip *iff* $P = Q$	equality test	(i)
POLDUP	Q,P	$Q = P$	duplicate	(j)
POLCHS	P	$P = -P$	change sign	(k)

Table IV (Continued)

(ii) *Alternatives for Added Convenience and/or Efficiency*

POLSMP	Q,C,P	$Q=C*P$	scalar multiply	(l)
POLSMO	C,P	$P=C*P$	scalar multiply and overwrite	(m)
POLOMP	Q,M,P	$Q=M*P$	one-term multiply	(n)
POLOMO	M,P	$P=M*P$	one-term multiply and overwrite	(o)
POLSAD	Q,C,P	$Q=C+P$	scalar add	(p)
POLSAO	C,P	$P=C+P$	scalar add and over-write	(q)
POLADO	P,Q	$P=P+Q$	add and overwrite	(r)
POLDFO	P,X	$P=\partial P/\partial X$	differentiate and over-write	(s)

(iii) *Explanation of Symbols*

F,G,P,Q,R = polynomials (symbolic addresses of pointers)
C = scalar (symbolic address of scalar)
M = monomial (symbolic address of pointer)
X = variable (specified in the manner indicated by the last previous VARTYP declaration – see Section 3.5.2)
LISTP = list of polynomials
LISTV = list of variables.

Table IV shows the basic set of instructions in ALPAK. As one can see the language macros require a fairly detailed knowledge of the working of the assembly language and the hardware. A simple program in ALPAK to obtain

$$R = P + \partial Q/\partial Y$$

where P, Q, R are polynomials is given in Figure 8.

The following program computes $R=P+\partial Q/\partial Y$.

```
        POLBEG   10000
        VARTYP   NAM
FMT     POLCVF   (X, 12, Y, 12, Z, 12)
        POLRDP   P, FMT
        POLRDP   Q, FMT
        POLDIF   DQDY, Q, Y
        POLADD   R, P, DQDY                          (73)
        POLPRT   R, –, (R=P+DQ/DY)
        TRA      ENDJOB
P       PZE
Q       PZE
R       PZE
        END
```

Fig. 8. Program in ALPAK.

The arithmetic of the program follows from the instruction set easily. Other statements mean: POLBEG 10000 reserves 10000 locations. VARTYP NAM is a mode for differentiation of whole series rather than individual terms. POLCVF (X, 12, Y, 12, Z, 12) is the format indicating a precision of 12 bits each for X, Y, Z. POLRDP P, FMT reads polynomial P according to the format FMT given in statement above. POLPRT R, $-$, ($R=R+DQ/DY$) is a printing command with title $R=R+DQ/DY$. ($-$) is a command to skip three spaces between title and text. The next three statements reserve space for P, Q, R.

Assembly language systems of this sort can be tremendously efficient but they are not available to a large class of possible users because of the restrictions already mentioned. Furthermore, assembly language is almost always endemic to a particular machine model and, hence, may not be transportable to another. There has been a valiant attempt in recent years to improve this situation by having the same memory structure and instruction set in different lines of computers, but communication in assembly language is virtually ruled out among machines of different manufacture.

The designers of ALPAK quickly realized this and decided to make their system more universally available by re-casting it into the FORTRAN mold. The evolution of ALPAK from an assembly language processor to a higher level processor embedded in FORTRAN is an example of the inevitable progressive simplification and generalization of most viable systems. The most recent example of this discussed later is Jeffery's pre-processor for TRIGMAN. The last step of putting a program into a universal language is quite often an insignificant portion of the total labor involved. Thus, the time spent in perfecting ALTRAN, the FORTRAN version of ALPAK, was

ALPAK A	6.5 man-yr
ALPAK B	2.5 man-yr
ALTRAN	0.5 man-yr
TOTAL	9.5 man-yr

Table V gives the basic ALTRAN commands. In that table A, B, C, F, G, X are polynomials and K is an integer. The problem of Figure 8 is more simply written in ALTRAN as

$$R=P+\mathrm{DIFF}(Q, Y).$$

TABLE V
Some basic commands in ALTRAN

Operations	
Addition	$A=B+C$
Subtraction	$A=B-C$
Multiplication	$A=B*C$
Division	$A=B/C$
Integral exponentiation	$A=B**K$
Substitution	$G=F(X=P,\ Y=Q)$
Differentiation	$G=\mathrm{DIFF}(F, X)$
Greatest Common Divisor	$A=\mathrm{GCD}(B, C)$

Figure 9 illustrates in a very elementary fashion what ALTRAN does when it operates upon polynomials.

We come now to the general-purpose high-level compiler called FORMAC (Sammet, 1965) which, as originally written for IBM 700 and 7000 computers, was made superset to FORTRAN. Table VI describes the early FORMAC language. In this language expressions can be of the most general type. For example, the differential formula for Legendre functions

$$P_n(x) = \frac{1}{2^n n!} \frac{\mathrm{d}^n}{\mathrm{d}x^n} (x^2 - 1)^n$$

Addition	$(x+y)+(x-y)=2x$
Subtraction	$(x+y)-(x-y)=2y$
Multiplication	$(x+y)(x-y)=x^2-y^2$
Division	$(x^2-y^2)\div(x+y)=x-y$
Integral Exponentiation	$(x+y)^3=x^3+3x^2y+3xy^2+y^3$

Substitution

$$\left\{\begin{aligned} f(x, y, z) &= xy+z \\ p &= x+y \\ q &= x-y \\ r &= x^2+y^2 \\ f(p, q, r) &= pq+r \\ &= (x+y)(x-y)+(x^2+y^2) \\ &= (x^2-y^2)+(x^2+y^2) \\ &= 2x^2 \end{aligned}\right.$$

Differentiation $\quad \frac{\partial}{\partial y}(2x+5xy^2-3y^3)=10xy-9y^2$

Greatest Common Divisor $\quad \mathrm{GCD}(x^2-y^2, x^2+2xy+y^2)=x+y$

Given: $\quad \dfrac{x^2-y^2}{x^2+2xy+y^2}$

Divide numerator and denominator by GCD $\quad \dfrac{x^2-y^2}{x^2+2xy+y^2}=\dfrac{(x^2-y^2)/(x+y)}{(x^2+2xy+y^2)/(x+y)}=\dfrac{x-y}{x+y}$

Fig. 9. Elementary operation in ALTRAN.

can be written in PL/I FORMAC as

```
LET (P("N")=DERIV((X**2-1)**"N", X,
"N")/2**"N"*FAC("N"))
```

where 'N' is not a literal, but in the modern version of FORMAC which is superset to

PL/I can take on the integral values 0, 1, 2,..., as they might be assigned in a DO-loop. The Cambridge Polish notation was previously explained. The FORTRAN version of FORMAC represents expressions in 'delimiter Polish' notation which is a variation of Cambridge Polish. In this notation a delimiter, say], defines the scope of variary

TABLE VI

The FORMAC language

(1) *Four Declarative Statements*

ATOMIC	– declare basic variables.
DEPEND	– declare implicit dependence relations
PARAM	– declare parametric pairs for SUBST and EVAL.
SYMARG	– declare subroutine arguments as FORMAC variables; flag program beginning.

(2) *Fifteen Executable Statements*

(a) Statements Yielding FORMAC Variables

LET	– construct specified expressions.
SUBST	– replace variables with expressions.
EXPAND	– remove parentheses.
COEFF	– obtain coefficient of variable or its power.
PART	– separate expressions into terms, factors, exponents.
ORDER	– specify sequencing of variables within expressions.

(b) Statements Yielding FORTRAN Variables

EVAL	– evaluate expressions.
MATCH	– compare two expressions for equivalence or identity.
FIND	– determine dependence relations.
CENSUS	– count words, terms, or factors.

(c) Miscellaneous Statements

BCDCON	– convert to BCD form from internal form.
ALGON	– convert to internal form from BCD form.
AUTSIM	– control arithmetic done during automatic simplification.
ERASE	– eliminate expressions no longer needed.
FMDMP	– symbolic dump.

operators (operators which operate on two or more operands). Actually, the precise internal representation is not important to us, but it is important to recognize that the LET statement above results in the creation of a string of symbols and not a numerical result. All FORMAC statements result in internal operations therefore on strings. All FORTRAN or PL/I statements produce numbers in the usual way (depending on which version is used) and the statements can be completely mixed.

Figure 10 is part of a program to compute Lagrange's f and g series (Sconzo, 1965),

$$f = \sum_{n=0}^{\infty} f_n \tau^n / n!, \qquad g = \sum_{n=0}^{\infty} g_n \tau^n / n!.$$

```
$IBFMC TS02
        SYMARG
        DIMENSION F(100), G(100), OUTPUT(22)

C       FORMAC DATA DECLARATIONS.
        ATOMIC MU, SIGMA, EPSI, T
        DEPEND (MU, SIGMA, EPSI / T)
        LET F(1)=1.
        LET G(1)=0.
        DO 13 1=1, 100
C       CODE TO GENERATE F(I+1) AND G(I+1) FROM F(I) AND G(I)
      5 LET F(I+1)=SUBST FMCDIF(F(I), T, 1)-MU*G(I), LIST
        LET F(I+1)=EXPAND F(I+1)
     10 LET G(I+1)=SUBST(FMCDIF(G(I), T, 1)+F(I)), LIST
        LET G(I+1)=EXPAND G(I+1)
   LIST PARAM (FMCDIF(MU, T, 1), -3.*MU*SIGMA), (FMCDIF(SIGMA,
        T, 1), EPSI-2.*SIG
        1MA**2), (FMCDIF(EPSI, T, 1), -SIGMA*MU-SIGMA*2.*EPSI)
C       CODE TO EDIT THE EXPRESSIONS FOR OUTPUT.
        CALL GROUP (F(I+1), SIGMA)
        LET F(I+1)=ORDER F(I+1), DEC, FUL, (SIGMA)
        CALL GROUP (G(I+1), SIGMA)
        LET G(I+1)=ORDER G(I+1), DEC, FUL, (SIGMA)
C       CODE TO OUTPUT THE EXPRESSIONS.
        WRITE(6, 210) I
    210 FORMAT(1H110X2HF(13, 3H)=)
        J=O
    120 LET J=BCDCON F(I+1), OUTPUT, 12
        WRITE(6, 200)(OUTPUT(K), K=2, 12)
        IF(J.NE.0) GO TO 120
        WRITE(6, 240) I
    240 FORMAT(1H110X2HG(13, 3H)=)
    130 LET J=BCDCON G(I+1), OUTPUT, 12
        WRITE(6, 200)(OUTPUT(K), K=2, 12)
        IF(J.NE.0) GO TO 130
     13 CONTINUE
        STOP
    200 FORMAT(5×21A6)
        END
$IEDIT                SYSCKI
$IBFTC  TS02
$IEDIT
```

Fig. 10. Partial program for computing *f* and *g* series.

The solution is based upon the recursion relations

$$f_n = \dot{f}_{n-1} - (1/r^3)\, g_{n-1}, \qquad g_n = f_{n-1} + \dot{g}_{n-1}$$

where $f_0 = 1$, $g_0 = 0$ and we put

$$\mu = 1/r^3, \qquad \sigma = s/r^3, \qquad \varepsilon = \dot{s}/r^2, (s = x\dot{x} + y\dot{y} + z\dot{z})$$
$$\dot{\mu} = -3\mu\sigma, \qquad \dot{\sigma} = \varepsilon - 2\sigma^2, \qquad \dot{\varepsilon} = -\sigma(\mu + 2\varepsilon)$$

MU, SIGMA, EPSI and T are declared as atomic variables (unnecessary in PL/I FORMAC) and dependence of μ, σ, ε on t is declared in the next statement. After $F(1)$ and $G(1)$ are set, the first FORMAC statement within the DO essentially says: find the first derivative of f_i with respect to t, subtract μg_i and substitute into that expression for f_{i+1} from LIST (four statements below),
i.e., $\dot{\mu} = -3\mu\sigma$, $\dot{\sigma} = \varepsilon - 2\sigma^2$, $\dot{\varepsilon} = -\sigma\mu - \sigma \cdot 2 \cdot \varepsilon$. For purposes of simplification this is expanded and the series at g_{i+1} is similarly developed. That is the entire program. The rest of the program provides that the output be given in a fairly readable form. Figure 11 shows partial output from this program.

```
          F( 1)=
    0$
          F( 2)=
  -MU$
          F( 3)=
   3*SIGMA*MU$
          F( 4)=
 -15*SIGMA**2*MU+3*EPSI*MU*MU**2$
          F( 5)=
  105*SIGMA**3*MU+SIGMA*(-45*EPSI*MU-15*MU**2)$
          F( 6)=
-945*SIGMA**4*MU+SIGMA**2*(630*EPSI*MU+210*MU**2)-24*
EPSI*MU**2-45
*EPSI**2*MU-MU**3$
```

Fig. 11. Partial output for f and g series program.

The series was run as far as $n = 27$ as a test case and took less than 19 min on an IBM 7094.

Figure 12 gives the coefficients to 12th order as published in *Astron. J.* For purposes of printing it is necessary to rewrite the computer output in the ordinary mathematical notation of the journals. There are currently two ways around this problem. PL/I FORMAC pays particular attention to readability.
Consider the PL/I FORMAC program in Figure 13 which computes Legendre poly-

$$
\begin{aligned}
f_0 &= 1\\
f_1 &= 0\\
f_2 &= -\mu\\
f_3 &= 3\sigma\mu\\
f_4 &= -15\sigma^2\mu + 3\varepsilon\mu + \mu^2\\
f_5 &= 105\sigma^3\mu - \sigma(45\varepsilon\mu + 15\mu^2)\\
f_6 &= -945\sigma^4\mu + \sigma^2(630\varepsilon\mu + 210\mu^2) - 24\varepsilon\mu^2 - 45\varepsilon^2\mu - \mu^3\\
f_7 &= 10395\sigma^5\mu - \sigma^3(9450\varepsilon\mu + 3150\mu^2) + \sigma(882\varepsilon\mu^2 + 1575\varepsilon^2\mu + 63\mu^3)\\
f_8 &= -135135\sigma^6\mu + \sigma^4(155925\varepsilon\mu + 51975\mu^2)\\
&\quad - \sigma^2(24570\varepsilon\mu^2 + 42525\varepsilon^2\mu + 2205\mu^3) + 117\varepsilon\mu^3 + 1575\varepsilon^3\mu\\
&\quad + 1107\varepsilon^2\mu^2 + \mu^4\\
f_9 &= 2027025\sigma^7\mu - \sigma^8(2837835\varepsilon\mu + 945945\mu^2)\\
&\quad + \sigma^3(644490\varepsilon\mu^2 + 1091475\varepsilon^2\mu + 65835\mu^3)\\
&\quad - \sigma(10935\varepsilon\mu^3 + 99225\varepsilon^3\mu + 74385\varepsilon^2\mu^2 + 255\mu^4)\\
f_{10} &= -34459425\sigma^8\mu + \sigma^6(56756700\varepsilon\mu + 18918900\mu^2)\\
&\quad - \sigma^4(17027010\varepsilon\mu^2 + 28378350\varepsilon^2\mu + 1891890\mu^3)\\
&\quad + \sigma^2(599940\varepsilon\mu^3 + 4365900\varepsilon^3\mu + 3421440\varepsilon^2\mu^2 + 21120\mu^4)\\
&\quad - 498\varepsilon\mu^4 - 99225\varepsilon^4\mu - 85410\varepsilon^3\mu^2 - 15066\varepsilon^2\mu^3 - \mu^5\\
f_{11} &= 654729075\sigma^9\mu - \sigma^7(1240539300\varepsilon\mu + 413513100\mu^2)\\
&\quad + \sigma^5(465404940\varepsilon\mu^2 + 766215450\varepsilon^2\mu + 54864810\mu^3)\\
&\quad - \sigma^3(27027000\varepsilon\mu^3 + 170270100\varepsilon^3\mu + 137837700\varepsilon^2\mu^2\\
&\quad + 1201200\mu^4) + \sigma(114444\varepsilon\mu^4 + 9823275\varepsilon^4\mu + 9058500\varepsilon^3\mu^2\\
&\quad + 2023758\varepsilon^2\mu^3 + 1023\mu^5)\\
f_{12} &= -13749310575\sigma^{10}\mu + \sigma^8(29462808375\varepsilon\mu + 9820936125\mu^2)\\
&\quad - \sigma^6(13315121820\varepsilon\mu^2 + 21709437750\varepsilon^2\mu + 1640268630\mu^3)\\
&\quad + \sigma^4(1122971850\varepsilon\mu^3 + 6385128750\varepsilon^3\mu + 5298643350\varepsilon^2\mu^2\\
&\quad + 58108050\mu^4) - \sigma^2(12072060\varepsilon\mu^4 + 638512875\varepsilon^4\mu\\
&\quad + 618918300\varepsilon^3\mu^2 + 159729570\varepsilon^2\mu^3 + 195195\mu^5) + 2031\varepsilon\mu^5\\
&\quad + 9823275\varepsilon^5\mu + 9951525\varepsilon^4\mu^2 + 2480958\varepsilon^3\mu^3 + 164610\varepsilon^2\mu^4 + \mu^6\\
g_0 &= 0\\
g_1 &= 1\\
g_2 &= 0\\
g_3 &= -\mu\\
g_4 &= 6\sigma\mu\\
g_5 &= -45\sigma^2\mu + 9\varepsilon\mu + \mu^2\\
g_6 &= 420\sigma^3\mu - \sigma(180\varepsilon\mu + 30\mu^2)\\
g_7 &= -4725\sigma^4\mu + \sigma^2(3150\varepsilon\mu + 630\mu^2) - 54\varepsilon\mu^2 - 225\varepsilon^2\mu - \mu^3\\
g_8 &= 62370\sigma^5\mu - \sigma^3(56700\varepsilon\mu + 12600\mu^2)\\
&\quad + \sigma(3024\varepsilon\mu^2 + 9450\varepsilon^2\mu + 126\mu^3)\\
g_9 &= -945945\sigma^6\mu + \sigma^4(1091475\varepsilon\mu + 259875\mu^2)\\
&\quad - \sigma^2(111510\varepsilon\mu^2 + 297675\varepsilon^2\mu + 6615\mu^3) + 243\varepsilon\mu^3 + 11025\varepsilon^3\mu\\
&\quad + 4131\varepsilon^2\mu^2 + \mu^4\\
g_{10} &= 16216200\sigma^7\mu - \sigma^5(22702680\varepsilon\mu + 5675670\mu^2)\\
&\quad + \sigma^3(3617460\varepsilon\mu^2 + 8731800\varepsilon^2\mu + 263340\mu^3)\\
&\quad - \sigma(35100\varepsilon\mu^3 + 793800\varepsilon^3\mu + 371790\varepsilon^2\mu^2 + 510\mu^4)\\
g_{11} &= -310134825\sigma^8\mu + \sigma^6(510810300\varepsilon\mu + 132432300\mu^2)\\
&\quad - \sigma^4(113513400\varepsilon\mu^2 + 255405150\varepsilon^2\mu + 9459450\mu^3)\\
&\quad + \sigma^2(2589840\varepsilon\mu^3 + 39293100\varepsilon^3\mu + 21116700\varepsilon^2\mu^2 + 63360\mu^2)\\
&\quad - 1008\varepsilon\mu^4 - 893025\varepsilon^4\mu - 457200\varepsilon^3\mu^2 - 50166\varepsilon^2\mu^3 - \mu^5\\
g_{12} &= 6547290750\sigma^9\mu - \sigma^7(12405393000\varepsilon\mu + 3308104800\mu^2)\\
&\quad + \sigma^5(3587023440\varepsilon\mu^2 + 7662154500\varepsilon^2\mu + 329188860\mu^2)\\
&\quad - \sigma^3(145945800\varepsilon\mu^3 + 1702701000\varepsilon^3\mu + 1005404400\varepsilon^2\mu^2\\
&\quad + 4804800\mu^4) + \sigma(355608\varepsilon\mu^4 + 98232750\varepsilon^4\mu + 60350400\varepsilon^2\mu^2\\
&\quad + 9227196\varepsilon^2\mu^3 + 2046\mu^5)
\end{aligned}
$$

Fig. 12. Coefficients f_n, g_n to $n = 12$.

Listing of deck LEGR

```
//LEGR    JOB   FORMAC, FORMAC, MSGLEVEL=I
//JOBLIB  DD    DSNAME=SYSI. SYSTEM. LINKLIB, DISP=OLD
//        EXEC  FORMAC
//SYSIN     DD  *
      LEGR: PROCEDURE OPTIONS(MAIN);
      FORMAC-OPTIONS;
      OPTSET(LINELENGTH=72);
      OPTSET(EXPND);
            PUT PAGE;
/*GENERATE LEGENDRE POLYNOMIALS BY METHOD 1 */
            DO N=0 TO 10;
            LET(P("N")=DERIV((X**2-1)**"N", X, "N")/(2**"N"*
            *FAC("N")) );
            END:
/* GENERATE LEGENDRE POLYNOMIALS BY METHOD 2 */
            LET( Q(0)=1; Q(1)=X );
            DO N=2 TO 10;
            LET( N="N";
                  Q(N)=(2*N-1)/N*X*Q(N-1)-(N-1)/N*Q(N-2)  ):
            END:
/*CHECK THAT P(N)=Q(N) AND PRINT OUT RESULTS */
             PUT LIST( 'LEGENDRE POLYNOMIALS'); PUT SKIP(3);
             DO N=0 TO 10;
             LET( N="N" );
             IF IDENT(P(N); Q(N)) THEN PRINT-OUT(P(N));
                                  ELSE STOP;
             END;
             END LEGR:
/*
```

The program computes $P_n(x)$:

(1) By means of the differential formula

$$P_n(x) = \frac{1}{2^n n!} \frac{\mathrm{d}^n}{\mathrm{d}x^n} (x^2 - 1)^n$$

(2) By means of the recurrence relation

$$Q_n(x) = \frac{2n-1}{n} \times Q_{n-1}(x) - \frac{n-1}{n} Q_{n-2}(x)$$

with $Q_0(x) = 1$; $Q_1(x) = x$.

The program checks that $P_n(x) = Q_n(x)$ and prints out the results.

Fig. 13. FORMAC program to compute Legendre polynomials.

nomials according to the differential formula and well-known recurrence relations, compares them and prints out the first term. The results are shown in Figure 14, which is probably acceptable in many publications as to quality.

The other choice is to use the method of photocomposition as was done by Gerard *et al.* (1965) when they computed Newcomb operators in the analytical development of the planetary disturbing function, using a powerful special-purpose program which used array rotation for representation of polynomials like ALPAK. A photocomposition-machine (Photon' was used by Izsak and his associates) is a machine which has a variety of type fonts which might include mathematical symbols, Greek letters, etc. A command to the machine selects a particular character from a particular type font and it passes optically with pre-selected magnification to a pre-selected position onto a photographic emulsion. Thus, at great speed, and under computer control, it is possible to prepare and compose a sheet which can then be developed and later printed by photographic offset or other methods. Today, photocomposition of type is a

$P(0) = 1$

$P(1) = X$

$P(2) = 3/2\ X^2 - 1/2$

$P(3) = -3/2\ X + 5/2\ X^3$

$P(4) = -15/4\ X^2 + 35/8\ X^4 + 3/8$

$P(5) = 15/8\ X - 35/4\ X + 63/8\ X^5$

$P(6) = 105/16\ X^2 - 315/16\ X^4 + 231/16\ X^6 - 5/16$

$P(7) = -35/16\ X + 315/16\ X^3 - 693/16\ X^5 + 429/16\ X^7$

$P(8) = -315/32\ X^2 + 3465/64\ X^4 - 3003/32\ X^6 + 6435/128\ X^8 + 35/128$

$P(9) = 315/128\ X - 1155/32\ X^3 + 9009/64\ X^5 - 6435/32\ X^7 + 12155/128\ X$

9

--

$P(10) = 3465/256\ X^2 - 15015/128\ X^4 + 45045/128\ X^6 - 109395/256\ X^8 +$

$46189/256\ X^{10} - 63/256$

Fig. 14. Legendre polynomial output.

widely accepted method of preparing copy for publication. Many newspapers in the U.S.A., for example, are composed in this way. Figure 15 shows partial output from Izsak's program using photocomposition.

$$
\begin{aligned}
\Sigma_{6,\,1}^{n,\,k} = {} & 0.42191319\mathrm{k} + 0.13270400\mathrm{k}^2 - 0.58263887\mathrm{k}^3 - 0.52940538\mathrm{k}^4 - 0.17751735\mathrm{k}^5 \\
& - 0.260^41666 \times 10^{-1}\mathrm{k}^6 - 0.13888888 \times 10^{-2}\mathrm{k}^7 \\
& + (-0.22146267 - 0.15382270 \times 10\mathrm{k} - 0.13371744 \times 10\mathrm{k}^2 - 0.21294488\mathrm{k}^3 \\
& + 0.11219618\mathrm{k}^4 \\
& + 0.40104166 \times 10^{-1}\mathrm{k}^5 + 0.34722222 \times 10^{-2}\mathrm{k}^6)\mathrm{n} \\
& + (0.10800021 \times 10 + 0.23384766 \times 10\mathrm{k} + 0.13933919 \times 10\mathrm{k}^2 + 0.27007379\mathrm{k}^3 \\
& + 0.77610213 \times 10^{-10}\mathrm{k}^4 - 0.31249999 \times 10^{-2}\mathrm{k}^5)\mathrm{n}^2 \\
& + (-0.78080512 - 0.97878689\mathrm{k} - 0.35248481\mathrm{k}^2 - 0.35156250 \times 10^{-1}\mathrm{k}^3 \\
& + 0.86805555 \times 10^{-3}\mathrm{k}^4)\mathrm{n}^3 \\
& + (0.20079210 + 0.14930555\mathrm{k} + 0.27018229 \times 10^{-1}\mathrm{k}^2 + 0.43402778 \times 10^{-3}\mathrm{k}^3)\mathrm{n}^4 \\
& + (-0.21755642 \times 10^{-1} - 0.83007813 \times 10^{-2}\mathrm{k} - 0.39062499 \times 10^{-3}\mathrm{k}^2)\mathrm{n}^5 \\
& + (0.94401041 \times 10^{-3} + 0.10850694 \times 10^{-3}\mathrm{k})\mathrm{n}^6 - 0.10850694 \times 10^{-4}\mathrm{n}^7 \\
\Sigma_{5,\,2}^{u,\,k} = {} & -0.16601565 \times 10^{-1}\mathrm{k} - 0.71250000\mathrm{k}^2 - 0.94023438\mathrm{k}^3 - 0.20800781\mathrm{k}^{-4} \\
& + 0.11588541\mathrm{k}^5 \\
& + 0.46875000 \times 10^{-1}\mathrm{k}^6 + 0.41666666 \times 10^{-2}\mathrm{k}^7 \\
& + (0.15107422 - 0.61458333\mathrm{k} + 0.57226563\mathrm{k}^2 + 0.14643554 \times 10\mathrm{k}^3 \\
& + 0.55013020\mathrm{k}^4 \\
& + 0.25520833 \times 10^{-1}\mathrm{k}^5 - 0.62499999 \times 10^{-2}\mathrm{k}^6)\mathrm{n} \\
& + (0.56184895 + 0.17567708 \times 10\mathrm{k} + 0.26090495\mathrm{k}^2 - 0.50292968\mathrm{k}^3 \\
& - 0.12500000\mathrm{k}^4 \\
& + 0.10416666 \times 10^{-2}\mathrm{k}^5)\mathrm{n}^2 \\
& + (-0.12462239 \times 10 - 0.16437174 \times 10\mathrm{k} - 0.30875651\mathrm{k}^2 + 0.63802083 \times 10^{-1}\mathrm{k}^3 \\
& + 0.26041666 \times 10^{-2}\mathrm{k}^4)\mathrm{n}^3 \\
& + (0.59261068 + 0.37565104\mathrm{k} + 0.15950520 \times 10^{-1}\mathrm{k}^2 - 0.13020833 \times 10^{-2}\mathrm{k}^3)\mathrm{n}^4 \\
& + (-0.85774739 \times 10^{-1} - 0.18066406 \times 10^{-1}\mathrm{k} - 0.13020833 \times 10^{-3}\mathrm{k}^2)\mathrm{n}^5 \\
& + (0.33528645 \times 10^{-2} + 0.19531249 \times 10^{-3}\mathrm{k})\mathrm{n}^6 - 0.32552083 \times 10^{-4}\mathrm{n}^7
\end{aligned}
$$

Fig. 15. Partial output of Newcomb operators program.

FORMAC is a powerful general-purpose algebraic manipulator written at a high level and has tremendous convenience. The phenomenon of expression pulsation is the enlargement of expressions during various manipulative phases. Most of the time the expressions then simplify and reduce to manageable proportions. However, in carrying out consequential problems, users of the FORMAC system, especially in dynamical astronomy have found that one runs out of space very quickly. Of course, there are provisions for releasing space that is no longer needed (the ERASE instruction, for example), but the problem arises even under the most careful attention to space conservation. Because of the great generality of the compiler, the system itself must occupy a large share of memory.

In passing, mention should be made of two ALGOL systems. The first, FORMULA ALGOL (Perlis, 1965), is superset to ALGOL and provides a language which can be used to construct a compiler of one's own design. Available to the user are list processing and string manipulative techniques, special expression forms such as formula expressions. Thus, it devolves entirely on some designer to construct a suitable system. He would have to be conversant with many aspects of computer science but he could build a powerful compiler. This has not been done, nor have any consequential problems been solved to my knowledge.

The second system, which has no name, is described by van de Riet (1968) in two short volumes entitled 'Formula Manipulation in ALGOL 60'. Two systems are described: the simple and the general system. Both systems are entirely coded in ALGOL so that any computer system capable of processing ALGOL is capable of compiling the manipulator. Furthermore, the user can suitably modify the system to satisfy some of the special requirements he may have in particular applications. Or – since the general system occupies much storage space and requires longer run-times, parts of the system not needed may be omitted. Obviously, this assumes an intimate knowledge of the compiler and must be done with great caution. Clearly, this would be true of all manipulators, especially those written in universal languages like FORTRAN, ALGOL, PL/I. Van de Riet recommends, as do all compiler builders, that to achieve the optimal running efficiency, it is highly desirable that the most frequently used, and most time consuming, procedures be hand-coded in machine language. A number of applications have been carried out but, as far as is known, no problems in dynamical astronomy.

This brings us to most recent times. It was recognized a decade ago (Davis, 1963) that the most useful compilers for applications in celestial mechanics would be those specially designed to deal with Poisson series, but implementation of such systems did not begin until 1965 with the work of Danby, Deprit and Rom, and in 1966 with the work of Barton. Eaily specifications for such systems are shown in Table VII (Davis, 1963; Deprit, 1967; Barton, 1966).

Barton's work called attention to the exciting possibilities of specialized systems. The compiler was written in less than six hours (debugging time was not given) and carried out Delaunay's development of the lunar disturbing function to the 8th order in 7 min on Cambridge's Titan computer. Among the great advantages of the mechanized computer approach to solving such problems is the direct utilization of the basic equations. When work is done by hand, short cuts are almost always devised. From a programming standpoint, providing there is no loss of precision (in the case of numerical problems), it is simpler, more direct and more easily suitable to iterative machine processes to deal with basic equations themselves rather than their transformations. Thus, the disturbing function **R**, the cosine of the elongation S of the Moon from the Sun in the barycentric system, the radius vector **r**, the mean anomaly **l**, the semi-major axis **a** and $r^2(\mathrm{d}f/\mathrm{d}l)$ are given by:

TABLE VII

Specifications for poisson series manipulators

Davis	Deprit	Barton
Function MACROS		
1. Substitute	1. Creation of a series	1. Addition
2. Simplify	2. Annihilation of a series	2. Subtraction
3. Differentiate	3. Ordering of terms	3. Negation
4. Integrate	4. Insertion of new terms in series	4. Multiplication by a rational number
Series MACROS		
5. 'Multinomiate'	5. Selection of particular terms	5. Selection of a particular term
6. 'Fourierate'	6. Fusion of several series	6. Differentiation with respect to any of 14 variables
7. Series Multiply	7. Addition of series (especially $f \leftarrow f + \alpha g$)	7. Integration with respect to any of 14 variables
8. Series Add, Subtract	8. Multiplication of series (especially $f \leftarrow f + gh$)	8. Multiplication
9. Combine	9. Differentiation with respect to any of three variables	9. Substitution for a polynomial variable into another series
10. Series Differentiate and Integrate	10. Numerical evaluation of series	10. Substitution for a harmonic variable into another series along with a truncated Taylor series

$$R = m' \frac{a^2}{a'^2}\left(\frac{a'}{r'}\right)\left[\left(\frac{r}{a}\right)^2\left(\frac{a'}{r'}\right)^2 P_2(S) + \left(\frac{a}{a'}\right)\left(\frac{r}{a}\right)^3\left(\frac{a'}{r'}\right)^3 P_3(S) + \cdots\right], \quad (2)$$

$$S = (1-\gamma^2)\cos(f+g+h-f'-g'-h') + \\ + \gamma^2 \cos(f+g-h+f'+g'+h'), \quad (3)$$

$$r = a(1 - e\cos E), \quad (4)$$

$$l = E - e\sin E, \quad (5)$$

$$a = r\,\mathrm{d}E/\mathrm{d}l, \quad (6)$$

$$r^2\,(df/dl) = a^2\,(1-e^2)^{1/2}, \quad (7)$$

where the other symbols have their usual meanings.

The object is to express R as a function of $\gamma, a, e, l, g, h, a', e', l', g', h'$. The program then executes the following sequence of steps.

(1) Kepler's Equation (5) is solved by direct iteration to give $E(e, l)$.

(2) $E(e, l)$ is substituted into (4) to give r/a.

(3) $a/r = \mathrm{d}E/\mathrm{d}l$ is obtained from (6) by differentiation.

(4) $\mathrm{d}f/\mathrm{d}l$ is obtained from (7) and integrated to give f.

(5) f'(true longitude of the Sun) is obtained similarly.

(6) f and f' are substituted into (3)

(7) After a'/r' is obtained, substitute a'/r', a/r and S into (2) to give R.

And now we come to contemporaneous systems of the last few years. Only those that have been, or are planning to be, used on consequential problems are discussed here. (The reader is referred to Kovalevsky (1968) for one very useful approach to these problems. It is a (5, 4)-system where the sums $P_i = \sum N_j x_1^{j_1} x_2^{j_2} \ldots x_k^{j_k}$ are treated as a table of numbers N_j whose position in the table are determined by a unique combination of the $j_1, j_2, \ldots, j_k$. Thus, Poisson series may be regarded as pure trigonometric series whose coefficients are these tables. An algebra of operations is constructed for series manipulators. This technique has many advantages of speed but limitations of space.) We shall briefly describe four other systems.

(1) MAO (Rom, 1970)

(2) Broucke's Programming System (Broucke, 1969)

(3) ESP (Rom, 1971)

(4) TRIGMAN (Jeffreys, 1970).

The first version of MAO (Mechanized Algebraic Operations) was designed by Danby *et al.* (1965) (q.v.). It is a (10,6)-system, i.e., there are 10 polynomial parameters and 6 trigonometric arguments written in FORTRAN and assembly language. As with other systems, compactness of equations is not sought before the problem is tackled but rather the user must organize the problem into a sequence of routine steps. This generally means developing recursive procedures to be later fed to the system. The system, as are all others described, is fundamentally a Poisson series manipulator which in MAO is represented as:

$$\sum_{i_1}\sum_{i_2}\cdots\sum_{i_m}\sum_{j_1}\sum_{j_2}\cdots\sum_{j_n} p_1^{i_1}p_2^{i_2}\cdots p_n^{i_m} \times C_{i_1, i_2, \dots, i_m}^{j_1, j_2, \dots, j_n} \times$$

$$\times \begin{matrix}\cos\\ \sin\end{matrix}(j_1t_1 + j_2t_2 + \cdots + j_nt_n) \tag{8}$$

where all the i's and j's are positive, zero or negative, except for j_1 which is always non-negative. The principal stratagem is to partition the power series as sequences of coefficients each of which is itself treated as a Poisson series. Thus, series may be partitioned into blocks which are homogeneous and whose dimensions are known. The great advantage of this approach is that each block may be treated as an expression. Great convenience is introduced by having all the terms in a block with the same powers in the small parameter. This approach thus easily lends itself to the storage of series, or partitions of series, on outside media such as tapes and disk. Only those blocks are kept in core storage that are currently active, since core memory is the most valuable resource in the system.

The user is saved from the requirement of allocating space for each series his problem requires by simply allocating a FORTRAN variable to each series. Thus, storage management is handled by the processor which utilizes a simple type of algorithm for dynamic storage of series. All series in core are kept contiguous, the currently active one being at the 'bottom'. Every time series operations occur, the active series is moved to the bottom and all the others are moved up. Some of these moves are done automatically by the system, others carried out by the MOVE(A) command (see Table VIII). The dynamic core storage has an extension to disk for the more inactive series. The language, summarized by CALL's in Table VIII, may be supplemented by all the capabilities of FORTRAN– Since the 360 FORTRAN compiler distinguishes subroutine arguments by *value* or by *name* (address, location or argument), the dummy arguments in CALL's must be enclosed in slashes. This insures that the subroutine itself does not allocate space to the dummy variables but uses already existing location for them.

Figure 16 shows how each term of a series is stored. Six words of four bytes each are used (8 bits = 1 byte). As seen in Equation (8), all i's and j's can be positive or negative except j_1 which must be $\geqslant 0$. The high order bit in each $i_k, j_k (k \neq 1)$ is the sign bit, $0 \equiv +, 1 \equiv -$. If the number is negative, it is stored as a two's complement. Thus, the range for the i's and j's is $-64_{10}(40_{16})$ to $+63_{10}(3F_{16})$. Since j_1 is always understood to be positive, its high order digit serves as a flag such that $0 \equiv$ cosine term, $1 \equiv$ sine term. Hence the range of j_1 is from 0 to $+63_{10}(3F_{16})$. Figure 17 shows the structure of the i and j bytes. This judicious method of packing the indices in the words allows for packed word arithmetic, i.e., additions and subtractions of four bytes at a time may take place without the time-consuming operations of packing and unpacking. The techniques for doing so are elaborated by Rom in his paper.

Because core storage is always at a premium, the user may ERASE a series which is

TABLE VIII

Mao language

Calling sequence	Operation
CALL SPACE(n)	open up dynamic storage
CALL DEFINE(A)	set up new series; $A = 0$
CALL ANNEX(A, a, iw, sc)	augment A
CALL LIST(A*,'× × × × × × × ×', n)	print terms of A
CALL MOVE(A)	place A on bottom of stack
CALL ERASE(A)	remove A from storage
CALL PXCA(A, B)	change identifiers; $A = B$
CALL ORDER(A)	sort terms of A in natural order
CALL TSORT(A)	sort terms of A according to trigonometric arguments
CALL HEXIN(A)	read D from hexadecimal cards
CALL HEXOUT(A)	punch A in hexadecimal
CALL NEWDSK	fresh start on disk
CALL DISKON	restart disk
CALL LOCK	set disk in read-only mode
CALL UNLOCK	set disk in read-write mode
CALL CLOSED	close disk file
CALL SEEK(A_*)	point to read A_* from disk
CALL SEEK(0)	point to write on disk
CALL WRITER(A, B_*, P_*, I)	write A, and point to P_*
CALL READP(A, P_*)	read A, and point to P_*
CALL SREAD(A, B_*, P_*, I)	read A by searching, and point to P_*
CALL SCAN(A, α)	remove from A terms whose absolute value is less than α
n = ITERM(A)	set n equal to term count of A
CALL PROD(A_*, B_*, C, α)	$\alpha_* A_* B + C \rightarrow C$
CALL ACUM(A_*, B, α)	$\alpha_* A + B \rightarrow B$
CALL PDP(A_*, B, n)	$\partial A/\partial p_n \rightarrow B$
CALL PDT(A_*, B, m)	$\partial A/\partial t_m \rightarrow B$
CALL PMULT(A, n, i)	$p_n{}^{i} * A \rightarrow A$
CALL FETCH(A, α, iw, sc)	unpack next term of A
CALL RPLACE(A, α, iw, sc)	replace current term in A
CALL ADD(A, α, iw, sc)	store term in A; combine like terms

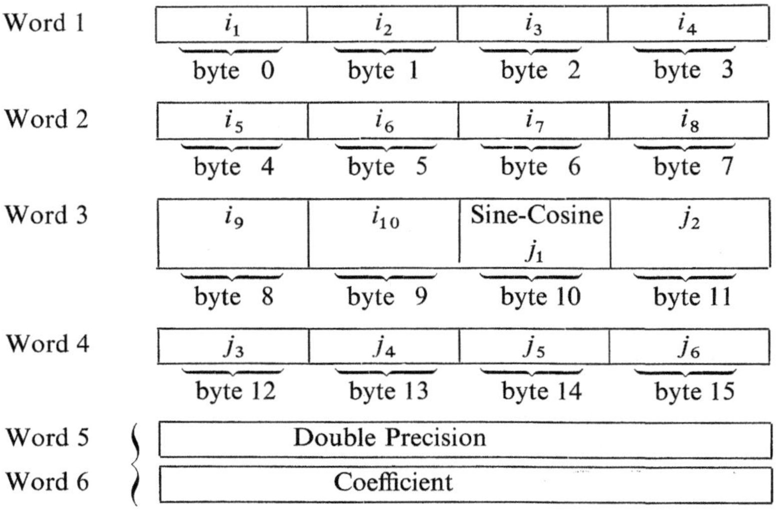

Fig. 16. Storage of a term.

no longer needed. MAO does this automatically for any series that it 'knows' can be erased and closes up the vacated space by its dynamic relocation subroutines.

The MAO-1970 system, unlike its 1965 prototype which was experimental in nature and coded mostly in FORTRAN, is a highly optimized system designed to carry out consequential problems in dynamical astronomy. Mention has already been made of packed word arithmetic. Further, frequently used or time-consuming routines are hand-optimized in assembly code to take advantage of any particular features of a machine. Finally, optimum algorithms are sought for the most efficient execution in time and space of the subroutines.

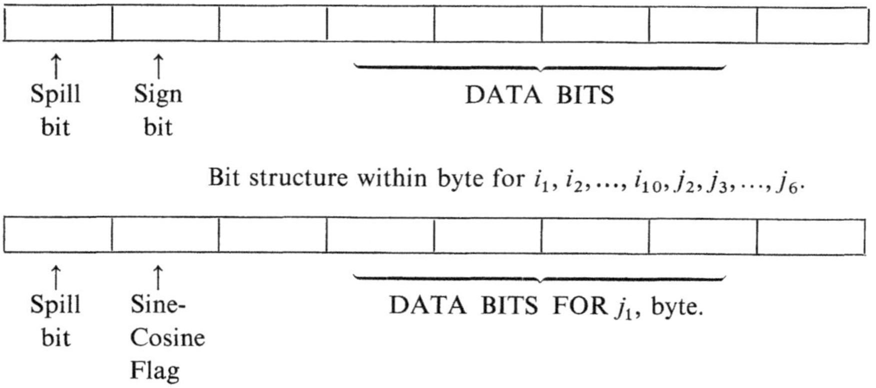

Fig. 17. Structure of the *i* and *j* bytes.

As an example of an efficient algorithm, table-look is a very frequently used subroutine within the PROD and ADD calls. Time consumption can be enormous after multiplying two series. In the particular method of multiplying and series storage it was observed that the current product terms are usually near the last one which was looked up. If a record is kept of the position of the last term looked up, one has a pointer to start a searching process in the neighborhood of the last term looked up. Figure 18 illustrates this process. If the header is reached before the term is found, a sequential downward search is started. If the bottom of a series is reached, a sequantial upward search is made.

A large number of consequential problems have been solved using MAO. Consult the references in Rom's paper of 1970.

Boucke and Garthwaites' system (1969) is a successor to the early MAO system described above with an overall design influence by van Flandern (private communication), Barton (1966, 1967, 1968) and Kovalevsky (1959, 1968). Two separate systems were developed: a (3, 3)-system and a (6, 6)-system, both written in FORTRAN IV.

One of the principal features of these systems is management of storage so that coefficients are stored in any order and those that are too small will be discarded and hence will not occupy valuable space.

Figure 19 illustrates the technique for series storage.

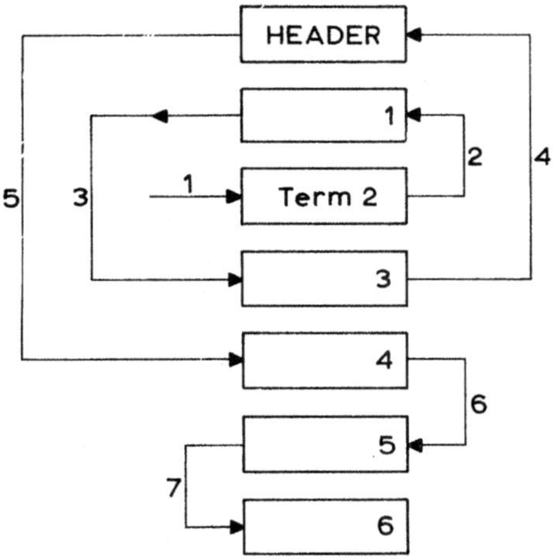

Fig. 18. Table look-up algorithm.

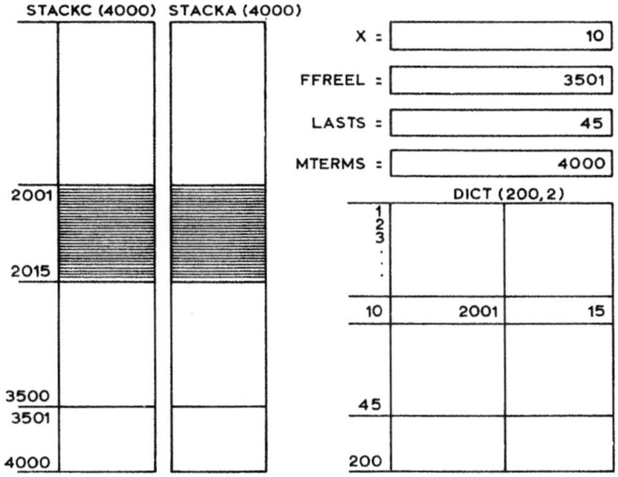

Fig. 19. Series storage.

When a series X is defined, it is assigned a location in the dictionary (10 in this case). DICT is a stack, made up of two columns, the first giving the location of the coefficients of the series, 2001, and the second giving the number of terms. STACKC contains these coefficients in double precision while STACKA contains the indices and arguments in packed form. As in MAO, each series to be used must first be defined. Further, the same kind of dynamic storage algorithm is used so that active series are always at the bottom. On ERASING a series from the stacks, vacated spaces are eliminated by relocation upward of all the series below it, but empty rows are left in the dictionary. These are later refilled as new series are defined.

Because it is necessary to reserve space in named COMMON for the dimensions of

DICT of STACKA and STACKC, the user must have an *a priori* estimate of the number of series as well as the maximum number of terms to be reserved for each problem. As in MAO, the most crucial subroutines are optimized in machine language, for example the sort and multiplication.

This system is provided with control over the size of the smallest coefficients allowed. The tolerance is a given quantity EPSLN in a named COMMON area so that appropriate subroutines can automatically refer to it. Furthermore, individual orders of magnitude can be placed on any individual index or on the *total* order of a polynomial term x and the ranges of the i's and the j's can be controlled.

The generalization of MAO is ESP (Echeloned Series Processor). Poisson series, as defined in equation (8) above, do not contain literal divisors, though, obviously, monomial terms may have negative exponents. In the Lunar Theory, integration of series gives rise to such divisors which may then become more complicated after other operations have taken place, e.g., multiplication. In order to accommodate series with literal divisors, Rom introduces a tree structure for these series arranged as shown in Figure 20. The top echelon contains trigonometric terms T_i, the middle echelon the

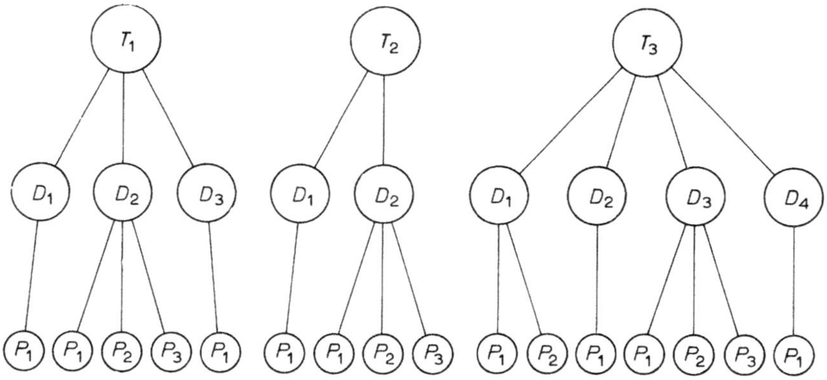

Fig. 20. Echeloned series.

divisors D_j, and the bottom echelon the monomials P_m. The general representation of Echeloned Series E is then given by

$$E = \sum_i T_i \sum_j D_j \sum_m P_m ,$$

where

$$T_i = \frac{\cos}{\sin} (i_1 l_1 + i_2 l_2 + \cdots i_6 l_6) ,$$

$$D_j = \frac{1}{\prod_k (\alpha_{j,k} n_1 + \beta_{j,k} n_2 + \gamma_{j,k} n_3)^{\beta_{j,k}}} ,$$

and

$$P_m = \text{coeff} \times p_1^{\gamma_1} \times p_2^{\gamma_2} \times \cdots p_{12}^{\gamma_{12}} .$$

It is clear that the l's, n's, and p's are literals, all other indices are integers. Moreover, $\delta \geqslant 0$ and α, β, γ are constructed to be relatively prime.

The same dynamic storage algorithm of MAO is used globally in ESP. However, the block structure has changed significantly. Conceptually, Echeloned Series are stored in parenthesized form according to the following expression:

$$T_1 \{D_1 [P_1 + P_2 + \cdots] + D_2 [P_1 + P_2 + \cdots] + \cdots\} + \\ + T_2 \{D_1 [P_1 + P_2 + \cdots] + D_2 [P_1 + P_2 + \cdots] + \cdots\} + \cdots$$

with obvious liberties taken with the subscripts. The principal operations on Echeloned Series are performed on triplets (T_i, D_j, P_k) in which the subscript k varies most frequently, then j, then i, which is suited to bringing out the terms

$$T_1, D_1, P_1, P_2, \ldots, D_2, P_1, P_2, \ldots, T_2, D_1, P_1, \ldots . \tag{9}$$

This type of triplet algebra requires that one be able to insert or extract terms in any echelon. ESP deals with this contingency by employing two structure modes: *un-listed* and *listed*. Figure 21 illustrates the un-listed mode and reproduces the sequence

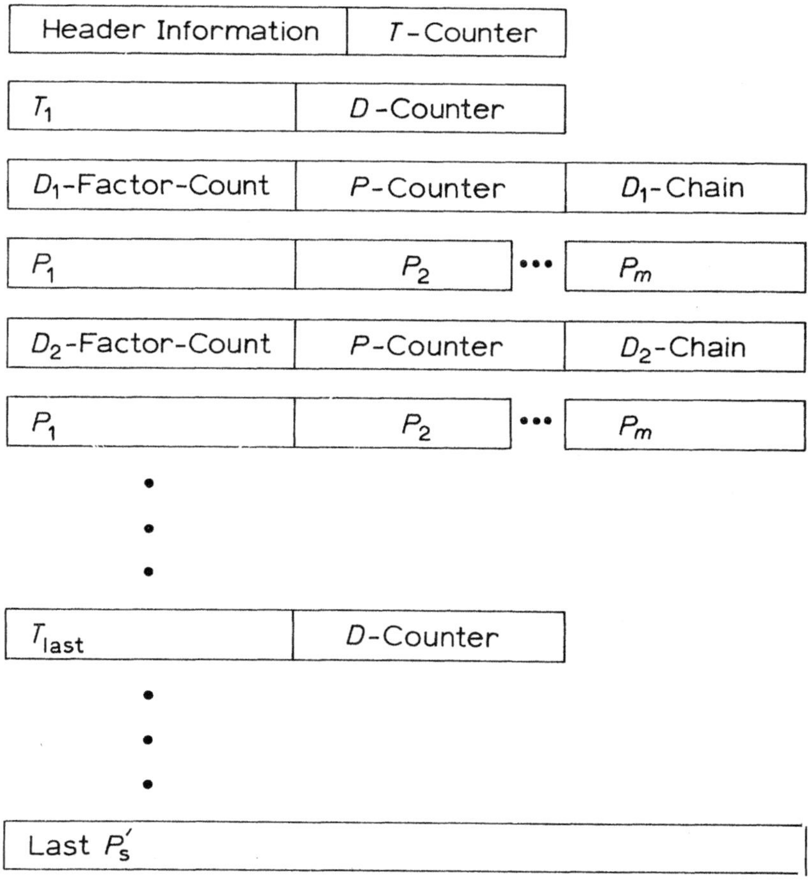

Fig. 21. *Un-listed* storage of echeloned series.

(9) above. The T-terms and P-terms are handled as in MAO (and have the same structure). A D-term is a sequence of words of 4-bytes, the first three representing the multiples of the arguments and the last, the power of the divisor. The absence of a D-term indicates a divisor of 1.

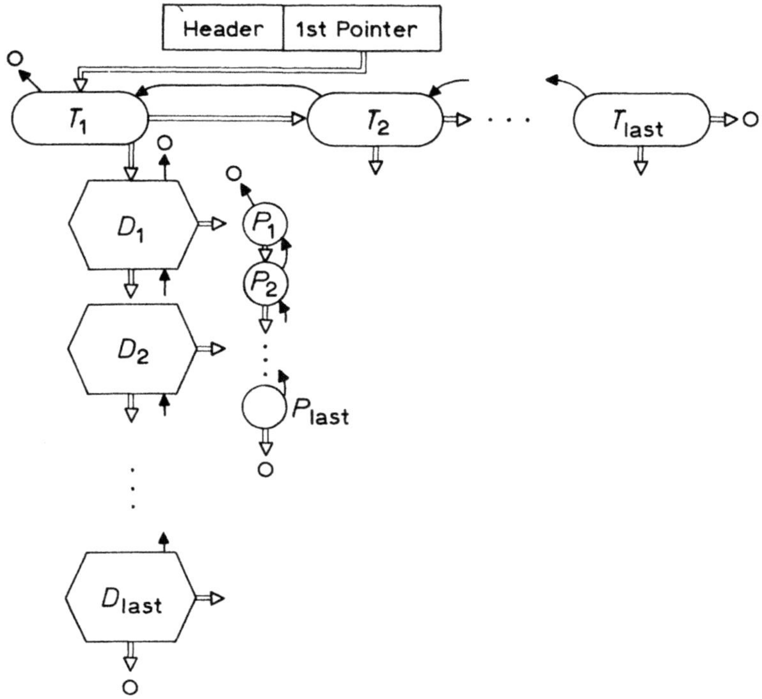

Fig. 22. *Listed* storage of echeloned series.

The *listed* mode is shown in Figure 22, and is the usual associative list structure described earlier. In the figure double line arrows indicate backwards pointers. The elements in the *listed* mode have the same arrangement as in the *un-listed* mode. However, all counters (except for D) are replaced by pointers. Thus, from a particular element, one can locate the element's 'ancestor' or 'offspring' on the same echelon, or possibly the first 'offspring' one echelon below.

In this way the list structure permits the insertion (or deletion) of elements anywhere. The dual pointer system simplifies the table look-up, as only a one-direction search is necessary. The *listed* mode is needed for construction of Echeloned Series, but since it is time and space consuming, provisions are made to destroy the list structure with the UNLIST command. The inverse command LIST is also available. The entire language is shown in Table IX.

As can be seen, most of the CALL's are the same as those of MAO. The most important deviation from MAO are the subroutines which allow the user to manipulate individual terms. These may be used to build other operators in FORTRAN, a very powerful tool, indeed.

TRIGMAN departs from all other Poisson series manipulators in that it is fully dedicated to list processing structures and techniques. The subroutines are written in

TABLE IX

The ESP language

Calling Sequence	Operation
CALL SPACE(N)	open up dynamic storage
CALL DEFINE(X)	define a series X
CALL ANNEXT(X, it, sc)	annex trigonometric element
CALL ANNEXD(X, d, nf)	annex divisor chain
CALL ANNEXP(X, c, p)	annex monomial element
CALL PRINT(X,'× × × × × × × ×', n)	print X
CALL ERASE(X)	remove X from storage
CALL PXCA(X, Y)	change identifiers; $X = Y$
CALL HEXIN(X)	read X from hex. cards
CALL HEXOUT(X)	punch X in hexadecimal
CALL NEWDSK	fresh start on disk
CALL DISKON	restart disk
CALL LOCK	set disk in read only mode
CALL UNLOCK	set disk in read-write mode
CALL CLOSED	close disk file
CALL SEEK(DX)	point to read DX
CALL SEEK(0)	point to write a series
CALL WRITEP(X, DX, PN, l)	write X and point to PN
CALL READP(X, PN)	read X and point to PN
CALL SREAD(X, DX, PN, l)	search for X based on l, and read
CALL REWIND(io)	rewind tape unit io
CALL SEARCH(lb, io, k)	position tape io by searching for label lb
CALL LABEL(lb, io)	write label lb on io
CALL TIN(X, io)	read series X from io
CALL TOUT(X, io)	write series X on io
CALL WEF(io)	write end of file on io
CALL UNLOAD(io)	rewind and unload tape io
CALL PROD(X, Y, Z, α, ε)	$\alpha * X * Y + Z = Z$, α & ε optional
CALL ACUM(X, Y, α, ε)	$\alpha * X + Y = Y$, ε optional
CALL PDP(X, Y, i)	$\partial X / \partial P_i = Y$
CALL PDT(X, Y, i)	$\partial X / \partial T_i = Y$
CALL ENTER(n, X_1, X_2, ...)	bring series X_n, $1 \leqslant 4 \leqslant$ from disk if not in core
CALL REMOVE	erase all that was read with ENTER
CALL FETCH(X, n_1, n_2, i, iw, α)	extract elements sequentially from X
CALL LIST(X)	construct *listed* mode for X
CALL UNLIST(X, ε)	construct *un-listed* mode for X ε is a treshold on numeric coefficients
CALL TSTORE(X, iw, lp)	store trigonometric element in *listed* X
CALL DSTORE(X, l_p, nf, d, d_p)	store divisor in *listed* X under trigonometric element l_p
CALL PSTORE(X, d_p, p, α, p_p)	store monomial element and numeric coefficient in *listed* X under divisor d_p
CALL TPOINT(X, iw, i_p)	assembly language equivalent of TSTORE
CALL SPOINT(X, l_p, nf, d, s_p)	assembly language equivalent of DSTORE
CALL PPOINT(X, s_p, p, α, p_p)	assembly language equivalent of PSTORE

FORTRAN primarily for compatibility with other machines (as well as interpretation) and individual features are machine-dependent on word structure, but Jefferys indicates these routines can be easily modified. TRIGMAN started out as a (9,10)-system but was extended to a (29,10)-system with negative exponents and able to handle rational coefficients.

Series are stored in blank common and a pointer is the FORTRAN subscript of the first word of the next cell. Two or more pointers are permitted so that trees of arbitrary complexity can be constructed. A pointer value of zero indicates the last element of a sequence. To locate a given cell, one starts at the series name which permits one to follow all the pointers to the desired cell. The existence of several pointers permits pointing from several directions to the same subchain, reducing time and space.

The data structures (lists) in TRIGMAN are:

(1) Polynomial (one pointer per cell gives a linear array)

(2) Poisson Series (a 'down' pointer chains together trigonometric terms and a 'right' pointer chains together the polynomial terms)

(3) *Name List* points to a polynomial structure, a Poisson series or a working list in the process of being constructed.

Thus, if a programmer names a series, a pointer is created which points to the FORTRAN variable naming the series, and another pointer is set in the FORTRAN variable to the entry in the *Name List*. Because of this structure, it is unnecessary for the user to define his series variables (as is the case, e.g., in MAO). Whenever a FORTRAN name is used in a subroutine call, the *Name List* is looked up. If the address of the FORTRAN variable is there, the series is obtained, otherwise, a new series with that name and value zeıo is created.

The usual list-processing techniques of storage management are used in TRIGMAN. Available to the system is a *free space list*. Any time a cell is needed, the system points to a free cell in the free space. It is possible during operations on series that individual cells will be lost to the system and these become unavailable because pursuing pointers through the name list or the free space list bypass them. In time the free space is exhausted. An efficient method for 'garbage collection' comes from the original LISP system and places a flag in every cell that can be reached beginning with the first cell of the name list and following every branch. Thus, after every active cell has a flag, it is removed, if not, it is added to the new free space list. Jefferys wrote the pointer routines described in FORTRAN so that they could be easily modified for other computers.

Tables X and XI show the TRIGMAN language. As far as storage management is concerned, associative techniques are at a disadvantage both in storage space and fetch-times because of the larger system overhead. Apparently, the principal advantage is the flexibility and complex data structures that it can handle. An example is the implementation of mixed real and rational arithmetic.

Jeffreys sees as an important application of TRIGMAN the pursuit of quadratically convergent perturbation methods which may prove to be useful in those instances

TABLE X

Commands in trigman

	Commands in the TRIGMAN language	
Truncation	CALL NORDER(*pvar*, n)	Truncates terms of order $n \geqslant 0$, where nk is degree of a polynomial variable.
	CALL MAXSET(m)	Truncates all terms with degree $m \geqslant 0$.
	CALL TOLSET(*tol*)	Floating point consts. are truncated if $< tol$ (where $tol \geqslant 0$).
Scalar def.	CALL RATDEF(m, n, *scalar*)	*scalar* is defined as the rational no. m/n or a FORTRAN fl. point number.
Input and Output	CALL INPUT(SER)	Reads a series from cards giving it the FORTRAN variable name SER.
	CALL INPUT(SER, *file*)	Reads input where a file has already been named.
	CALL OUTPUT(*option*, *option*, ...)	*option* may be any of the following: 5HALGOL. ALGOL-style output. 7HFORTRAN. FORTRAN-style output. 6HOUTPUT. Output to printer. 5HPUNCH. Output to punch. nLxxxxx. Output to file xxxxx. 9HNOCOMPILE. Readable output (not compilable). 7HCOMPILE. FORTRAN or ALGOL compatible output.
	CALL OUTPUT(*name*, SER, *option*, *option*, ...)	Same as above except SER is given the name *name*.
Basic series operations	CALL CLEAR(A)	Sets A = zero series and frees space in core.
	CALL CONST (*scalar*, A)	Creates a constant series with value = *scalar*.
	NONZERO(A)	= 0 if A = 0. = 1 if $A \neq 0$.
	CALL TRIGADD(A, B, C)	C = A + B series.
	CALL TRIGSUB(A, B, C)	C = A − B series.
	CALL TSCMPY(*scalar*, A, B)	B = *scalar* * A series.
	CALL TRIGMPY(*scalar*, A, B, C)	C = *scalar* * A * B series.
	CALL RAISE(SER, n, ANS)	ANS = SER **n, where $n \geqslant 0$.
	CALL SUBX(A, *pvar*, B, C)	C = B → *pvar* in A series.
	CALL SUBX(A, *pvar*, B, C, D)	C = B → *pvar* in A series. Every appearacne of 1/*pvar* → D series.
	CALL SUBX(A, *pvar*, B, C, D, n)	C = B → *pvar* in A series. Every appearance of 1/*pvar***n → D series.
	CALL SUBYSC(A, *tvar*, SIN, COS, B)	B = SIN(*tvar*) or COS(*tvar*) → SIN, COS in A
	CALL TDIFINT(A, *tvar*, n, B, REM)	B = $d^nA/d(tvar)^n$, where REM = all terms in A which do not depend upon *tvar*.
	CALL PDIFINT(A, *pvar*, n, B, REM)	B = $d^nA/d(pvar)^n$, where REM = all terms in A which would have become logarithms if integrated

Table X (Continued)

	Commands in the TRIGMAN language	
Taylor series Expansions, etc.	CALL RCPRCL(*scalar*, A, B)	B = 1/(1 + *scalar* * A) series.
	CALL BINOM(*scalar*, A, i, j, B)	B = (1 + *scalar* * A)** *i/j* for rational powers.
	CALL BINOM(*scalar*, A, *pwr*, O, B)	B = (1 + scalar * A)** *pwr* for fl. pt. powers.
	CALL SUBXEPS(A, *pvar*, B, C)	C = *pvar* → *pvar* + B in A series.
	CALL SUBYEPS(A, *tvar*, B, C)	C = *tvar* → *tvar* + B in A series.
	CALL SUBY(A, *tvar*, *const*, *arg*, B, C)	C = *tvar* → *const* + *arg* + B in A series, where *const* = fl. pt. no. and *arg* = a Hollerith constant representing a new angular argument.
House-keeping	CALL INIT(i, j, k)	Initialization subroutine: *i* = length of blank common for user's use; *j* = maximum no. of polynomial variables = 29; *k* = maximum no. of trig. variables = 10.

TABLE XI

Commands in the TRIGMAN language

Other available subroutines

CALL TERM(*scalar*, *poly*, *sincos*, *trigarg*, SER)
Example:
CALL TERM(− 0.3, 11HA**2*b*C** − 1., 3HCOS, 7H − 2*L + G., AUGMENT) adds $-0.3A^2BC^{-1}\cos(-2L+G)$ to the series AUGMENT.

CALL NEWTERM(*scalar*, *poly*, *sincos*, *trigarg*, SER)
SER is put = 0 before adding the new term.

CALL PWRLIMS(SER, *pvar*, m, n) — SER is searched for appearances of *pvar*.
m = Min(0, min power of *pvar*)
n = Max(0, max power of *pvar*).

CALL GETXPWR(SER, *pvar*, k, ANS) — ANS = *pvar* **k (....series....)

CALL FOURIER(SER, *tvar*, n, *list*, SIN, COS)
Initially set $n = -1$. Then, n is set = the largest multiple of *tvar* found. SIN and COS are the *coefficients of* sin(n * *tvar*) and cos(n * *tvar*). *list* is an unused FORTRAN variable name needed by the subroutine.

CALL GMESS, CALL NOGMESS	These subroutines turn messages from the Garbage Collector on or off.
CALL REALNLY, CALL REALRAT	The first subroutine produces a real mode in the calculation of coefficients, the second produces rational terms.
CALL SETXNAM(*pvar*, *pvar'*)	Changes the name of the p-variable from the unprimed to the primed.
CALL SETYNAME(*tvar*, *tvar'*)	Changes the name of the t-variable from the unprimed to the primed.

Data Format in TRIGMAN

⟨name⟩	=	⟨series⟩	$	FORTRAN-style
⟨name⟩	:=	⟨series⟩	;	ALGOL-style

For example:

S5 = (2*E − 1/4*E**3)SIN(L) + 5/4*E**2*SIN(2*L) $ FORTRAN-style

or, S5 : = (2*E − 1/4*E ↑ 3)SIN(L) + 5/4*E ↑ 2*SIN(2*L); ALGOL-style

like Lunar Theory where series converge slowly. He illustrates the convenience of the language by expanding the Lunar disturbing function to the sixth order of the eccentricities and inclinations. The program took 'an afternoon' to write, ran for 30 s on the CDC 6600 computer and agreed perfectly with Delaunay's literal series.

In a recent paper entitled 'A Precompiler for the Formula Manipulation System TRIGMAN', Jefferys has eliminated the CALL's to his system by writing a preprocessor in the string processing language SNOBOL. The program to compute the Legendre polynomials in E*COS(U) is on the left in Figure 23 and is written directly in the TRIGMAN language. The program on the right is written in the new precompiler.

As can be seen, there is a remarkable improvement in readability. (Compare this program with Figure 13 for a similar program in FORMAC). The pre-compiler, called TRIGRUN receives a FORTRAN-like program like that shown and outputs a program acceptable to the local FORTRAN compiler. It is superset to FORTRAN and has three new statement types:

(1) Declarations (POLY, TRIG, SERIES)
(2) Input/Output statements for series
(3) Assignment statements for series.

Furthermore, to eliminate the usual two-phase procedure where a preprocessor exists, a single control is available which will fetch and execute the TRIGRUN compiler and then load and execute the FORTRAN compiler, adding additional pre-compiled decks and finally, loading and executing the program. However, this depends intimately on the installation operating system and hardware.

```
  DIMENSION P (11)
  CALL INIT (0, 1, 1)
  CALL TERM (1.0, 2HE., 3HCOS, 2HU., X)
  CALL CONST (1.0, P (1))                    POLY E
  CALL TRIGADD (ZERO, X, P (2))              TRIG U
  DO 1 I = 2, 10                             SERIES P (11)
  CALL RATDEF (2*I − 1, I, COEFF)            P (1) = 1
  CALL TRIGMPY (COEFF, X, P (I), TEMP)       P (2) = E * COS (U)
  CALL RATDEF (I − 1, I, COEFF)              DO 1 I = 2, 10
  CALL TSCMPY (COEFF, P (I − 1), TEMP2)      P(I + 1) = (2*I − 1)/I*P(I) − (I − 1)/I*P(I–1)
  CALL TRIGSUB (TEMP, TEMP2, P (I + 1))    1 OUTPUT P (I + 1)
1 CALL OUTPUT (1HP, P (I + 1))               END
  END
```

Fig. 23. Programs in TRIGMAN and its pre-compiler.

Four new functions have been designed:

(1) DERIV (partial derivative of a series with respect to a variable)
(2) INTEGRAL (integral of a series with respect to a variable)
(3) SPLIT (all the terms factored by a variable to a given power)
(4) SUBST (substitute one series into another)

An interesting compiler problem is encountered in handling DO-loops because the last statement of the DO may be expanded into several CALL's. The last of the pre-compiled statements must therefore contain the control number of the DO. However, it may be that the program may need to transfer control to the first of expanded statements. TRIGRUN deals with this situation by replacing the statement number *within* the DO with a very large statement number and using the first of the expanded statements. Jefferys gives this example:

```
PRE-COMPILER              OUTPUT
DO 1 I=1,10         DO 99999 I=1,10
    :                      :
1 X=A+B−C           1    CALL TRIGADD (A, B, T0001)
                    9999 CALL TRIGSUB (T0001, C, X)
                         CONTINUE
```

One of the principal motivations in writing the pre-compiler was to make the system more widely available to non-specialists. As in the case of ALPAK and ALTRAN, once the main work has been done in designing the 'heart' of the system, the cost of improving it as far as convenience, legibility and clarity are concerned is small. While there is this gain, there is still another trade-off in a degradation of running time (and some loss of space).

A still further improvement in making the system easier to use is to have the entire system execute interpretively, that is, to execute effectively one step at a time. Among the great advantages of such a mode of operation are built in diagnostics, debugging and tracing facilities. The cost in execution time is bound to be enormous. However, one of the most promising aspects of an interpretive system is its application to *conversational algebraic manipulation.* FORMAC has been quite successful in the interpretive mode using a CRT as input/output. An interpretive TRIGMAN may turn out to be very exciting.

References

Abrahams, P. W.: 1968, *Adv. Computers* **9**, Academic Press, p. 51.

Baker, Marks, and Tobey: 'PL 1 FORMAC Course Notes', Federal Systems Division, IBM, Gaithersburg, Maryland.

Barton, D.: 1966, *Astron. J.* **71**, 438–442.

Barton, D.: 1967, *Astron J.* **72**, 1281–7.

Barton, D., Bourne, S. R., and Burgess, C. J.: 1968, *Computer J.* **10**, 293–8.

Bender, B.: 1966, 'MANIP: A Computer System for Algebra and Analytic Differentation', ACM Symposium, March 1966, Washington, D.C. Institute for Environmental Research, Environmental Services Administration, Boulder, Colorado.

Bobrow, D. G.: 1968, *Symbol Manipulation: Languages and Techniques*, North-Holland Publ. Co.

Broucke, R. and Garthwaite, K.: 1969, *Celes. Mech.* **1**, 271–84.

Brown: 1966, 'The ALTRAN Language and the ALPAK System for Symbolic Algebra on a Digital Computer', Bell Tel. Labs.

Brown, Leagus: 'Oedipus Reference Manual', Bell Tel. Labs., Murray Hill, N.J.

Brown, Leagus: ALPAK-B: 'A New Version of the ALPAK System for Symbolic Algebra on a Digital Computer', Bell Tel. Labs., Murray Hill, N.J.

Brown, Hyde, and Tague: 1964, 'The ALPAK System for Non-numerical Algebra on a Digital Computer', Monograph 4869, Bell Telephone System Tech. Publs.

Campbell, J. A. and Jefferys, W. H.: 1970, *Celes. Mech.* **2**, 467–73.

Chapront, Ghertzman, and Kovalevsky: 1967, *Coll. Méc. Cél.*, Moscow.

Collins: 1966, PM, 'A System for Polynominal Manipulation', RC-1526, IBM Watson Research Center, Yorktown Heights, N.Y.

Danby, J. M. A., Deprit, A., and Rom, A.: 1965, 'The Symbolic Manipulation of Poisson Series', Math. Note No. 432, Document D1-82-0481, Boeing Scientific Research Laboratories.

Davis, M. S.: 1958, *Astron. J.* **63**, 462–4.

Davis, M. S.: 1963, in W. F. Freiberger and W. Prager (eds.), *Applications of Digital Computers*, Ginn and Co., Boston, pp. 85–96. Round Table Discussion. Celestial Mechanics Conference.

Davis, M. S.: 1968, *Astron. J.* **73**, 195–202.

Deprit, A.: 1971, *Celes. Mech.* **3**, 312–9.

Deprit, A. and Rom, A.: 1967, 'Computerized Expansions in Elliptic Motion', Boeing Scientific Research Laboratories (Seattle, Washington). Doc. D1-82-0601.

Deprit, A. and Rom, A.: 1970, *Celes. Mech.* **2**, 166–206.

Deprit, A., Henrard, J., and Rom, A.: 1970, *Science* **168**, 1569–70.

Deprit, A., Henrard, J., and Rom, A.: 1970, 'Anal. Lunar Eph. I. Def. of the Main Problem'. Boeing Scientific Research Laboratories (Seattle, Washington). Doc. D1-82-0963.

Deprit, A., Henrard, J., and Rom, A.: 1970, 'Anal. Lunar Ephemeris: The Mean Motions', Math. and Info. Science Report 21, Doc. D1-82-0990, Boeing Scientific Research Laboratories (Seattle, Washington).

'FORMAC Symbolic Mathematics Interpreter', IBM Corporation, Prog. Info. Dept., 40 Saw Mill River Rd., Hawthorne, N.J. 10532, U.S.A.
Forte, A.: 1967, *Snobol 3 Primer*, MIT Press.
Foster, J. M.: *List Processing*, Am. Elsevier Publ. Co.
Fox, L.: 1966, *Advances in Programming and Non-numerical Computation*, Pergamon Press, p. 145.
Gerard, Izsak, and Barnett: 1965, *Comm. ACM* **8**, 1.
Gotlieb and Novak: ALGEM – 'An Algebraic Manipulator', Dept. of Comp. Science, University of Toronto, Toronto, Canada.
Herget, P. and Musen, P.: 1959, *Astron. J.* **64**, 11–20.
Hoffleit, D.: *Catalogue of Bright Stars*, Yale University Observatory, New Haven, Conn., 3rd rev. ed.
Izsak, Gerard, Efimba, and Barnett: 1964, 'Construction of Newcomb Operator on a Digital Computer', Smithsonian Institution Astrophysical Observatory, Special Report No. 140.
Izsak, Benima, and Mills: 'Analytical Development of the Planetary Disturbing Function on a Digital Computer', Smithsonian Institution Astrophysical Observatory, No. 164.
Jefferys, W. H.: 1970, *Celes. Mech.* **2**, 474–80.
Jefferys, W. H.: 1971, *Celes. Mech.* **3**, 390–94.
Jefferys, W. H.: 1971, *Comm. of the ACM* **14**, 8.
Jefferys, W. H.: 'A Pre-Compiler for the Formula Manipulation System TRIGMAN', Dept. of Astronomy, Univ. of Texas, Austin, Texas, U.S.A.
Jirauch and Westerwick: 1964, 'A Use of Digital Computers for Algebraic Computation', Tech. Doc. Report RTD-TDR-63-4200, Air Force Flight Dynamic Lab. Wright-Patterson AFB, Ohio.
Korsvold, K.: 'An On-Line Program for Non-Numerical Algebra', Intern Report – E. 81, Norwegian Defence Research Establishment.
Kovalevsky, J.: 1959, *Bull. Astron.* **23**, 1–89.
Kovalevsky, J.: 1968, *Astron. J.* **73**, 203–9.
Lapidus and Goldstein: 1965, *Comm. of the ACM* **8**, 501–7.
LISP: 1962, *1.5 Programmer's Manual*, MIT Press.
Mangeney and Chapront: 1967, 'Application du Programme d'Opérations sur les Séries Littérales à la Théorie de la Lune', Bureau des Longitudes, Paris.
Information Processing Language V – Manual, Rand Corporation, 1964, Prentice-Hall, Inc.
McIlroy and Brown: 'The ALTRAN Language for Symbolic Algebra on a Digital Computer', Bell Tel. Labs., Murray Hill, N.J.
Neidleman, I. D.: 1967, *Comm. of the ACM* **1**, 167–8.
Pakin, S.: 1968, 'APL/360 Reference Manual', Science Research Associates, Inc., 259, E. Erie St., Chicago, Ill. 60611.
Perlis and Iturriaga: 1964, *Comm. of the ACM* **7**, 2.
van de Riet, R. P.: 1968, 'Formula Manipulation in Algol 60', Mathematisch Centrum, Amsterdam.
Rom, A.: 1970, *Celes. Mech.* **1**, 301–19.
Rom, A.: 1971, *Celes. Mech.* **3**, 331–45.
Rom, A. R. M.: *Comm. of the ACM.*
Rosen and Saul (eds.): 1967, *Prog. Syst. and Languages*, McGraw-Hill.
Sammet, J. E.: 1967, *Adv. Computers* **8**, Academic Press, p. 51.
Sammet, J. E.: 1968, 'Symbol Manipulation Languages and Techniques', Revised Annotated Description Based Bibliography on the Use of Computers for Non-numerical Mathematics, North-Holland Publ. Co.
Sammet, J. E.: 1969, *Programming Languages: History and Fundamentals*, Prentice-Hall, Inc.
Sconzo, Le Schack, and Tobey: 1965, *Astron. J.* **70**, 269–71.
Scott, N. R.: 1960, *Analog and Digital Computer Technology*, McGraw Hill.
Sharaf, S. G.: 1955, *Trans. Inst. Theoret. Astron. Leningrad* **4**, 3.
SICSAM Bulletins: 'Special Interest Committee on Symbolic and Algebraic Manipulations', Association for Computing Machinery.
Surkan, A. J.: 'On-Line Realization of Non-Numeric Algebraic Operations of Recuisively Polynomial Coefficient Series of Orbital Motion', RC-2048 ($10482), Research Division, IBM, Yorktown Heights, N.Y.
Tobey, R. G.: 1969, 'Proceedings of the 1968 Summer Institute on Symbolic Mathematical Computation', IBM Boston Prog. Center, 545 Technology Square, Cambridge, Mass.
Westerwick and Brown: 'A FORTRAN IV Program to Derive the Equations of Motions of Systems',

Tech. Report AFFDL-TR-65-194, Air Force Flight Dynamics, Lab., Wright-Patterson AFB, Ohio.

Technical Reports from IBM

Bleiweiss, *et al.*: 1966, 'A Time-Shared Algebraic Desk Calculator Version of FORMAC', TROO. 1415.
Bond: 1966, 'History, Features and Commentary on FORMAC', TROO. 1426.
Bond, *et al.*: 1965, 'FORMAC – An Experimental Formula Manipulation Compiler', TROO. 1192-1.
Bond, *et al.*: 1965, 'Implementation of FORMAC', TROO. 1260.
Sammet: 1965, 'An Overall View of FORMAC', TROO. 1367.
Sammet: 1966, 'An Annotated Description Based Bibliography on the Use of Computers for Non-numerical Mathematics', TROO. 1427.
Sammet and Bond: 1964, 'Intro. to FORMAC', TROO. 1127.
Sconzo, *et al.*: 1965, 'Symbolic Computation of *f* and *g* Series by Computer', TROO. 1262-1.
Tobey: 1965, 'Eliminating Monotonous Mathematics with FORMAC', TROO. 1365.
Tobey, *et al.*: 1965, 'Automatic Simplification in FORMAC', TROO. 1343.
Tobey: 'Experience with FORMAC Algorithm Design', TROO. 1413.

The Use of Electronic Computers for Analytical Development in Celestial Mechanics: 1968, *Astron. J.* **73**, 3.
W. J. Eckert 195; M. S. Davis 195; J. Kovalevsky 203; A. Deprit and A. Rom 210; J. Chapront and L. Mangenay-Ghertzman 214; A. R. Le Schack and P. Scorzo 217.

ELEMENTARY DYNAMICAL ASTRONOMY

P. HERGET
Cincinnati Observatory, Cincinnati, Ohio, U.S.A.

Abstract. The essentials of Newtonian mechanics are presented, insofar as they apply to practical dynamical problems in the solar system.

The contents of these lectures will be presented in outline form. They are based upon simple Newtonian Mechanics, Differential Equations, Vector Analysis, and Numerical Calculus.

1. The Problem of Two Bodies

The force of gravitation of one body of mass M upon a second body of mass m is

$$\mathbf{F} = -k^2 \frac{mM}{r^3} \mathbf{r} = m \frac{\mathrm{d}^2\mathbf{r}}{\mathrm{d}t^2}$$

where $\mathbf{r}$ is the position vector of mass m relative to mass M.

Thus,

$$\frac{\mathrm{d}^2\mathbf{r}}{\mathrm{d}t^2} = \frac{\mathrm{d}\mathbf{v}}{\mathrm{d}t} = -\frac{k^2 M}{r^3} \mathbf{r},$$

and the solution gives a conic section for the locus of $\mathbf{r}$, with M at one focus. There are six scalar constants of integration, called elements, which may be represented in various ways. For example, the simplest set is the components of $\mathbf{r}_0$ and $\mathbf{v}_0$ at t_0. A set which is geometrically more descriptive is: i, Ω, ω, e, a, M_0 at t_0. (Bauschinger, 1906; Gauss, 1963; Herget, 1948a; Stracke, 1929; Watson, 1869.)

The problem of preliminary orbits requires that the elements with respect to the Sun be determined upon the basis of three or more observations of $\boldsymbol{\varrho}^*$ $(\cos\delta \cos\alpha, \cos\delta \sin\alpha, \sin\delta)$ as seen from the Earth. Lambert made some important advances in studying this problem, Olbers devised a good practical method for finding the parabolic orbits of comets, and Gauss developed an unsurpassed general solution after the discovery of Ceres in 1801. LaPlace also developed a method which has practical advantages in certain cases. The basic equation connecting the heliocentric orbit and the observations is $\mathbf{r} + \mathbf{R} = \varrho\boldsymbol{\varrho}^*$, where $\mathbf{R}$ is the solar coordinate vector from the observer to the origin.

Lambert's equation states that

$$KV^2\varrho = \left(\frac{1}{r^3} - \frac{1}{R^3}\right) R \sin F$$

where K is the geodesic curvature of the apparent path across the sky, V is the angular velocity of the object along its apparent path, and F is the angle from the Sun to the great circle defined by the tangent to the apparent path. These quantities are not

B. D. Tapley and V. Szebehely (eds.), Recent Advances in Dynamical Astronomy, 392–395.

easily derived from a few observations. Otherwise it would be possible to solve for r and ϱ from this equation in the form $\varrho = A + B/r^3$ and the equation $r^2 = \varrho^2 - 2[\boldsymbol{\varrho}^* \cdot \mathbf{R}]\varrho + \mathbf{R}^2$. (Herget, 1948b).

The method of Gauss is based upon the formula which he devised for computing the sector-triangle ratio, $Y(i, j)$, given $\mathbf{r}_i$, $\mathbf{r}_j$, and $(t_j - t_i)$. Thus

$$\frac{(t_3 - t_2)\, Y(3,1)}{(t_3 - t_1)\, Y(3,2)} [\varrho_1 \boldsymbol{\varrho}_1^* - \mathbf{R}_1] + \frac{(t_2 - t_1)\, Y(3,1)}{(t_3 - t_1)\, Y(2,1)} [\varrho_3 \boldsymbol{\varrho}_3^* - \mathbf{R}_3] = [\rho_2\, \boldsymbol{\varrho}_2^* - \mathbf{R}_2] = \mathbf{r}_2 .$$

If we multiply through this equation by $\cdot(\boldsymbol{\varrho}_2^* \times \boldsymbol{\varrho}_3^*)$ and $\cdot(\boldsymbol{\varrho}_1^* \times \boldsymbol{\varrho}_2^*)$, successively, we get

$$c_1 [\boldsymbol{\varrho}_1^* \cdot \boldsymbol{\varrho}_2^* \times \boldsymbol{\varrho}_3^*]\, \varrho_1 = c_1 [\mathbf{R}_1 \cdot \boldsymbol{\varrho}_2^* \times \boldsymbol{\varrho}_3^*] - [\mathbf{R}_2 \cdot \boldsymbol{\varrho}_2^* \times \boldsymbol{\varrho}_3^*] + c_3 [\mathbf{R}_3 \cdot \boldsymbol{\varrho}_2^* \times \boldsymbol{\rho}_3^*]$$

$$c_3 [\boldsymbol{\varrho}_1^* \cdot \boldsymbol{\varrho}_2^* \times \boldsymbol{\varrho}_3^*]\, \varrho_3 = c_1 [\mathbf{R}_1 \cdot \boldsymbol{\varrho}_1^* \times \boldsymbol{\varrho}_2^*] - [\mathbf{R}_2 \cdot \boldsymbol{\varrho}_1^* \times \boldsymbol{\varrho}_2^*] + c_3 [\mathbf{R}_3 \cdot \boldsymbol{\varrho}_1^* \times \boldsymbol{\varrho}_2^*]$$

Gauss' method contains essentially only two unknowns. This has led to the method of Variation of Geocentric Distances, which is well adapted to electronic computer operations, and it enables one to include all available observations at one time (Herget, 1965a).

LaPlace's method applies numerical differentiation to the observations in order to solve the differential equations, and it is equivalent to a Taylor's series expansion about some point, t_0.

$$\mathbf{r} = \varrho \boldsymbol{\varrho}^* - \mathbf{R}$$

$$\mathbf{v} = \varrho' \boldsymbol{\varrho}^* + \varrho \boldsymbol{\varrho}^{*\prime} - \mathbf{R}'$$

$$\frac{\mathrm{d}^2 \mathbf{r}}{\mathrm{d}t^2} = \varrho'' \boldsymbol{\varrho}^* + 2\varrho' \boldsymbol{\varrho}^{*\prime} + \varrho \boldsymbol{\varrho}^{*\prime\prime} - R'' = (\mathbf{R} - \varrho \boldsymbol{\varrho}^*)\, k^2 M\, r^{-3} .$$

For each observation

$$\boldsymbol{\varrho}^* = \boldsymbol{\varrho}_0^* + (\Delta t)\, \boldsymbol{\varrho}_0^{*\prime} + \frac{(\Delta t)^2}{2!} \boldsymbol{\varrho}_0^{*\prime\prime} + \frac{(\Delta t)^3}{3!} \boldsymbol{\varrho}_0^{*\prime\prime\prime} + \cdots$$

and the unknown derivatives on the right hand side can be found approximately from three or more observations. Thus

$$\varrho [\boldsymbol{\varrho}^* \times \boldsymbol{\varrho}^{*\prime} \cdot \boldsymbol{\varrho}^{*\prime\prime}] = \mathbf{R}'' \cdot \boldsymbol{\varrho}^* \times \boldsymbol{\varrho}^{*\prime} + [\mathbf{R} \cdot \boldsymbol{\varrho}^* \times \boldsymbol{\varrho}^{*\prime}]\, k^2 M / r^{-3} .$$

This is the same as $\varrho = A + B/r^3$ in Lambert's equation. However, LaPlace's method has the advantage that it is equally applicable to a more complicated case than the differential equations for the Problem of Two Bodies. We shall show an application to the outer satellites of Jupiter. (Herget, 1965b.)

2. Planetary Perturbations

The complete differential equations for a body moving in the solar system under the simultaneous gravitational attraction of the Sun and all the planets is

$$\frac{d^2x}{dt^2} = -k^2(M+m)\frac{x}{r^3} + \sum_1^i m_i\left(\frac{x_i - x}{\Delta_i^3} - \frac{x_i}{r_i^3}\right),$$

and similarly for y and z, where the center of the Sun is taken as the origin. One method of solution is the direct numerical integration of this system of simultaneous differential equations, and this is usually known as Cowell's method. Because the mass of the Sun, M, is greater than 1000 m_i, it is possible to solve the complete problem by various perturbation methods, otherwise known as the Variation of Arbitrary Constants. This may be done in several different ways: (1) by direct numerical integration of the differential equations for the perturbations, (2) by expressing the perturbing forces in terms of Fourier Series and integrating the equations analytically or with numerical coefficients in each special case, (3) by expressing the disturbing forces in terms of Chebyshev polynomials over a limited range of time and integrating that set of equations. (Brouwer and Clemence, 1961; Carpenter, 1966; Herget, 1962a; Musen, 1954.)

The principal considerations in Cowell's method include a good start-up procedure, the predictor-corrector extrapolation, and the automatic control of the interval size. Since the coordinates of the disturbing planets may be known in advance, there is the option in a large electronic computer to read them from a magnetic tape, or to compute them *ab initio* by the same numerical integration procedure, so that they are available in high speed memory as one goes along. (Herget, 1962b; Schubert and Stumpff, 1966.)

Special perturbation methods for the Variation of Elements are based upon the equation for the total differential:

$$\frac{dE}{dt} = \frac{\partial E}{\partial t} + \frac{\delta E}{\delta t},$$

where the $\partial/\partial t$ represents any variation of E due to the pasage of time in the instantaneous, osculating, elliptic orbit, and the $\delta/\delta t$ is any variation due to the perturbing forces. For example,

$$\frac{d\mathbf{r}}{dt} = \mathbf{v} = \frac{\partial \mathbf{r}}{\partial t} + \frac{\delta \mathbf{r}}{\delta t} = \mathbf{v} + 0,$$

because the variation of $\mathbf{r}$ in the osculating elliptic orbit is also $\mathbf{v}$, and therefore $\delta\mathbf{r}/\delta t = 0$. However, since

$$\frac{d\mathbf{v}}{dt} = -\frac{\mathbf{r}}{r^3} + \mathbf{F} = \frac{\partial \mathbf{v}}{\partial t} + \frac{\delta \mathbf{v}}{\delta t}$$

we get $\delta\mathbf{v}/\delta t = F$, because in the elliptic orbit $\partial\mathbf{v}/\partial t = -\mathbf{r}/\mathbf{r}^3$. Consider $\mathbf{r}\times\mathbf{v} = \mathbf{c}$.

$$\frac{\delta \mathbf{c}}{\delta t} = \frac{\delta \mathbf{r}}{\delta t} \times \mathbf{v} + \mathbf{r} \times \frac{\delta \mathbf{v}}{\delta t} = 0 + \mathbf{r} \times \mathbf{F}.$$

The perturbations of the usual set of elements (i, Ω, ω, e, a, M_0) will get into difficulties if the values of either i or e are nearly zero, so that special considerations must be given to the selection of a suitable set of elements. (Cohen and Hubbard, 1962; Herrick, 1965.)

3. Differential Corrections

The initial elements may not be expected to agree exactly with all subsequent observations, and therefore differential corrections to the initial elements can be derived from the residuals. This process is based upon the equation for the total differential of a function of many variables.

$$\mathrm{d}F = \sum \frac{\partial F}{\partial E_i} \Delta E_i = \Delta F\,(\text{Obs.-Calc.})$$

The problem is to find $\partial F/\partial E_i$ for whatever quantity, F, has been observed. Then the $\Delta E_i's$ become the unknowns in the system of linear equations, which is usually solved by least squares. Again one must give special consideration to the choice of a set of $E_i's$, otherwise the determinant of the coefficients may have two or more nearly equal columns and the solution will be indeterminate. (Eckert and Brouwer, 1937; Herget, 1940.)

4. Variational Equations

Whenever the perturbations are very small, $\partial F/\partial E_i$ may be represented with sufficient accuracy by the value which it would have in the unperturbed elliptic orbit. However, when the elliptic elements are subjected to more severe perturbations, then $\partial F/\partial E_i$ accumulates an additional effect over a long period of time, due to the changing values of the elements. In such a case it becomes necessary to obtain the true values of $\partial F/\partial E_i$ by integrating the Variational Equation. An example of this is the case of the outer satellites of Jupiter, which are strongly perturbed by the Sun. (Herget, 1968.)

References

Bauschinger, J.: 1906, *Bahnbestimmung der Himmelskoerper*, Leipzig.
Brouwer, D. and Clemence, G.: 1961, *Methods of Celestial Mechanics*, New York.
Carpenter, L.: 1966, 'Planetary Perturbations in Chebyshev Series', NASA TN D-3168.
Cohen, C. and Hubbard, C.: 1962, *Astron. J.* **67**, 10.
Eckert, W. and Brouwer, D.: 1937, *Astron. J.* **46**, 125.
Gauss, K.: 1963, *Theoria Motus* (transl. by C. H. Davis), Dover.
Herget, P.: 1940, *Astron. J.* **48**, 105.
Herget, P.: 1948a, *Computation of Orbits*, Cincinnati, Ch. 6.
Herget, P.: 1948b, *Computation of Orbits*, Cincinnati, p. 37.
Herget, P.: 1962a, *Astron. J.* **67**, 16.
Herget, P.: 1962b, *Astron. J.* **67**, 89.
Herget, P.: 1965a, *Astron. J.* **70**, 1.
Herget, P.: 1965b, *Astron. J.* **70**, 98.
Herget, P.: 1968, *Astron. J.* **73**, 737.
Herrick, S.: 1965, *Astron. J.* **70**, 309.
Musen, P.: 1954, *Astron. J.* **59**, 262.
Schubert, J. and Stumpff, P.: 1966, *Veröff. Astron. Rechen-Inst.*, No. 18.
Stracke, G.: 1929, *Bahnbestimmung der Planeten und Kometen*, Berlin.
Watson, J.: 1869, *Theoretical Astronomy*, Philadelphia.

STATISTICAL ORBIT DETERMINATION THEORY

B. D. TAPLEY*

Dept. of Aerospace Engineering and Engineering Mechanics,
The University of Texas at Austin, Austin, Tex., U.S.A.

Abstract. Recent advances in methods for estimating the state of a continuous non-linear dynamic system using discrete observations are reviewed. Particular attention is given to a comparison of batch and sequential algorithms. The stability and convergence characteristics as well as questions related to the computation effort are considered. Finally, numerical results obtained in applying the recently developed Dynamic Model Compensation algorithm are compared with results obtained using a batch algorithm using range-rate data obtained during the Apollo 10 and 11 lunar orbit missions. These results indicate a distinct improvement in estimation accuracy using the Dynamic Model Compensation algorithm.

1. Introduction

The problem of estimating the state of any nonlinear dynamical system influenced by random forces using discrete observations which are subject to random error has received considerable attention during the last decade. The conventional solution to the problem involves linearizing the nonlinear equations about a reference solution and, then, applying linear estimation techniques. The estimate of the deviation from the reference solution can be obtained, then, by either a batch data processing algorithm in which a large batch of observations are used to obtain the estimate of the state at some reference epoch or the state may be estimated sequentially by processing each observation as it is made.

Errors of four basic types influence the accuracy of the linear estimates:

(1) errors due to the linearization assumptions;

(2) errors introduced in the computational procedures;

(3) errors which occur in the observation process; and

(4) errors due to inaccuracies in the mathematical model used to describe the dynamical process.

While the errors which occur in the observation process are important in determining the overall accuracy of any estimation procedure, they will not be considered in this discussion. Instead, attention will be confined to the effects of the errors encountered in the computation and modeling procedures. In complex orbit determination programs, the errors which occur in the computational process and the errors due to inaccurate modeling are closely related and careful consideration of the errors is required to distinguish the particular cause. With a complex mathematical model, it is difficult to obtain an accurate numerical solution to the relations which define the estimation procedure. On the other hand, if the mathematical model is simplified to alleviate computational problems, important physical effects may be neglected and the estimation procedure may diverge.

* Professor and Chairman, Department of Aerospace Engineering and Engineering Mechanics.

B. D. Tapley and V. Szebehely (eds.), Recent Advances in Dynamical Astronomy, 396–425.

There have been a comparatively few studies related to the problem of orbit determination in the presence of unmodeled dynamic errors (as contrasted with a considerable bibliography of literature which considers the effects of the first three error sources). Since the unmodeled accelerations will vary, generally, with time, the effect of the unmodeled accelerations can be compensated for most easily by using a sequential or Kalman-Bucy type of filter. Furthermore, if the extended form of the sequential estimation procedure is used, where the estimate at a time, t_i, is used to define the reference trajectory for propagating the estimate to the next observation epoch, t_j, then the errors due to the linearization assumptions will be minimized.

These concepts along with several other pertinent points are discussed in the subsequent sections. In Section 2, the general formulation of the orbit determination problem for a near-Earth satellite is discussed, and the problem is reduced to a linear state estimation problem. In Section 3, the 'batch' minimum variance (or 'least squares') solution to the problem is discussed. In Section 4, an alternate procedure for solving the normal equations associated with the batch estimate is described. In Section 5, the sequential estimation algorithm (or Kalman-Bucy filter) is derived directly from the minimum-variance batch algorithm. In Section 6, the problem of estimating the state in the presence of unmodeled accelerations is discussed and the application of the Dynamic Model Compensation algorithm to the lunar orbit determination problem is described. In Section 7, a summary of the conclusions discussed in the previous sections is given.

2. Problem Formulation

The equations which govern the motion of a vehicle moving in a central force field can be expressed as

$$\dot{\mathbf{r}} = \mathbf{v}, \qquad \dot{\mathbf{v}} = \frac{\mu\mathbf{r}}{r^3} + \mathbf{R}(\mathbf{r}, t) \tag{2.1}$$

where $\mathbf{r}$ is the position vector, $\mathbf{v}$ is the velocity vector, μ is the gravitational parameter, and $\mathbf{R}$ is the vector of perturbations. Equations (2.1) can be expressed in first order form as follows.

$$\begin{aligned} \dot{\zeta} &= f(\zeta, \alpha, t) \\ \dot{\alpha} &= 0 \end{aligned} \tag{2.2}$$

where α is a q-vector of all constant parameters in Equations (2.1) whose values are to be estimated during the orbit determination procedure. The six-vectors ζ and $f(\zeta, \alpha, t)$ are defined as

$$\zeta^T = [\mathbf{r}^T : \mathbf{v}^T] \qquad f^T(\zeta, \alpha, t) = [\mathbf{v}^T : (-\mu\mathbf{r}/r^3 + \mathbf{R})^T]. \tag{2.3}$$

A. THE STATE EQUATIONS

As defined, the state vector is composed of all dependent variables or constant parameters required to define the time rate of change of the state of the dynamical system. With this definition, the n-dimensional state vector, X, can be expressed as follows

$$X^T = [\zeta^T : \alpha^T] \tag{2.4}$$

and the state equations become,

$$\dot{X} = F(X, t); \quad X(t_0) = X_0 \tag{2.5}$$

where $F^T(X, t) = [f^T(\zeta, \alpha, t) \vdots 0^T]$. With X_0, in Equation (2.5), specified, the motion of the dynamical system will be determined uniquely by the solution to Equation (2.5). This solution can be expressed formally as

$$X(t) = \theta(X_0, t_0, t). \tag{2.6}$$

In the usual orbit determination problem, X_0 will not be known perfectly and, consequently, the true solution, $X(t)$, will differ from the nominal solution, $X^*(t)$, obtained with the specified initial state, X_0^*. As a consequence, observations of the motion must be made to determine the true motion.

B. THE OBSERVATION-STATE RELATIONSHIP

Usually, the state vector, $X(t)$, cannot be observed directly. Instead, the observations will be a non-linear function of the state. Furthermore, the observations are usually influenced by random error which occurs in the process of making the observations. As a consequence of these points, the observation-state relationship can be expressed as

$$Y_i = G(X_i, t_i) + \varepsilon_i \tag{2.7}$$

where Y_i is a p-vector of observations of the state X_i at the time epoch t_i, $G(X_i, t_i)$ is a p-vector of non-linear functions relating the state and the observations, and ε_i is a p-vector of observation errors. As an example, the observation may be the range between a tracking station and the satellite and in this case, Equation (2.7) would be expressed as

$$R_i = [\boldsymbol{\varrho} \cdot \boldsymbol{\varrho}]_i^{1/2} + \varepsilon_{R_i}$$

where $\boldsymbol{\varrho} = \mathbf{r} - \mathbf{r}_s$ and $\mathbf{r}_s$ is the location of the tracking station. For this case, $p = 1$ and Equation (2.7) will be a scalar observation-state relation at the time t_i. The quantity R_i will be obtained as the result of some measurement process.

In general, more than one quantity may be observed at a given epoch, i.e., Y_i may consist of measurements of range, range-rate, azmuth, elevation, etc., where the observed quantities are measured from some topocentric tracking station.

The following comments can be made regarding the observation-state relationship expressed in Equation (2.7).

(1) Usually, Y_i will be of smaller dimension than X_i, i.e., $p \leqslant n$ for the orbit determination problem.

(2) Furthermore, even if $p = n$, X_i cannot be determined from Equation (2.7) unless $\varepsilon_i = 0.$, i.e., unless the observations are perfect.

As a consequence, a large number of observations are obtained and the information available for estimating the initial state can be expressed as

$$Y_i = G(X_i, t_i) + \varepsilon_i, \quad i = 1, \dots, l. \tag{2.8}$$

Since $X_i = \Theta(X_0, t_0, t_i)$ from Equation (2.6), it follows that Equations (2.8) will represent $l \times p$ equations in the $(l \times p \times n)$ unknowns, i.e., the unknown components of the observation error and the initial state. Since the number of equations are fewer than the number of unknowns in Equations (2.8), some additional criterion must be adopted for choosing X_0. The additional criterion is usually selected to minimize the error in the estimate of the initial state. Depending on the criterion, the estimate of X_0 may be based on a least squares criterion or a statistical based criterion such as the minimum variance or maximum likelihood estimation criterion may be used.

In either of these cases, solution for an estimate of X_0 requires solving, iteratively, the nonlinear system of equations given by Equations (2.8) for X_0. The difficulties associated with obtaining an estimate of X_0 using Equations (2.6) and (2.8) have led to the use of linear estimation techniques where they can be applied.

C. LINEARIZATION OF THE NON-LINEAR PROBLEM

If the true value of the state, $X(t)$ and some initially assumed value of the state, $X^*(t)$, are sufficiently close throughout some time interval of interest, $t_0 \leqslant t \leqslant t_l$, a Taylor series expansion about the initially assumed, or nominal, trajectory at each point in time can be used to linearize the non-linear problem. To accomplish this, let

$$\begin{aligned} x(t) &= X(t) - X^*(t); \qquad t_0 \leqslant t \leqslant t_l \\ y_i &= Y_i - G(X_i^*, t_i); \qquad i = 1, \ldots, l. \end{aligned} \tag{2.9}$$

Then substituting Equations (2.9) into Equations (2.5) and (2.8) leads to

$$\begin{aligned} \dot{X} &= \dot{X}^* + \left[\frac{\partial F}{\partial X}\right]^* [X - X^*] + \cdots, \\ Y_i &= G(X_i^*, t_i) + \left[\frac{\partial G}{\partial X}\right]^* [X_i - X_i^*] + \cdots + \varepsilon_i, \qquad i = 1, \ldots, l. \end{aligned} \tag{2.10}$$

Now, if terms of $0[(X_i - X_i^*)^2]$ are neglected and if the definitions

$$A(t) = [\partial F/\partial X]^*, \quad \tilde{H}(t_i) = [\partial G/\partial X]_i^*, \tag{2.11}$$

are introduced into Equations (2.10), Equations (2.9) can be used to express Equations (2.10) as

$$\begin{aligned} \dot{x} &= A(t)\,x, \qquad x(t_0) = x_0 \\ y_i &= \tilde{H}_i x_i + \varepsilon_i, \qquad i = 1, \ldots, l. \end{aligned} \tag{2.12}$$

The unknown state, X_0, is replaced by the unknown state deviation, x_0, and the non-linear estimation problem is replaced by an estimation problem in which the observations are related in a linear manner to the state and the state is propagated by a system of linear equations with time dependent coefficients. The quantity to be estimated, $x(t)$, is the deviation from the reference solution, $X^*(t)$. Equations (2.5) must be integrated with some initial value, X_0^*, to obtain a reference solution for evaluating $A(t)$ and $\tilde{H}_i$.

D. REDUCTION OF OBSERVATIONS TO A SINGLE EPOCH

To reduce $(l \times p + l \times n)$ unknowns in Equations (2.12) to $(l \times p + n)$ unknowns, all of the state variables, x_i, are expressed as a function of the state at a single epoch, say x_k. To this end, note that the solution to the first of Equations (2.12) can be expressed as

$$x_i = \Phi(t_i, t_k)\, x_k \tag{2.13}$$

where $\Phi(t_i, t_k)$ is the state transition matrix. See for example, Coddington and Levinson (1955). The state transition matrix will satisfy the following properties:

(1) $\Phi(t_0, t_0) = I = \Phi(t_k, t_k)$

When the time intervals, t_i and t_k, coincide, the state transition matrix reduces to the $n \times n$ identity matrix.

(2) $\Phi(t_2, t_0) = \Phi(t_2, t_1)\, \Phi(t_1, t_0)$

This follows from the uniqueness of Newtonian motion.

(3) $\Phi(t_2, t_1) = \Phi^{-1}(t_1, t_2)$

This property follows from the reversability of Newtonian mechanics. The state transition matrix satisfies the following differential equation (Coddington and Levinson, 1955):

$$\dot{\Phi}(t, t_k) = A(t)\, \Phi(t, t_k), \qquad \Phi(t_k, t_k) = I \tag{2.14}$$

where $A(t)$ is defined in Equation (2.11). Finally, it can be shown (Coddington and Levinson, 1955) that the inverse of the state transition matrix satisfies the differential equation

$$\dot{\Phi}^{-1}(t, t_k) = \Phi^{-1}(t, t_k)\, A(t), \qquad \Phi^{-1}(t_k, t_k) = I. \tag{2.15}$$

Using Equation (2.13), the system of Equations (2.12) can be expressed as

$$\begin{aligned} x_i &= \Phi(t_i, t_k)\, x_k \\ y_i &= \tilde{H}_i x_i + \varepsilon_i, \qquad i = 1, \ldots, l. \end{aligned} \tag{2.16}$$

Now, if the first of Equations (2.16) is used to express each state, X_i, in terms of the state at some general epoch t_k, the following system of equations is obtained.

$$\begin{aligned} y_1 &= \tilde{H}_1 \Phi(t_1, t_k)\, x_k + \varepsilon_1 \\ y_2 &= \tilde{H}_2 \Phi(t_2, t_k)\, x_k + \varepsilon_2 \\ \vdots \quad & \quad \vdots \qquad\qquad\qquad \vdots \\ y_l &= \tilde{H}_l \Phi(t_l, t_k)\, x_k + \varepsilon_l. \end{aligned} \tag{2.17}$$

Now defining $H_i = \tilde{H}_i \Phi(t_i, t_k)$ and

$$y = \begin{bmatrix} y_1 \\ y_2 \\ \vdots \\ y_l \end{bmatrix} \quad H = \begin{bmatrix} \tilde{H}_1 \Phi(t_1, t_k) \\ \tilde{H}_2 \Phi(t_2, t_k) \\ \vdots \\ \tilde{H}_l \Phi(t_l, t_k) \end{bmatrix} \quad \varepsilon = \begin{bmatrix} \varepsilon_1 \\ \varepsilon_2 \\ \vdots \\ \varepsilon_l \end{bmatrix} \tag{2.18}$$

where y and ε are m-vectors where $m = l \times p$ and H is an $m \times n$ matrix, the system represented by Equations (2.17) can be expressed as

$$y = Hx_k + \varepsilon \,. \tag{2.19}$$

Equations (2.19) represents a system of m equations in $m+n$ unknowns where $m>n$. For the case $\varepsilon=0$, any n of Equations (2.19) that are independent can be used to determine x_k. For the general case, $\varepsilon\neq 0$ and some further criterion, such as least squares, maximum likelihood or minimum variance must be specified to determine x_k. In the following discussion the minimum variance criteria will be selected and the best linear unbiased minimum variance estimate will be obtained.

3. Minimum Variance Estimate

The minimum variance criterion is widely used in developing solutions to estimation problems because of the simplicity in its use. It has the additional advantage that the complete statistical description of the random errors in the problem is not required. Rather, only the first and second moments of the conditional density function are required. This information is expressed in the mean and covariance matrix associated with the random error. For further discussion of minimum variance estimation methods see Liebelt (1967) and Lewis and Odell (1971).

A. THE NORMAL CASE

If it is assumed that the observation error ε_i is random with zero mean and specified covariance, i.e.

$$E[\varepsilon_i] = 0 \qquad E[\varepsilon_i\varepsilon_i^T] = R_i \tag{3.1}$$

then the estimation of the state can be formulated in the following problem statement:

Given: The system of state-propagation equations and observation state equations

$$\begin{aligned} x_i &= \Phi(t_i, t_k)\, x_k \\ y_i &= \tilde{H}_i x_i + \varepsilon_i \,, \qquad i = 1, \ldots, l \end{aligned} \tag{3.2}$$

where $E[\varepsilon_i]=0$; $E[\varepsilon_i\varepsilon_j^T]=R_i\delta_{ij}$ for all (i,j) and $E[(x_k-\bar{x}_k)\varepsilon_j^T]=0$ for all (i,j),

Find: The best linear unbiased minimum variance estimate, $\hat{x}_k$, of the state x_k.

The solution to this problem procedes as follows. Using the state transition matrix, reduce Equations (3.2), to the following form

$$y = Hx_k + \varepsilon \tag{3.3}$$

where

$$E[\varepsilon] = \begin{bmatrix} E[\varepsilon_1] \\ E[\varepsilon_2] \\ \vdots \\ E[\varepsilon_l] \end{bmatrix} = 0 \qquad E[\varepsilon\varepsilon^T] = \begin{bmatrix} R_1 & & & 0 \\ & R_2 & & \\ & & \ddots & \\ 0 & & & R_l \end{bmatrix}. \tag{3.4}$$

Generally, $R_1=R_2=\cdots=R_l$, but this is not a necessary restriction in the following argument. Furthermore, the more general case of time-correlated observation errors where $E[\varepsilon_i\varepsilon_j^T]=R_{ij}$ and where the off-diagonal zero's in the definition of the R matrix will be replaced by non-zero quantities can be treated within the framework of the following discussion.

From the requirements of the problem statement the estimate is to be the best linear, unbiased minimum variance estimate. The consequences of each of these requirements are examined in the following steps.

(1) *Linear*: The requirement of a linear estimate implies that the estimate is to be made up of a linear combination of the observations, i.e.,

$$\hat{x}_k = My. \tag{3.5}$$

The $(n \times m)$ matrix M is unspecified and is to be selected to obtain the best estimate

(2) *Unbiased*: If the estimate is unbiased, then

$$E[\hat{x}] = x. \quad (3.6$$

Substituting, Equations (3.5) and (3.3) into Equation (3.6) leads to the following requirement

$$E[My] = E[MHx_k + \varepsilon_k] = x_k.$$

But, since $E[\varepsilon_k]=0$, this reduces to

$$MHx_k = x_k$$

from which the following constraint on M is obtained.

$$MH = I. \tag{3.7}$$

That is, if the estimate is to be unbiased the linear mapping matrix M must satisfy Equation (3.7).

(3) *Minimum Variance*: If the estimate is unbiased, then the covariance matrix can be expressed as

$$\begin{aligned} P_k &= E[(\hat{x}_k - E[\hat{x}_k])(\hat{x}_k - E[\hat{x}_k])^T] = \\ &= E[(\hat{x}_k - x_k)(\hat{x}_k - x_k)^T]. \end{aligned} \tag{3.8}$$

Hence, the problem statement requires the $\hat{x}_k$ be selected to minimize Equation (3.8) while satisfying Equations (3.6) and (3.7). Substituting Equation (3.6) into Equation (3.8) leads to the following result.

$$\begin{aligned} P_k &= E[(My - x_k)(My - x_k)^T] = \\ &= E[\{M(Hx_k + \varepsilon) - x_k\}\{M(Hx_k + \varepsilon) - x_k\}^T] = \\ &= E[M\varepsilon\varepsilon^T M^T] \end{aligned}$$

since $MH-I=0$. It follows then that the cavariance matrix can be written as

$$P_k = MRM^T \tag{3.9}$$

and M is to be selected to satisfy Equation (3.7). To involve the constraint imposed by

Equation (3.7) and to keep the constrained relation for P_k symmetric, Equation (3.7) is adjoined to Equation (3.9) in the following form.

$$P_k = MRM^T + \Lambda^T [I - MH]^T + (I - MH)\,\Lambda$$

where Λ is a $n \times n$ matrix of unspecified Lagrange multipliers. Then, for a minimum of P, it is necessary that the first variation vanish.

$$\delta P_k = 0 = (MR - \Lambda^T H^T)\,\delta M^T + \delta M\,[RM^T - H\Lambda] + \\ + \delta\Lambda^T [I - MH]^T + (I - MH)\,\delta\Lambda\,.$$

Now if δP_k is to vanish for arbitrary δM and $\delta\Lambda$, the coefficients of each term in the variation must vanish and, hence, the following conditions must be satisfied.

$$MH - I = 0\,, \qquad MR - \Lambda^T H^T = 0\,. \tag{3.10}$$

From the second of these conditions

$$M = \Lambda^T H^T R^{-1} \tag{3.11}$$

since R is assumed to be positive definite. Substituting Equation (3.11) into the first of Equations (3.10) leads to the following result

$$\Lambda^T (H^T R^{-1} H) = I\,.$$

Now, since the coefficient of Λ^T will be full rank if R^{-1} is positive definite, it follows that

$$\Lambda^T = (H^T R^{-1} H)^{-1}$$

and in view of Equation (3.11),

$$M = (H^T R^{-1} H)^{-1}\, H^T R^{-1}\,. \tag{3.12}$$

This is the value of M which satisfies the unbiased and minimum variance requirements. Substitution of Equation (3.12) into Equation (3.9) leads to the following expression for the covariance matrix.

$$P_k = (H^T R^{-1} H)^{-1}\,.$$

With Equations (3.12) and (3.5), the best linear unbiased minimum variance estimate of x_k is given as

$$\hat{x}_k = (H^T R^{-1} H)^{-1}\, H^T R^{-1} y\,. \tag{3.13}$$

Note that computation of the estimate, $\hat{x}_k$, requires inverting an $n \times n$ matrix and for a large dimension system, the computation of this inverse can impose considerable difficulty. The solution given by Equation (3.13) will agree with the weighted least squares solution if the weighting matrix, W, used in the least squares approach is equal to the inverse of the observation noise covariance matrix, i.e., if $W = R^{-1}$. The minimum variance estimate given in Equation (3.13) will agree with the maximum likelihood estimate, if the observation errors are assumed to be distributed normally with zero mean and covariance R.

B. PROPAGATION OF THE ESTIMATE

If the estimate at a time t_j is obtained by using Equation (3.13), the estimate at any later time prior to the next observation point can be determined as follows. Let

$$\begin{aligned}\hat{x}_j &= E\left[x_j \mid y_1, y_2, \ldots, y_j\right] \\ \bar{x}_k &= E\left[x_k \mid y_1, y_2, \ldots, y_j\right].\end{aligned} \tag{3.14}$$

That is, $\bar{x}_k$, $t_k > t_j$, is the best estimate of x_k based on the observations available up to t_j. From the first of Equations (3.2), it follows that

$$E\left[x_k \mid y_1, \ldots, y_j\right] = \Phi(t_k, t_j)\, E\left[x_j \mid y_1, \ldots, y_j\right].$$

In view of Equations (3.14), the a priori estimate of x_k, i.e., $\bar{x}_k$, is given by the following expression.

$$\bar{x}_k = \Phi(t_k, t_j)\, \hat{x}_j. \tag{3.15}$$

The expression for propagating the covariance matrix can be obtained as follows.

$$\bar{P}_k \underline{\underline{\Delta}} E\left[(x_k - \bar{x}_k)(x_k - \bar{x}_k)^T \mid y_1, \ldots, y_j\right]. \tag{3.16}$$

In view of Equation (3.15), Equation (3.16) becomes

$$\bar{P}_k = E\left[\Phi(t_k, t_j)(x_j - \hat{x}_j)(x_j - \hat{x}_j)^T \Phi^T(t_k, t_j) \mid y_1, \ldots, y_j\right].$$

Since the state transition matrix is deterministic, it follows that

$$\bar{P}_k = \Phi(t_k, t_j)\, P_j \Phi^T(t_k, t_j) \tag{3.17}$$

where $P_j = E[(x_j - \hat{x}_j)(x_j - \hat{x}_j)^T \mid y_1, \ldots, y_j]$. Equations (3.15) and (3.17) can be used to determine the estimate of the state at t_k based on an estimate at t_j.

C. MINIMUM VARIANCE ESTIMATE WITH A PRIORI INFORMATION

If an estimate and the associated covariance matrix are obtained at a time t_j, and an additional observation or observation sequaence is obtained at a time t_k, the estimate and the observation can be combined in a straightforward manner to obtain the new estimate $\hat{x}_k$. The estimate $\hat{x}_j$ and P_j are propagated forward to t_k, using the following equations

$$\bar{x}_k = \Phi(t_k, t_j)\, \hat{x}, \qquad \bar{P}_k = \Phi(t_k, t_j)\, P_j \Phi^T(t_k, t_j). \tag{3.18}$$

The problem to be considered can be stated as follows:

Given $\bar{x}_k$, P_k and $y_k = \tilde{H}_k x_k + \varepsilon_k$, where $E[\varepsilon_k] = 0$, $E[\varepsilon_k \varepsilon_j^T] = R_k \delta_{kj}$ and $E[(x_j - \hat{x}_j)\varepsilon_k^T] = 0$, find the best linear minimum variance unbiased estimate.

The solution to the problem can be obtained by reducing it to the previously solved problem. To this end note that if $\hat{x}_j$ is unbiased, $\bar{x}_k$ will be unbiased since $E[\bar{x}_k] = \Phi E[\hat{x}_j]$. Hence, $\bar{x}_k$ can be interpreted as an observation and the following relations will hold

$$\begin{aligned} y_k &= \tilde{H}_k x_k + \varepsilon_k \\ \bar{x}_k &= x_k + \eta_k \end{aligned} \tag{3.19}$$

where

$$E[\varepsilon_k] = 0\,, \quad E[\varepsilon_k \varepsilon_k^T] = R\,, \quad E[\eta_k] = 0 \quad \text{and} \quad E[\eta_k \eta_k^T] = \bar{P}_k\,.$$

Now, if the following definitions are used

$$y = \begin{bmatrix} y_k \\ \cdots \\ \bar{x}_k \end{bmatrix} \quad H = \begin{bmatrix} \tilde{H}_k \\ \cdots \\ I \end{bmatrix} \quad \varepsilon = \begin{bmatrix} \\ \cdots \\ \eta \end{bmatrix} \quad R = \begin{bmatrix} R_k & \vdots & 0 \\ \cdots & \cdots & \cdots \\ 0 & \vdots & \bar{P}_k \end{bmatrix}. \tag{3.20}$$

Equations (3.19) can be expressed as $y = Hx + \varepsilon$ and the solution obtained in the previous section can be applied to obtain the following estimate for $\hat{x}_k$.

$$\hat{x}_k = (H^T R^{-1} H)^{-1}\, H^T R^{-1} y\,.$$

In view of the definitions in Equation (3.20),

$$\hat{x}_k = \left([\tilde{H}_k^T \vdots I] \begin{bmatrix} R_k^{-1} & \vdots & 0 \\ \cdots & \vdots & \cdots \\ 0 & \vdots & \bar{P}_k^{-1} \end{bmatrix} \begin{bmatrix} \tilde{H}_k^T \\ \cdots \\ I \end{bmatrix} \right)^{-1} \left([\tilde{H}_k^T \vdots I] \begin{bmatrix} R_k^{-1} & \vdots & 0 \\ \cdots & \vdots & \cdots \\ 0 & \vdots & \bar{P}_k^{-1} \end{bmatrix} \begin{bmatrix} y_k \\ \\ \bar{x}_k \end{bmatrix} \right)$$

or in expanded form,

$$\hat{x}_k = (\tilde{H}_k^T R_k^{-1} \tilde{H}_k + \bar{P}_k^{-1})^{-1} (\tilde{H}_k^T R_k^{-1} y_k + \bar{P}_k^{-1} \bar{x}_k)\,. \tag{3.21}$$

The following remarks are pertinent in regard to Equation (3.21).

(1) The vector y_k may be only a single observation or it may include an entire batch of observations.

(2) The a priori estimate, $\bar{x}_k$, may represent the estimate based on a priori initial conditions or the estimate based on the reduction of a previous batch of data.

(3) As in the previous solution, an $n \times n$ matrix must be inverted and if the dimension n is large, this inversion can lead to computational problems.

An alternate approach to solving either Equation (3.13) or (3.21) is discussed in the following section.

4. An Alternate Solution of the Normal Equations

In many practical applications, the solution to either Equations (3.13) or (3.21) requires the inversion of an $n \times n$ matrix. Frequently, when the dimension n is large, the matrix to be inverted, i.e., $(H^T R^{-1} H)$, will be ill-conditioned. As a consequence, a recent line of inquiry into alternate methods for solving the system of linear equations has been investigated. Discussions of these approaches are given in Businger and Golub (1968), Householder (1958) and Kamiski *et al.* (1971). The general approach used in these investigations is outlined in the following sections.

A. SOLUTION OF THE NORMAL EQUATIONS

The application of the least squares or the minimum variance estimation criteria, leads to the following equation which must be solved for the best estimate, $\hat{x}_k$.

$$(H^T R^{-1} H)\,\hat{x}_k = H^T R^{-1} y\,. \tag{4.1}$$

Providing $(H^T R^{-1} H)$ is full rank, $\hat{x}_k$ can be obtained by direct matrix inversion. If the dimensions of $\hat{x}_k$ are large, this inversion process may be quite time consuming.

A more convenient procedure is to use the Cholesky decomposition (Fadeeva, 1959) to separate $(H^T R^{-1} H)$ into two triangular matrices S^T and S such that

$$S^T S = H^T R^{-1} H \tag{4.2}$$

where S is an upper triangular $n \times n$ matrix. Then $\hat{x}_k$ can be determined without matrix inversion by applying a forward substitution, followed by a backward substitution.

To this end, let

$$b = S\hat{x}_k\,. \tag{4.3}$$

Then, in view of Equation (4.2), Equation (4.1) can be expressed as

$$S^T b = c \tag{4.4}$$

where $c = H^T R^{-1} y$. In component form, Equation (4.4) can be expressed as

$$\begin{bmatrix} S_{11} & 0 & 0 & \ldots 0 \\ S_{12} & S_{22} & 0 & \ldots 0 \\ S_{13} & S_{23} & S_{33} & \ldots 0 \\ \vdots & & & \vdots \\ S_{1n} & S_{2n} & S_{3n} & \ldots S_{nn} \end{bmatrix} \begin{bmatrix} b_1 \\ b_2 \\ b_3 \\ \vdots \\ b_n \end{bmatrix} = \begin{bmatrix} c_1 \\ c_2 \\ c_3 \\ \vdots \\ c_n \end{bmatrix}. \tag{4.5}$$

By expanding Equation (4.5), the following relations are obtained.

$$\begin{aligned} &S_{11} b_1 = c_1 \\ &S_{12} b_1 + S_{22} b_2 = c_2 \\ &S_{13} b_1 + S_{23} b_2 + S_{33} b_3 = c_3 \\ &\vdots \\ &S_{1n} b_1 + S_{2n} b_2 + S_{3n} b_3 \cdots + S_{nn} b_n = c_n\,. \end{aligned} \tag{4.6}$$

Equations (4.6) can be solved for the b_i as follows

$$\begin{aligned} &b_1 = (c_1/S_{11}) \\ &b_2 = (c_2 - S_{12} b_1)/S_{22} \\ &b_3 = (c_3 - S_{13} b_1 - S_{23} b_2)/S_{33} \\ &\vdots \\ &b_n = (c_n - S_{1n} b_1 - S_{2n} b_2 - \cdots - S_{nn} b_n)/S_{nn}\,. \end{aligned} \tag{4.7}$$

The general term in Equations (4.7) can be expressed as

$$\begin{aligned} &b_1 = c_1/S_{11} \\ &b_2 = (c_2 - S_{12} b_1)/S_{22} \\ &b_i = (c_i - \sum_{j=1}^{i-1} S_{ji} b_j)/S_{ii}\,; \qquad i = 3, \ldots, n\,. \end{aligned} \tag{4.8}$$

Then by recalling the definition $S\hat{x}=b$, components of $\hat{x}$ can be determined by inverse substitution as follows.

$$\begin{aligned}
\hat{x}_n &= b_n/S_{nn} \\
\hat{x}_{n-1} &= (b_{n-1} - S_{n-1,n}\hat{x}_n)/S_{n-1,n-1} \\
\hat{x}_{n-2} &= (b_{n-2} - S_{n-2,n}\hat{x}_n - S_{n-2,n-1}\hat{x}_{n-1})/S_{n-2,n-2} \\
&\vdots \\
\hat{x}_2 &= (b_2 - S_{23}\hat{x}_3 - S_{24}\hat{x}_4 - \cdots - S_{2,n}\hat{x}_n)/S_{22} \\
\hat{x}_1 &= (b_1 - S_{12}\hat{x}_2 - S_{13}\hat{x}_3 - \cdots - S_{1n}\hat{x}_n)/S_{11}\,.
\end{aligned} \tag{4.9}$$

The general term in Equations (4.9) can be expressed as

$$\begin{aligned}
\hat{x}_n &= \bar{x}_n/S_{nn} \\
\hat{x}_{n-1} &= (b_{n-1} - S_{n-1,n}\hat{x}_n)/S_{n-1,n-1} \\
\hat{x}_i &= (b_i - \sum_{j=i+1}^{n} S_{ij}\hat{x}_j)/S_{ii}\,.
\end{aligned} \tag{4.10}$$

Hence, by Equations (4.8) and (4.10), the solution to Equations (4.1) can be obtained without inverting the matrix $(H^T R^{-1} H)$. Only the recursive scalar divisions are required.

B. THE CHOLESKY DECOMPOSITION

The upper triangular matrix S can be computed by using the Cholesky decomposition (Fadeeva, 1959) which states that any symmetric positive definite $n \times n$ matrix P may be written in factored form as

$$\begin{bmatrix} P_{11} & P_{12} \ldots P_{1n} \\ P_{12} & P_{22} \ldots P_{2n} \\ \vdots & \\ P_{1n} & P_{2n} \ldots P_{nn} \end{bmatrix} = \begin{bmatrix} S_{11} & 0 & \ldots 0 \\ S_{21} & S_{22} & \ldots 0 \\ \vdots & \vdots & \ddots \vdots \\ S_{n1} & S_{n2} & \ldots S_{nn} \end{bmatrix} \begin{bmatrix} S_{11} & S_{21} & \ldots S_{n1} \\ 0 & S_{22} & \ldots S_{n2} \\ \vdots & \vdots & \ddots \vdots \\ 0 & 0 & S_{nn} \end{bmatrix}$$

In correspondence with the scalar case, S is often called the square root, or the Cholesky square root of P. Cholesky gave the folowing recursive algorithm for computing S:

$$\begin{aligned}
S_{ii} &= \sqrt{P_{ii} - \sum_{j=1}^{i-1} S_{ij}^2}, \quad \text{for} \quad i = 1, \ldots, n \\
S_{ij} &= \begin{cases} 0; j < i \\ \dfrac{1}{S_{ii}}\left[P_{ij} - \sum_{k=1}^{i-1} S_{ik}S_{jk}\right]; & j = i+1, n \end{cases}
\end{aligned} \tag{4.11}$$

Example: As an example consider the following matrix

$$P = \begin{bmatrix} 1 & 2 & 3 \\ 2 & 8 & 2 \\ 3 & 2 & 14 \end{bmatrix}.$$

Using Equations (4.11), it follows that

$$\begin{array}{lll} S_{11} = 1 & S_{21} = 0 & S_{31} = 0 \\ S_{12} = 2 & S_{22} = \sqrt{8 - (2)^2} = 2 & S_{32} = 0 \\ S_{13} = 3 & S_{23} = [2 - (2)(3)]/2 = -2 & S_{33} = 14 - 9 - 4 = 1 . \end{array}$$

Hence

$$S = \begin{bmatrix} 1 & 2 & 3 \\ 0 & 2 & -2 \\ 0 & 0 & 1 \end{bmatrix}$$

and

$$S^T S = \begin{bmatrix} 1 & 0 & 0 \\ 2 & 2 & 0 \\ 3 & -2 & 1 \end{bmatrix} \begin{bmatrix} 1 & 2 & 3 \\ 0 & 2 & -2 \\ 0 & 0 & 1 \end{bmatrix} = \begin{bmatrix} 1 & 2 & 3 \\ 2 & 8 & 2 \\ 3 & 2 & 14 \end{bmatrix} = P .$$

5. Sequential Estimation Algorithm

In the previous section, an alternate approach to inverting the $(n \times n)$ matrix $(H^T R^{-1} H)$ was presented. In this section, a second approach is discussed in which the observations are processed as soon as they are received. An advantage of the sequential processing algorithm presented in the following discussion is that the matrix to be inverted will be of the same dimension as the observation error covariance. Hence, for a single scalar observation, only a scalar division will be required to obtain the estimate of x_k. The algorithm presented was developed originally by Swerling in 1958 (Swerling, 1959), but the treatment which has received a more popular aclaim is that due to the work of Kalman and Bucy (1961). The sequential estimation algorithm developed in the subsequent discussion is referred to universally as the Kalman-Bucy filter.

A. THE SEQUENTIAL ESTIMATION ALGORITHM

Recall that with an estimate x_j and P_j can be propagated forward to a time t_k by the relations,

$$\begin{aligned} \bar{x}_k &= \Phi(t_k, t_j)\, \hat{x}_j \\ \bar{P}_k &= \Phi(t_k, t_j)\, P_j \Phi^T(t_k, t_j) \end{aligned} \tag{5.1}$$

and with an additional observation at t_k,

$$y_k = H_k x_k + \varepsilon_k \tag{5.2}$$

where $E[\varepsilon_k] = 0$ and $E[\varepsilon_k, \varepsilon_j^T] = R_k \delta_{kj}$, the best estimate of x_k is obtained in Equation (3.21) as

$$\hat{x}_k = (H_k^T R_k^{-1} H_k + \bar{P}_k^{-1})^{-1} (H_k^T R_k^{-1} y_k + \bar{P}_k^{-1} \bar{x}_k) \tag{5.3}$$

where $\tilde{H}_k = H_k$ is used in Equation (3.21). The primary computational problems are associated with computing the $(n \times n)$ matrix inverse in Equation (5.3). Recall that in the original derivation, it is shown that the quantity to be inverted is the covariance matrix P_k associated with estimate $\hat{x}_k$. That is,

$$P_k = (H_k^T R_k^{-1} H_k + \bar{P}_k^{-1})^{-1}. \tag{5.4}$$

From Equation (5.4), it follows that

$$P_k^{-1} = H_k^T R_k^{-1} H_k + \bar{P}_k^{-1}. \tag{5.5}$$

Pre-multiplying each side of Equation (5.5) by P_k and then post-multiplying by $\bar{P}_k$ leads to the following expression.

$$\bar{P}_k = P_k H_k^T R_k^{-1} H_k \bar{P}_k + P_k \tag{5.6}$$

or

$$P_k = \bar{P}_k - P_k H_k^T R_k^{-1} H_k^T \bar{P}_k.$$

Now if Equation (5.6) is post-multiplied by the quantity $H_k R^{-1}$, the following expression is obtained.

$$\begin{aligned}\bar{P}_k H_k^T R_k^{-1} &= P_k H_k^T R_k^{-1} [H_k \bar{P}_k H_k^T R_k^{-1} + I] = \\ &= P_k H_k^T R_k^{-1} [H_k \bar{P}_k H_k^T + R_k] R_k^{-1}.\end{aligned}$$

Now, solving for the quantity $P_k H_k R_k^{-1}$ leads to

$$P_k H_k^T R_k^{-1} = \bar{P}_k H_k^T [H_k \bar{P}_k H_k^T + R_k]^{-1}. \quad (5.8$$

This relates the a priori covariance matrix $\bar{P}_k$ to the a posteriori covariance matrix P_k. If Equation (5.8) is used to eliminate $P_k H_k^T R_k^{-1}$ in Equation (5.7),/the following result is obtained.

$$P_k = \bar{P}_k - \bar{P}_k H_k^T [H_k \bar{P}_k H_k^T + R_k]^{-1} H_k \bar{P}_k. \tag{5.9}$$

Note that Equation (5.9) is an alternate way of computing the inverse in Equation (5.4). In Equation (5.9), the matrix to be inverted is of dimension $p \times p$, e.g., the same as the observation error covariance matrix. If only a scalar observation is involved, the inverse is obtained through only one scalar division. The identity in Equation (5.9) is referred to in the literature as the Schurr identity or the inside out rule. If the weighting matrix, K_k, is defined as

$$K_k = \bar{P}_k H_k^T [H_k \bar{P}_k H_k^T + R_k]^{-1} \tag{5.10}$$

then Equation (5.9) can be expressed in the compact form

$$P_k = [I - K_k H_k] \bar{P}_k. \tag{5.11}$$

Now, if Equation (5.11) is substitued into Equation (5.3), the sequential form for computing x_k can be obtained, i.e.,

$$\hat{x}_k = [I - K_k H_k] \bar{P}_k [H_k^T R_k^{-1} y_k + \bar{P}_k^{-1} \bar{x}_k].$$

Re-arranging leads to

$$\hat{x}_k = [I - K_k H_k] \bar{x}_k + [I - K_k H_k] \bar{P}_k H_k^T R_k^{-1} y_k. \tag{5.12}$$

It is a straightforward algebraic exercise to show that the coefficient of y_k reduces to K_k, i.e.

$$[I - K_k H_k] \bar{P}_k H_k^T R_k^{-1} = K_k .$$

Substituting this result into Equation (5.12) leads to the following result.

$$\hat{x}_k = \bar{x}_k + K_k [y_k - H_k \bar{x}_k] . \tag{5.13}$$

Equation (5.13), along with Equations (5.1), (5.10) and (5.11) can be used in a recursive fashion to compute the estimate of $\hat{x}_k$, incorporating the observation, y_k.

The algorithm for computing the estimate sequentially can be summarized then as follows:

Given: $\hat{x}_{k-1}$, $\bar{P}_{k-1}$ and X_{k-1}^*.

(1) Integrate from t_{k-1} to t_k,

$$\dot{X}^* = F(X^*, t), \qquad X(t_{k-1}) = X_{k-1}^*$$
$$\dot{\Phi}(t, t_{k-1}) = A(t)\,\Phi(t, t_{k-1}), \qquad \Phi(t_{k-1}, t_{k-1}) = I .$$

(2) Compute

$$\bar{x}_k = \Phi(t_k, t_{k-1})\,\hat{x}_{k-1}$$
$$\bar{P}_k = \Phi(t_k, t_{k-1})\,P_{k-1}\Phi^T(t_k, t_{k-1})$$
$$y_k = Y_k - G(X_k^*, t_k)$$
$$H_k = [\partial G/\partial X]^* .$$

(3) Compute

$$K_k = \bar{P}_k H_k^T [H_k \bar{P}_k H_k^T + R_k]^{-1}$$
$$P_k = [I - K_k H_k]\,\bar{P}_k$$
$$\hat{x}_k = \bar{x}_k + K_k [y_k - H_k \bar{x}_k] .$$

(4) Replace k with $k+1$ and return to (1).

As pointed out previously, the only inverse required is the inverse required in calculating the weighting matrix K_k and this inverse will be of the size of the state noise covariance matrix. The estimate of the state of the nonlinear system is given by $\hat{X}_k = X_k^* + \hat{x}_k$. One disadvantage of this algorithm lies in the fact that, if the true state and the reference state are not close together, then the linearization assumption leading to Equations (2.12) may not be valid and the estimation process may diverge. A second unfavorable characteristic of the sequential estimation algorithm is that the state estimate covariance matrix will approach zero as the number of observations becomes large.

Examination of the estimation algorithm will indicate that, as $P_k \to 0$, the estimation procedure will become insensitive to the observations and the estimate will diverge due either to errors introduced in the linearization procedure, computational errors or errors due to an incomplete mathematical model. Two modifications which improve these deficiencies are discussed in the following sections.

B. EXTENDED SEQUENTIAL ESTIMATOR

In order to minimize the effects of the errors due to the neglect of the higher order terms in the linearization procedure leading to Equations (2.12), the extended form of the sequential estimation algorithm is used. The method is given a complete discussion in Jazwinski (1969) where it is referred to as the extended Kalman-Bucy filter. The Extended Sequential Estimation algorithm is obtained by nothing that the best estimate of the true state of the vehicle is given as

$$\hat{X}_k = X_k^* + \hat{x}_k . \tag{5.14}$$

The effects of the errors introduced in the linearization process will be minimized when x_k is as small as possible. Since the true state is related to the nominal state by

$$X_k = X_k^* + x_k \tag{5.15}$$

and since that tacit assumption is made that $\hat{X}_k$ will be nearer to X_k than X_k^*, the nominal trajectory can be rectified at each observation point by using the estimate of the state as the new nominal for propogating the estimate forward. That is, assume that

$$X(t) = \bar{X}(t) + x(t), \qquad t_k \leqslant t \leqslant t_{k+1} \tag{5.16}$$

where the reference solution is the a priori mean solution obtained by integrating the following differential equations

$$\dot{\bar{X}} = F(\bar{X}, t); \qquad \bar{X}(t_k) = X_k^* + \hat{x}_k . \tag{5.17}$$

Since the estimate of the state deviation, $\hat{x}_k$, is included in the initial conditions $\bar{X}_k$, it follows that the initial condition for mapping forward the state deviation, i.e., $\bar{x}_{k+1} = \Phi(t_{k+1}, t_k)\, x_k$, will be zero. Hence, this relation will lead to

$$\bar{x}_{k+1} = 0, \qquad t_{k+1} > t_k . \tag{5.18}$$

Furthermore, the a priori observation residual becomes

$$\bar{r}_k = y_k - \tilde{H}_k \bar{x}_k . \tag{5.19}$$

In view of Equation (2.19) and (5.18), Equation (5.19) becomes

$$r_k = Y_k - G(\bar{X}_k, t_k). \tag{5.20}$$

Since the covariance relation represents the growth of the covariance matrix associated with the estimate at time t_k, it follows that $\bar{P}_{k+1}$ will be governed by the same equations as for the sequential case, i.e., Equation (3.17).

With these assumptions, the computational algorithm for the Extended Sequential Estimation procedure can be summarized as follows:

Given: $\bar{X}_k = X_k^* + \hat{x}_k$ and P_k.

(1) Integrate to t_{k+1}

$$\begin{aligned} &\dot{\bar{X}} = F(\bar{X}, t), \qquad \bar{X}(t_k) = \hat{X}_k \\ &\dot{\Phi}(t, t_k) = A(t)\, \Phi(t, t_k), \qquad \Phi(t_k, t_k) = I . \end{aligned} \tag{5.21}$$

(2) Obtain the observation Y_{k+1} and calculate

$$\begin{aligned} H_{k+1} &= [\partial G(\bar{X}_{k+1}, t_{k+1})/\partial X] \\ \bar{P}_{k+1} &= \Phi(t_{k+1}, t_k)\, P_k \Phi^T(t_{k+1}, t_k) \\ y_{k+1} &= Y_{k+1} - G(\bar{X}_{k+1}, t_{k+1}). \end{aligned} \tag{5.22}$$

(3) Calculate:

$$\begin{aligned} K_{k+1} &= \bar{P}_{k+1} H_{k+1}^T [H_{k+1} \bar{P}_{k+1} H_{k+1}^T + R_{k+1}]^{-1}. \\ P_{k+1} &= [I - K_{k+1} H_{k+1}]\, \bar{P}_{k+1} \\ \hat{X}_{k+1} &= \bar{X}_{k+1} + K_{k+1} [Y_{k+1} - G(\bar{X}_{k+1}, t_{k+1})]. \end{aligned} \tag{5.23}$$

(4) Replace $k+1 \rightarrow k+2$ and return to (1).

Note that in the estimate of X_{k+1}, i.e., $\hat{X}_{k+1}$, the difference between the observation and the a priori estimate is used. Compare this with the linearized observation residual, $\bar{r}_{k+1} = Y_{k+1} - G(X^*_{k+1}\; t_{k+1}) - \tilde{H}_{k+1}\bar{x}_{k+1}$ which is used in the standard sequential estimation algorithm. It is obvious that the linearization effects are minimized if the extended form of the estimation algorithm is used, since if X^*_k is replaced by $\bar{X}_k$, then $\bar{x}_k$ will be zero and $\bar{x}_{k+1}$ will be zero. Furthermore, the errors involved in propagating the state forward from t_k to t_{k+1} will be minimized also, since the propagation of $\bar{x}_k$ is incorporated in the integration of the non-linear equations used to determine the reference solution, i.e., Equations (5.23).

C. STATE NOISE COMPENSATION ALGORITHM

In addition to the effects of the non-linearities, the effects of errors in the dynamical model can lead to divergence in the estimate. See, for example, the discussion in Schlee *et al.* (1967). As pointed out previously, for a sufficiently large number of observations the value of the covariance matrix $\bar{P}_k$ will asymptotically approach zero and the estimation algorithm will be insensitive to any further observations. One approach to eliminating this divergence is to recognize that the linearized equations for propagating the estimate of the state are in error. The first of Equations (2.12) is replaced by the following system of linear equations

$$\dot{x} = A(t)\, x + m(t) \tag{5.24}$$

where $m(t)$ is an n-vector which represents the error in the linear system equations which describe the state-deviation, $x(t)$. The vector, $m(t)$, may include such effects as determanistic model errors, numerical integration errors due to truncation and/or roundoff or random process-noise. In the frequently used approach for correcting for the errors due to $m(t)$. the matrix $m(t)$ is approximated by the following relation:

$$m(t) = u(t) \tag{5.25}$$

where $u(t)$ is a random process which satisfies the following a priori statistics

$$E[u(t)] = 0 \qquad E[u(t)\, u^T(\tau)] = Q(t)\, \delta(t - \tau) \tag{5.26}$$

where $\delta(t-\tau)=0$, for $t \neq \tau$ and $\int_{\tau-}^{\tau+} \delta(t-\tau)\mathrm{d}t = 1$. Using Equation (5.26), the equation for propagating the a priori estimate of x_k, $\bar{x}_k = E[x_k \mid y_1, \ldots, y_{k-1}]$ can be expressed as

$$\dot{\bar{x}} = A(t)\,\bar{x} \tag{5.27}$$

which is the same result as that obtained with the first of Equations (2.12). However, the equation for the covariance matrix will not be the same. Recall that

$$\bar{P}(t) = E[(x(t) - \bar{x}(t))(\bar{x}(t) - \bar{x}(t))^T \mid y_1 \ldots y_{k-1}],\ t > t_{k-1}. \tag{5.28}$$

Differentiating this quantity leads to

$$\dot{\bar{P}} = E[\Delta x\,\Delta\dot{x}^T + \Delta\dot{x}\,\Delta x^T \mid y_1, \ldots, y_{k-1}], \qquad t > t_{k-1} \tag{5.29}$$

where $\Delta x = x - \bar{x}$. Now from Equation (5.24) and Equation (5.27),

$$\Delta\dot{x} = A(t)\,\Delta x + u(t). \tag{5.30}$$

Substituting (5.30) in Equation (5.29) and using the result (Kalman and Bucy, 1961) that

$$E[\Delta x(t)\,u^T(t)] = Q(t)/2 \tag{5.31}$$

leads to the following differential equation for propagation of the state estimate covariance matrix.

$$\dot{\bar{P}} = A(t)\,\bar{P} + \bar{P}A^T(t) + Q(t). \tag{5.33}$$

Equation (5.33) is an $n \times n$ matrix Riccati equation, whose solution must be integrated with the initial conditions $\bar{P}(t_k) = P_k$ to obtain $\bar{P}_{k+1}$. As an alternative to integrating Equation (5.33), the following integral form can be used.

$$\bar{P}_{k+1} = \Phi(t_{k+1}, t_k)\,P_k\Phi^T(t_{k+1}, t_k) + \int_{t_k}^{t_{k+1}} \Phi(t, \tau)\,Q(\tau)\,\Phi^T(t, \tau)\,d\tau \tag{5.34}$$

where the $2n \times n$ system of equations to be integrated can be expressed as

$$\begin{aligned} &\dot{\Phi}(t, t_k) = A(t)\,\Phi(t, t_k), \qquad \Phi(t_k, t_k) = I \\ &\dot{\Gamma}(t_{k+1}, t) = \Phi(t_{k+1}, t)\,Q(t)\,\Phi^T(t_{k+1}, t), \qquad \Gamma(t_{k+1}, t_{k+1}) = 0. \end{aligned} \tag{5.35}$$

Then, the solution to Equation (5.34) becomes

$$\bar{P}_{k+1} = \Phi(t_{k+1}, t_k)\,P_k\Phi(t_{k+1}, t_k) + \Gamma_k. \tag{5.36}$$

If the time interval $\Delta t = t_{k+1} - t_k$ is sufficiently small, then the quadrature represented by the second term in Equation (5.34) can be replaced by $\bar{Q} = Q_{avg}\Delta t$. In this case, only the first of Equations (5.35) must be integrated.

The comparison of Equation (5.33) with Equations (5.34) indicates the following.

(1) Since $P(t)$ is symmetric, only $n(n+1)/2$ of the $n \times n$ system of equations represented by Equation (5.33) must be integrated. However, the $[n(n+1)/2]$-equations are coupled and must be integrated as a single first order system of dimension $(n^2+n)/2$.

(2) The $2n \times n$ system represented by Equations (5.34) can be separated into an $n \times n$ system of differential equations and an $n \times n$ quadrature. Furthermore, the $n \times n$ system of equations represented by the first of Equation (5.34) can be integrated as a sequence of $n \times 1$ column vectors.

Hence, the comparison between the two methods indicates that fewer equations must be integrated in obtaining the solution for $\bar{P}(t)$ with Equation (5.33), but that the integration of these equations is more difficult than the integrations associated with the larger system represented by Equations (5.36).

D. THE Q-MATRIX ALGORITHM

Either Equation (5.33) or Equation (5.36) can be used with the previously developed algorithms to obtain a sequential estimation procedure in which the effects of state noise are accounted for by the inclusion of the Q-matrix. For instance, the Extended Sequential Estimation Algorithm given in Equations (5.21), (5.22) and (5.23) would be modified by replacing Equations (5.21) with the equations

$$\begin{aligned} \dot{\bar{X}} &= F(\bar{X}, t), \quad \bar{X}(t_k) = \hat{X}_k \\ \dot{\bar{P}} &= A(t) P + PA^T(t) + Q(t), \quad \bar{P}(t_k) = P_k \end{aligned} \tag{5.37}$$

and second of Equations (5.22) is eliminated if Equation (5.33) is used. The remaining equations are unchanged. If Equation (5.35) is used, Equation (5.21) is replaced with

$$\begin{aligned} \dot{\bar{X}} &= F(\bar{X}, t), \quad \bar{X}(t_k) = \hat{X}_k \\ \dot{\Phi}(t, t_k) &= A(t)\, \Phi(t, t_k), \quad \Phi(t_k, t_k) = I \\ \dot{\Gamma}(t_{k+1}, t) &= \Phi(t_{k+1}, t)\, Q(t)\, \Phi^T(t_{k+1}, t), \quad \Gamma(t_{k+1}, t_{k+1}) = 0 \end{aligned} \tag{5.38}$$

and the second of Equations (5.22) is replaced by Equation (5.36). The remaining equations are unchanged. If the assumption $\Gamma_k = Q_{a-g}$ can be used the last of Equations (5.38) can be omitted and the only change in Equations (5.21) through (5.23) will be the replacement of the second of Equations (5.22) with Equation (5.36).

The advantage of using the Q-matrix compensated sequential estimation algorithm lies in the fact that the asymptotic value of $\bar{P}(t)$ will approach a non-zero value determined by the magnitude of Q. That is, for certain values of $Q(t)$, the increase in the state error covariance matrix $\bar{P}(t)$ during the interval between observations will balance the decrease in the covariance matrix which occurs at the observation point; see the second of Equations (5.23). In this situation, the estimation procedure will always be sensitive to new observations and, in particular, large observation residuals will have an effect on the estimate of the state.

6. Estimation in the Presence of Unmodeled Accelerations

Precise determination of the orbit of a near-earth satellite is a primary requirement for the utilization of a satellite in many navigation, geodetic and oceanographic studies. The successful completion of the scientific aspects of the planetary exploration and lunar exploration missions relies also on such precise determination of the orbit. The determination and prediction of the orbit of a satellite in a near planetary gravitational field is a difficult task. For example, in the near earth environment the satellite is influenced by the non-spherical effects of the Earth's gravitational field, atmospheric resistance, solar radiation pressure, lunar solar perturbations, etc. Similar remarks could be made concerning the lunar satellite. In order to have an accurate determination of a near-Earth satellite's orbit, an accurate knowledge of the geopotential must be available. However, the determination of the coefficients of the geopotential by conventional least squares estimation procedures is limited by the effects af atmospheric drag. For some orbits, the specific effects of the atmospheric resistance are difficult to determine since the atmospheric density and hence the drag undergo large unmodeled fluctuations. Such fluctuations as well as the effects of other uncertainties in the mathematical model lead to a discrepancy between the model forces used to describe the motion of a near-Earth satellite and the actual forces to which it is subjected. The effects of these unmodeled forces ultimately lead to disagreement between the observed and the predicted motions. In some cases, the disagreement is sufficiently large to constitute a divergence of the estimation procedure. As a consequence, a consideration of the modification of the methods discussed in the previous sections for reducing observation data is required with the objective of developing an estimation theory which compensates for the effects of the unmodeled forces.

As pointed out in the previous sections, there are two basic procedures for processing data obtained during a tracking period to obtain an estimate of the orbit of a space vehicle. The methods have been referred to as sequential processors and batch processors. If the observations are made at discrete points in time, the sequential orbit determination procedure processes each observation at the time point at which it is received. The batch processor, on the other hand, will store the entire sequence of observations until the observation period is completed. Then the estimate of the trajectory at some reference epoch is obtained by processing all of the data simultaneously. There are advantages and disadvantages associated with each method. When the mathematical model gives a precise description of the motion, the sequential processor is best suited to the problem of determining the orbit in real time; while the batch processor is well suited to postflight data analysis for extremely large batches.

The nature of the errors which occur when the batch processor is used is discussed in Anderle *et al.* (1969); Gapcynski *et al.* (1968); Hamer and Johnson (1969) and Squires *et al.* (1969), while the errors which occur using the sequential processor are discussed in Schlee *et al.* (1967). The correction for the effects of unmodeled forces can be implemented most easily using the sequential estimation algorithm. The algorithms which account for noise in the state equations, i.e., the Q-matrix compensation algorithms discussed in the previous section are derived by adding a term to the matrix equation for propagating the covariance matrix associated with the state estimate (Potter and Dicker, 1970 and Sorenson, 1966). This method of accounting for the effects of the unmodeled accelerations suffers from two significant disadvantages: (1) The accuracy with which the estimate can be obtained depends on the value chosen for the additive covariance. If the term is too large, undue emphasis will be placed on the most recent observations. If the additive term is too small, the state estimate covariance matrix can become too small and the estimate can diverge. (2) Such an estimation algorithm does not yield any direct information concerning the values of the unmodeled accelerations. In Ingram (1971a, b); Tapley (1971, 1972), a method for estimating the state of a spacecraft in the presence of unmodeled accelerations is presented. The proposed method processes the data in a sequential fashion and has the advantage that, in addition to obtaining a more precise estimate of the state of the spacecraft, the values of the unmodeled acceleration are determined as a function of time. The method is statistically categorized as a first order auto regressive estimation procedure in which the unmodeled acceleration in each of the three components of the equations of motion is modeled as a first order Gauss-Markov Process (Feller, 1966). The correlation times are treated as unknown parameters and are estimated also during the orbit determination procedure. In Ingram (1971a, b) and Tapley (1971) the method is applied to the problem of estimating the unmodeled accelerations acting on a lunar satellite. Range rate data obtained during the Apollo 10 and 11 missions is processed and in some regions the range rate residuals are improved by a factor of approximately 10 over the residual obtained with the conventional NASA-MSC orbit determination procedure. The estimated values of the unmodeled accelerations obtained in this study are repeatable from revolution to revolution within a given mission and from mission to mission when the spacecraft covers the same ground track. In addition, the variations in the unmodeled acceleration show a high physical correlation with the location of reported lunar mascons (Sjogren *et al.*, 1971). The method is referred to as the dynamic model compensation (DMC) estimation algorithm and in Tapley and Hagar (1972), it is shown that the algorithm can be successfully applied to the problem of estimating purely random components of unmodeled accelerations such as those associated with noisy fluctuations in a low thrust space vehicle. In the following section, the assumptions used in developing this estimation algorithm, along with the results obtained in the application of the algorithm to the orbit determination data from the Apollo 10 and 11 missions are briefly reviewed. For more extensive discussion of the algorithm and of the applications, Ingram (1971a, b) and Tapley (1971, 1972) should be consulted.

A. THE DMC ESTIMATION ALGORITHM

The equations which describe the motion of a nonlinear dynamical system can be expressed as

$$\dot{r} = v, \qquad \dot{v} = a_m(r, v, t) + m(t) \tag{6.1}$$

where r is a 3-vector of position components, v is a 3-vector of velocity components, and $a_m(r, v, t)$ is a 3-vector of modeled acceleration components. The three-vector $m(t)$ designates the unmodeled acceleration components and it represents the effects of all accelerations which are not accounted for in the mathematical model used to describe the physical process. Some of these effects, such as those due to parametric uncertainties, can be handled by treating certain parameters in the modeled accelerations in Equations (6.1) as unknown, but constant, parameters and estimating their value during the orbit determination process. This procedure is discussed in Section 2. See, for example, Equation (2.2). However, in situations where a large number of such parameters are required, as in the description of the Earth or Lunar gravitational fields, this approach will lead to computational difficulties. See Anderle *et al.* (1969); Gapcynski *et al.* (1968); Hamer and Johnson (1969) and Squires *et al.* (1969). Other effects such as radiation pressure, translational forces due to unbalanced attitude jets, fluctuations in atmospheric drag, anomalies in propulsion systems, etc. depend on such a large number of parameters that they can be approximated best as a continuous dynamic process.

There have been a number of investigations which have considered the problem of state estimation in the presence of such unmodeled forces. In particular, the effects of errors in the mathematical model on the accuracy and stability of the linear sequential estimation procedures are discussed in Potter and Dicker (1970); Schlee *et al.* (1967) and Sorenson (1966). In (Schlee *et al.* 1967), the question of divergence of the estimation procedure due to errors in the dynamic model is considered, as there have been a number of methods proposed for alleviating the problem of estimation divergence. These methods include the addition of a constant matrix to the state error covariance matrix to account for noise in the differential equations which govern the state process (16), specification of a minimum bound on the estimation error covariance matrix (17) and the techniques of using a finite (10) or exponentially time-decaying (16) data set. While these methods improve the estimation accuracy in the presence of the unmodeled accelerations, they do not provide any direct information on the nature of the unmodeled forces.

In the following discussion, a sequential estimation method, which compensates for unmodeled effects in the differential equations which describe the motion of an orbiting vehicle, is described. The algorithm, referred to as the Dynamic Model Compensation (DMC) algorithm, has two advantages: (1) the compensation for the effects of the unmodeled accelerations will yield an improved estimate of the vehicle state in real time estimation problems and (2) the method will yield information which can be used in post-flight data analysis to improve the mathematical model. The 'unmodeled' accelerations, $m(t)$ in Equations (6.1), are assumed to consist of the superposition of a time correlated component and a purely random component. In developing the estimation algorithm, the unmodeled acceleration, $m(t)$ is approximated as an adaptive first order Gauss-Markov process, $\varepsilon(t)$, which satisfies the following vector differential equation

$$\dot{\varepsilon}(t) = B\varepsilon(t) + u(t) \tag{6.2}$$

where $\varepsilon(t)$ is a three-vector and $u(t)$ is a 3-vector of Gaussian noise whose components are assumed to be described by the a priori statistics:

$$E[u(t)] = 0 \qquad E[u(t)\, u^T(\tau)] = q(t)\, \delta(t - \tau). \tag{6.3}$$

The coefficient matrix, B, is defined by the components $B_{ij} = \beta_i \delta_{ij}$ $(i, j = 1, 2, 3)$

where the β_i are assumed to be unknown parameters whose values are to be determined during the estimation process by combining the equation $\dot{\beta}=0$ where $\beta^T=[\beta_1\beta_2\beta_3]$ with Equation (6.2). If $\varepsilon(t)$ is substituted for $m(t)$ in Equation (6.1) and if the state vector $X(t)$ is defined as

$$X^T = [r^T : v^T : \varepsilon^T : \beta^T] \tag{6.4}$$

then Equations (6.1), (6.2) and the relations $\dot{\beta}=0$ can be combined to obtain the following differential equations for the state:

$$\dot{X} = F(X, u, t), \quad X(t_0) = X_0 \tag{6.5}$$

where $F^T=[v^T:(a+\varepsilon)^T:(B\varepsilon+u)^T:0]$ and where the initial conditions, X_0, are unknown.

For $t>t_i$, where t_i is some reference epoch, the solution to Equations (6.5) can be expressed in integral form as follows (Ingram, 1971a):

$$\begin{aligned} r(t) &= r_i + v_i\Delta t + \int_{t_i}^{t} a(\tau)[t-\tau]\,\mathrm{d}\tau \\ v(t) &= v_i\Delta t + \int_{t_i}^{t} a(\tau)\,\mathrm{d}\tau \\ \varepsilon(t) &= E(t)\,\varepsilon_i + l_i \\ \beta(t) &= \beta_i \end{aligned} \tag{6.6}$$

where $\Delta t=t-t_i$ and $a(t)=a_m(t)+\varepsilon(t)$. The matrices $E(t)$ and l_i are defined as

$$E(t) = \begin{bmatrix} \alpha_x & 0 & 0 \\ 0 & \alpha_y & 0 \\ 0 & 0 & \alpha_z \end{bmatrix} \tag{6.7}$$

$$l_i^T = [\sigma_{x_i}\sqrt{1-\alpha_{x_i}^2}u_{x_i} \,\vdots\, \sigma_{y_i}\sqrt{1-\alpha_{y_i}^2}u_{y_i} \,\vdots\, \sigma_{z_i}\sqrt{1-\alpha_{z_i}^2}u_{z_i}] \tag{6.8}$$

where $q_x=\sigma_x\sqrt{T_x/2\alpha_x}=\exp[-(t-t_i)/T_x]$ with similar definitions for σ_y, σ_z, α_y and α_z.

Using the definition in Equation (6.4), the solution to Equations (6.5) can be expressed as

$$X(t) = \theta(X_i, t_i, t) + \eta_i, \quad t \geqslant t_i \tag{6.9}$$

where $\eta^T=[\eta_r^T:\eta_v^T:\eta_\varepsilon^T:0]$. The components of the state noise matrix, η_i, are due to the purely random components of the unmodeled accelerations and can be defined as follows:

$$\eta_r = \int_{t_i}^{t} l_i(t-\tau)\,\mathrm{d}\tau, \quad \eta_v = \int_{t_i}^{t} l_i\,\mathrm{d}\tau, \quad \eta_\varepsilon = l_i. \tag{6.10}$$

In view of Equations (6.3), the random process η_i will satisfy the conditions

$$E[\eta_i] = 0 \qquad E[\eta_i \eta_j]^T = Q_i \delta_{ij} \tag{6.11}$$

where δ_{ij} is the Kronecker delta. Equation (6.9) is used to propagate the state vector from an observation point t_i to an observation point t_j.

The relationship between the p-dimensional observation vector, Y_i, the p-vector of observation noise, v_i, and the state at the time, t_i, is

$$Y_i = G(X_i, t_i) + v_i . \tag{6.12}$$

In the following discussion, it is assumed that the observation noise satisfies the following conditions:

$$E[v_i] = 0, \quad E[v_i v_j]^T = R_i \delta_{ij}, \quad E[v_i X_j^T] = 0. \tag{6.13}$$

The problem considered then is posed as follows: Given the relation for propagating the state, Equation (6.6), the observation state relation, Equation (6.12), the sequence of observations Y_i, $i = 1, \ldots, k$, and the statistics on the state noise, Equation (6.3), and the observation noise, Equation (6.13), find the best estimate, in the minimum variance sense, of the state, $\hat{X}_k$, at the time t_k.

Under the conditions given in the problem statement, the estimate at time, t_k, can be obtained using the extended sequential estimation algorithm given by Equations (5.21) through (5.23) where the Q-matrix modification discussed in either Equation (5.37) or (5.38) is used. The application of the Dynamic Model Compensation Algorithm to the problem of estimating the motion of a lunar satellite is discussed in the following section.

B. THE DETERMINATION OF THE ORBIT OF A LUNAR SATELLITE

The equations which describe the motion of a lunar satellite can be expressed as

$$\ddot{\bar{r}} = \bar{a}_c + \bar{a}_N + \bar{m} \tag{6.14}$$

where the 3×1-matrices, $\bar{a}_c$, $\bar{a}_N$ and $\bar{m}$, represent the acceleration due to the central body, the perturbing acceleration due to the N-bodies included in the mathematical model and the acceleration due to the unmodeled forces, respectively.

For a lunar satellite, the central body acceleration is due to the modeled gravitational effects of the Moon. The Lunar gravitational potential can be expressed as

$$V = \frac{-\mu_m}{r} \sum_{n=0}^{\infty} \sum_{m=0}^{n} \left(\frac{R_p}{r}\right)^n P_{nm}(\sin\phi)\left[C_{nm}\cos m\lambda + S_{nm}\sin m\lambda\right] \tag{6.15}$$

where μ_m is the gravitational parameter for the Moon, R_p is the equatorial radius of the Moon, $P_{nm}(\sin\phi)$ is the Legendre associated function and C_{nm} and S_{nm} are coefficients which represent the mass distribution of the Moon. Furthermore, r, ϕ and λ are selenographic spherical coordinates where r is the radial distance from the origin of a selenographic coordinate system, λ is the selenographic longitude measured clockwise from the x-axis in the lunar equatorial plane and ϕ is the selenographic latitude, measured positive toward the positive z-axis. The positive x-axis coincides with the zero-selenographic longitude and the x-y plane coincides with the lunar equatorial plane. The components of the central body acceleration can be expressed, then, as

$$\bar{a}_c = \tilde{R} \begin{bmatrix} -(1/r)\, \partial V/\partial r \\ -(1/r \cos\phi)\, \partial V/\partial\lambda \\ -\partial V/\partial\phi \end{bmatrix} \tag{6.16}$$

where R is the matrix which transforms the acceleration from the rotating body-fixed selenocentric system to a Mean of 1950 system. The rotation matrix, $\tilde{R}$, includes the effects of the lunar libration and sway as well as the effects of the precession and nutation of the Earth-Moon orbit plane (de Sulima, 1970). The position of the Moon with respect to the Earth and the Sun, as well as its orientation with respect to the Mean of 1950 inertial coordinate system, is obtained from the JPL DE Ephemeris tape (Devine, 1967).

In the investigation described here, the only perturbing effects included, in addition to the lunar effects, are the point mass effects of the Earth and the Sun. Hence, $\bar{a}_n$ in Equation (6.15) is modeled as

$$\bar{a}_N = -\mu_e\left[\frac{\bar{r}-\bar{r}_e}{|\bar{r}-\bar{r}_e|^3} + \frac{\bar{r}_e}{r_e^3}\right] - \mu_s\left[\frac{\bar{r}-\bar{r}_s}{|\bar{r}-\bar{r}_s|^3} - \frac{\bar{r}_s}{r_s^3}\right] \tag{6.17}$$

where μ_e and μ_s are the gravitational constants of the Earth and the Sun, respectively, and $\bar{r}_e$ and $\bar{r}_s$ are the position vectors of the Earth and Sun with respect to the Moon in the Mean of 1950 reference system.

The unmodeled acceleration will be estimated in the Mean of 1950 system; hence, it can be described as an additive term in Equation (1) which can be expressed as follows:

$$\dot{\bar{r}} = \bar{v}, \qquad \dot{\bar{v}} = \bar{a}_c + \bar{a}_N + \bar{m} \tag{6.18}$$

where $\bar{a}_c$ and $\bar{a}_N$ are obtained from Equations (6.16) and (6.17) respectively.

In the study described here, the potential in Equation (6.16) was truncated to include numerical values for only the $C_{2,0}$, $C_{2,2}$, $C_{3,0}$, $C_{3,1}$ and $C_{3,3}$. The potential model resulting from this approximation is designated by the NASA Manned Spacecraft Center as LM-1 (Wollenhaupt, 1969) and the values adopted for the harmonic coefficients are: $C_{2,0} = -2.07108 \times 10^{-4}$, $C_{2,2} = 0.20716 \times 10^{-4}$, $C_{3,0} = 0.21 \times 10^{-4}$, $C_{3,1} = 0.34 \times 10^{-4}$, and $C_{3,3} = 0.02583 \times 10^{-4}$. The coefficients in the LM-1 model have been specified to obtain optimal results when tracking data from an Apollo-type Lunar satellite is processed with a batch processor (Wollenhaupt, 1969). In view of the limited number of coefficients included in the Lunar potential, it is reasonable to expect that the unmodeled acceleration terms, $\bar{m}(t)$, will be due predominately to the neglected effects of the Lunar gravitational potential. In the following section, results obtained by using the algorithm to process tracking data obtained during the Apollo 10 and 11 missions are discussed.

C. APPLICATION TO THE APOLLO 10 AND 11 MISSIONS

This section examines the results which were obtained by processing Apollo 10 and 11 doppler data. The method of generating the doppler data is described in Ingram (1971a) and de Sulima (1970). The numerical results obtained by estimating position, velocity and the unknown accelerations are compared with the results obtained by using a batch processor. The batch processing technique is a standard weighted least squares method and is described in de Sulima (1970).

The Apollo 10 and 11 data tape and the results obtained by using a batch processor were supplied by the NASA Manned Spacecraft Center. The data tape contains doppler data which is averaged over a 6 second time interval. As a result, the observation noise is about 0.15 cps. In order that the comparison of the residual pattern obtained with the Dynamic Model Compensation sequential processor and the residual obtained with the NASA MSC batch processor be meaningful, it is necessary that given the same initial state vector, the two programs generate essentially the same numerical value for a doppler observation. The two programs use the same method of computing the doppler observable and the same ephemeris tape, JPL DE 19. Numerical experiments have shown that the two programs generate the same residual pattern to within 0.002 cps. At each observation time, there are between 4 and 6 stations simultaneously tracking the satellite. It requires about 70 min for a lunar satellite to travel across the front face of the Moon. During this time, approximately 700 vectors of 5 observations each are obtained, for a total of approximately 3500 observations.

D. RESIDUAL PATTERNS

As a means of evaluating the DMC algorithm, it is instructive to compare the residual curve obtained using a batch processor with the results obtained using the DMC algorithm. The residual computed by the batch processor corresponds to the difference between the observed and computed values of the doppler data after the batch processor has iterated to convergence. The residual curve computed by the DMC algorithm is a predicted residual. That is, it is the residual which results from propagating the best estimate of the state vector to an observation time and computing the observation without including the observation obtained at that time. The residual associated with the batch processor corresponds to the residual computed by including the entire sequence of observations. Unless stated otherwise the initial state vector for the SPSN corresponds to the converged solution of the batch processor.

Figure 1 compares the residual curves of the SPSN and the batch processor for the Merritt Island (MIL) station for the first 8 min of Revolution 5 of the Apollo 11 mission. The characteristics illustrated in Figure 1 are reasonable representative of the residual curve for all revolutions. The batch processor has segments of the revolution over which its residual pattern is definitely periodic. To a lesser extent, the results obtained by the SPSN also manifests this characteristic.

The residual curve for Revolution 6 of the Apollo 11 Mission is plotted in Figure 1

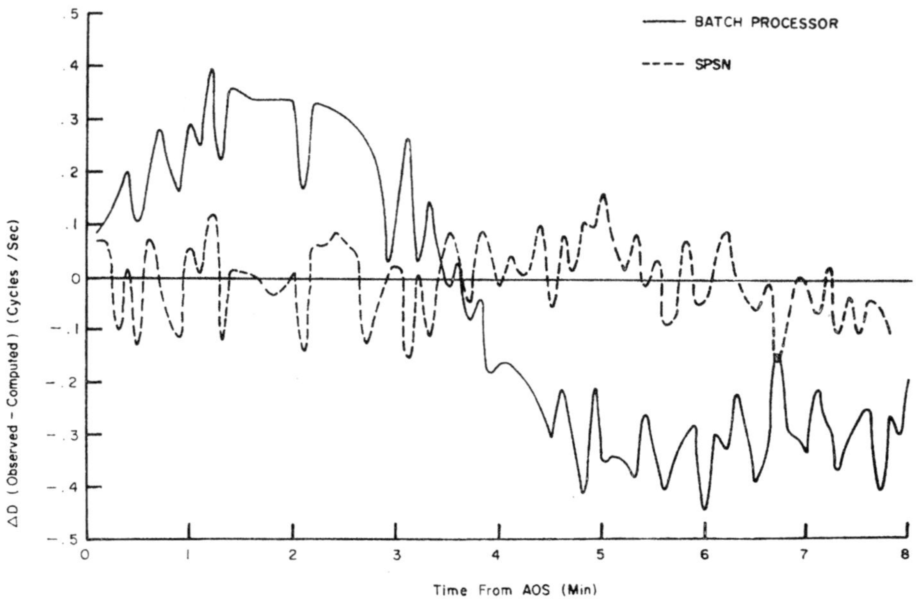

Fig. 1. Doppler residual comparison – Apollo 11, Rev. 5, Merritt Island Tracking Station.

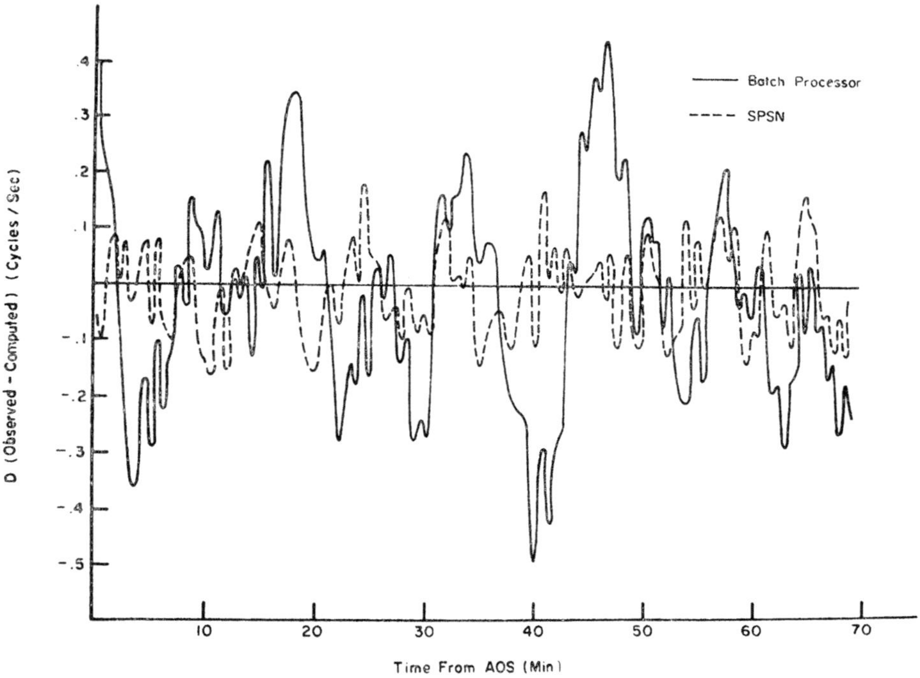

Fig. 2. Doppler residual comparison – Apollo 11, Rev. 6, Honeysuckle Tracking Station.

where the abbreviation AOS stands for acquisition of signal. It should be noted that every fifth point associated with the Honeysuckle (HSK) station is plotted. The points are connected to indicate the nature of the curve. The solid line represents results obtained with the batch processor while the dashed line represents the SPSN results. From an examination of Figure 2, it is clear that the residual curve obtained with the SPSN has an amplitude which is much less than that obtained with the batch processor. In fact, the amplitude of the resudial pattern for the batch processor peaks at about 0.45 cps while the maximum amplitude associated with SPSN is about 0.15 cps.

E. UNMODELED ACCELERATIONS

The unmodeled accelerations are estimated in a Mean of 1950 selenocentric inertial coordinate system. However, the estimated values are interpreted, most easily, in a spherical coordinate system which has unit vectors $\bar{\varepsilon}_r$ in the radial direction, $\bar{\varepsilon}_\lambda$ in the positive longitude direction and $\bar{\varepsilon}_\phi$ in the positive latitude direction. In Figure 3, the radial components of the unmodeled accelerations for Revolutions 6 and 7 in the Apollo 11 mission are shown. Figure 3 indicates that the estimated values of the

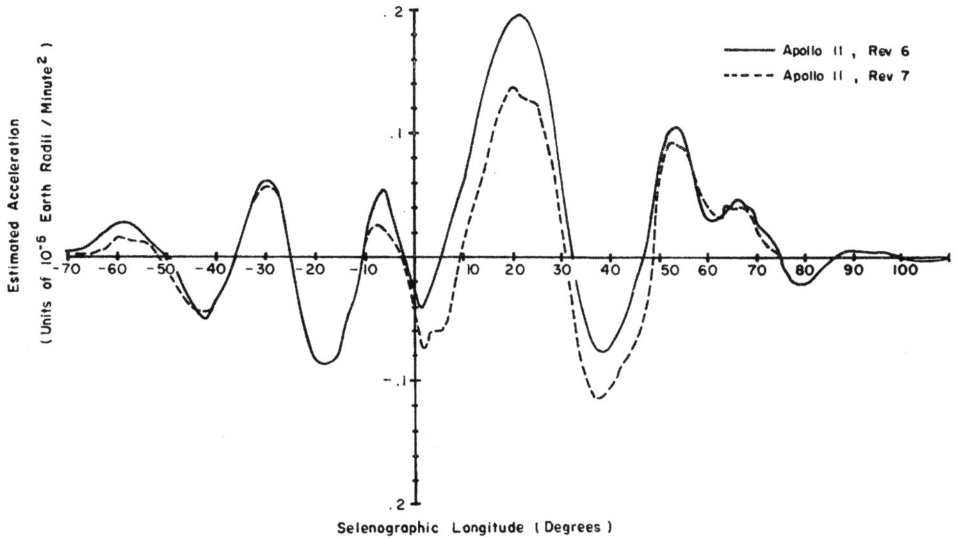

Fig. 3. Comparison of radial accelerations – Apollo 11, Revs. 6 and 7.

unmodeled accelerations are repeatable from revolution to revolution within a given mission. It is further shown in Ingram (1971a, b) and Tapley (1971) that the estimated values of the unmodeled accelerations are repeatable from mission to mission when the same ground track is covered. Finally, the peaks in the radial accelerations can be correlated with the reported location of lunar mascons (Sjogren *et al.*, 1971). In particular, the accelerations peaks at $-30°$, $20°$, $40°$, $50°$ and $60°$ selenographic longitude can be correlated with the locations of the mascons shown in Figure 4. The data shown in Fig. 4 is abstracted from the map of lunar mascons reported in Sjogren *et al.* (1971).

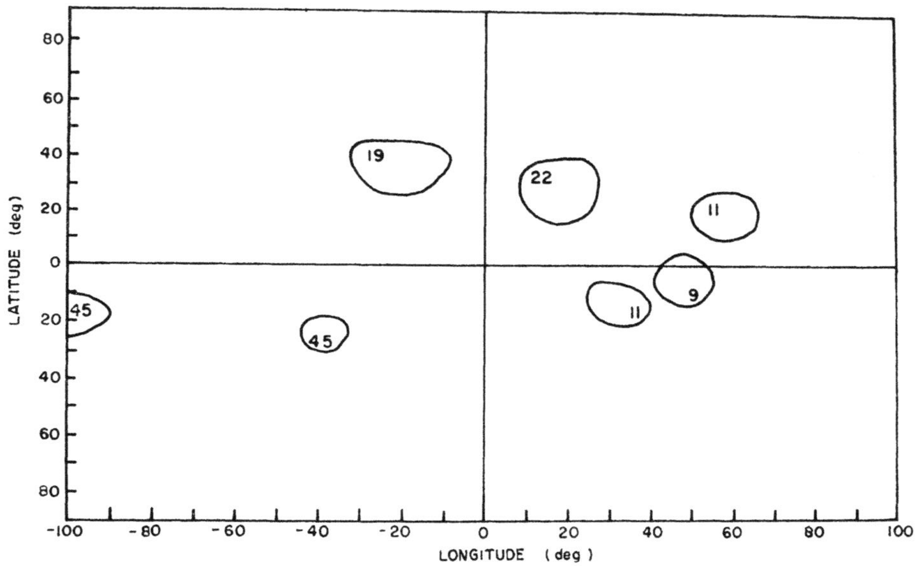

Fig. 4. Location of Lunar mascons.

Based on the numerical results presented in the previous section, it is concluded that the DMC algorithm provides a significant improvement over the conventional batch or sequential method of estimating the state in the presence of unmodeled accelerations.

7. Summary

The previous sections have reviewed the algorithms for determining the orbit of a space vehicle. The relationship between the batch and sequential estimation algorithms is given and the topic of divergence of the sequential estimation algorithm is considered. The concept of dynamic model compensation (DMC) is introduced and the extended sequential estimation algorithm is used to obtain an estimation procedure for the determination of the orbit of a lunar satellite in the presence of unmodeled accelerations. Finally, results obtained during the reduction of tracking data from the Apollo 10 and 11 missions are described. It is determined that the DMC algorithms give a more accurate estimate of the state than the conventional batch or sequential estimation algorithms. In addition, examination of the numerical results leads to the conclusion that the estimated values of the unmodeled accelerations are repeatable from orbit to orbit within a given mission and from mission to mission when the same ground track is covered. Finally, the variations in the magnitude of the radial component of the unmodeled acceleration shows a high correlation with the reported location of Lunar surface mascons.

Acknowledgement

This investigation was supplied by AFOSR under Grant No. 72-2233.

References

Anderle, R. J., Malyevac, P. A., and Green, H. L. Jr.,: 1969, 'Effect of Neglected Gravity Coefficients on Computed Satellite Orbits and Geodetic Parameters', *AIAA/AGU Symposium on Astrodynamics and Related Planetary Sciences*, Washington, D.C., April 21–25.

Businger, P. and Golub, G. H.: 1968, *Numerische Mathematik* 7.

Coddington, E. A. and Levinson, N.: 1955, *Theory of Ordinary Differential Equations*, McGraw-Hill.

de Sulima, T. H.: 1970, 'Houston Operations Predictor/Estimator (HOPE) Engineering Manual', TRW Note No. 70-FMT-792, Houston, Texas.

Devine, C. J.: 1967, 'JPL Development Ephemeris Number 19', Technical Report 32-1181, Je Propulsion Laboratory, Pasadena, California.

Fadeeva, V. N.: 1959, *Computational Methods of Linear Algebra*, Dover Pub., New York, N.Y., pp. 81–84.

Feller, W.: 1966, *An Introduction to Probability Theory and Its Applications*, Vol. II, John Wiley and Sons, New York, pp. 95–98.

Gapcynski, J. P., Blackshear, W. T., and Campton, H. R.: 1968, 'The Lunar Gravitational Field as Determined from the Tracking Data of the Lunar Orbiter Series of Spacecraft', *AAS/AIAA Astrodynamics Specialist Conference*, AAS Paper No. 68–132, Jackson, Wyoming, September.

Hamer, H. A. and Johnson, K. G.: 1969, 'Effect of Gravitational-Model Selection on the Accuracy of Lunar Orbit Determination from Short Data Arcs', NASA TN D-5105, Langley Research Center, March.

Householder, A. S.: 1958, *Assoc. Comp. Mech.* **5**, 339–342.

Ingram, D. S.: 1971, 'Orbit Determination in the Presence of Unmodeled Accelerations', Applied Mechanics Research Laboratory, Report No. AMRL-1022, The University of Texas at Austin (Also, Ph.D. Dissertation).

Ingram, D. S. and Tapley, B. D.: 1971b, 'Lunar Orbit Determination in the Presence of Unmodeled Accelerations', *AAS/AIAA Astrodynamics Specialist Conference* 1971, AAS Paper No. 71–37.

Jazwinski, A. H.: 1969, *Stochastic Process and Filtering Theory*, Academic Press, New York, N.Y.

Kalman, R. E. and Bucy, R. S.: 1961, *J. Basic Eng.* **83D**, 95–108.

Kamiski, P. G., Bryson, A. E., and Schmidt, S. F.: 1971, 'Discrete Square Root Filtering: A Survey of Current Techniques', IEEE Trans. on Automatic Control, Vol. AC-16, No. 6, pp. 727–735.

Lewis, T. V. and Odell, P. L.: 1971, *Estimation in Linear Models*, Prentice Hall.

Liebelt, P. B.: 1967, *An Introduction to Optimal Estimation*, Addison, Wesley.

Potter, J. E. and Dicker, J. C.: 1970, *SIAM Journal of Control* **8**, 513–526.

Schlee, S. F., Standish, C. J., and Toda, N. F.: 1967, *AIAA J.* **5**, 1114–1120.

Sjogren, W. L., *et al.*: 1971, *Moon* **2**, 338–353.

Sorenson, H. W.: 1966, in C. T. Leondes (ed.) *Adv. Control Systems* Vol. **3**, 219–292.

Squires, R. K., *et al.*: 1969, 'Response of Orbit Determination Systems to Model Errors', C-643-69-503, Goddard Space Flight Center, Greenbelt, Maryland.

Swerling, P.: 1959, *J. Astronaut Sci.* **6**, 46–52.

Tapley, B. D. and Hagar, Jr., H.: 1972, 'Navigation Strategy and Filter Design for Solar Electric Missions', Applied Mechanics Research Laboratory, Tech. Report AMRL-1040.

Tapley, B. D. and Ingram, D. S.: 'Orbit Determination in the Presence of Unmodeled Accelerations'. Proceedings of the Second Symposium on Nonlinear Estimation Theory, San Diego, California.

Wollenhaupt, W. R.: 1969, 'Lunar Gravitational Model for Apollo 12', MSC Internal Note 69-FM49-323, NASA/MSC, Houston, Texas.

ORBIT DETERMINATION BY SOLVING A BOUNDARY VALUE PROBLEM

M. SCHNEIDER

Institute for Astronomical and Physical Geodesy, Technical University Munich, Munich, F.R.G.

Abstract. Using Hammersteins method for solving nonlinear integral equations a procedure for a combined preliminary and definitive orbit determination of artificial earth satellites has been developed. A computer programme has been written which can at present use photographic and distance measurements. Orbit determination is done with respect to an inertial frame of reference. The usual reductions (polar motion, parallactic refraction etc.) can be applied. The model of the force field is now restricted to the gravitational field of the Earth (spherical harmonics expansion up to any order and degree), but will be extended to include air drag (CIRA 72), lunisolar attraction, radiation pressure and major geodynamical effects.

1. Introduction

Orbit determination plays an important part in the dynamical use of artificial satellites, i.e. in the determination of field parameters by analyzing orbital arcs. A study group formed at the Technical University of Munich is mainly interested in the determination of parameters scaling the Earth's gravitational field. For this purpose a series of orbit determination programmes has been developed which allow the transformation of observations of geodetically useful earth satellites into empirical orbits in a suitably chosen frame of reference. These empirical orbits form the basis of the dynamical use outlined in Sigl *et al.* (1970).

Here the very first part of a combined preliminary/definitive orbit determination procedure is dealt with.

2. Principle of the Orbit Determination Procedure

The orbit determination procedure to be given might be characterized as an extended Gaussian one. This is due to the fact that the orbit determination problem is basically formulated as a (two-point-) boundary-value problem.

The motion of a satellite (mass point m_s) with respect to an inertial reference frame is governed by the equation of motion

$$m_s \frac{d^2\mathfrak{r}}{dt^2} = \mathfrak{F}(t; \mathfrak{r}, \dot{\mathfrak{r}}), \tag{1}$$

where $\mathfrak{r}$ denotes the position vector of m_s, $\mathfrak{F}$ is the resulting force acting upon the satellite and t is the time.

A solution $\mathfrak{r}=\mathfrak{r}(t)$ of the equation of motion (1) satisfying the boundary values

$$\begin{aligned} t_A &: \mathfrak{r}_A = \mathfrak{r}(t_A) \\ t_B &: \mathfrak{r}_B = \mathfrak{r}(t_B) \end{aligned} \tag{2}$$

B. D. Tapley and V. Szebehely (eds.), Recent Advances in Dynamical Astronomy, 426–428.

can be given by first transforming the (two-point-) boundary-value-problem (1), (2) into the integral equation

$$\mathfrak{r}(t) = \bar{\mathfrak{r}}(t) - \frac{T^2}{m_s} \int_0^1 K^I(t, \tau)\, \mathfrak{F}(\tau; \mathfrak{r}, \dot{\mathfrak{r}})\, \mathrm{d}\tau ,$$

where time counting is normalized to $t_A = 0 \leqslant t, \tau \leqslant t_B = 1$ ($T = t_B - t_A$ serves as the time unit) and where

$$\bar{\mathfrak{r}}(t) = \mathfrak{r}_A + (\mathfrak{r}_B - \mathfrak{r}_A)\, t \tag{3}$$

and $K^I(t, \tau)$ is the kernel of the integral equation (see Schneider, 1972a).

This integral equation can be solved by using a method proposed by Hammerstein. The basic idea is to expand the solution to the eigenfunctions $\bar{\varphi}_\nu^I(t)$ of the kernel $K^I(t, \tau)$, i.e.

$$\mathfrak{r}(t) = \bar{\mathfrak{r}}(t) + \sum_{\nu=1}^{\infty} \mathfrak{c}_\nu \bar{\varphi}_\nu^I(t) \qquad (0 \leqslant t \leqslant 1) \tag{4}$$

and to determine the coefficients $\mathfrak{c}_\nu$ from the condition equations, given in Schneider (1972a),

$$\mathfrak{c}_\nu = -\frac{T^2}{\lambda_\nu} \int_0^1 \bar{\varphi}_\nu^I(\tau)\, \frac{\mathfrak{F}(\tau, \mathfrak{r}, \dot{\mathfrak{r}})}{m_s}\, \mathrm{d}\tau \qquad (\nu = 1, 2, \ldots \infty) \tag{5}$$

where λ_ν are the eigenvalues of the kernel. The righthand sides of these equations are functions of the coefficients $\mathfrak{c}_1, \mathfrak{c}_2, \ldots$ due to the dependence of $\mathfrak{F}$ on $\mathfrak{r}$ and $\dot{\mathfrak{r}}$.

The representation of the solution $\mathfrak{r}(t)$ in the form (4) is equivalent to an orbital theory which is related to the observations through the geometrical relation

$$\mathfrak{q}(t) = \mathfrak{r}(t) - \mathfrak{R}(t), \tag{6}$$

where $\mathfrak{q}(t)$ is the topocentric position vector joining the observer (at time t at the inertial position $\mathfrak{R}$) and the satellite at time t.

An orbit determination programme can now be established by combining the observation Equations (6) for all observations obtained at times $t_i (t_A \leqslant t_i (i = 1, 2, \ldots k) \leqslant t_B)$ and Equations (4) and (5). Details of this approach can be found in Ilk (1972), where the results of various applications are also given.

3. Concluding Remarks

An interesting feature of the outlined orbit determination procedure might be mentioned here. One is not restricted to a special force field $\mathfrak{F}$ as in many of the Gaussian approaches.

It is possible to use any model for $\mathfrak{F}$. Also it is not necessary to have an orbital theory available. The expansion (4) is due to a theorem by Mercer a uniformly converging representation of the solution $\mathfrak{r}(t)$in the time interval $[t_A, t_B]$ to be analyzed.

The result of an orbit determination following these lines is a completely defined expansion (4). If one is interested in the velocity at some time t_k with $t_A \leqslant t_k \leqslant t_B$ one can successfully apply a procedure given in Schneider (1972b).

References

Ilk, K. H.: 1972, 'Zur vorläufigen Bahnbestimmung künstlicher Erdsatelliten', BMBW-FB W 72–16 Munich.

Sigl, R. *et al.*: 1970, 'Anwendung der Hammersteinschen Methode der unendlich vielen Variablen auf Probleme der Satellitengeodäsie und Himmelsmechanik', BMBW-FB W 70–33, Munich.

Schneider, M.: 1972a, 'Determinierung von Bewegungsproblemen durch zeitliche Randwerte', BMBW-FB W 72–17, Munich.

Schneider, M.: 1972b, 'Über die Lösung von Anfangswertproblemen der Punktmechanik mit Hilfe von Randwertproblemen', BMBW-FB W 72–23, Munich.

LONG-TERM MOTION IN A RESTRICTED PROBLEM OF ROTATIONAL MOTION*

J. E. COCHRAN

Assistant Professor of Aerospace Engineering. Auburn University, Auburn, Ala.. U.S.A.

Abstract. A method of general perturbations, based on the use of Lie series to generate approximate canonical transformations, is applied to study the long-term effects of gravity-gradient torque and orbital evolution on the rotational motion of a triaxial, rigid satellite. The center of mass of the satellite is constrained to move in an elliptic orbit about an attracting point mass. The orbit, which has a constant inclination, is constrained to precess and spin with constant rates. The method of general perturbations is used to obtain the Hamiltonian for the nonresonant secular and long-period rotational motion of the satellite to second order in n/ω_0, where n is the orbital mean motion of the center of mass and ω_0 is a reference value of the magnitude of the satellite's rotational angular velocity. The differential equations derivable from the transformed Hamiltonian are integrable and the solution for the long-term motion may be expressed in terms of Jacobian elliptic functions and elliptic integrals. Geometrical aspects of the long-term rotational motion are discussed and a comparison of theoretical results with observations is made.

1. Introduction

Since the advent of artificial satellites, there has been renewed interest in obtaining analytical theories for the rotational motions of rigid bodies about their centers of mass when the centers of mass are constrained to orbit attracting primary bodies. Earlier works, dealing with natural bodies, include LaPlace's (1829) and Tisserand's (1891) investigations of the rotational motions of the Earth and the Moon.

Among the many recent works on the subject are those of Colombo (1964), Beletskii (1965) and Holland and Sperling (1969). However, the studies most closely related to the one discussed here are those made by Crenshaw and Fitzpatrick (1968) and Hitzl and Breakwell (1969).

Crenshaw and Fitzpatrick developed a first-order, gravity-gradient theory for the complete rotational motion of a rapidly spinning, uniaxial, rigid body, the center of mass of which was required to move in a uniformly precessing, circular orbit about an attracting point mass. They used the theory of canonical transformations to obtain a solution to the unperturbed, free-Eulerian motion, derived a set of differential equations analogous to Lagrange's planetary equations, and integrated these to first order. Hitzl and Breakwell used canonical transformation theory to study the rotational motion of a rapidly tumbling triaxial, rigid satellite, the center of mass of which was constrained to move in a fixed elliptic orbit about an attracting point mass, by applying an averaging procedure to the perturbing Hamiltonian. They studied the nonresonant and internally near-resonant effects of the gravity-gradient perturbations.

The mathematical model discussed here extends that of Hitzl and Breakwell to

* This paper is a slightly modified version of a previously published paper (Cochran, 1972) by the author.

B. D. Tapley and V. Szebehely (eds.), Recent Advances in Dynamical Astronomy, 429–453.

include effects of orbital evolution. That is, the plane of the orbit of the satellite's center of mass is constrained to precess and spin (movement of apsidal line) at constant rates, $\dot{\Omega}$ and $\dot{\omega}$, respectively.

Our restricted problem of rotational motion is studied using a slight modification of a new theory of general perturbations introduced by Hori (1966.)* The method of general perturbations is used to obtain the Hamiltonian for the nonresonant secular and long-period rotational motion of the triaxial satellite to second order in n/ω_0, where n is the orbital mean motion of the center of mass and ω_0 is a reference value of the satellite's rotational angular speed. The differential equations derivable from from the transformed Hamiltonian are integrable in terms of Jacobian elliptic functions and elliptic integrals. Geometrical aspects of the long-term rotational motion are discussed and, as an application of the theory, a comparison of theoreticalresults with observations is made.

2. Coordinate Systems

It is convenient at the outset to define certain coordinate systems which will be used in what follows. In Figure 1, five orthogonal coordinate systems with their common origin O at the center of mass of the triaxial satellite are shown. The $OXYZ$ coordiante system is a nonrotating system to which rotational motion about O is referred. The $Ox^oy^oz^o$ system is associated with the orbit of the attracting point mass, P, about O. It may be obtained from the $OXYZ$ system by positive rotations through the angles, Ω and I, the longitude of the ascending node and the inclination, respectively, of the orbit of P. The z^o-axis is directed along the normal to the orbital plane. The $Oxyz$ system is associated with the rotational angular momentum, $\mathbf{H}$, of the satellite about O. The z-axis of this system is directed in the sense of $\mathbf{H}$ and the x-axis lies along the line of intersection of the XY plane and the plane normal to $\mathbf{H}$ through O, as shown in Figure 1b. The $Oxyz$ system may be obtained from the $OXYZ$ system by rotations through the angles, ψ^* and θ^*. The $Ox_Hy_Hz_H$ system is associated with both the vector $\mathbf{H}$ and the orbital system $Ox^oy^oz^o$. It may be obtained from the $Ox^oy^oz^o$ system by rotations through the angles, ψ_H and θ_H. Finally, the $Ox'y'z'$ coordinate system (see Figure 1c) is such that its axes are principal axes for the triaxial satellite at its center of mass. It may be obtained from the $OXYZ$, $Oxyz$, and $Ox_Hy_Hz_H$ systems by rotations though the angles, (ψ, θ, ϕ), (ϕ^*, θ', ϕ'), and (ϕ_H, θ', ϕ'), respectively.

3. The Hamiltonian

It is a straightforward matter to obtain the dynamical Hamiltonian $\mathscr{H}$ for our problem using the Eulerian angles, (ψ, θ, ϕ), as generalized coordinates. However, if this is done (see Fitzpatrick, 1970), the form of $\mathscr{H}_0$, the part of the Hamiltonian consisting of the

* A theory similar to Hori's has been set forth by Deprit (1970) and much controversy has arisen as to the exact connection between the two theories. See, for example, Campbell and Jefferys (1970), Mersman (1971), and Henrard (1971).

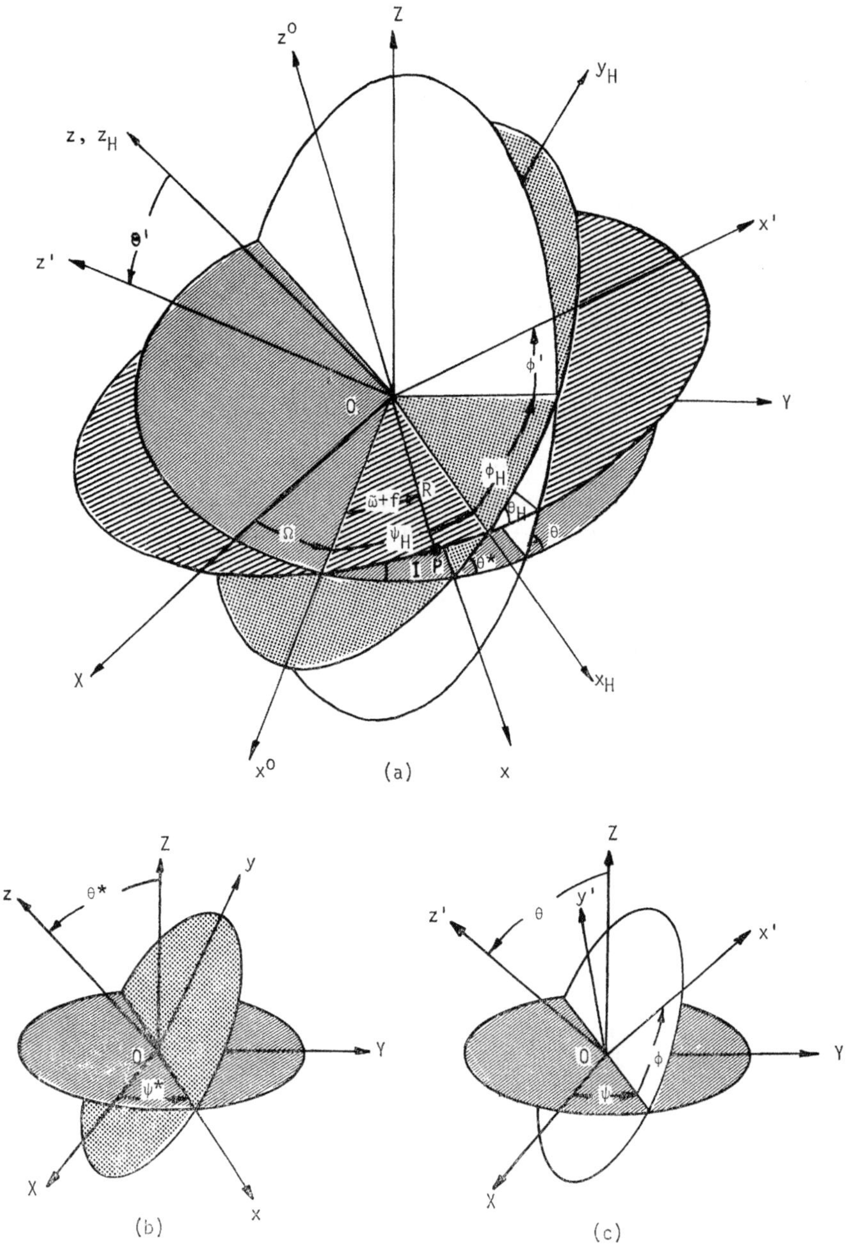

Fig. 1. Coordinate systems.

rotational kinetic energy of the satellite about O, is not a in very simple form. Deprit (1967), following Tisserand to some extent, obtained a simpler form for $\mathscr{H}_0$ using canonical transformations. The same technique may be applied to the triaxial Hamiltonian, $\mathscr{H}$, expressed in terms of ψ, θ, ϕ, and their conjugate moments, to obtain the transformed hamiltonian

$$\mathscr{H}^* = \frac{1}{2}\left[\frac{\sin^2 \phi'}{A} + \frac{\cos^2 \phi'}{B}\right](P^2_{\phi^*} - P^2_{\phi'}) + \frac{1}{2C} P^2_{\phi} + V,$$

where A, B, and C denote the principal moments of inertia of the satellite about the x'-, y'-, and z'-axes, respectively, and $P_{\phi^*}=h=|H|$, $P_{\phi'}=h\cos\theta^*$, and $P_{\psi^*}=h\ \cos\theta^*$ are the momenta conjugate to ψ^*, ϕ', and ϕ^*, respectively. The term V which represents that part of the potential energy due to the gravity-gradient torque which depends explicitly on the orientation of the satellite is given by

$$V = \frac{3}{2}\frac{GM_P}{R^3}\left[(A - B)\cos^2\gamma + (C - B)\cos^2\chi\right]. \tag{1}$$

In Equation (1), G is the universal gravitational constant, M_P is the mass of P, R is the distance from O to P, and γ and χ are the angles between the line segment OP and the x'- and z'-axes, respectively. Also, for an elliptic orbit,

$$R = a(1 - e^2)/(1 + e\cos f),$$

where a, e, and f denote the semi-major axis, eccentricity, and true anomaly, respectively.

In this paper, the orbit of P is assumed to precess at a constant rate $\dot{\Omega}$. For Earth satellites, $\dot{\Omega}$ is $\mathcal{O}(10^{-3})$ compared to n, so that it is convenient to treat the orbital precession as well as the gravity-gradient torque as perturbation. This may be done by referring the rotational motion to the $OXYZ$ system using the angles Ω, I, ψ_H, θ_H, ϕ_H, θ' and ϕ'. The angles Ω, I, ψ_H, θ_H and ϕ_H can be introduced through a canonical transformation.

By referring to the spherical triangle in Figure 2, $\sin\phi^*$, $\cos\phi^*$, $\sin\psi^*$ and $\cos\psi^*$

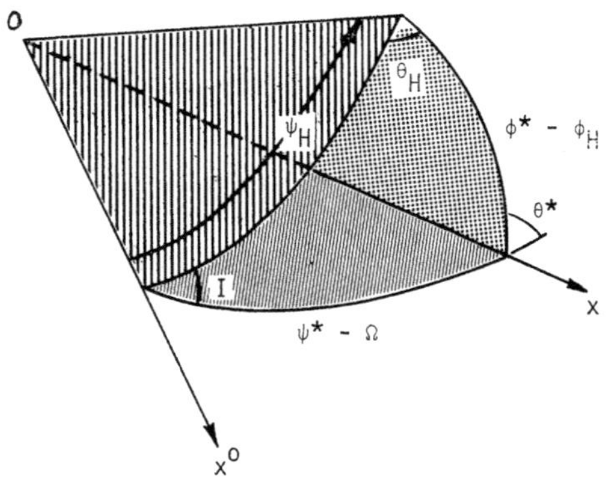

Fig. 2. Spherical triangle for the transformation $(\phi', \phi^*, \psi^*, P_{\phi'}, P_{\phi^*}, P_{\psi^*}) \to (\phi', \phi_H, \psi_H, P_{\phi'}, P_{\phi_H}, P_{\psi_H})$.

may be replaced by functions of Ω, I, ψ_H, θ_H and ϕ_H as follows:

$$\begin{aligned}\sin\phi^* &= a_1 \sin\phi_H + b_1 \cos\phi_H\\ \cos\phi^* &= a_1 \cos\phi_H - b_1 \sin\phi_H\\ \sin\psi^* &= c_1 \sin\Omega + d_1 \cos\Omega\\ \cos\psi^* &= c_1 \cos\Omega - d_1 \sin\Omega,\end{aligned}$$

where

$$\begin{aligned}a_1 &= (\cos I - \cos\theta^* \cos\theta_H)/\sin\theta^* \sin\theta_H\\ b_1 &= \sin I \sin\psi_H/\sin\theta^*\\ c_1 &= (\cos\theta_H - \cos\theta^* \cos I)/\sin\theta^* \sin I\\ d_1 &= \sin\theta_H \sin\psi_H/\sin\theta^*\\ \cos\theta^* &= \cos\theta_H \cos I - \sin\theta_H \sin I \cos\psi_H .\end{aligned}$$

Furthermore, by using the differential identity,

$$\mathrm{d}\phi^* = \mathrm{d}\phi_H - \cos\theta^* (\mathrm{d}\psi^* - \mathrm{d}\Omega) - \cos\theta_H \, \mathrm{d}\psi_H ,$$

obtained from the elementary spherical trigonometry of Figure 2, we find that the differential condition,

$$P_{\psi^*} \, \mathrm{d}\psi^* + P_{\phi^*} \, \mathrm{d}\phi^* - \mathscr{H}^* \, \mathrm{d}t = P_{\psi_H} \, \mathrm{d}\psi_H + P_{\phi_H} \, \mathrm{d}\phi_H - \mathscr{H}_H \, \mathrm{d}t ,$$

which is sufficient for a canonical transformation, is satisfied by

$$\begin{aligned}P_{\psi_H} &= P_{\phi^*} \cos\theta_H\\ P_{\phi_H} &= P_{\phi^*}\\ \mathscr{H}_H &= \mathscr{H}^* - \dot{\Omega} \left(P_{\psi_H} \cos I - \sqrt{P_{\phi_H}^2 - P_{\psi_H}^2} \sin I \cos\psi_H\right).\end{aligned} \tag{2}$$

In the last of Equations (2), the new Hamiltonian $\mathscr{H}_H$ is to be formed as indicated, after expressing $\mathscr{H}^*$ in terms of the variables $(\phi', \phi_H, \psi_H, P_{\phi'}, P_{\phi_H}, P_{\psi_H})$ and the time t. It turns out that the angle Ω does not appear in $\mathscr{H}_H$, and since $\dot{\Omega}$, the constant rate of orbital precession, is small, the second term in $\mathscr{H}_H$ will be treated as a perturbing Hamiltonian along with $V(\phi', \phi_H, \psi_H, P_{\phi'}, P_{\phi_H}, P_{\psi_H}, t)$.

For sufficiently small values of the eccentricity, the functions of f which appear in V may readily be expressed as functions of the mean anomaly, $M = n(t - t_0)$. The resulting Hamiltonian will, because of M and $\tilde{\omega}$, be nonautonomous, but by treating M and $\tilde{\omega}$ as additional coordinates and thereby artificially increasing the order of our system, the explicit dependence of $\mathscr{H}_H$ on t may be removed. The reason for doing this is that we desire an autonomous Hamiltonian as a starting point for application of the perturbation scheme which we shall employ here.

In addition to eliminating t, we shall introduce the dimensionless variables $P_1 \equiv \equiv P_{\phi'}/I_0\omega_0$, $P_2 \equiv P_{\phi_H}/I_0\omega_0$, and $P_3 \equiv P_{\psi_H}/I_0\omega_0$, where I_0 has the units of a moment of inertia and ω_0 is an angular speed such that, at $t = t_0$, $h = I_0\omega_0$ and the rotational kinetic energy, $\mathscr{T} = \frac{1}{2} I_0 \omega_0^2$. We also introduce the notation, $\bar{A} \equiv A/I_0$, $\bar{B} \equiv B/I_0$, $\bar{C} \equiv C/I_0$, $Q_1 \equiv \phi'$, $Q_2 \equiv \phi_H$, $Q_3 \equiv \psi_H$, $Q_4 \equiv M$, and $Q_5 \equiv \tilde{\omega}$. Using these definitions, removing the explicit time dependence from $\mathscr{H}_H$, dividing by $I_0\omega_0^2$, and letting F denote the negative of the result, we have

$$F = F_0 + \varepsilon F_1 + \varepsilon^2 F_2,$$

where

$$F_0 = -\frac{1}{2}\left[\left(\frac{\sin^2 Q_1}{\bar{A}} + \frac{\cos^2 Q_1}{\bar{B}}\right)(P_2^2 - P_1^2) + P_1^2/\bar{C}\right]$$

$$F_1 = -P_4$$

$$F_2 = -\frac{3}{2}\left(\frac{a}{R}\right)^3 [(\bar{A} - \bar{B})\cos^2\gamma + (\bar{C} - \bar{B})\cos^2\chi] +$$

$$+\left(\frac{\dot{\Omega}}{n}\right)\left(\frac{\omega_0}{n}\right)[P_3 \cos I - \sqrt{P_2^2 - P_3^2}\sin I \cos Q_3] - \left(\frac{\dot{\tilde{\omega}}}{n}\right)\left(\frac{\omega_0}{n}\right)P_5 .$$

and $(a/R)^3 \cos^2\gamma$ and $(a/R)^3 \cos^2\chi$ are functions of the Q_j, $j=1, 2, 3, 4, 5$, and P_j, $j=1, 2, 3$, and do not contain $t^* \equiv t\omega_0$, the new independent variable, explicitly. The small parameter $\varepsilon \equiv n/\omega_0$ has appeared naturally during the introduction of dimensionless variables and inertia parameters.

The problem embodied by F will be approached as a standard, general perturbation problem treating F_1 and F_2 as perturbing Hamiltonians, and using the method of general perturbations which will now be described.

4. Method of General Perturbations

We will study the triaxial problem using a method of general perturbations which is based on the use of Lie series to generate approximate, direct canonical transformations. The procedure we will follow is essentially that developed by Hori (1966).

A convenient way to introduce the method is by considering the following autonomous, canonical, differential system:

$$\begin{aligned} \frac{\mathrm{d}P_j}{\mathrm{d}\tau} &= \frac{\partial S}{\partial Q_j}; \; P_j(0) = x_j \\ \frac{\mathrm{d}Q_j}{\mathrm{d}\tau} &= -\frac{\partial S}{\partial P_j}; \; Q_j(0) = y_j \quad (j = 1, 2, 3, \ldots, n), \end{aligned} \tag{3}$$

where

$$S = S(P_1, P_2, \ldots, P_n, Q_1, Q_2, \ldots, Q_n; \varepsilon)$$

is an arbitrary function except that it must be such that a solution to the system (3) exists in a domain, $0 \leqslant |\tau| \leqslant \delta$, $0 \leqslant \varepsilon \leqslant \delta$, and should have a convergent Taylor series expansion about $\varepsilon = 0$.† In Equations (3) ε is not a true variable, but a small positive constant and τ, the independent variable, is not necessarily the time.

† Since we are more concerned with obtaining a 'formal' solution, no attempt to prove a particular function S has such properties will be made.

For convenience, we let

$$\mathbf{x} \equiv (x_1 x_2 x_3 \dots x_n)^T, \qquad \mathbf{y} \equiv (y_1 y_2 y_3 \dots y_n)^T$$
$$\mathbf{P} \equiv (P_1 P_2 P_3 \dots P_n)^T, \qquad \mathbf{Q} \equiv (Q_1 Q_2 Q_3 \dots Q_n)^T$$

where the superscript T denotes the transpose. Furthermore, a function $f(P_1, P_2, \dots, \dots, P_n, Q_1, Q_2, Q_3, \dots, Q_n; \varepsilon)$ will be denoted by $f(\mathbf{P}, \mathbf{Q}; \varepsilon)$ and the notation

$$\frac{\partial S}{\partial \mathbf{P}} = \left(\frac{\partial S}{\partial P_1} \frac{\partial S}{\partial P_2} \cdots \frac{\partial S}{\partial P_n}\right)^T$$

will be used.

The power series solution to (3) is of particular interest. To implement obtaining such a solution the additional notation

$$\begin{aligned} D_s^0 v &\equiv v \\ D_s^1 v &\equiv \{v, S\} \\ D_s^2 v &\equiv \{\{v, S\}, S\} \\ &\vdots \\ D_s^k &\equiv \{D^{k-1} v, S\}, \end{aligned}$$

where $\{v, S\}$ denotes the Poisson bracket of v, a differentiable function of the P_j and Q_j, and of S, is adopted. Then, the power series solution to (3) may be written as

$$\begin{aligned} \mathbf{P} &= \mathbf{x} + \sum_{k=1}^{\infty} \frac{\tau^k}{k!} D_s^k \mathbf{x} \\ \mathbf{Q} &= \mathbf{y} + \sum_{k=1}^{\infty} \frac{\tau^k}{k!} D_s^k \mathbf{y}. \end{aligned} \tag{4}$$

In Equations (4), $S = S(\mathbf{x}, \mathbf{y}; \varepsilon)$.

It may also be easily shown that an indefinitely differentiable function, $f(\mathbf{P}, \mathbf{Q}) = f(P_1, P_2, \dots P_n, Q_1, Q_2, \dots Q_n)$, has the series representation

$$f(\mathbf{P}, \mathbf{Q}) = f(\mathbf{x}, \mathbf{y}) + \sum_{k=1}^{\infty} \frac{\tau^k}{k!} D_s f. \tag{5}$$

Furthermore, because the system (3) is autonomous, we have

$$\begin{aligned} \mathbf{x} &= \mathbf{P} + \sum_{k=1}^{\infty} \frac{(-\tau)^k}{k!} D_s^k \mathbf{P} \\ \mathbf{y} &= \mathbf{Q} + \sum_{k=1}^{\infty} \frac{(-\tau)^k}{k!} D_s^k \mathbf{Q} \end{aligned} \tag{6}$$

and

$$f(\mathbf{x}, \mathbf{y}) = f(\mathbf{P}, \mathbf{Q}) + \sum_{k=1}^{\infty} \frac{(-\tau)^k}{k!} D_{sJ}^k f(\mathbf{P}, \mathbf{Q}), \tag{7}$$

where $S = S(\mathbf{P}, \mathbf{Q}; \varepsilon)$.

Equations (4) represent infinitely many canonical transformations and Equations (6) represent the corresponding inverse transformations. The particular transformation we will use is that obtained from (4) by letting $\tau = \varepsilon$,

$$\begin{aligned} \mathbf{P}(\varepsilon) &= \mathbf{x} + \sum_{k=1}^{\infty} \frac{\varepsilon^k}{k!} D_s^k \mathbf{x} \\ \mathbf{Q}(\varepsilon) &= \mathbf{y} + \sum_{k=1}^{\infty} \frac{\varepsilon^k}{k!} D_s^k \mathbf{y}. \end{aligned} \tag{8}$$

Equations (8), along with the corresponding transformation equation,

$$f(\mathbf{P}, \mathbf{Q}) = f(\mathbf{x}, \mathbf{y}) + \sum_{k=1}^{\infty} \frac{\varepsilon^k}{k!} D_{sJ}^k f(\mathbf{x}, \mathbf{y}), \tag{9}$$

for an arbitrary function $f(\mathbf{P}, \mathbf{Q})$, form the basis of our perturbation method.

Let S be expressed in the form $S = S_1(\mathbf{P}, \mathbf{Q}) + \varepsilon S_2(\mathbf{P}, \mathbf{Q}) + \varepsilon^2 S_3(\mathbf{P}, \mathbf{Q}) + \cdots$ and let $F(\mathbf{P}, \mathbf{Q}; \varepsilon) = F_0(\mathbf{F}, \mathbf{Q}) + \varepsilon F_1(\mathbf{P}, \mathbf{Q}) + \varepsilon^2 F_2(\mathbf{P}, \mathbf{Q}) + \cdots$ denote the Hamiltonian for an autonomous dynamical system which has a solution when $\varepsilon = 0$. Then, if we transform from the canonical set $(\mathbf{P}, \mathbf{Q})$ to the set $(\mathbf{x}, \mathbf{y})$ using the autonomous transformation Equations (8), we have $F^*(\mathbf{x}, \mathbf{y}, \varepsilon) = F(\mathbf{P}(\mathbf{x}, \mathbf{y}; \varepsilon), Q(\mathbf{x}, \mathbf{y}; \varepsilon); \varepsilon)$ as the new Hamiltonian. It then follows, from (9) and the expressions for S and $F(\mathbf{P}, \mathbf{Q}; \varepsilon)$, that

$$F^*(\mathbf{x}, \mathbf{y}, \varepsilon) = \sum_{k=0}^{\infty} \varepsilon^k F_k^*, \tag{10}$$

where

$$\begin{aligned} F_0^* &= F_0(\mathbf{x}, \mathbf{y}) \\ F_1^* &= F_1(\mathbf{x}, \mathbf{y}) + \{F_0, S_1\} \\ F_2^* &= F_2(\mathbf{x}, \mathbf{y}) + \tfrac{1}{2}\{F_1 + F_1^*, S_1\} + \{F_0, S_2\} \\ &\vdots \end{aligned} \tag{11}$$

We have given only the first three terms of F^* explicitly, since only these will be used here.

For a particular $F(\mathbf{P}, \mathbf{Q}; \varepsilon)$, it is our aim to choose the functions S_k in such a manner that the differential equations associated with F^* are more tractable than the original canonical equations. If F_1 contains both momenta and coordinates, the partial differential equations which are obtained by expanding the Poisson brackets in Equations (11) may be complicated, thus making the choices of the S_k difficult. To simplify the brackets $\{F_0, S_k\}$ and implement the choices of the S_k, we shall introduce an auxiliary transformation into Equations (11). This transformation is defined by a complete integral of the Hamilton-Jacobi equation,

$$F_0\left(-\frac{\partial \bar{S}}{\partial \mathbf{y}}, \mathbf{y}\right) + \frac{\partial \bar{S}}{\partial t} = 0. \tag{12}$$

The complete integral $\bar{S} = \bar{\alpha}_1 t + \bar{S}_1(\bar{\boldsymbol{\alpha}}, \mathbf{y})$, where $\bar{\boldsymbol{\alpha}} = (\bar{\alpha}_1\, \bar{\alpha}_2 \ldots \bar{\alpha}_n)^T$ and the $\bar{\alpha}_j$ are new canonical variables, defines the transformation

$$\begin{aligned} \mathbf{x} &= -\frac{\partial \bar{S}}{\partial \mathbf{y}} \\ t + \bar{\beta}_1 &= -\frac{\partial \bar{S}_1}{\partial \bar{\alpha}_1} \\ \bar{\beta}_j &= -\frac{\partial \bar{S}_1}{\partial \bar{\alpha}_j} \quad (j = 2, 3, \ldots, n), \end{aligned} \tag{13}$$

and the new Hamiltonian, $K^*(\bar{\boldsymbol{\alpha}}, \bar{\boldsymbol{\beta}}, t; \varepsilon)$ is given by

$$K^*(\bar{\boldsymbol{\alpha}}, \bar{\boldsymbol{\beta}}, t; \varepsilon) = \sum_{k=1}^{\infty} \varepsilon^k K_k^*, \tag{14}$$

where

$$K_k^* = F_k^*(\mathbf{x}(\bar{\boldsymbol{\alpha}}, \bar{\boldsymbol{\beta}}, t), \mathbf{y}(\bar{\boldsymbol{\alpha}}, \bar{\boldsymbol{\beta}}, t)).$$

Since a Poisson bracket is invariant under a canonical tran sformation, by using Equations (13) and the fact that $F_0 = -\bar{\alpha}_1$, we find that Equat ions (11) become

$$\begin{aligned} K_1^* &= F_1(\mathbf{x}(\bar{\boldsymbol{\alpha}}, \bar{\boldsymbol{\beta}}, t), \mathbf{y}(\bar{\boldsymbol{\alpha}}, \bar{\boldsymbol{\beta}}, t)) - \frac{\partial S}{\partial \bar{\beta}_1} \\ K_2^* &= F_2(\mathbf{x}(\bar{\boldsymbol{\alpha}}, \bar{\boldsymbol{\beta}}, t), \mathbf{y}(\bar{\boldsymbol{\alpha}}, \bar{\boldsymbol{\beta}}, t)) + \tfrac{1}{2}\{F_1 + F_1^*, S_1\} - \frac{\partial S_2}{\partial \bar{\beta}_1}. \\ &\vdots \end{aligned} \tag{15}$$

We now try to choose the S_k successively, in such a manner that our transformed dynamical system is easier to solve than the original system. If the Hamiltonian $F(\mathbf{P}, \mathbf{Q}; \varepsilon)$ is periodic in the Q_j with period 2π, an acceptable choice of S_1 is

$$S_1 = \int F_{1_p}\, \mathrm{d}\bar{\beta}_1, \tag{16}$$

where F_{1_p} is the part of $F_1(x(\bar{\boldsymbol{\alpha}}, \bar{\boldsymbol{\beta}}, t), y(\bar{\boldsymbol{\alpha}}, \bar{\boldsymbol{\beta}}, t))$ which contains $\bar{\beta}_1$. In general, we will let the subscript s denote the part of a function of $\bar{\boldsymbol{\alpha}}$, $\bar{\boldsymbol{\beta}}$, and t which does not contain $\bar{\beta}_1$ explicitly, while the subscript p will denote the remainder of that same function.

Then, along with (16), we have, through second order in ε,

$$
\begin{aligned}
K_1^* &= F_{1s} \\
S_2 &= \int [F_2 + \tfrac{1}{2}\{F_1 + F_1^*, S_1\}]_p \, \mathrm{d}\bar{\beta}_1 \\
K_2^* &= F_{2s} + \tfrac{1}{2}\{F_1 + F_1^*, S_1\}_s .
\end{aligned} \tag{17}
$$

If K_1^* and K_1^* are chosen according to (17) and F is periodic in the Q_j with period 2π, they will not contain t or $\bar{\beta}_1$ explicitly. Thus, through second order in ε, we have

$$K^* (\bar{\boldsymbol{\alpha}}, -, \bar{\beta}_2, \bar{\beta}_3, \ldots, \bar{\beta}_n, -) = \text{const.}, \tag{18}$$

where a dash is used in place of a variable to emphasize its explicit absence from K^*, and since $\bar{\beta}_1$ is ignorable, we also have

$$\dot{\bar{\alpha}}_1 = \frac{\partial K^*}{\partial \bar{\beta}_1} = 0 ,$$

or

$$\bar{\alpha}_1 = \text{const.}, \tag{19}$$

which is a new integral of the system. The remaining canonical equations are

$$
\begin{aligned}
\dot{\bar{\alpha}}_j &= \frac{\partial K^*}{\partial \bar{\beta}_j} \; (j = 2, 3, \ldots, n) \\
\dot{\bar{\beta}}_j &= -\frac{\partial K^*}{\partial \bar{\alpha}_j} \; (j = 1, 2, \ldots, n) .
\end{aligned} \tag{20}
$$

If Equations (20) are integrable, the problem is solved through second order in ε. If they are not integrable, but the equations,

$$
\begin{aligned}
\dot{\bar{\boldsymbol{\alpha}}} &= \varepsilon \frac{\partial K_1^*}{\partial \bar{\boldsymbol{\beta}}} \\
\dot{\bar{\boldsymbol{\beta}}} &= -\varepsilon \frac{\partial K_1^*}{\partial \bar{\boldsymbol{\alpha}}} ,
\end{aligned} \tag{21}
$$

are integrable, then the procedure just described may be applied to the Hamiltonian $K^*(\bar{\boldsymbol{\alpha}}, \bar{\boldsymbol{\beta}}; \varepsilon)$ in another effort to obtain additional integrals and/or integrable equations.

It may be seen that each time the method we have outlined is applied, two canonical transformations, one approximate transformation via Lie series and one exact, auxiliary transformation using the solution to the Hamilton-Jacobi equation for the 'unperturbed' problem are made. The method given here differs from Hori's method (Hori, 1966) in the use of the auxiliary Hamilton-Jacobi equation and the invariance of the Poisson bracket under such a transformation to obtain the simplified Equations (15).

It should be noted that the auxiliary transformation may not always be needed. If F_0 is a function of only the P_j, then the brackets $\{F_0, S_k\}$ will not be very complicated and the S_k can easily be chosen so that one or more of the y_j are eliminated from F^*.

5. Long-Term Motion in the Triaxial Problem

A. TRANSFORMATIONS

In applying the perturbation method just described to the triaxial problem, we first make a second-order canonical transformation defined by

$$\begin{aligned} \mathbf{P} &= \mathbf{x} + \varepsilon \{\mathbf{x}, S_1\} + \varepsilon^2 [\{\mathbf{x}, S_2\} + \tfrac{1}{2}\{\{\mathbf{x}, S_1\}, S_1\}] \\ \mathbf{Q} &= \mathbf{y} + \varepsilon \{\mathbf{y}, S_1\} + \varepsilon^2 [\{\mathbf{y}, S_2\} + \tfrac{1}{2}\{\{\mathbf{y}, S_1\}, S_1\}]. \end{aligned} \tag{22}$$

This gives us the new Hamiltonian,

$$F^*(\mathbf{x}, \mathbf{y}; \varepsilon) = F_0^* + \varepsilon F_1^* + \varepsilon^2 F_2^*, \tag{23}$$

where

$$\begin{aligned} F_0^* &= -\tfrac{1}{2}\left[\left(\frac{\sin^2 y_1}{\bar{A}} + \frac{\cos^2 y_1}{\bar{B}}\right)(x_2^2 - x_1^2) + x_1^2/\bar{C}\right] \\ F_1^* &= -x_4 + \{F_0, S_1\} \\ F_2^* &= -\frac{3}{2}\left(\frac{a}{R}\right)^3 [(\bar{A} - \bar{B})\cos^2\gamma + (\bar{C} - \bar{B})\cos^2\chi]|_{\substack{P=x \\ Q=y}} + \\ &\quad + \left(\frac{\dot{\Omega}}{n}\right)\left(\frac{\omega_0}{n}\right)[x_3 \cos I - \sqrt{x_2^2 - x_3^2}\sin I \cos y_3] - \\ &\quad - \left(\frac{\dot{\tilde{\omega}}}{n}\right)\left(\frac{\omega_0}{n}\right) x_4 + \tfrac{1}{2}\{F_1 + F_1^*, S_1\} + \{F_0, S_2\}. \end{aligned} \tag{24}$$

The second and third of Equations (24) are to be simplified by choosing S_1 and S_2, respectively. Clearly, the second equation is simple if we choose $S_1 \equiv 0$, and this choice simplifies the third equation. The choice of S_2 is more difficult.

Since F_0 contains x_1 *and* y_1, the expanded form of the Poisson bracket $\{F_0, S_2\}$ in terms of x_j and the y_j is complicated. To simplify this form, we introduce an auxiliary canonical transformation which is defined by the complete integral,

$$\bar{S} = \bar{\alpha}_1 t^* - \sum_{j=2}^{5} \bar{\alpha}_j y_j - \int_{y_{10}}^{y_1} x_1 \, \mathrm{d}y_1, \tag{25}$$

of the Hamilton-Jacobi equation,

$$-\left[\frac{\sin^2 y_1}{2\bar{A}} + \frac{\cos^2 y_1}{2\bar{B}}\right]\left[\left(\frac{\partial \bar{S}}{\partial y_2}\right)^2 - \left(\frac{\partial \bar{S}}{\partial y_1}\right)^2\right] - \left(\frac{\partial \bar{S}_1}{\partial y_1}\right)^2 / 2\bar{C} + \frac{\partial \bar{S}}{\partial t^*} = 0. \tag{26}$$

In (25), $\bar{\alpha}_1 = -F_0(\mathbf{x}, \mathbf{y})$ and $\bar{\alpha}_j, j = 2, 3, 4, 5$, are new canonical variables, while

$$x_1 = \sqrt{\bar{C}\,\frac{(a' + b' \sin^2 y_1)}{(c' + d' \sin y_1)}}, \tag{27}$$

where

$$\begin{aligned} a' &= \bar{A}(2\bar{B}\bar{\alpha}_1 - \bar{\alpha}_2^2) \\ b' &= \bar{\alpha}_2^2(\bar{A} - \bar{B}) \\ c' &= \bar{A}(\bar{B} - \bar{C}) \\ d' &= \bar{C}(\bar{A} - \bar{B}) \end{aligned}$$

and y_{10} will be specified in what follows. The plus sign is taken on the radical in (27), since by proper labeling of the principal axes we may make $0 \leqslant \theta' < \pi 2$ and since $\bar{\theta}' \equiv \cos^{-1}(x_1/x_2)$ should also be in this range for small perturbations in P_1 and P_2.†

The transformation equations derived from $\bar{S}(\bar{\alpha}, \bar{\beta}, t)$, according to (13), in addition to (27) are

$$\begin{aligned} x_j &= \bar{\alpha}_j \quad (j = 2, 3, 4, 5) \\ t^* + \bar{\beta}_1 &= I_1 \\ y_2 - \bar{\beta}_2 &= I_2 \\ y_j &= \bar{\beta}_j \quad (j = 3, 4, 5), \end{aligned} \tag{28}$$

where

$$\begin{aligned} I_1 &= \int_{y_{10}}^{y_1} \frac{\bar{A}\bar{B}\sqrt{\bar{C}}\,\mathrm{d}y_1}{\sqrt{(a' + b' \sin^2 y_1)(c' + d' \sin^2 y_1)}} \\ I_2 &= \int_{y_{10}}^{y_1} \frac{\sqrt{\bar{C}}[\bar{B} + (\bar{A} - \bar{B}) \cos y_1]\,\mathrm{d}y_1}{\sqrt{(a' + b' \sin^2 y_1)(c' + d' \sin^2 y_1)}}. \end{aligned} \tag{29}$$

The integral I_1 may be simplified by making a change of variable defined by $U \tan y_1 = \cot \zeta$, where $U = [\bar{A}(\bar{B} - \bar{C})/\bar{B}(\bar{A} - \bar{C})]^{1/2}$. Setting $y_{10} = -\pi/2$ so that $\zeta_0 = 0$, from (29), we get

$$u = \int_0^{\zeta} \frac{\mathrm{d}\zeta}{\sqrt{1 - k^2 \sin^2 \zeta}}, \tag{30}$$

where

$$\begin{aligned} u &= \lambda(t^* + \bar{\beta}_1) \\ \lambda &= \sqrt{\frac{(\bar{B} - \bar{C})(2\bar{B}\bar{\alpha}_1 - \bar{\alpha}_2^2)}{\bar{A}\bar{B}\bar{C}}} \\ k^2 &= \frac{(\bar{A} - \bar{B})(\bar{\alpha}_2^2 - 2C\bar{\alpha}_1)}{(\bar{B} - \bar{C})(2A\bar{\alpha}_1 - \bar{\alpha}_2^2)}. \end{aligned}$$

From (30) and the theory of elliptic functions, it follows that $\sin \zeta = \mathrm{sn}\, u$. We note that $-\bar{\beta}_1$ is the value of t^* when $y_{10} = -\pi/2$.

Using the above results and formulae in Whittaker and Watson (1963), we find that,

† Only motion which does not correspond to a separatrix polhode will be considered here. See, for example, MacMillan (1960).

if $\alpha^2 \equiv \bar{C}(\bar{A}-\bar{B})/\bar{A}(\bar{B}-\bar{C})$ and $k^2 \operatorname{sn} ia \equiv -\alpha^2$, where $i=\sqrt{-1}$, then

$$I_2 = \left[\frac{\bar{\alpha}_2}{\bar{A}} - iZ(ia)\,\lambda\right](t^* + \bar{\beta}_1) + (i/2)\ln\left[\frac{\theta(u+ia)}{\theta(u-ia)}\right]. \tag{31}$$

In Equation (31), $Z\,(ia)$ is Jacobi's Zeta-function and $\theta(u \pm ia)$ are Theta-functions. The coefficient of $(t^*+\bar{\beta}_1)$ in Equation (31) is the mean rate of precession of the z'-axis about the z-axis if $\varepsilon=0$ and $\dot{\Omega}=0$.

Using the above results, and setting

$$\bar{R} = \sqrt{\frac{\bar{C}(2\bar{A}\bar{\alpha}_1 - \bar{\alpha}_2^2)}{\bar{\alpha}_2^2(\bar{A}-\bar{C})}},$$

we may write (27) as

$$x_1 = \bar{\alpha}_2 \bar{R}\,\operatorname{dn} u\,. \tag{32}$$

Also, since $\sin\zeta = \operatorname{sn} u$ and $\cos\zeta = \operatorname{cn} u$, we have

$$\begin{aligned} \sin y_1 &= -\operatorname{cn} u/\sqrt{1+\alpha^2\operatorname{sn}^2 u} \\ \cos y_1 &= U^{-1}\operatorname{sn} u/\sqrt{1+\alpha^2\operatorname{sn}^2 u}\,. \end{aligned} \tag{33}$$

The solution for the unperturbed problem, embodied in Equations (28), (31), (32), and (33), corresponds exactly to that given by Whittaker (1965) for a freely spinning, triaxial, rigid body. If $\varepsilon=0$ and $\dot{\Omega}=0$, $y_1=\phi'$ is the angle of spin, $y_2=\phi_H$ is angle of precession, and $\bar{\theta}'=\theta'$ is the nutation angle of the satellite.

When the equations for the transformation $(\mathbf{x}, \mathbf{y}) \to (\bar{\boldsymbol{\alpha}}, \bar{\boldsymbol{\beta}})$ are substituted into the last two of Equations (24), we get

$$\begin{aligned} F_1^* &= -\bar{\alpha}_4 - \frac{\partial S_1}{\partial \bar{\beta}_1} \\ F_2^* &= F_2(\mathbf{x}(\bar{\boldsymbol{\alpha}}, \bar{\boldsymbol{\beta}}, t^*), \mathbf{y}(\bar{\boldsymbol{\alpha}}, \bar{\boldsymbol{\beta}}, t^*)) + \tfrac{1}{2}\{F_1 + F_1^*, S_1\} - \frac{\partial S_1}{\partial \bar{\beta}_1}\,. \end{aligned} \tag{34}$$

As pointed out earlier, the obvious choice for S_1 is $S_1 \equiv 0$; for, then we have, from (34),

$$\begin{aligned} F_1^* &= \bar{\alpha}_4 \\ F_2^* &= F_2(\mathbf{x}(\bar{\boldsymbol{\alpha}}, \bar{\boldsymbol{\beta}}, t^*), \mathbf{y}(\bar{\boldsymbol{\alpha}}, \bar{\boldsymbol{\beta}}, t^*)) - \frac{\partial S_1}{\partial \bar{\beta}_1}, \end{aligned}$$

In principle, the explicit substitution of the transformation Equations (27) and (28) into F^* can be carried out using the expressions

$$\begin{aligned} \cos\gamma = {} & \cos\mu(\cos y_2 \cos y_1 - \sin y_2 \sin y_1 \cos\bar{\theta}') - \\ & - \cos\bar{\theta}_H \sin(\sin y_2 \cos y_1 + \cos y_2 \sin y_1 \cos\bar{\theta}') + \\ & + \sin\bar{\theta}_H \sin\mu \sin\bar{\theta}' \sin y_1 \end{aligned}$$

and

$$\cos\chi = \cos\mu \sin y_2 \sin\bar{\theta}' + \cos\bar{\theta}_H \sin\mu \cos y_2 \sin\bar{\theta}' + \sin\bar{\theta}_H \sin\mu \cos\bar{\theta}'\,.$$

where $\mu \equiv y_3 - (\tilde{\omega} + f)$, $\cos \bar{\theta}' \equiv x_1/x_2$, and $\cos \bar{\theta}_H \equiv x_3/x_2$. This will not be done here; however, it is fairly easy to show that $\cos^2 \gamma$ and $\cos^2 \chi$ have the forms

$$\cos^2 \gamma = \tfrac{1}{4}(1 - 3\cos^2 \bar{\theta}_H)\sin^2 \bar{\theta}' \sin^2 y_1 + \tfrac{1}{4}(1 - \sin^2 \bar{\theta}' \sin^2 y_1) \times \times \sin^2 \bar{\theta}_H \cos^2 \mu + \sum_{j=-2}^{2} \sum_{j=-2}^{2} \sum_{k=-2}^{2} B_{ijk} \cos(iy_1 + ky_2)$$

and

$$\cos^2 \chi = \sum_{j=-2}^{2} \sum_{k=-2}^{2} C_{jk} \cos(j\mu + ky_2),$$

where $B_{000} = \frac{1}{4}(1 + \cos^2 \bar{\theta}_H)$, $C_{00} = \frac{1}{4}[(1 + \cos^2 \bar{\theta}_H) + (1 - 3\cos^2 \bar{\theta}_H)\cos^2 \bar{\theta}']$, $B_{i20} = 0$, $C_{20} = \frac{1}{4}\sin^2 \bar{\theta}_H (1 - 3\cos^2 \bar{\theta}')$, and B_{ij0} and C_{j0} are zero for $j \neq 2$.

Now, y_2 is monotonically increasing with t^*, and if the possibility of internal resonance (Hitzl and Breakwell, 1969) is not considered, the parts of $\cos^2 \gamma$ and $\cos^2 \chi$ which are free of $\bar{\beta}_1$ are those parts of the terms outside the summation signs which are free of $\bar{\beta}_1$, plus those parts of B_{000}, C_{00} and C_{20} free of $\bar{\beta}_1$. To determine the parts of these terms which are free of $\bar{\beta}_1$, we must consider only the functions $\cos^2 \bar{\theta}'$ and $\sin^2 \bar{\theta}' \sin^2 y_1$.

We have $\cos^2 \bar{\theta}' = R^2 \mathrm{d}n^2 u$ and $\sin^2 y_1 = \mathrm{cn}^2 u/(1 + \alpha^2 \mathrm{sn}^2 u)$. Hence, by using $\cos^2 \bar{\theta}' = = 1 - \sin^2 \bar{\theta}'$, $\mathrm{dn}^2 u = 1 - k^2 \mathrm{sn}^2 u$ and the identity †

$$\mathrm{d}n^2 u \equiv E/K + Z'(u),$$

where $Z'(u) = \mathrm{d}Z(u)/\mathrm{d}u$ is periodic in u, we find that the parts of $\cos^2 \bar{\theta}'$ and $\sin^2 \bar{\theta}' \sin^2 y_1$ free of $\bar{\beta}_1$ are

$$\langle \cos^2 \bar{\theta}' \rangle = \bar{R}^2 E/K \tag{35}$$

and

$$\langle \sin^2 \bar{\theta}' \sin^2 y_1 \rangle = \bar{Q}^2 (E - k'^2 K)/k^2, \tag{36}$$

respectively. In (36), $\bar{Q}^2 = \bar{B}(\bar{\alpha}_2^2 - 2\bar{C}\bar{\alpha}_1)/[\bar{\alpha}_2^2(\bar{B} - \bar{C})]$ and $k'^2 = 1 - k^2$.

Using (35) and (36) along with the expressions for $\cos^2 \gamma$ and $\cos^2 \chi$, we obtain

$$K^*(\bar{\alpha}, \bar{\beta}_3, \bar{\beta}_4, \bar{\beta}_5; \varepsilon) = -\varepsilon\bar{\alpha}_4 - \varepsilon^2 \left(\frac{\dot{\tilde{\omega}}}{n}\right)\left(\frac{\omega_0}{n}\right)\alpha_5 + \varepsilon^2 [K_\Omega - \langle\!\langle V_G \rangle\!\rangle],$$

where

$$K_\Omega = \left(\frac{\dot{\Omega}}{n}\right)\left(\frac{\omega_0}{n}\right)[\bar{\alpha}_3 \cos I - \sqrt{\bar{\alpha}_2^2 - \bar{\alpha}_3^2} \sin I \cos \bar{\beta}_3]$$

$$\langle\!\langle V_G \rangle\!\rangle = \frac{3}{8}\left(\frac{a}{R}\right)^3 \{(\bar{A} + \bar{C} - 2\bar{B})(1 + \cos^2 \bar{\theta}_H) + (1 - 3\cos^2 \bar{\theta}_H)\Delta + + (\bar{A} + \bar{C} - 2\bar{B})(1 - \cos^2 \bar{\theta}_H)\cos^2 \bar{\mu}\} \tag{37}$$

$$\Delta = \bar{B}[(E/K)(\bar{B} - \bar{C})/\bar{B} - 1](2\bar{A}\bar{\alpha}_1/\bar{\alpha}_2^2 - 1) + \bar{A} - \bar{B}$$

† Here we have used the facts that $E(u) = \int \mathrm{d}n^2 u \, \mathrm{d}u$ and $E(u) = (E/K)u + Z(u)$, where $E(u)$ is the incomplete elliptic integral of the second kind of K and E are complete elliptic integrals of the first and second kinds, respectively.

and

$$\bar{\mu} \equiv \bar{\beta}_3 - [\bar{\beta}_5 + f(\bar{\beta}_4)].$$

The generating function S_2 is given by

$$S_2 = -\int [V_G(\mathbf{x}(\bar{\boldsymbol{\alpha}}, \bar{\boldsymbol{\beta}}, t^*), \mathbf{y}(\bar{\boldsymbol{\alpha}}, \bar{\boldsymbol{\beta}}, t^*)) - \langle\langle V_G \rangle\rangle] \, d\bar{\beta}_1, \tag{38}$$

where $V_G = (1/n^2)V$. Exact analytical evaluation of the integral in (38) appears to be a very formidable task and has not been done; however, an approximate analytical evaluation has been given by Cochran (1970). The numerical evaluation of the short-period perturbations derivable from S_2 involves only numerical quadratures, and is not considered in this paper. Only secular and long-period perturbations will be discussed.

We note that K^* contains neither $\bar{\beta}_1$ nor $\bar{\beta}_2$, so that the new canonical equations are

$$\begin{aligned}
\frac{d\bar{\alpha}_j}{dt^*} &= 0 \quad (j = 1, 2) \\
\frac{d\bar{\alpha}_j}{dt^*} &= \frac{\partial K^*}{\partial \bar{\beta}_j} \quad (j = 3, 4, 5) \\
\frac{d\bar{\beta}_j}{dt^*} &= \frac{\partial K^*}{\partial \bar{\alpha}_j} \quad (j = 1, 2, 3, 4, 5).
\end{aligned} \tag{39}$$

The integrals $\bar{\alpha}_1$ = constant and $\bar{\alpha}_2$ = constant, of Equations (39) express the facts that the average values of the rotational kinetic energy of the satellite are the magnitude of the satellite's rotational angular momentum, respectively, are constant.

Equations (39) do not appear to be completely integrable. To obtain equations which are integrable, we make the approximate transformation,

$$\begin{aligned}
\bar{\boldsymbol{\alpha}} &= \boldsymbol{\xi} + \varepsilon \frac{\partial S_1^*}{\partial \boldsymbol{\eta}} \\
\bar{\boldsymbol{\beta}} &= \boldsymbol{\eta} - \varepsilon \frac{\partial S_1^*}{\partial \boldsymbol{\xi}}
\end{aligned} \tag{40}$$

The function S_1^* is at this point arbitrary and the new Hamiltonian is

$$K^{**}(\xi, \eta; \varepsilon) = \varepsilon K_1^{**} + \varepsilon_2 K_2^{**}$$

where

$$\begin{aligned}
K_1^{**} &= -\xi_4 \\
K_2^{**} &= K_2^*(\boldsymbol{\xi}, \boldsymbol{\eta}) - \frac{\partial S_1^*}{\partial \eta_4}.
\end{aligned} \tag{41}$$

Since the second of Equations (41) is a simple partial differential eqaution, no auxiliary transformation will be made. Instead, we let $K_{2\,p}^*$ denote the part of K_2^* which contains η_4 and choose

$$S_1^* = \int K_{2_p}^* \, d\eta_4. \tag{42}$$

To $\mathcal{O}(e^3)$, S_1^* is given by

$$\begin{aligned} S_1^* = f_1 &[(9/8)\, e \sin \eta_4 + (27/32)\, e^2 \sin 2\eta_4 + \\ &+ (3/16)\, e \sin(2\eta_3 - 2\eta_5 - \eta_4)] + \\ &+ f_2 \{[-3/16 + (15/32)\, e^2] \sin(2\eta_3 - 2\eta_5 - 2\eta_4) - \\ &- (7/16)\, e \sin(2\eta_3 - 2\eta_5 - 3\eta_4) - \\ &- (51/64)\, e^2 \sin(2\eta_3 - 2\eta_5 - 4\eta_4)\}, \end{aligned}$$

where

$$\begin{aligned} f_1 &= (\bar{A} + \bar{C} - 2\bar{B})(1 + \cos^2 \hat{\theta}_H) + (1 - 3\cos^2 \hat{\theta}_H)\, \Delta \\ f_2 &= (\bar{A} + \bar{C} - 2\bar{B} - 3\Delta)(1 - \cos^2 \hat{\theta}_H). \end{aligned}$$

In the expressions for $f_{1,2}$, Δ is to be obtained by replacing $\bar{\alpha}_1$ by ξ_1 and $\bar{\alpha}_2$ by ξ_2 in the last of Equations (37). Also, we have adopted the notation $\cos\hat{\theta}_H \equiv \xi_3/\xi_2$.

The perturbations in the variables $\bar{\alpha}_j$ and $\bar{\beta}_j$, which may be computed using Equations (40), may be termed 'quasi-long period perturbations' since they are perturbations with periods of order $2\pi/n$. We also note that the amplitudes of these perturbations are of order ε.

Using the choice, (42), of S_1^*, we have, from the last of Equations (41),

$$K_2^{**} = K_\Omega^*(\boldsymbol{\xi}, \boldsymbol{\eta}) - \langle\langle\langle V_G \rangle\rangle\rangle - (\tilde{\omega}/n)(\omega_0/n)\, \xi_5,$$

where

$$\begin{aligned} K_\Omega^*(\boldsymbol{\xi}, \boldsymbol{\eta}) &= \left(\frac{\dot{\Omega}}{n}\right)\left(\frac{\omega_0}{n}\right)[\xi_3 \cos I - \sqrt{\xi_2^2 - \xi_3^2} \sin I \cos \eta_3] \\ \langle\langle\langle V_G \rangle\rangle\rangle &= \tfrac{3}{8}(1 - e^2)^{-3/2}[(\bar{A} + \bar{C} - 2\bar{B})(1 + \cos^2 \hat{\theta}_H) + \\ &\quad + (1 - 3\cos^2 \hat{\theta}_H)\, \Delta]. \end{aligned} \tag{43}$$

In the second of Equations (43), $(1-e^2)^{-3/2}$ is the part of $(a/R)^3$ which does not contain $\eta_4 = M$. Note that the rotation of the apsidal line of the orbit of P does not affect the long-term rotational motion of the satellite.

The new Hamiltonian K^{**} does not contain η_1, η_2, η_4, or η_5 and the new canonical equations are

$$\begin{aligned} \frac{d\xi_j}{dt^*} &= 0 \quad (j = 1, 2, 4, 5) \\ \frac{d\xi_3}{dt^*} &= \frac{\partial K^{**}}{\partial \eta_3} \\ \frac{dn}{dt^*} &= -\frac{\partial K^{**}}{\partial \xi}. \end{aligned} \tag{44}$$

Hence, ξ_1, ξ_2, ξ_4, and ξ_5 are constant. Also, from the last two of Equations (44), we immediately obtain

$$\begin{aligned} \eta_4 &= nt + \eta_{40} = M \\ \eta_5 &= \dot{\tilde{\omega}} t + \eta_{50} = \tilde{\omega}. \end{aligned} \tag{45}$$

These last results represent merely the recovery of our assumed variations for M and $\tilde{\omega}$.

The rest of Equations (44) can also integrated. Their solution, which will now be considered, determines the long-term rotational motion of the satellite.

B. LONG-TERM MOTION

Since a large portion of the previous work on problems of rotational motion dealt with uniaxial satellites, during the course of the work which led to this paper, a set of canonical variables which may be related to uniaxial variables was introduced for comparative purposes. The variables in this set are

$$\begin{aligned} L_1 &= \sqrt{\frac{\bar{B}\bar{C}(2A\xi_1 - \xi_2^2)}{\bar{A}(\bar{B}-\bar{C})}} \\ L_j &= \xi_j \quad (j = 2, 3, 4, 5) \\ l_1 &= \bar{\lambda}(t^* + \eta_1) \\ l_2 &= (\xi_2/\bar{A})(t^* + \eta_1) + \eta_2 \\ l_j &= \eta_j \quad (j = 3, 4, 5), \end{aligned} \tag{46}$$

where

$$\hat{\lambda} = (\bar{B} - \bar{C}) L_1/\bar{B}\bar{C}.$$

For the unperturbed, uniaxial ($\bar{A} = \bar{B}$) case, $l_1 = \phi'$, $l_2 = \phi_H$, $l_3 = \psi_H$, $L_1 = P_{\phi'}/I_0\omega_0$, $L_2 = P_{\phi_H}/I_0\omega_0$, and $L_3 = P_{\psi_H}/I_0\omega_0$.

The transformation Equations (46) involve t^*, so that, when expressed in terms of the l_j and the L_j, the transformed Hamiltonian, $\hat{F}$, is given by

$$\hat{F} = -\tfrac{1}{2}(\bar{B} - \bar{C}) L_1^2/\bar{B}\bar{C} - L_2^2/2\bar{A} + K^{**}(\boldsymbol{\xi}(\mathbf{L}), \boldsymbol{\eta}(\mathbf{L}, \mathbf{l}); \varepsilon),$$

and may be written explicitly in the form,

$$\begin{aligned} \hat{F} = &-\tfrac{1}{2}(\bar{B} - \bar{C}) L_1^2/\bar{B}\bar{C} - L_2^2/2\bar{A} - \varepsilon L_4 - \varepsilon^2 \left(\frac{\dot{\tilde{\omega}}}{n}\right)\left(\frac{\omega_0}{n}\right) L_5 + \\ &+ \varepsilon^2 \left[a_0 L_3 + b_0 \sqrt{L_2^2 - L_3^2} \cos l_3 + c_0 L_3^2 + d_0\right], \end{aligned} \tag{47}$$

where

$$\begin{aligned} a_0 &= \left(\frac{\dot{\Omega}}{n}\right)\left(\frac{\omega_0}{n}\right) \cos I \\ b_0 &= \left(\frac{\dot{\Omega}}{n}\right)\left(\frac{\omega_0}{n}\right) \sin I \\ c_0 &= -\tfrac{3}{8}(1 - e^2)^{-3/2} (\bar{B} + \bar{C} - 2\bar{A} + 3\Delta_0)/L_2^2 \\ d_0 &= -\tfrac{3}{8}(1 - e^2)^{-3/2} \Delta_0 \\ \Delta_0 &= \Delta - \bar{A} + \bar{B}. \end{aligned} \tag{48}$$

Since t^* is absent from $\hat{F}$, we have the first integral, $\hat{F} = \text{constant}$, which may be expressed in the form

$$\hat{a}L_3^2 + \hat{b}L_3 + \hat{c} = -\sqrt{L_2^2 - L_3^2}\cos l_3\,. \tag{49}$$

where $\hat{a} = -C_o/b_o$, $\hat{b} = -a_o/b_o$, and $\hat{c}$ is an arbitrary constant. Note that since l_1 and l_2 do not appear in $\hat{F}$, L_1 and L_2 are constant. Level curves for the long-term motion of a typical triaxial satellite (Pegasus A) are presented in Figure 3.

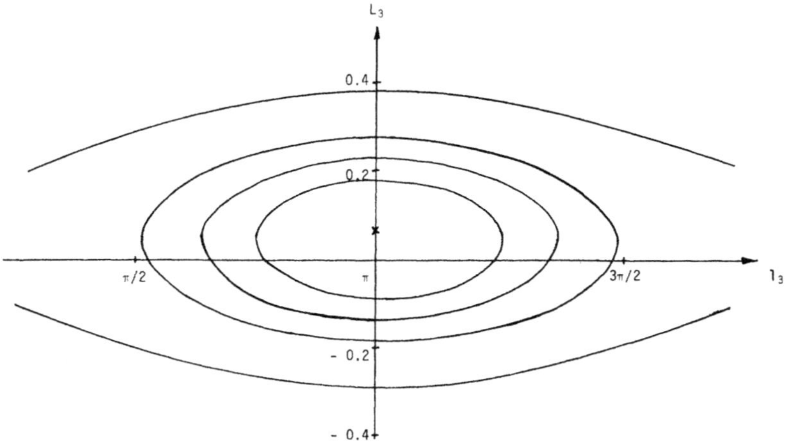

Fig. 3. Level curves, $F(L_3, l_3) = \text{constant}$.

To obtain a geometrical integral (49), we let $\mathbf{i}^o$, $\mathbf{j}^o$, and $\mathbf{k}^o$ denote unit vectors directed along the positive x^o-, y^o-, and z^o-axes, respectively, and define a vector $\mathbf{L}$ by

$$\mathbf{L} \equiv L_{x^o}\mathbf{i}^o + L_{y^o}\mathbf{j}^o + L_{z^o}\mathbf{k}^o = \sqrt{L_2^2 - L_3^2}\sin l_3\mathbf{i}^o - \\ - \sqrt{L_2^2 - L_3^2}\cos l_3\mathbf{j}^o + L_3\mathbf{k}^o\,. \tag{50}$$

Then, (49) becomes

$$L_{y^o} = \hat{a}L_{z^o}^2 + \hat{b}L_{z^o} + \hat{c}\,. \tag{51}$$

Equation (51) represents a family of parabolic cylinders which open in either the positive or negative y^o-direction. Thus, the vector, $\mathbf{L}$, which may be construed as the averaged rotational angular momentum (nondimensionalized) of the satellite, must change in such a way that the projections of its terminus onto the $y^o z^o$-plane form a family of parabolas.

Since L_2 is constant and is the magnitude of $\mathbf{L}$, we have

$$L_{x^o}^2 + L_{y^o}^2 + L_{z^o}^2 = L_2^2 = \text{const}\,. \tag{52}$$

Equation (52) represents a family of spheres with origins at O.

For any particular problem, with given initial conditions which define the initial state of the rotational motion of the satellite, we have, according to (51) and (52), a unique parabolic cylinder and a unique sphere. The intersection(s) of these two surfaces is (are) the locus (possible loci) of the terminus of **L**. A typical example is shown in Figure 4. If the parabolic cylinder penetrates the sphere as shown in Figure 5b, there are two possible loci. The 'occupied' locus may be determined from initial conditions.

To determine when the terminus of **L** occupies a given point on the line of inter-

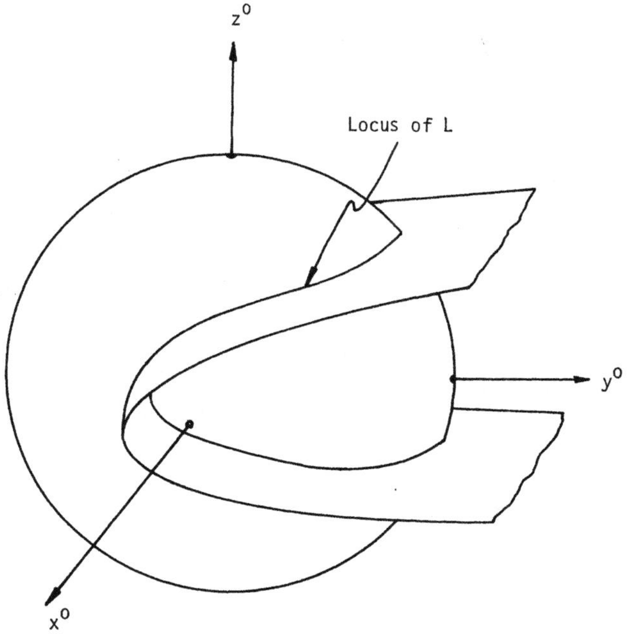

Fig. 4. Geometrical representation of the integrals $L_{y^o} = \hat{a}L_{z^o}^2 + \hat{b}L_{z^o} + \hat{c}$ and $L_{x^o}^2 + L_{y^o}^2 + L_{z^o}^2 = L_2^2$.

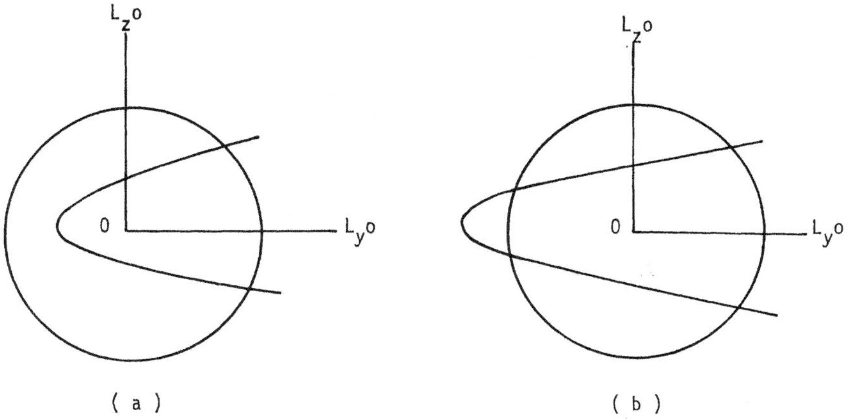

Fig. 5. Projections of the integrals $L_{y^o} = \hat{a}L_{z^o}^2 + \hat{b}L_{z^o} + \hat{c}$ and $L_{x^o}^2 + L_{y^o}^2 + L_{z^o}^2 = L_2^2$ onto the $L_{y^o}L_{z^o}$-plane.

section of the sphere and parabolic cylinder, we must obtain an integral involving t^* explicitly. From (47) and (50), we have

$$\frac{dL_3}{dt^*} = \varepsilon^2 b_o L_{x^o}, \tag{53}$$

where $L_{x^o}^2 = L_2^2 - L_{y^o}^2 - L_3^2$. Using (51), (52), and (53), we get

$$\frac{dL_3}{dt^*} = \pm\, \varepsilon^2 b_o \sqrt{L_2^2 - L_3^2 - (\hat{a}L_3^2 + \hat{b}L_3 + \hat{c})^2}, \tag{54}$$

which, as Holland and Sperling (1969) have pointed out, indicates that L_3 is a function of elliptic functions of t^*. It has not, to the author's knowledge, been pointed out, however, that the roots of the quartic equation $g(L_3) = L_{x^o}^2 = L_2^2 - L_3^2 - (\hat{a}L_3^2 + \hat{b}L_3 + \hat{c}) = 0$ are the values of L_3 at which the projections of the parabolic cylinder and the sphere onto the y^o-plane ($x^o = 0$) intersect. This is an important piece of information, since an understanding of the nature of the roots of $g(L_3) = 0$ is necessary for the integration of (54).

Excluding the rare cases in which the parabolic cylinder and the sphere are tangent at a point,† by geometry (see Figures 5a and 5b), it is obvious that, for real motion to occur, $g(L_3) = 0$ must have either two or four real roots. It follows then, excluding the rare cases as stated above, that there must be two possible forms of the solution to (54).

Case 1. Two Distinct Real Roots

When $g(L_3) = 0$ has two real roots, z_1 and z_2, $z_1 > z_2$, and a pair of complex roots, $z_3 = a_1 + ib_1$ and $z_3^* = a_1 - ib_1$ (see Figure 5a), we find that (see Byrd and Friedman, 1954, p. 153)

$$L_3 = \frac{A_1 + A_2 \operatorname{cn} \hat{u}}{A_3 + A_4 \operatorname{cn} \hat{u}} \tag{55}$$

where

$$A_1 = z_1 G_1 + z_2 G_2, \quad A_2 = z_2 G_1 - z_1 G_2$$
$$A_3 = G_1 + G_2, \quad A_4 = G_1 - G_2$$
$$G_1 = \sqrt{(z_1 - a_1)^2 + b_1^2}, \quad G_2 = \sqrt{(z_2 - a_1)^2 + b_1^2}$$

In Equation (55), we have also introduced the notation

$$\hat{u} = \hat{\omega}(t^* - C_3), \tag{56}$$

where

$$\hat{\omega} = \varepsilon^2 c_o \sqrt{G_1 G_2}, \tag{57}$$

and C_3 is an arbitrary constant. Furthermore, the modulus $\hat{k}$ of the Jacobian elliptic function $\operatorname{cn} \hat{u}$ is given by

$$\hat{k}^2 = [(z_1 - z_2)^2 - (G_1 - G_2)^2]/4G_1 G_2. \tag{58}$$

† These cases correspond to stable or unstable orientations of L.

We may also determine l_3, as a function of t^* for this case. To do this, we use Equations (49), (50), (53), (56), and (57) to get

$$\tan l_3 = -\hat{a}(\mathrm{d}L_3/\mathrm{d}\hat{u})\sqrt{G_1G_2}/[-(\hat{a}L_3^2 + \hat{b}L_3 + \hat{c})], \tag{59}$$

where

$$\mathrm{d}L_3/\mathrm{d}\hat{u} = -[A_2A_3 - A_1A_4]\,\mathrm{dn}\,\hat{u}\,\mathrm{cn}\,\hat{u}/(A_3 + A_4\,\mathrm{cn}\,\hat{u})^2,$$

the quadrant of l_3 being governed by the signs of the numerator and denominator of the right-hand side of (59).

Case 2. Four Distinct Real Roots

If the parabola intersects the circle at four distinct points (see Figure 5b), the quartic equation $g(L_3)=0$, has four distinct real roots, say, $z_1 > z_2 > z_3 > z_4$. The solution for L_3 in this case takes the form, (see Byrd and Friedman, 1954, p. 97 and 133)

$$L_3 = \frac{B_1 + B_2\,\mathrm{sn}^2\,\hat{u}}{B_3 + B_4\,\mathrm{sn}^2\,\hat{u}}, \tag{60}$$

where

$$\hat{u} = \hat{\omega}(t^* - C_3)$$
$$\hat{\omega} = \varepsilon^2(c_o/2)\sqrt{(z_1 - z_3)(z_2 - z_4)}.$$

The modulus, $\hat{k}$, of $\mathrm{sn}^2\,\hat{u}$ is given by

$$\hat{k}^2 = [(z_1 - z_2)(z_3 - z_4)(z_1 - z_3)(z_2 - z_4)],$$

and, depending on between which two roots L_3 librates, the B_j have two possible forms. If $z_1 \leqslant L_3 \leqslant z_2$, then $B_1 = z_1(z_2 - z_4)$, $B_2 = z_4(z_1 - z_2)$, $B_3 = z_2 - z_4$, and $B_4 = z_1 - z_2$. If $z_4 \leqslant L_3 \leqslant z_3$, $B_1 = z_4(z_1 - z_3)$, $B_2 = z_1(z_3 - z_4)$, $B_3 = z_1 - z_3$, and $B_4 = z_3 - z_4$. Values of l_3 may be obtained from

$$\tan l_3 = -(\hat{a}/2)(\mathrm{d}L_3/\mathrm{d}\hat{u})\sqrt{(z_1 - z_3)(z_2 - z_4)}/[-(\hat{a}L_3^2 + \hat{b}L_3 + \hat{c})],$$

where (61)

$$\mathrm{d}L_3/\mathrm{d}\hat{u} = 2[B_2B_3 - B_1B_4]\,\mathrm{cn}\,\hat{u}\,\mathrm{sn}\,\hat{u}\,\mathrm{dn}\,\hat{u}/(B_3 + B_4\,\mathrm{sn}^2\,\hat{u})^2,$$

and L_3 is given by (60).

Now, $L_3 = L_2\cos\hat{\theta}_H$ and $l_3 = \hat{\psi}_H$, and the angles $\hat{\theta}_H$ and $\hat{\psi}_H$ define the orientation of L in space. We will use these results in the next section to predict the orientation of the rotational angular momentum of an artificial satellite as a function of time.

The coordinates l_1 and l_2 exhibit secular and long-period perturbations. By using the integral (55) and the Hamiltonian $\hat{F}$, we may write

$$\frac{\mathrm{d}l_1}{\mathrm{d}t^*} = \hat{n}_1 + D_1L_3^2 \tag{62}$$

$$\frac{\mathrm{d}l_2}{\mathrm{d}t^*} = \hat{n}_2 + D_2 L_3^2 + D_3/(L_2 + L_3) + D_4/(L_2 - L_3),$$

where

$$\hat{n}_1 = [(\bar{B} - \bar{C})/\bar{B}\bar{C}]\, L_1 + \varepsilon^2\, \tfrac{3}{8} (1 - e^2)^{-3/2} \frac{\partial \Delta_o}{\partial L_1}$$

$$\hat{n}_2 = L_2/\bar{A} - \varepsilon^2\, \tfrac{3}{8} (1 - e^2)^{-3/2} \left[\frac{\partial \Delta_o}{\partial L_2} + c_o L_2\right]$$

$$D_1 = -\, \varepsilon^2\, \tfrac{9}{8} (1 - e^2)^{-3/2} \frac{\partial \Delta_o}{\partial L_1} / L_2^2$$

$$D_2 = -\, \varepsilon^2\, \tfrac{9}{8} (1 - e^2)^{-3/2} \left[\frac{\partial \Delta_o}{\partial L_2} / L_2^2 - 2c_o / L_2\right]$$

$$D_3 = -\, \varepsilon^2 b_o (\hat{a} L_2^2 - \hat{b} L_2 + \hat{c})/2$$

$$D_4 = -\, \varepsilon^2 b_o (\hat{a} L_2^2 + \hat{b} L_2 + \hat{c})/2$$

$$\frac{\partial \Delta_o}{\partial L_1} = (2\Delta_o - \Delta^*)/L_1$$

$$\frac{\partial \Delta_o}{\partial L_2} = -\, (2\Delta_o - \Delta^*)/L_2$$

$$\Delta^* = [(\bar{A} - \bar{B})\, (\bar{A} - \bar{C})/\bar{A}]\, \frac{(E - K)\, K - E\, (E - k'^2 K)/k'^2}{k^2 K^2}.$$

If either of the solutions, (55) or (60), for L_3 is substituted into Equations (62), l_1 and l_2 may be obtained by quadrature. In fact, analytical expressions for l_1 and l_2 may be obtained (Cochran, 1970). These expressions involve elliptic integrals of the second and third kinds as well as elliptic functions, and are not very well suited for computations. However, by merely knowing the forms of the solutions for l_1 and l_2, we can conclude that the angles of spin and precession of the satellite will experience both secular and long-period perturbations.

Since the potential V is composed of periodic functions, it is not surprising that we do not find secular perturbations in momenta L_j. Furthermore, referring to the transformations $(\mathbf{P}, \mathbf{Q}) \to (\mathbf{x}, \mathbf{y})$, $(\mathbf{x}, \mathbf{y}) \to (\boldsymbol{\xi}, \boldsymbol{\eta})$ and $(\boldsymbol{\xi}, \boldsymbol{\eta}) \to (\mathbf{L}, \mathbf{l})$ it is apparent that the P_j, to the order of the present theory, do not experience secular perturbations. Hence the angles $\theta' = \cos^{-1}(P_1/P_2)$ and $\theta_H = \cos^{-1}(P_3/P_2)$, are not secularly perturbed by the gravity-gradient torque.

6. Comparison of Theoretical Results and Observations

The theoretical results derived in the preceding section have been used to predict the long-term changes in the rotational motion of the Pegasus A satellite. This satellite 'tumbled' rapidly, so the assumption $\omega \gg n$ is valid. However, it was not spin-stabilized against gravity-gradient and other environmental torques. These torques produced decidedly non-Eulerian type motion, which was inferred from data obtained from on-board Sun and horizon sensors.

During the period of time we will consider, Pegasus A was essentially spinning about its axis of maximum moment of inertia, so that its angular momentum and body-fixed z'-axis were colinear. The data obtained by the on-board sensors therefore may be used to describe changes in the rotational angular momentum. Since the data presented here (Holland, 1969) was obtained by *statistically* averaging acquired data over six-hour time periods and since the theoretical results were obtained by an *analytical* averaging process, what we compare here are two descriptions of **L**, the averaged rotational angular momentum of the satellite.

The analytical descriptions of $\hat{\theta}_H = \cos^{-1}(L_3/L_2)$ and $\hat{\psi}_H = l_3$ were obtained using Equations (55) and (59). The initial conditions used were $h = 5.842 \times 10^5$ kg-m^2 min^{-1}, $\psi_H = 310°$, $\theta_H = 88°$, $\phi_H = \phi' = \theta' = 0$. Other pertinent data used were $A = 1.03068 \times 10^5$ kg-m^2, $B = 3.33455 \times 10^5$ kg-m^2, $C = 3.94992 \times 10^5$ kg-m^2, $n = 3.71°$ min^{-1}, $e = 0.1617$, $I = 31.74°$ and $\dot{\Omega} = -6.152°$ day^{-1}. The initial conditions for $\hat{\theta}_H$ and $\hat{\psi}_H$ were corrected for quasi-long period discrepancies using Equations (40) and (42).

Theoretical and observed time histories of $\hat{\theta}_H$ and $\hat{\psi}_H$ are presented in Figures 6 and 7. The small black triangles denote observed values while theoretical results are shown as solid lines. Excellent agreement can be observed for the time period of 17 days covered by the data. Deviations of no more than 10° in $\hat{\theta}_H$ and $\hat{\psi}_H$ are noted.

The predicted period of the motion of the terminus of **L** on the sphere (Equation 52)

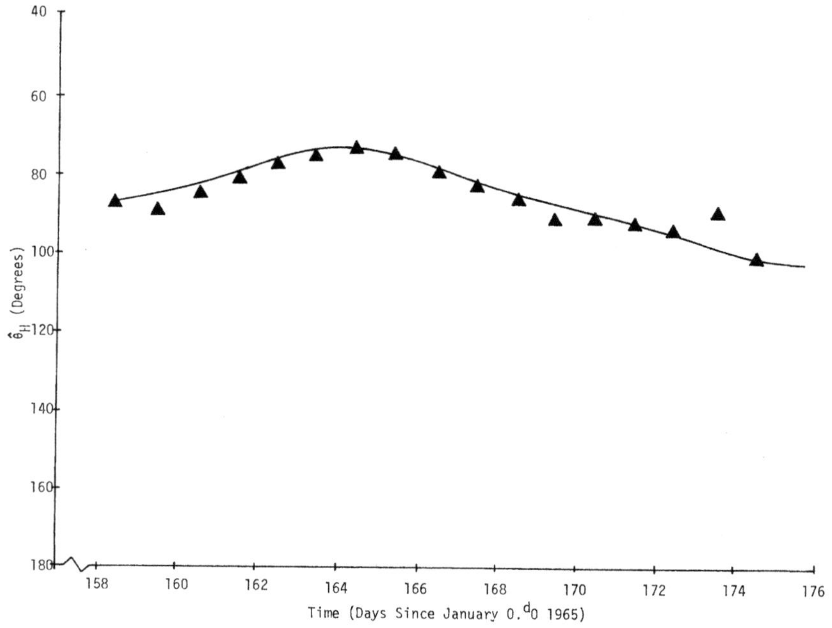

Fig. 6. Predicted and observed motion in $\hat{\theta}_H$

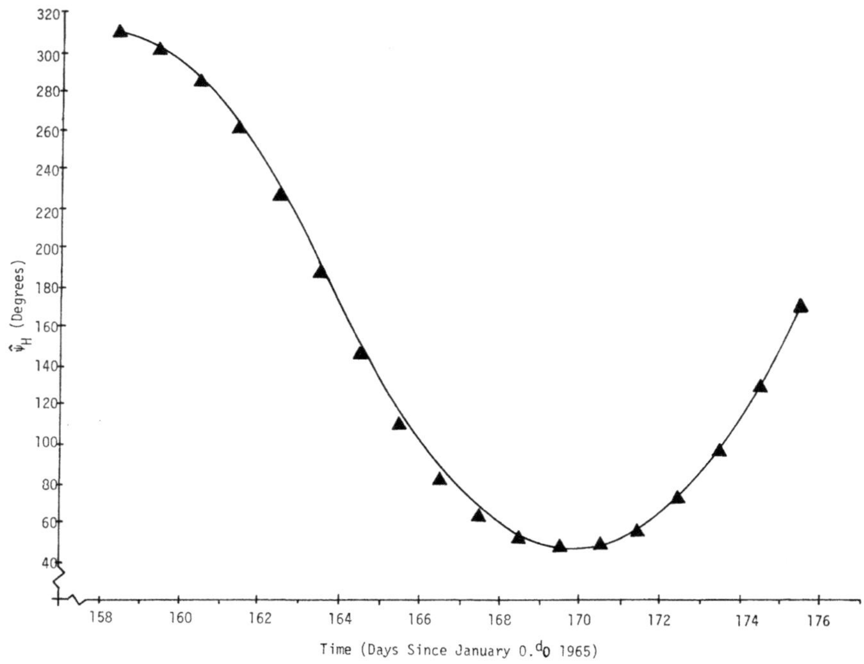

Fig. 7. Predicted and observed motion in $\hat{\psi}_H$.

for the case examined is 22 days. Agreement of theory with observations during this time period is very good; however, since forces of other than a gravitational origin affect the satellite's motion, good agreement cannot usually be maintained for time periods longer than one or two predicted periods of the motion of **L**. In particular, it has been observed that the **L**-vector for the satellite Pegasus A passed from the state given in Figures 6 and 7 into a state in which circulation in $\hat{\psi}_H$ occurred. Meanwhile, h remained essentially constant. From the theory developed here, we conclude that the parabolic cylinder (Equation (49)) must have been altered or at least shifted so that it completely penetrated the sphere. These changes are, of course, not predicted by the present theory.

Another change observed in the rotational motion of Pegasus A was the transition from rotation, essentially about the axis of minimum moment of inertia, to rotation about the axis of maximum moment of inertia. As previously stated, the corresponding secular change in the Eulerian angle θ' is not predicted by a gravitational theory.

Acknowledgements

The support of this work by the National Science Foundation through its Graduate Fellowship Program as well as financial support from the United States Air Force*

* Contract No. AFOSR 69-1744A.

and the National Aeronautics and Space Administration* is gratefully acknowledged. The author also expresses his thanks to Dr B. D. Tapley** and Dr Phillip M. Fitzpatrick† for their encouragement and suggestions during the course of this investigation.

References

Beletskii, V. V.: 1965, 'Motion of an Artificial Satellite about Its Center of Mass', NASA TT F-429.

Byrd, P. F. and Friedman, M. D.: 1954, *Handbook of Elliptic Integrals for Engineers and Physicists*, Springer-Verlag, Berlin.

Campbell, J. A. and Jefferys, W. H.: 1970, *Celes. Mech.* **2**, 467.

Cochran, J. E.: 1970, Dissertation, University of Texas, Austin.

Cochran, J. E.: 1972, *Celes. Mech.* **6**, 2, 213.

Colombo, G.: 1964, *Acad. Press Appl. Math.* **7**, Academic Press, New York, 175.

Crenshaw, J. W. and Fitzpatrick, P. M.: 1968, *AIAA J.* **6**, 2140.

Deprit, A.: 1967, *Am. J. Phys.* **35**, 424.

Fitzpatrick, P. M.: 1970, *Principles Celes. Mech.*, Academic Press, New York, 348.

Henrad, J.: 1970, *Celes. Mech.* **3**, 107.

Hitzl, D. L. and Breakwell, J. V.: 1971, *Celes. Mech.* **3**, 346.

Holland, R. L.: 1969, Private communication.

Holland, R. L. and Sperling, H. J.: 1969, *Astron. J.* **74**, 490.

Hori, G.: 1966, *Publ. Astron. Soc. Japan* **18**, 287.

LaPlace, P. S.: 1829, *Mecanique Céleste*, Vol. II, Chelsa Publishing Company, New York, Fifth Book, Chapters I–II.

MacMillan, W. D.: 1960, *Dynamics of Rigid Bodies*, Dover Publications, New York, Chapter VII.

Mersmann, W. A.: 1970, *Celes. Mech.* **3**, 81.

Tisserand, F.: 1891, *Traité de Mécanique Céleste*, Vol. II, Gauther-Villars, Paris, Chapters XXII–XXIII.

Whittaker, E. T.: 1965, *A Treatise on the Analytical Dynamics of Particles and Rigid Bodies*, Cambridge University Press, New York, 144.

Whittaker, E. T. and Watson, G. N.: 1927, *Modern Analysis*, Cambridge University Press, New York, 522.

* Contract No. NAS8-20175.

** Chairman, Dept. of Aerospace Engineering and Engineering Mechanics, The University of Texas, Austin, Texas.

† Professor, Dept. of Mathematics, Auburn University, Auburn, Alabama.

UNIFORM THEORY OF A ROTATING RIGID BODY WITH DYNAMICAL SYMMETRY

M. VITINS

Seminar für Angewandte Mathematik, Zürich, Switzerland

Abstract. The existing numerical or analytical theories of the rotational motion of a rigid body based on the use of elements have a common drawback that singularities appear when the inclination of the body approaches some critical value. The elements which will be derived in this paper, being based on the Euler parameters, avoid this difficulty. A complete set of elements valid for all inclinations are introduced. The key to these regular elements is the use of a redundant set of elements.

1. Representation of Rotations

The Euler parameters were first used in classical mechanics by Felix Klein. These parameters are used for representing rotations. Let a coordinate system be rotated about a unit vector (l_1, l_2, l_3) with an angle φ. The Euler parameters are then defined by

$$u_1 = l_1 \sin\frac{\varphi}{2}, \quad u_2 = l_2 \sin\frac{\varphi}{2}, \quad u_3 = l_3 \sin\frac{\varphi}{2}, \quad u_4 = \cos\frac{\varphi}{2}. \tag{1}$$

Clearly the magnitude of the 4-vector $\mathbf{u}$ is unity,

$$u_1^2 + u_2^2 + u_3^2 + u_4^2 = 1. \tag{2}$$

Let $\mathbf{x}$ be any point in the physical space and let $\mathbf{z}$ represent the same point with respect to a rotated coordinate system attached to a rigid body, whose center of mass is placed in the origin of the coordinate systems. Then we have

$$\mathbf{x} = A(u)\,\mathbf{z}, \tag{3}$$

where

$$A(u) = \begin{bmatrix} u_1^2 - u_2^2 - u_3^2 + u_4^2 & 2(u_1u_2 - u_3u_4) & 2(u_1u_3 + u_2u_4) \\ 2(u_1u_2 + u_3u_4) & -u_1^2 + u_2^2 - u_3^2 + u_4^2 & 2(-u_1u_4 + u_2u_3) \\ 2(u_1u_3 - u_2u_4) & 2(u_1u_4 + u_2u_3) & -u_1^2 - u_2^2 + u_3^2 + u_4^2 \end{bmatrix}.$$

The matrix A, together with Equation (2), is a classical representation of an orthogonal matrix with unity determinant as a function of four parameters and is known as Euler parametrisation. Any attempt to replace these four parameters by three other quantities without admitting singularities is doomed to failure.

2. Free Rotational Motion

Denoting the vector of turning with respect to the principle axes of the body system

B. D. Tapley and V. Szebehely (eds.), Recent Advances in Dynamical Astronomy, 454–456.

by $\boldsymbol{\omega}$, the equations of free motion of a symmetrical rigid body are given by

$$\dot{\mathbf{u}} = \tfrac{1}{2} B(\omega)\,\mathbf{u}\,, \tag{4}$$

$$\begin{aligned} &\dot{\omega}_1 + m\omega_3\omega_2 = 0 \\ &\dot{\omega}_2 - m\omega_3\omega_1 = 0 \\ &\dot{\omega}_3 = 0 \end{aligned} \tag{5}$$

where

$$B(\omega) = \begin{bmatrix} 0 & \omega_3 & -\omega_2 & \omega_1 \\ -\omega_3 & 0 & \omega_1 & \omega_2 \\ \omega_2 & -\omega_1 & 0 & \omega_3 \\ -\omega_1 & -\omega_2 & -\omega_3 & 0 \end{bmatrix}$$

and m is a constant

$$m = \frac{I_3 - I_2}{I_1} = \frac{I_3 - I_1}{I_2}.$$

I_1, I_2 and I_3 are the moments of inertia and $I_1 = I_2$. Equations (5) are the usual Euler equations. Their solution is easily obtained by realizing that they are a system of linear coupled differential equations

$$\begin{bmatrix} \omega_1 \\ \omega_2 \end{bmatrix} = \begin{bmatrix} \cos\psi_2 & -\sin\psi_2 \\ \sin\psi_2 & \cos\psi_2 \end{bmatrix} \begin{bmatrix} a_1 \\ a_2 \end{bmatrix}, \qquad \omega_3 = \text{const.}, \qquad \psi_2 = m\omega_3 t. \tag{6}$$

The initial values of the constants a_1, a_2 and ω_3 are always well defined.

Inserting Equation (6) into Equation (4) a final solution of the form

$$\mathbf{u} = C\mathbf{b}$$

is found, where b_i are four further constants and where

$$C = \frac{1}{2\varrho_2}$$

$$\begin{bmatrix} -2\varrho_2 \sin\varphi_1 & 2\varrho_2 \cos\varphi_1 & a_1 \sin\varphi_2 + a_2 \cos\varphi_2 & -a_2 \sin\varphi_2 + a_1 \cos\varphi_2 \\ -2\varrho_2 \cos\varphi_1 & -2\varrho_2 \sin\varphi_1 & -a_1 \cos\varphi_2 + a_2 \sin\varphi_2 & a_2 \cos\varphi_2 + a_1 \sin\varphi_2 \\ a_1 \sin\varphi_2 + a_2 \cos\varphi_2 & a_2 \sin\varphi_2 - a_1 \cos\varphi_2 & 2\varrho_2 \sin\varphi_1 & 2\varrho_2 \cos\varphi_1 \\ -a_1 \cos\varphi_2 + a_2 \sin\varphi_2 & -a_2 \cos\varphi_2 - a_1 \sin\varphi_2 & 2\varrho_2 \cos\varphi_1 & -2\varrho_2 \sin\varphi_1 \end{bmatrix}$$

$$\nu = \sqrt{(m+1)^2\,\omega_3^2 + a_1^2 + a_2^2}\,, \qquad \varrho_2 = \frac{m+1}{2}\,\omega_3 + \tfrac{1}{2}\nu$$

$$\psi_1 = \nu t\,,$$

$$\varphi_1 = \tfrac{1}{2}(\psi_1 - \psi_2), \qquad \varphi_2 = \tfrac{1}{2}(\psi_1 + \psi_2).$$

Without loss of generality we may assume that $\varrho_2 > 0$. The matrix C has the important properties that it is orthogonal and that its determinant cannot vanish. Hence the initial conditions of b_1, b_2, b_3 and b_4 are always well defined.

The solution is thus determined by the seven elements $a_1, a_2, \omega_3, b_1, b_2, b_3, b_4$ which are constants and by the two linearly increasing elements ψ_1 and ψ_2; hence the total order is nine. Due to the redundancy of the coordinates u_i necessary for regularization, the elements satisfy the restriction

$$\frac{\nu}{\varrho_2}\left(b_1^2 + b_2^2 + b_3^2 + b_4^2\right) = 1,$$

which is found by casting Equation (2) into elements.

Finally it must be pointed out that the differential equations for these elements in perturbed motion will be void of singularities provided the torques are free of singularities. These element equations can be used to determine the gravity-gradient perturbations of a unaxial satellite tumbling at a frequency much greater than its orbital rate.

More details concerning the proposed elements will be presented in a forthcoming thesis.

Acknowledgement

The author is grateful to Prof. E. Stiefel for his continued interest and support.

ATTITUDE STABILITY OF A SATELLITE IN THE RESTRICTED THREE-BODY PROBLEM

W. J. ROBINSON

Dept. of Mathematics, University of Bradford, Bradford, Yorkshire, England

Abstract. The satellite is usually regarded as a point mass in the restricted problem of three bodies. In this paper, the satellite is regarded as an axially symmetric rigid body with its center of mass placed at L_4. A condition for a stable attitude is found when the body is at rest relative to the primaries with the axis of symmetry lying in the orbital plane.

1. Introduction

In the short space allowed for this presentation, it is necessary to choose a subject which is not too far ranging. The present calculation was carried out as part of a check on some work of a more general nature. The outcome was that a result was derived which had previously been obtained by Kane and Marsh (1971). In addition a further result was obtained which is believed to be new.

2. Description of the System

The system consists of three bodies. Two of them are relatively heavy with masses m_1 and m_2. These are regarded as point masses and are referred to as the primaries. They rotate in a plane in circular orbits about their common mass center O with constant angular velocity ω. The unit of mass is chosen so that $m_1+m_2=1$ and the unit of distance is chosen so that the distance between the primaries is one unit of length. The unit of time is chosen so that $\omega=1$. It then follows that the gravitational constant also has the value unity.

Rectangular cartesian coordinate axes OX, OY, OZ are chosen so that OZ is the axis of rotation, m_1 has coordinates $(m_2, 0, 0)$ and $OXYZ$ is a righthanded frame. The primaries are then at rest relative to this frame which itself rotates about OZ with angular velocity $\omega=1$.

The third body, referred to as the satellite, has mass m, the magnitude of m being regarded as insignificant in comparison with the masses m_1 and m_2. Relative to principal axes GX_3, GY_3, GZ_3 at its center of mass G, the satellite has moments of inertia A, A, C respectively, it being assumed that $A\neq C$.

Axes GX_0, GY_0, GZ_0 are parallel to OX, OY, OZ respectively. Euler angles ϕ, θ and ψ are chosen in a symmetrical manner as follows. The frame $GX_0Y_0Z_0$ is rotated through an angle ϕ about GX_0 to the position $GX_1Y_1Z_1$. This frame is rotated through an angle θ about GY_1 to occupy the position $GX_2Y_2Z_2$. Finally, this frame is rotated through an angle ψ about GZ_2 to occupy the position of the principal frame of axes $GX_3Y_3Z_3$.

In the present problem we assume that the satellite is free to turn about G which coincides with the point L_4, the equilateral point whose coordinates are $(\mu, \frac{1}{2}\sqrt{3}, 0)$

B. D. Tapley and V. Szebehely (eds.), Recent Advances in Dynamical Astronomy, 457–460.

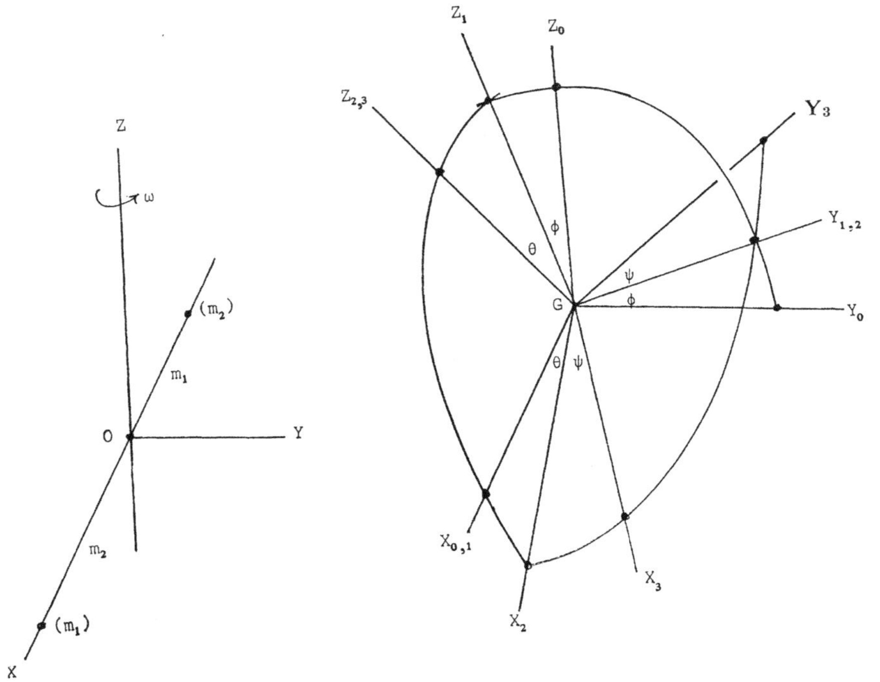

Fig. 1. Illustrating the coordinate frames used in the problem.

relative to the frame $OXYZ$, μ being defined by $\mu = \frac{1}{2}(m_2 - m_1)$. It is further assumed that $m_2 \geqslant m_1$ so that $\frac{1}{2} \geqslant \mu \geqslant 0$.

3. Equations of Rotation

From this point onwards components are specified with respect to the frame $L_4X_3Y_3Z_3$.

The angular velocity $\mathbf{\Omega}$ of the satellite is given by

$$\mathbf{\Omega} = [c_\psi(\dot{\phi}c_\theta - c_\phi s_\theta) + s_\psi(\dot{\theta} + s_\phi),\ - s_\psi(\dot{\phi}c_\theta - c_\phi s_\theta) + c_\psi(\dot{\theta} + s_\phi),\ \dot{\phi}s_\theta + c_\phi c_\theta + \dot{\psi}]$$

where $c\hat{\psi} = \cos\psi$, $s_\theta = \sin\theta$, $\dot{\phi}$ = time derivative of ϕ with respect to the local frame, and so on.

The couple about G acting on the satellite due to the gravitational attractions of the primaries has leading terms involving the moments of inertia. It is assumed that the linear dimensions of the satellite are so small compared to the unit of length that the third and higher moments of inertia are negligible compared to the second moments. Euler's equations of motion then take the form

$$A(Pc_\psi + Qs_\psi) = (A - C)\{c_\psi(\dot{\theta} + s_\phi) - s_\psi(\dot{\phi}c_\theta - c_\phi s_\theta)\} \times \times (\dot{\psi} + \dot{\phi}s_\theta + c_\theta c_\phi) + \tfrac{3}{2}(A - C)(SC_4C_\varphi - TS_4) \tag{1}$$

$$A(Qc_\psi - Ps_\psi) = (C - A)\{s_\psi(\dot{\theta} + s_\phi) + c_\psi(\dot{\phi}c_\theta - s_\theta c_\phi)\} \times \times (\dot{\psi} + \dot{\phi}s_\theta + c_\theta c_\psi) + \tfrac{3}{2}(C - A)(Ss_\psi c_\phi + Tc_\psi) \tag{2}$$

$$\dot{\psi} + \dot{\phi} s_\theta + c_\phi c_\theta = \text{const}. \tag{3}$$

where

$$P = \ddot{\phi} c_\theta - \dot{\theta}\dot{\phi} s_\theta + \dot{\theta}\dot{\psi} - \dot{\theta} c_\theta c_\phi + \dot{\phi} s_\theta s_\phi + \dot{\psi} s_\phi$$
$$Q = \ddot{\theta} - \dot{\phi}\dot{\psi} c_\theta + \dot{\phi} c_\phi + \dot{\psi} s_\theta c_\phi$$
$$S = \tfrac{3}{2} c_\theta s_\phi - \mu\sqrt{3} s_\theta$$
$$T = \mu\sqrt{3} c_{2\theta} s_\phi + \tfrac{1}{8} s_{2\theta} - \tfrac{3}{8} s_{2\theta} c_{2\phi}.$$

4. Special cases

One solution of the Equations (1), (2) and (3) is

$$\theta = \phi = 0, \qquad \dot{\psi} = \text{const}.$$

This is the case which is discussed by Kane and Marsh (1971). It will not be discussed further here.

Another solution of the equations is given by the conditions

$$\phi = \frac{\pi}{2}, \qquad s_{2(\theta+\alpha)} = 0, \qquad \dot{\psi} = 0$$

where 2α is the acute angle such that $\tan 2\alpha = \mu\sqrt{12}$. Two cases follow from these conditions according to whether $\theta = -\alpha$ or $\theta = \pi/2 - \alpha$.

Considering small oscillations about the equilibrium position in the first case, the following substitutions are made in Equations (1), (2) and (3)

$$\theta = \xi - \alpha, \qquad \phi = \eta + \frac{\pi}{2}, \qquad \dot{\psi} = \zeta$$

where ξ, η and ζ and their time derivatives are all regarded as first order small quantities which are functions of time. The equations of rotation, correct to the first order of small quantities reduce to the following

$$4Ac_\alpha \ddot{\eta} + \{(A - C)(13c_\alpha + 6\sqrt{3}\mu\, s_\alpha) + 4Cc_\alpha\}\,\eta = 0 \tag{4}$$
$$2A\ddot{\xi} + 3M(A - C)\,\xi = 0 \tag{5}$$
$$\dot{\zeta} - \dot{\eta} s_\alpha - n c_\alpha = 0 \tag{6}$$

where $M = (1 + 12\mu^2)^{1/2}$. For stability, Equation (5) demands that $A > C$, which condition also ensures the stability of the solutions of Equations (4). The stability of ζ follows from (6). The periods of the small oscillations are also easily deduced. Thus this position is stable as long as $A > C$, which means that the axis of minimum moment lies in the plane of OX and OY and is directed into the angle between m_1 and m_2 as shown in the diagram.

Considering the third position corresponding to

$$\phi = \frac{\pi}{2}, \qquad \theta = \frac{\pi}{2} - \alpha, \qquad \dot{\psi} = 0$$

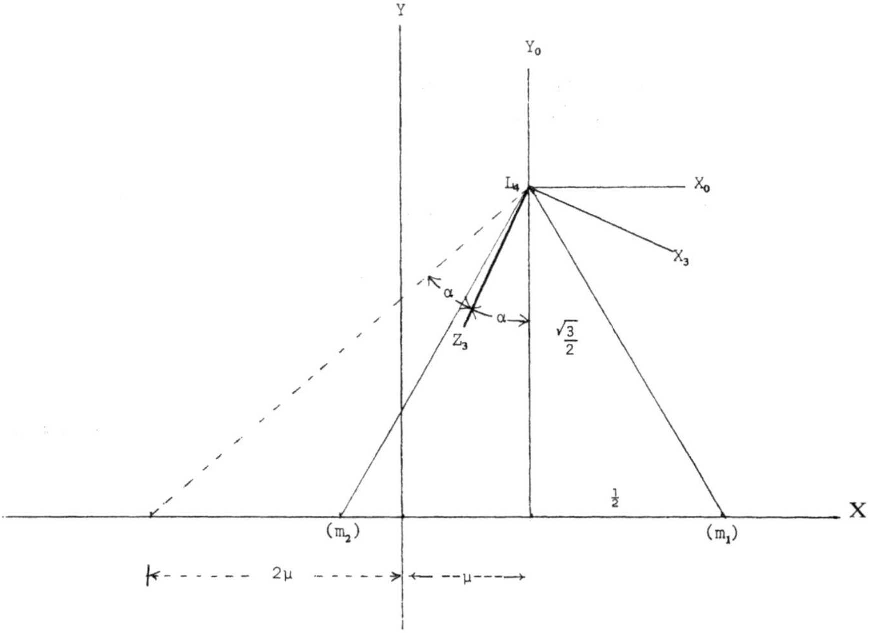

Fig. 2. Diagram showing position of stable equilibrium when $A > C$.

in the same way, it is found that this position is unstable for all values of A and C.

It may therefore be concluded that if $C < A$, there is a position of stable equilibrium with the axis of symmetry lying in the orbital plane as shown in Figure 2. If however $C > A$, there is no such position of stable equilibrium.

Reference

Kane, T. R. and Marsh, E. L.: 1971, *Celes. Mech.* **4**, 78–90.

ANALYTICAL TREATMENT OF STRAIGHT-LINE FLIGHTS INTO THE SUN AFTER PLANETARY FLY-BY

B. L. STANEK

Sagenbrugg, CH 6318 Walchwil, Switzerland

Abstract. Here the swing-by-technique is for once not used to fly to another planet or out of the solar system, but to realize a straight-line flight into the Sun. An analytic theory is given to represent the totality of these orbits for arbitrary swing-by-planets.

1. Introduction

A general transfer from a circular orbit 1 (Earth) to another orbit 2 (planet) is completely determined by the velocity vector at departure, v_1 and α_1, and at arrival, v_2 and α_2. In heliocentric coordinates (Figure 1), the equations are (Gravitational constant = 1, distances in AU):

$$v_1 \cos\alpha_1 = r v_2 \cos\alpha_2 \tag{1}$$

$$\frac{v_1^2}{2} - 1 = \frac{v_2^2}{2} - \frac{1}{r}. \tag{2}$$

Hence

$$v_2^2 = \frac{2\left(1 - \frac{1}{r}\right)\cos^2\alpha_1}{r^2\cos^2\alpha_2 - \cos^2\alpha_1}. \tag{3}$$

For every α_2, v_2 reaches a maximum if $\cos^2\alpha_1 = 1$, which means tangential departure. Therefore this condition determines the limiting curve for all arrival velocity vectors in the velocity plane. Introducing cartesian coordinates by

$$x = v_2 \cos\alpha_2, \qquad y = v_2 \sin\alpha_2, \tag{4}$$

this curve turns out to be a hyperbola (Figure 1)

$$x^2 \frac{r(r+1)}{2} - y^2 \frac{r}{2(r-1)} = 1. \tag{5}$$

$\cos\alpha_2 = 1/r$ in (3) yields $v_2 = \infty$. The asymptotes of the hyperbola therefore correspond to the tangents from the planet to Earth orbit.

The relative departure velocity Δv_1 at Earth is

$$\Delta v_1^2 = 1 + v_1^2 - 2v_1\cos\alpha_1. \tag{6}$$

Introducing (1) and (2) yields

$$\Delta v_1^2 = 3 - \frac{2}{r} + v_2^2 - 2rv_2\cos\alpha_2. \tag{7}$$

B. D. Tap ey and V. Szebehe y (eds.), Recent Advances in Dynamical Astronomy, 461–468.

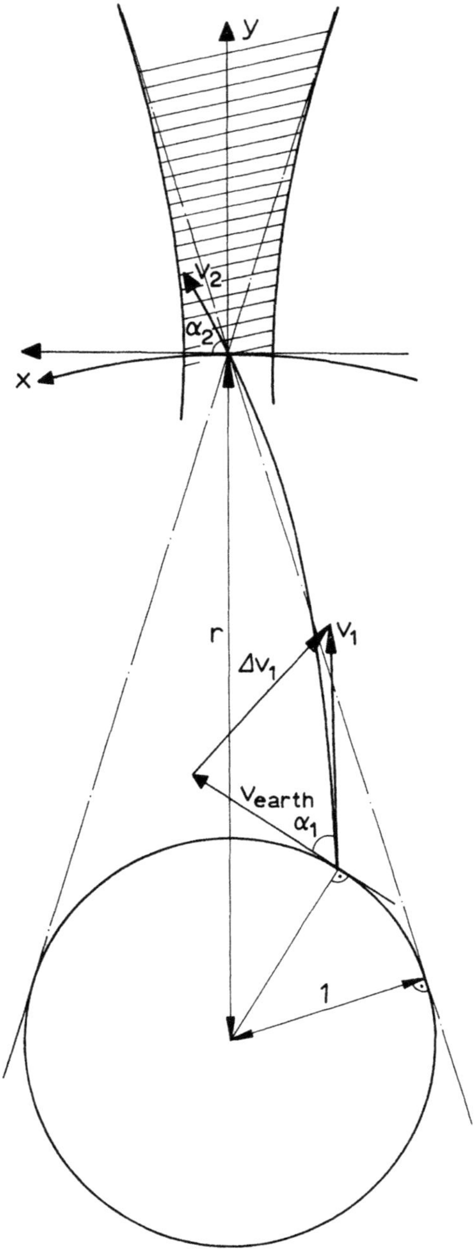

Fig. 1. Heliocentric orbital geometry.

$\Delta v_1(v_2, \alpha_2)$ is defined on the domain inside of the hyperbola in Figure 1. The outside arrival vectors cannot be realized by a one-impulse transfer from Earth. The curves of constant Δv_1^2 in the velocity plane (4) turn out to be circles of radius ϱ:

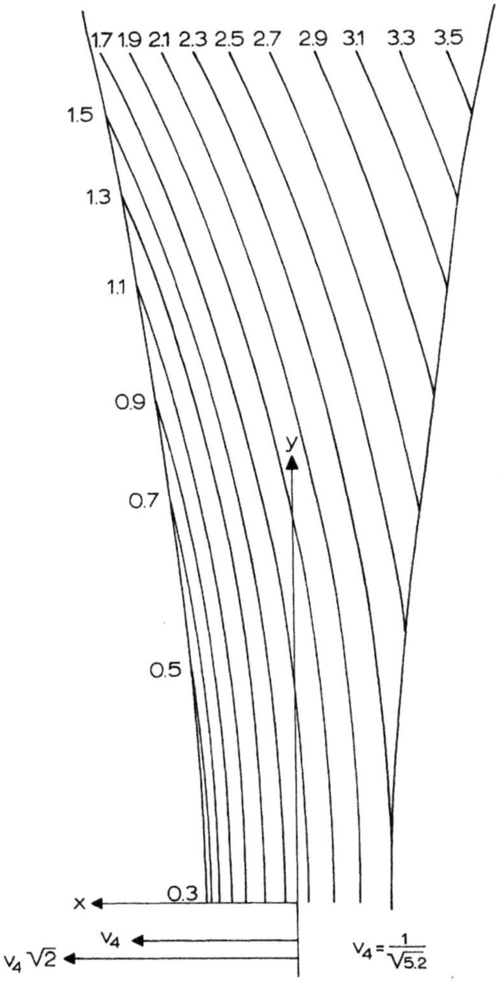

Fig. 2. Curves of constant relative departure velocity from Earth to Jupiter in the arrival velocity plane. (Units: EMOS = 29.78 km s^{-1})

$$(x - r)^2 + y^2 = \varrho^2, \tag{8}$$

where

$$\varrho^2 = \Delta v_1^2 + \frac{2}{r} - 3 + r^2. \tag{9}$$

The corresponding nomogram for $r = 5.2$ (Jupiter) is shown in Figure 2. Unit for Δv_1 is Earth mean orbital speed. It can be noted that near-tangential arrivals are not too costly in Δv_1, if an arrival velocity close to the left branch of the hyperbola (near-tangential departure!) is chosen.

2. Fly-by Conditions at Jupiter for a Straight Fall into the Sun

The vector $\mathbf{v}_p$ in Figure 3 represents the orbital velocity of the planet. It's length is

$$v_p = \frac{1}{\sqrt{r}}. \tag{10}$$

$\Delta v_2 = \mathbf{v}_2 - \mathbf{v}_p$ is the relative arrival velocity at the planet. Encounter with the planet turns the vector $\Delta \mathbf{v}_2$ into $\Delta \mathbf{v}_2^*$.

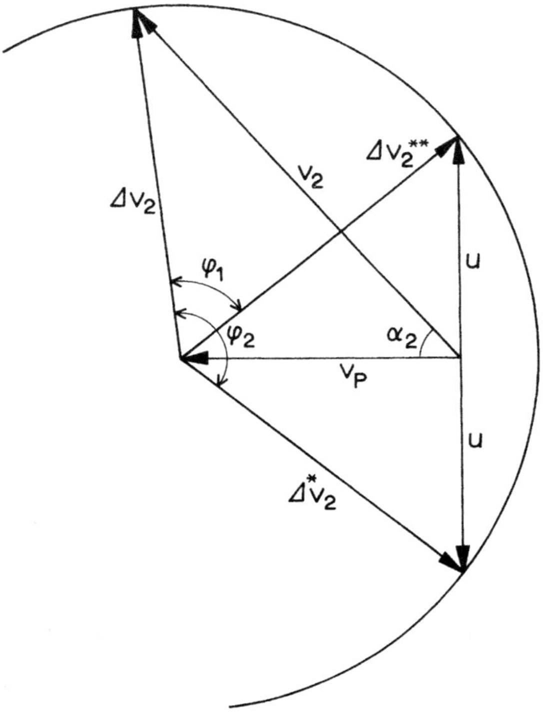

Fig. 3. Vector geometry in the arrival velocity plane.

There are in general two turning angles φ_1 and φ_2 for which the heliocentric post-flyby velocity is radial, that means plus or minus **u**. Simple geometry in Figure 3 yields

$$\cos \varphi_{1,2} = \frac{1}{\Delta v_2^2} \left[v_p^2 - v_p v_2 \cos \alpha_2 \mp v_2 \sin \alpha_2 \sqrt{v_2^2 - 2 v_p v_2 \cos \alpha_2} \right], \tag{11}$$

$$\Delta v_2^2 = v_2^2 + v_p^2 - 2 v_p v_2 \cos \alpha_2 . \tag{12}$$

A formula for the hyperbolic two-body motion gives the relation of $\varphi_i (i = 1, 2)$ to the planet's mass M and periapsis distance R of the fly-by:

$$q = \frac{R}{M} = \frac{1}{\Delta v_2^2}\left(\frac{1}{\sin\frac{\varphi_i}{2}} - 1\right) \equiv \frac{1}{\Delta v_2^2}\left(\sqrt{\frac{2}{1-\cos\varphi_i}} - 1\right). \tag{13}$$

Elimination of φ_i and Δv_2 in (13) by means of (11), (12) and (10) gives a relation $q(v_2, \alpha_2, r)$. For every r the function q is defined on a slightly smaller domain of the velocity plane between the hyperbola branches in Figure 1 than Δv_1. The further restriction comes from the condition $\Delta v_2 \geqslant v_p$. Smaller values of Δv_2 cannot lead to a straight-line flight. Figure 4 will show the domain.

3. Practical Examples

There does not exist a simple analytical equation for the curves of constant q as was the case for Δv_1. For $r = 5.2$, the nomogram was numerically computed and is shown in Figure 4. It corresponds to the second case φ_2 which indicates a fall into the Sun with inward looking initial velocity. The units for $q = R/M$ are Jupiter radii, because

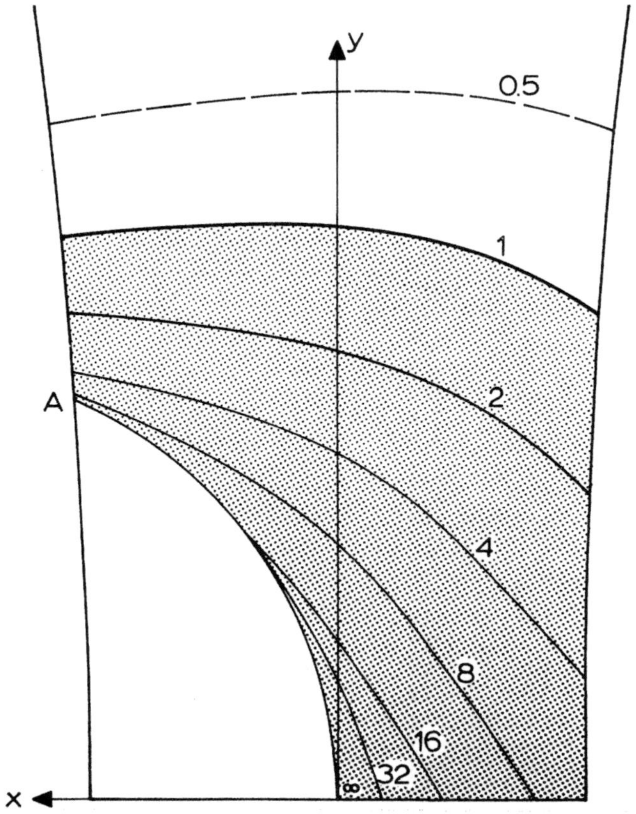

Fig. 4. Curves of constant flyby distance over Jupiter for subsequent fall into the Sun. (arrival velocity plane)

TABLE I

Δv_1-optimal fly-bys into the Sun as a function of r

r AU	Δv_1 km s^{-1}	V_2 km s^{-1}	α_2 degr.	R/M Jupiter	t_{E-p} d	t_{p-S} d	t_{E-S} d
1.0	29.780	59.560	0.0	0.000	0.000	64.569	64.596
1.1	24.800	53.082	20.8	0.157	17.406	74.492	91.898
1.2	21.315	48.115	27.8	0.320	27.895	84.878	112.773
1.3	18.824	44.193	32.2	0.485	38.001	95.706	133.707
1.4	17.007	41.015	35.4	0.652	48.082	106.959	155.040
1.5	15.653	38.379	37.9	0.820	58.207	118.621	176.827
1.6	14.627	36.150	39.8	0.991	68.386	130.678	199.064
1.7	13.836	34.234	41.5	1.164	78.617	143.119	221.736
1.8	13.217	32.564	42.8	1.339	88.898	155.931	244.829
1.9	12.727	31.092	44.0	1.516	99.230	169.104	268.334
2.0	12.335	29.780	45.0	1.697	49.563	182.628	232.191
2.1	12.019	28.602	45.9	1.879	119.988	196.495	316.483
2.2	11.762	27.536	46.7	2.065	130.504	210.697	341.200
2.3	11.551	26.566	47.4	2.253	141.094	225.224	366.318
2.4	11.378	25.677	48.1	2.444	151.747	240.071	391.818
2.5	11.235	24.860	48.7	2.638	162.478	255·231	417.709
2.6	11.116	24.104	49.3	2.835	173.287	270.697	443.984
2.7	11.018	23.403	49.8	3.035	184.197	286.463	470.660
2.8	10.936	22.751	50.3	3.237	195.194	302.524	497.719
2.9	10.869	22.141	50.7	3.442	206.289	318.875	525.163
3.0	10.812	21.570	51.1	3.650	217.479	335.510	552.989
3.1	10.766	21.034	51.6	3.861	228.783	352.424	581.207
3.2	10.728	20.530	51.9	4.075	240.183	369.614	609.797
3.3	10.697	20.054	52.3	4.291	251.702	387.074	638.777
3.4	10.671	19.604	52.6	4.510	263.323	404.801	668.124
3.5	10.651	19.177	53.0	4.731	275.061	422.791	697.851
3.6	10.636	18.773	53.3	4.955	286.912	441.039	727.952
3.7	10.623	18.388	53.6	5.182	298.875	459.543	758.418
3.8	10.615	18.022	53.9	5.411	310.963	478.298	789.261
3.9	10.609	17.673	54.1	5.643	323.164	497.302	820.466
4.0	10.605	17.340	54.4	5.877	335.483	516.551	852.034
4.1	10.603	17.021	54.6	6.113	347.917	536.042	883.959
4.2	10.603	16.716	54.9	6.352	360.473	555.773	916.246
4.3	10.605	16.424	55.1	6.594	373.154	575.740	948.894
4.4	10.608	16.144	55.3	6.837	385.955	595.940	981.895
4.5	10.613	15.875	55.6	7.083	398.862	616.371	1015.233
4.6	10.618	15.617	55.8	7.331	411.907	637.030	1048.938
4.7	10.624	15.368	56.0	7.581	425.051	657.916	1082.967
4.8	10.631	15.129	56.2	7.834	438.336	679.024	1117.360
4.9	10.639	14.898	56.4	8.089	451.737	700.354	1152.091
5.0	10.647	14.675	56.6	8.345	465.245	721.902	1187.147
5.1	10.655	14.460	56.8	8.604	478.897	743.667	1222.564
5.2	10.664	14.253	56.9	8.865	492.642	765.647	1258.289
5.3	10.674	14.052	57.1	9.129	506.518	787.839	1294.357
5.4	10.683	13.858	57.3	9.394	520.524	810.241	1330.764
5.5	10.693	13.671	57.4	9.661	534.632	832.851	1367.483

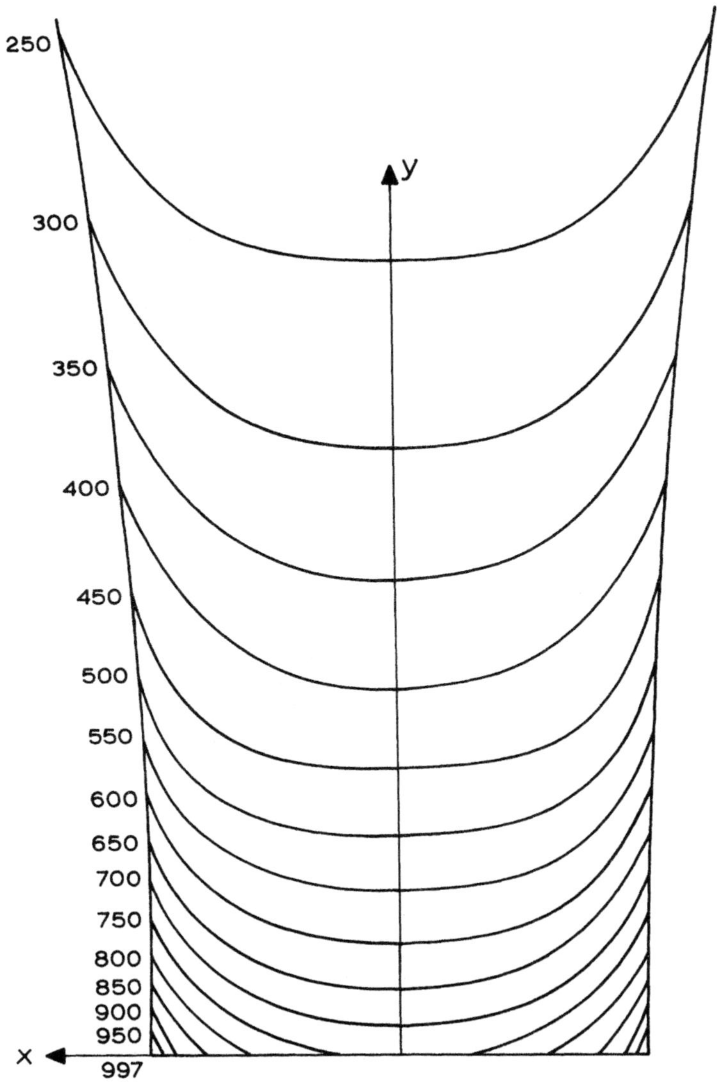

Fig. 5. Curves of constant flight time in days from Earth to Jupiter in the arrival velocity plane.

the planet's real mass M was chosen and the length also transformed from AU to Jupiter radii. Comparing Figure 2 and Figure 4, it becomes clear that the point A in Figure 4 is the one corresponding to a fall into the Sun achieved with the minimal departure velocity Δv_1 at Earth. The fly-by altitude is about 9 Jupiter radii. The question therefore arises how the parameters for this most favorable point A vary with r. The answer is shown in Table I. Apparently a planet with the size of Jupiter would be sufficient for fly-bys into the Sun, if it circles farther out than about 1.6 AU. Circling the Sun at approximately 4.15 AU, a flyby around such a 'Jupiter' would require a

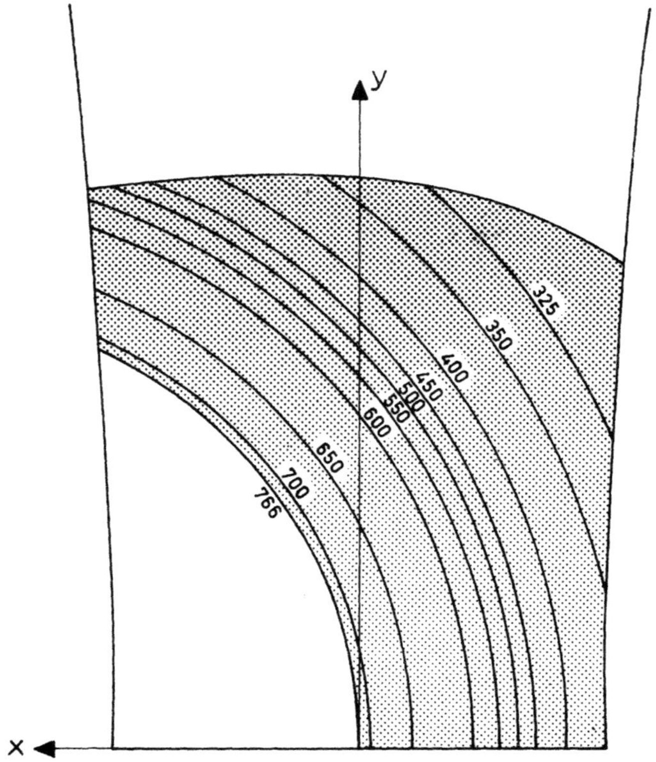

Fig. 6. Curves of constant fall time in days from Jupiter into the Sun. (arrival velocity plane)

minimum of $\Delta v_1 = 10.603$ km s^{-1} to reach the Sun, slightly less than the actual value of 10.664 km s^{-1} for $r = 5.2$, and considerably less than a direct flight into the Sun. The latter requires 29.78 km s^{-1}. In the last 3 columns, Table I also shows the flight times in days: Earth–planet, planet–Sun and the sum of both.

Similar investigations can be carried out for the inner planets, where $0 < r < 1$. The situation, however, is somewhat different: α_2 varies only between 45 and 90°, and even the 'cheapest' fall into the Sun via a fly-by requires 17.826 km s^{-1}. This would be for a planet at 0.364 AU, but having a mass/radius ratio of at least 1.86, the one of Jupiter! As additional information to the reader, Figures 5 and 6 show the transfer times to Jupiter and the subsequent fall times into the Sun.

ASTROPHYSICS AND SPACE SCIENCE LIBRARY

Edited by

J. E. Blamont, R. L. F. Boyd, L. Goldberg, C. de Jager, Z. Kopal, G. H. Ludwig, R. Lüst, B. M. McCormac, H. E. Newell, L. I. Sedov, Z. Švestka, and W. de Graaff

1. C. de Jager (ed.), *The Solar Spectrum. Proceedings of the Symposium held at the University of Utrecht, 26–31 August, 1963.* 1965, XIV + 417 pp.
2. J. Ortner and H. Maseland (eds.), *Introduction to Solar Terrestrial Relations. Proceedings of the Summer School in Space Physics held in Alpbach, Austria, July 15–August 10, 1963 and Organized by the European Preparatory Commission for Space Research.* 1965, IX + 506 pp.
3. C. C. Chang and S. S. Huang (eds.), *Proceedings of the Plasma Space Science Symposium, Held at the Catholic University of America, Washington, D.C., June 11–14, 1963.* 1965, IX + 377 pp.
4. Zdeněk Kopal, *An Introduction to the Study of the Moon.* 1966, XII + 464 pp.
5. Billy M. McCormac (ed.), *Radiation Trapped in the Earth's Magnetic Field. Proceedings of the Advanced Study Institute, Held at the Chr. Michelsen Institute, Bergen, Norway, August 16–September 3, 1965.* 1966, XII + 901 pp.
6. A. B. Underhill, *The Early Type Stars.* 1966, XIII + 282 pp.
7. Jean Kovalevsky, *Introduction to Celestial Mechanics*, 1967, VIII + 427 pp.
8. Zdeněk Kopal and Constantine L. Goudas (eds.), *Measure of the Moon. Proceedings of the Second International Conference on Selenodesy and Lunar Topography held in the University of Manchester, England, May 30–June 4, 1966.* 1967, XVIII + 479 pp.
9. J. G. Emming (ed.), *Electromagnetic Radiation in Space. Proceedings of the Third ESRO Summer School in Space Physics, held in Alpbach, Austria, from 19 July to 13 August, 1965.* 1968, VIII + 307 pp.
10. R. L. Carovillano, John F. McClay, and Henry R. Radoski (eds.), *Physics of the Magnetosphere. Based upon the Proceedings of the Conference held at Boston College, June 19–28, 1967.* 1968, X + 686 pp.
11. Syun-Ichi Akasofu, *Polar and Magnetospheric Substorms.* 1968, XVIII + 280 pp.
12. Peter M. Millman (ed.), *Meteorite Research. Proceedings of a Symposium on Meteorite Research held in Vienna, Austria, 7–13 August, 1968.* 1969, XV + 941 pp.
13. Margherita Hack (ed.), *Mass Loss from Stars. Proceedings of the Second Trieste Colloquium on Astrophysics, 12–17 September, 1968.* 1969, XII + 345 pp.
14. N. D'Angelo (ed.), *Low-Frequency Waves and Irregularities in the Ionosphere. Proceedings of the 2nd ESRIN-ESLAB Symposium, held in Frascati, Italy, 23–27 September, 1968.* 1969, VII + 218 pp.
15. G. A. Partel (ed.), *Space Engineering. Proceedings of the Second International Conference on Space Engineering, held at the Fondazione Giorgio Cini, Isola di San Giorgio, Venice, Italy, May 7–10, 1969.* 1970, XI + 728 pp.
16. S. Fred Singer (ed.), *Manned Laboratories in Space. Second International Orbital Laboratory Symposium.* 1969, XIII + 133 pp.
17. B. M. McCormac (ed.), *Particles and Fields in the Magnetosphere. Symposium Organized by the Summer Advanced Study Institute, held at the University of California, Santa Barbara, Calif., August 4–15, 1969.* 1970, XI + 450 pp.
18. Jean-Claude Pecker, *Experimental Astronomy.* 1970, X + 105 pp.
19. V. Manno and D. E. Page (eds.), *Intercorrelated Satellite Observations Related to Solar Events. Proceedings of the Third ESLAB/ESRIN Symposium held in Noordwijk, The Netherlands, September 16–19, 1969.* 1970, XVI + 627 pp.
20. L. Mansinha, D.E. Smylie, and A.E. Beck, *Earthquake Displacement Fields and the Rotation of the Earth. A NATO Advanced Study Institute Conference Organized by the Department of Geophysics, University of Western Ontario, London, Canada, June 22–28, 1969.* 1970, XI + 308 pp.
21. Jean-Claude Pecker, *Space Observatories.* 1970, XI + 120 pp.

22. L. N. Mavridis (ed.), *Structure and Evolution of the Galaxy. Proceedings of the Nato Advanced Study Institute, held in Athens, September 8–19, 1969.* 1971, VII + 312 pp.
23. A. Muller (ed.), *The Magellanic Clouds. A European Southern Observatory Presentation: Principal Prospects, Current Observational and Theoretical Approaches, and Prospects for Future Research. Based on the Symposium on the Magellanic Clouds, held in Santiago de Chile, March 1969, on the Occasion of the Dedication of the European Southern Observatory.* 1971, XII + 189 pp.
24. B. M. McCormac (ed.), *The Radiating Atmosphere. Proceedings of a Symposium Organized by the Summer Advanced Study Institute, held at Queen's University, Kingston, Ontario, August 3–14, 1970.* 1971, XI + 455 pp.
25. G. Fiocco (ed.), *Mesospheric Models and Related Experiments. Proceedings of the 4th ESRIN-ESLAB Symposium, held at Frascati, Italy, July 6–10, 1970.* 1971, VIII + 298 pp.
26. I. Atanasijević, *Selected Exercises in Galactic Astronomy.* 1971, XII + 144 pp.
27. C. J. Macris (ed.), *Physics of the Solar Corona. Proceedings of the NATO Advanced Study Institute on Physics of the Solar Corona, held at Cavouri-Vouliagmeni, Athens, Greece, 6–17 September 1970.* 1971, XII + 345 pp.
28. F. Delobeau, *The Environment of the Earth.* 1971, IX + 113 pp.
29. E. R. Dyer (general ed.), *Solar-Terrestrial Physics 1970. Proceedings of the International Symposium on Solar-Terrestrial Physics, held in Leningrad, U.S.S.R., 12–19 May 1970.* 1972, VIII + 938 pp.
30. V. Manno and J. Ring (eds.), *Infrared Detection Techniques for Space Research, Proceedings of the Fifth ESLAB-ESRIN Symposium, held in Noordwijk, The Netherlands, June 8–11, 1971.* 1972, XII + 344 pp.
31. M. Lecar (ed.), *Gravitational N-Body Problem, Proceedings of IAU Colloquium No. 10, held in Cambridge, England, August 12–15, 1970.* 1972, XI + 441 pp.
32. B. M. McCormac (ed.), *Earth's Magnetospheric Processes. Proceedings of a Symposium Organized by the Summer Advanced Study Institute and Ninth ESRO Summer School, held in Cortina, Italy, August 30–September 10, 1971.* 1972, VIII + 417 pp.
33. Antonin Rükl, *Maps of Lunar Hemispheres.* 1972, V + 24 pp.
34. V. Kourganoff, *Introduction to the Physics of Stellar Interiors.* 1973, XI + 115 pp.
35. B. M. McCormac (ed.), *Physics and Chemistry of Upper Atmospheres. Proceedings of Symposium Organized by the Summer Advanced Study Institute, held at the Untversity of Orléans, France, July, 31–August 11, 1972.* 1973, VIII + 389 pp.
36. J. D. Fernie (ed.), *Variable Stars in Globular Clusters and in Related Systems. Proceedings of the IAU Colloqim No. 21, held at the University of Toronto, Toronto, Canada, August 29–31, 1972.* 1973, IX + 234 pp.
37. R. J. L. Grard (ed.), *Photon and Particle Interaction with Surfaces in Space. Proceedings of the 6th ESLAB Symposium, held at Noordwijk, the Netherlands, 26–29 September, 1972.* 1973, XV + 577 pp.
38. Werner Israel (ed.), *Relativity, Astrophysics and Cosmology. Proceedings of the Summer School Held, 14–26 August, 1972, at the BANFF Centre, BANFF, Alberta, Canada.* 1973, IX + 323.

SOLE DISTRIBUTORS FOR U.S.A. AND CANADA:

Vols 2–6, and 8: Gordon & Breach Inc., 150 Fifth Ave., New York, N.Y. 10011

Vols 7 and 9–28: Springer Verlag New York, Inc., 175 Fifth Ave., New York, N.Y. 10011